"3D" Parametric and Nonparametric Description of Surface Topography in Manufacturing Processes

"3D" Parametric and Nonparametric Description of Surface Topography in Manufacturing Processes

Editors

Grzegorz Królczyk
Wojciech Kacalak
Michal Wieczorowski

MDPI • Basel • Beijing • Wuhan • Barcelona • Belgrade • Manchester • Tokyo • Cluj • Tianjin

Editors

Grzegorz Królczyk
Faculty of Mechanical
Engineering
Opole University of Technology
Opole
Poland

Wojciech Kacalak
Department of Technical and IT
Systems Engineering
Koszalin University of
Technology
Koszalin
Poland

Michal Wieczorowski
Division of Metrology and
Measurement Systems
Poznan University of
Technology
Poznan
Poland

Editorial Office
MDPI
St. Alban-Anlage 66
4052 Basel, Switzerland

This is a reprint of articles from the Special Issue published online in the open access journal *Materials* (ISSN 1996-1944) (available at: www.mdpi.com/journal/materials/special_issues/surf_topog_manuf).

For citation purposes, cite each article independently as indicated on the article page online and as indicated below:

LastName, A.A.; LastName, B.B.; LastName, C.C. Article Title. *Journal Name* **Year**, *Volume Number*, Page Range.

ISBN 978-3-0365-1404-8 (Hbk)
ISBN 978-3-0365-1403-1 (PDF)

Contents

About the Editors . . . . **vii**

Grzegorz Królczyk, Wojciech Kacalak and Michał Wieczorowski
3D Parametric and Nonparametric Description of Surface Topography in Manufacturing Processes
Reprinted from: *Materials* **2021**, *14*, 1987, doi:10.3390/ma14081987 . . . . **1**

Anna Bazan, Andrzej Kawalec, Tomasz Rydzak, Paweł Kubik and Adam Olko
Determination of Selected Texture Features on a Single-Layer Grinding Wheel Active Surface for Tracking Their Changes as a Result of Wear
Reprinted from: *Materials* **2020**, *14*, 6, doi:10.3390/ma14010006 . . . . **9**

Tarek Eseholi, François-Xavier Coudoux, Patrick Corlay, Rahmad Sadli and Maxence Bigerelle
A Multiscale Topographical Analysis Based on Morphological Information: The HEVC Multiscale Decomposition
Reprinted from: *Materials* **2020**, *13*, 5582, doi:10.3390/ma13235582 . . . . **29**

Wojciech Kapłonek, Tadeusz Mikolajczyk, Danil Yurievich Pimenov, Munish Kumar Gupta, Mozammel Mia, Shubham Sharma, Karali Patra and Marzena Sutowska
High-Accuracy 3D Optical Profilometry for Analysis of Surface Condition of Modern Circulated Coins
Reprinted from: *Materials* **2020**, *13*, 5371, doi:10.3390/ma13235371 . . . . **61**

Sunpreet Singh, Chander Prakash, Alokesh Pramanik, Animesh Basak, Rajasekhara Shabadi, Grzegorz Królczyk, Marta Bogdan-Chudy and Atul Babbar
Magneto-Rheological Fluid Assisted Abrasive Nanofinishing of -Phase Ti-Nb-Ta-Zr Alloy: Parametric Appraisal and Corrosion Analysis
Reprinted from: *Materials* **2020**, *13*, 5156, doi:10.3390/ma13225156 . . . . **81**

Pawel Pawlus, Rafal Reizer and Wieslaw Zelasko
Prediction of Parameters of Equivalent Sum Rough Surfaces
Reprinted from: *Materials* **2020**, *13*, 4898, doi:10.3390/ma13214898 . . . . **97**

Marzena Sutowska, Wojciech Kapłonek, Danil Yurievich Pimenov, Munish Kumar Gupta, Mozammel Mia and Shubham Sharma
Influence of Variable Radius of Cutting Head Trajectory on Quality of Cutting Kerf in the Abrasive Water Jet Process for Soda–Lime Glass
Reprinted from: *Materials* **2020**, *13*, 4277, doi:10.3390/ma13194277 . . . . **117**

Hong Xiao, Wei Han, Wenbin Tang and Yugang Duan
An Efficient and Adaptable Path Planning Algorithm for Automated Fiber Placement Based on Meshing and Multi Guidelines
Reprinted from: *Materials* **2020**, *13*, 4209, doi:10.3390/ma13184209 . . . . **139**

Pawel Pawlus, Rafal Reizer and Michal Wieczorowski
Conditions of the Presence of Bimodal Amplitude Distribution of Two-Process Surfaces
Reprinted from: *Materials* **2020**, *13*, 4037, doi:10.3390/ma13184037 . . . . **157**

Tomasz Bartkowiak, Michał Mendak, Krzysztof Mrozek and Michał Wieczorowski
Analysis of Surface Microgeometry Created by Electric Discharge Machining
Reprinted from: *Materials* **2020**, *13*, 3830, doi:10.3390/ma13173830 . . . . **171**

Tomasz Bartkowiak, Johan Berglund and Christopher A. Brown
Multiscale Characterizations of Surface Anisotropies
Reprinted from: *Materials* **2020**, *13*, 3028, doi:10.3390/ma13133028 . . . . **199**

Julie Marteau, Raphaël Deltombe and Maxence Bigerelle
Quantification of the Morphological Signature of Roping Based on Multiscale Analysis and
Autocorrelation Function Description
Reprinted from: *Materials* **2020**, *13*, 3040, doi:10.3390/ma13133040 . . . . **219**

**Kubilay Aslantas, Mohd Danish, Ahmet Hasçelik, Mozammel Mia, Munish Gupta, Turnad
Ginta and Hassan Ijaz**
Investigations on Surface Roughness and Tool Wear Characteristics in Micro-Turning of
Ti-6Al-4V Alloy
Reprinted from: *Materials* **2020**, *13*, 2998, doi:10.3390/ma13132998 . . . . **231**

Karol Grochalski, Michał Wieczorowski, Paweł Pawlus and Jihad H'Roura
Thermal Sources of Errors in Surface Texture Imaging
Reprinted from: *Materials* **2020**, *13*, 2337, doi:10.3390/ma13102337 . . . . **251**

Grzegorz Struzikiewicz and Andrzej Sioma
Evaluation of Surface Roughness and Defect Formation after The Machining of Sintered
Aluminum Alloy AlSi10Mg
Reprinted from: *Materials* **2020**, *13*, 1662, doi:10.3390/ma13071662 . . . . **269**

Piotr Krawiec, Leszek Różański, Dorota Czarnecka-Komorowska and Łukasz Warguła
Evaluation of the Thermal Stability and Surface Characteristics of Thermoplastic Polyurethane
V-Belt
Reprinted from: *Materials* **2020**, *13*, 1502, doi:10.3390/ma13071502 . . . . **283**

**Tejinder Pal Singh, Anil Kumar Singla, Jagtar Singh, Kulwant Singh, Munish Kumar Gupta,
Hansong Ji, Qinghua Song, Zhanqiang Liu and Catalin I. Pruncu**
Abrasive Wear Behavior of Cryogenically Treated Boron Steel (30MnCrB4) Used for Rotavator
Blades
Reprinted from: *Materials* **2020**, *13*, 436, doi:10.3390/ma13020436 . . . . **301**

Pawel Pawlus, Rafal Reizer and Michal Wieczorowski
Reverse Problem in Surface Texture Analysis—One-Process Profile Modeling on the Basis of
Measured Two-Process Profile after Machining or Wear
Reprinted from: *Materials* **2019**, *12*, 4169, doi:10.3390/ma12244169 . . . . **317**

About the Editors

Grzegorz Królczyk

Grzegorz Krolczyk is a Professor and Vice-Rector for Science and Development at Opole University of Technology. Author & co-author of 250 scientific publications (150 JCR papers), as well as nearly 30 studies and industrial applications. His main directions of scientific activity are analysis and improvement of manufacturing processes, surface metrology, and surface engineering. His research focuses on sustainable manufacturing as a tool for the practical implementation of the concept of social responsibility in the area of machining. A member of several scientific organizations, including Member of the Machine Building Committee of the Polish Academy of Sciences. In addition, he is a member of several editorial committees of scientific journals. He participated in advisory and opinion forming bodies, including Advisory team of the Minister of Science and Higher Education. Co-author of four patent applications, awarded many times for scientific activities in Poland and in the world.

Wojciech Kacalak

Head of the Department of Technical and IT Systems Engineering at the Koszalin University of Technology. Author of over 600 publications and 102 patents in the field of mechanical engineering. Research area: design of machine tools and technological devices, mechatronics, nanometrology, gears with adjustable clearance and adaptive kinematic modules for applications in micromechanisms and microrobotics, diagnostics, optimization and automation of precision grinding, processes of creating and shaping surface topography, metrology of surface layers, biomimetic bases in creating innovative internal structures and surface properties of technical elements. Member of the Machine Building Committee of the Polish Academy of Sciences. Promoter of 18 doctors of technical sciences.

Michal Wieczorowski

Head of Division of Metrology and Measurement Systems and Vice-Rector for Development and Cooperation with the Economy at Poznań University of Technology. Author of 6 books and over 250 publications. Scientific interest: topography analysis, nanometrology, coordinate measurement technique, optical scanning, photogrammetry, reverse engineering, computed tomography. Fulbright scholar at Northwestern University (USA), Visiting professor at Université de Valenciennes et du Hainaut-Cambrésis (France). Associate Editor of Measurement and Metrology and Measurement Systems. A Member of Engineering Academy in Poland and Committee on Machine Building of the Polish Academy of Sciences. Delegate for ISO TC 213:WG 10, WG 15 and WG 16.

Editorial

3D Parametric and Nonparametric Description of Surface Topography in Manufacturing Processes

Grzegorz Królczyk [1],* , Wojciech Kacalak [2] and Michał Wieczorowski [3]

1   Faculty of Mechanical Engineering, Opole University of Technology, 45-758 Opole, Poland
2   Faculty of Mechanical Engineering, Koszalin University of Technology, 75-900 Koszalin, Poland; wk5@tu.koszalin.pl
3   Faculty of Mechanical Engineering and Management, Poznan University of Technology, 60-965 Poznan, Poland; michal.wieczorowski@put.poznan.pl
*   Correspondence: g.krolczyk@po.opole.pl

Citation: Królczyk, G.; Kacalak, W.; Wieczorowski, M. 3D Parametric and Nonparametric Description of Surface Topography in Manufacturing Processes. *Materials* **2021**, *14*, 1987. https://doi.org/10.3390/ma14081987

Received: 1 April 2021
Accepted: 13 April 2021
Published: 15 April 2021

Publisher's Note: MDPI stays neutral with regard to jurisdictional claims in published maps and institutional affiliations.

Surface topography has a profound influence on the function of a surface. In industrial practice, the geometric product specification is an important issue in multiple applications [1,2]. The measurement and characterization of the geometric features of machined parts are important when trying to determine the functional properties of surfaces and also in the control of process parameters during manufacturing [3]. However, there are many other areas of science or engineering where surface topography is critical to function. The emerging aim in a novel science approach is to determine the functional parameters of the generated surfaces. A well known fact is that surface topography can be described by the 3D parameters of a representative area of the selected surface and it is a result of the interaction between tool and surface [4] or tribological interaction [5]. Functional parameters can give us information about the future length of product life [6]. Proper parameter analysis can give us an answer to the question which parameter should control the entire manufacturing process [7]. Surface topography measuring instruments have advanced options which push measurement technologies to their limits. Therefore, deeper insight and a more comprehensive understanding of the performance of surface topography measurement solutions is needed [8]. Parametric description of surface is applied in various fields of science and can give as information about fundamental and practical approach. Królczyk [9] presents the *Ssk-Sku* map(Skewness-Kurtosis map) to better understanding the mixing process. The mixing process of particulate materials is a random process. The paper presents that in micro scale of the whole process many dependencies and behavioural conditions of the grains with respect to each other are observed. Gogolin et al. [10] presents surface topography as a method of pipe systems inspections. The authors present differences in geometry and a flow simulation to show flow nature and make validation of process by the PIV method. Pluta et al. [11] used surface topography in the area of biocompatibility assessment of polymer-ceramic connective tissue replacements; 3D samples topography has been given a broader picture of the surface of the analysed composite biomaterials.

Among the many methods of surface topography assessment [12], parametric and nonparametric methods can be distinguished. Parametric methods include the description of the surface topography using individual parameters from the S (Height Parameters) group. The most frequently described parameters in the open literature are: average roughness (*Sa*), root mean square roughness (*Sq*), maximum peak height (*Sp*), maximum valley depth (*Sv*), maximum height of surface (*Sz*), kurtosis (*Sku*) and skewness (Ssk). These parameters give a certain image of the surface, while a complementary image of the surface is obtained by presenting grouped parameters such as the set *Sp*, *Sv* and *Sz* or *Ssk-Sku* maps with the *Sq* parameter. Kacalak et al. [13] presents a new methodology for assessing the state of the surface in grinding using new sets of parameters. The aim of the work was to characterize their machining potential. The nonparametric methods of description of surface topography are surface morphology assessing, analysis of direction of surface structure, Abbott–Firestone curve (AFC), power spectral density (PSD) and fractal analysis. Generally, important applications of parametric analysis are an evaluation of the surface of technological machine parts for different manufacturing processes. There are also parameters whose values for each or most surfaces are significantly different, which means that they have a high ability to distinguish surfaces in relation to specific topography features [14]. In addition, sets of roughness parameters that differentiate a given set of surfaces depending on the criterion used. It is, therefore, important to determine both the individual criteria of the ability to differentiate surface features as well as to determine their combined impact on the selection of parameters with high classification capacity.

Overall look at surface topography in manufacturing processes with its parametric and nonparametric description can be divided into three main areas: numerical analysis, measurement devices and applications. From that point of view, in the Special Issue, there are a number of paper discussing numerical aspects of assessment of surface topography. Pawlus et al. [15] showed predicted parameters of the sum surfaces. They analyzed the relationships among the parameters of two contacted surfaces on parameters of equivalent surface, for which ordinates are sums of ordinates of both surfaces. Surfaces of various types (one- and two-process, isotropic and anisotropic, random or periodic) were studied. During the parameters selection in the research presented in the article, the authors present that the two pairs of parameters—Sp/Sz and Sq/Sa—can better describe the shape of the probability ordinate distribution of the analysed surface topography than the Ssk-Sku map. Additionally, they found that the RMS height Sq and the RMS slope Sdq parameters predicted with very high accuracy. The authors of reference [16] developed a method of the one-process profile valley modelling based on the two-process profile. They show that the one-process random profile is characterized by the Gaussian ordinate distribution by the standard deviation of the profile height and the length of correlation, addressing the problem of estimation of the correlation length of this one-process profile. As it was shown, the correlation length of the base one-process profile can be obtained on the basis of the vertical truncation of the measured two-process profile. The proposed procedure was validated for two groups of surfaces. The average error of the correlation length estimation in the research was not higher than 7%, while the maximum error was not larger than 14%. What is very important, the presented method in future applications can be extended to the simulated texture of areal surface topography. In another paper, [17], the authors present the developed limiting conditions of the presence of bimodal ordinate distribution, discussing two-process surfaces with plateau and valley parts. They are created by superimpositions of two one-process textures of Gaussian probability height distributions. Based on that, it is expected that the resulting two-process surface would have bimodal height probability distribution, but typically, two-process textures have unimodal ordinate distribution. Generated stratified textures and measured two-process surfaces of cylinder liners were taken into consideration. It was proved that conditions depend on the material ratio at the plateau-to-valley transition (Smq parameter) and the ratio of heights of the plateau and valley surface parts (Spq/Svq parameters). The bimodal ratio increased when the Svq/Spq ratio increased. In addition, when the Smq parameter is not lower than 50%, unimodal amplitude distribution exists. The results are functionally important because of the high tribological significance of the material ratio curve.

These problems are closely related to anisotropy of surfaces, that can be effectively quantified using multiscale approach. An example of such research is presented in [18], where Bartkowiak et al. discussed surface anisotropy and multiscale analysis after milling. Topographies were studied on two milled steel surfaces, one convex with an evident large scale, cylindrical form anisotropy, the other nominally flat with smaller scale anisotropies; a µEDMed surface, an example of an isotropic surface; and an additively manufactured surface with pillar-like features. The paper presents two methods for multiscale quantification and visualization of anisotropy. One method shows anisotropy in horizontal coordinates, the second uses multiple bandpass filters. Curvature tensors contain the two principal curvatures, i.e., maximum and minimum curvatures, which are orthogonal and their directions, at each location. The authors analysed texture aspect ratios (Str) and texture directions (Std) parameters. Analysed multiscale methods show changes in anisotropy with the scale on surface measurements with markedly different anisotropies. Changes of anisotropy with scale categorically failed to be detected by traditional characterization methods, while multiscale methods proved to be very useful. Directions of principal curvatures superimposed on height maps also elucidate anisotropies at specific scales. Different scales show the multiscale nature of different sorts of anisotropies. Continuing multiscale approach to surface topography, Marteau et al. [19] present a study with the morphological signature of roping to link roughness results with five levels identified by a visual inspection. Roping or ridging is a visual defect affecting the surface of ferritic stainless steels, assessed using visual inspection of the surfaces. The researchers show multiscale analysis of surface roughness of the Str parameter and used the autocorrelation function for the quantification. The use of the isotropy with the Sq parameter led to a clear separation of the five levels of roping. To obtain a gradation description instead of a binary one, a methodology based on the use of the autocorrelation function was created, that consisted of a low-pass filtering, segmentation of the autocorrelation into four stabilized portions and computation of isotropy and the root mean square roughness Sq on the obtained quarters of function. Both methodologies can be used to quantitatively describe surface morphology of roping in order to improve our understanding of the roping phenomenon. In addition, Eseholi et al. [20] present multiscale topographical analysis based on morphologonal information. In the article, the researchers evaluated the effect of scale analysis, as well as the filtering process on the performances of an original compressed domain classifier in the field of material surface topographies classification. Each of the surface profiles has been multiscale analysed using a Gaussian Filter and decomposed into three multiscale filtered image types: Low-pass (LP), Band-pass (BP) and High-pass (HP) filtered versions, respectively. The images are lossless compressed using the High-effciency video coding (HEVC) standard. Compared to conventional roughness descriptors, the HEVC-MD descriptors increase surfaces discrimination from 65% to 81%. The results demonstrated that the robust compressed-domain topographies classifier is based on multiscale analysis methodologies.

Surface topography or surface meshing to be more precise, can be also used to create an efficient path planning algorithm. This is presented in publication written by Xiao et al. [21]. Generally, path planning algorithms for automated fiber placement are used to determine the directions of the fiber paths, while the quality of the fiber paths determines the efficiency and quality of the automated fiber placement process. The authors proposed a method of the datum direction vector via a guide-line update strategy to make the path planning algorithm applicable for complex surfaces. Sub-surface boundary splicing and surface topology reconstruction algorithm were proposed and both the computational complexity reduction and the efficiency improvement of the algorithm were analyzed. Additionally, an accuracy analysis on the proposed algorithm was performed, to investigate the relationship between the triangulation parameters and distance deviation, angle deviation and algorithm efficiency. The analysis indicated that choosing appropriate triangulation parameters is crucial for generation of the fiber path with high accuracy and efficiency.

The second area discussed in this Special Issue of Materials is related to surface measuring devices and conditions that needs to be maintained. Here, Kaplonek et al. [22] present non-contact measurement techniques as a support to X-ray-based methods. The authors presented wear of coins, as their wear affects utility values, qualifying them as a legal tender in a country. In this paper, they presented measured and analyzed surfaces with advanced high-accuracy optical profilometry methods. The obtained results confirm the validity of the applied high-accuracy measurement systems in such an application. Furthermore, the analysis can be a significant source of information regarding the condition of coins in the context of maintaining their functional properties. Discussing environmental conditions necessary to perform proper measurements, Grochalski et al. [23] present the effect of thermal phenomena on areal measurements of surface topography using stylus profilometry. The measurements were implemented under variable environmental conditions. The influence of internal heat sources from profilometer drives and their electronic components was analyzed. For this purpose, a thermal chamber was designed and built. On the obtained data, the authors proposed the time after which the correct topography measurement can be started. The value of elongation in individual axes of the profilometer is different and it very much depends on the construction of the device, type of drives used and their location. The largest impact on the imaging of the surface topography has the displacement of the probe in the direction of the Z-axis. This displacement directly translates into the obtained value of the height of the measured surface. The thermal and geometrical stabilization times should be precisely determined before beginning a 3D surface measurement. It was shown that performing thermal stabilization of the profilometer significantly reduced surface irregularity errors.

Now, concentration is owed to applications. The most common ones are related to machining. In this field, Aslantas et al. [24] present experiments and predict surface topography after micro-turning, as a micro-mechanical cutting method used to produce cylindrical parts with small diameter, where a second operation such as grinding may be difficult. The scientists describe empirical relations between technological cutting parameters and surface topography of titan alloy, using a multi-objective optimization. The research was developed using the RSM method, while the scanning electron microscope (SEM) analysis was done on the cutting tools to observe abrasion and crater wear mechanism. The overall results depict that the feed rate is the prominent factor that significantly affects the responses in micro-turning operation. Struzikiewicz and Sioma [25] present selected practical problems related to the surface quality after machining of AlSi10MG alloy powder made by laser sintering. Thanks to surface topography they found the occurrence of breaches on the machined surface, which negatively influence on the surface quality. The cause of breaches and deformations on the machined surface is probably the structure of the surface layer of the sintered aluminum and the method and conditions of combining material particles during the laser sintering process. Surface topography has been presented based on 3D microscopic analysis. The results of the research on the effect of cutting parameters on the values of parameters describing the surface quality are also presented. Taguchi's method was used in the research methodology. It is likely that there are areas with weaker material particle joints that were produced by melting and subsequently by combining metal powder particles. In another paper, Bartkowiak et al. [26] studied the state of surface topographies after electric discharge machining. The measured topographies consist of overlapping microcraters. For this, the authors used conventional ISO parameters, nonparametric motifs and multiscale analysis with curvature tensor. Motif analysis uses watershed segmentation which allows extraction and geometrically characterization of each crater. Curvature tensor analysis focuses on the characterization of principal curvatures and their function and their evolution with scale. Surfaces have been measured by focus variation microscopy. Strong correlations between the height of the crater, diameter, area and curvature was observed. The scientists proved a stronger correlation between conventional areal parameter related and heights dispersion. The approach presented in paper allows for extraction of information directly relating to the shape and size of topographic features. The results show experimentally that the microgeometry of surfaces created by EDM (Electrical Discharge Machining) is strongly affected by the discharge energy. Sutowska et al. [27] show surface topography after the impact of curvature of a shape cut out in a brittle material. The curvature of a shape, resulting from the size of the radius of the cutting head trajectory, is one of the key requirements necessary for ensuring the required surface quality of materials shaped by the abrasive water jet process. The manufacturing process used to generate the surface has been an abrasive water jet (AWJ). The results of the experiments confirmed that the effect of the curvature of the cut shape is an important factor from the efficiency point of view. The parameter used in this research was the total height of surface irregularities given by the St amplitude parameter. The obtained results of the experimental studies confirmed that the effect of the curvature of the cut shape is important from the point of view of the efficiency of the glass-based brittle material-cutting process using AWJ. An interesting way of machining is described in [28], where Singh et al. present the potential of magneto-rheological fluid-assisted abrasive finishing for generating precise surface topography of titan alloy for orthopedic applications. The corrosion performance of the finished samples has also been analyzed through simulated body fluid (SBF) testing. The corrosion analysis of the finished samples specified that the resistance against corrosion is a direct function of the surface finish. It has been found that the selected input process parameters significantly influenced the observed MR and Ra values at 95% confidence level. The morphological nonparametric analysis presented that the rough sites on the surface have provided the nuclei for corrosion mechanics that ultimately resulted in the shredding of the appetite layer. Overall, the results highlight that the MRF-AF process is highly suitable for producing nano-scale finishing of the biomedical implants made of high-strength β-phase Ti-Nb-Ta-Zr alloy. Abrasion and wear are connected with all the types of machining. The effect of cryogenic treatment and post tempering on the behavior of abrasive wear, in the presence of angular quartz sand of rotavator blade is presented in [29]. Rotavator blades are prone to significant wear because of the abrasive nature of sand particles. In this research, the authors show that cryogenic treatment has caused an improvement in the abrasive wear resistance and microhardness compared to untreated material due to enhancement in hardness. Cryogenic treatment has caused an improvement in the abrasive wear resistance and microhardness, compared to untreated material due to enhancement in hardness, the conversion of retained austenite into martensite and the precipitation of secondary carbides in boron steel after exposure to cryogenic temperature. The additional cost was incurred due to cryogenic treatment, but economic analysis justifies the additional cost of that operation. Tool wear analysis was analyzed by Bazan et al. [30], who presented a comparison of surface microgeometry of the grinding wheel by stylus and optical profilometry. In the article, the authors propose a new methodology for determining the average level of binder, which allows the definition of the cut-off level required to separate from the measurement data. This methodology allows one to track changes in characteristic parameters computed from measurements of surface topography in the above-mentioned areas due to different wear processes. Among the parameters for assessing the geometric features of the surface, there are many that do not differentiate individual surfaces in the analyzed set or do not distinguish a significant group. Measurements of the active surface microgeometry of the grinding wheel are commonly used to obtain a cloud of points representing the surface of the examined tool. The research was based on the analysis of data obtained from measurements of single-layer grinding wheels using the replica technique. Discussing applications, finally, Krawiec et al. [31] propose thermography as a non-contact diagnostic tool for assessing drive reliability. The researchers present research during the operation of the belt transmission with a heat-welded thermoplastic polyurethane V-belt. The V-belt temperature changes depending on the braking torque load at different values of the rotational speed of the active pulley, which were adopted as diagnostic characteristics. In this article, the surface morphology of the polyurethane belts has been evaluated based on microscopic tests. The surface topography of the samples was determined by scanning electron microscopy (SEM) and optical profilometry. On the microscopic images, there was a lack of traces of cracks and scratches on the surface, for deeper analysis the authors used observations performed on an electron microscope. It was found that the most favorable operating conditions occurred when the temperature values of active and passive connectors were similar and the temperature difference between them was small.

Already, this short presentation shows how important surface topography is for various fields of science. It also presents many different approaches to the assessment of asperities. This makes the 3D analysis of surfaces in micro scale a very important topic, aiming to create better mechanisms, systems and solutions for our everyday life. The the Industry 4.0 strategy is discussed and different sensors monitoring the states of machines and communicating with other ones are considered. Artificial intelligence will govern many problems of today which are too complicated for operators. Here, filtration methods and procedures are a great example, bearing in mind a large and growing number of options that can be chosen for certain applications. Still, all the new ideas do not change the fact that, in order to obtain proper results, surface topography simply must be manufactured.

Author Contributions: Conceptualization, G.K., W.K. and M.W.; methodology, G.K., M.W.; formal analysis, G.K., W.K. and M.W.; investigation, G.K., W.K. and M.W.; writing—original draft preparation, G.K., W.K. and M.W.; writing—review and editing, G.K., W.K. and M.W. All authors have read and agreed to the published version of the manuscript.

Funding: This research received no external funding.

Institutional Review Board Statement: Not applicable.

Informed Consent Statement: Not applicable.

Data Availability Statement: Data sharing is not applicable to this article.

Acknowledgments: In memory of Wojciech Kapłonek, Koszalin University of Technology, academic teacher and expert from surface topography.

Conflicts of Interest: No conflict of interest was found.

References

1. Krolczyk, G.M.; Nieslony, P.; Krolczyk, J.B.; Samardzicc, I.; Legutko, S.; Hloch, S.; Barrans, S.; Maruda, R.W. Influence of argon pollution on the weld surface morphology. *Measurement* **2015**, *70*, 203–213. [CrossRef]
2. Pawlus, P.; Reizer, R.; Wieczorowski, M. A review of methods of random surface topography modeling. *Tribol. Int.* **2020**, *152*, 106530. [CrossRef]
3. Niemczewska-Wójcik, M.; Wójcik, A. The multi-scale analysis of ceramic surface topography created in abrasive machining process. *Measurement* **2020**, *166*, 108217. [CrossRef]
4. Kapłonek, W.; Nadolny, K.; Zieliński, B.; Plichta, J.; Pimenov, D.Y.; Sharma, S. The role of observation-measurement methods in the surface characterization of X39Cr13 stainless-steel cutting blades used in the fish processing industry. *Materials* **2020**, *13*, 5796. [CrossRef]
5. Wójcik, A.; Frączek, J.; Niemczewska-Wójcik, M. The relationship between static and kinetic friction for plant granular materials. *Powder Technol.* **2020**, *361*, 739–747. [CrossRef]
6. Gupta, M.K.; Song, Q.; Liu, Z.; Pruncu, C.I.; Mia, M.; Singh, G.; Lozano, J.A.; Carou, D.; Khan, A.M.; Jamil, M.; et al. Machining characteristics based life cycle assessment in eco-benign turning of pure titanium alloy. *J. Clean. Prod.* **2020**, *251*, 119598. [CrossRef]
7. Pawlus, P.; Reizer, R.; Wieczorowski, M. Characterization of the shape of height distribution of two-process profile. *Measurement* **2020**, *153*, 107387–107395. [CrossRef]
8. Feng, X.; Senin, N.; Su, R.; Ramasamy, S.; Leach, R. Optical measurement of surface topographies with transparent coatings. *Opt. Lasers Eng.* **2019**, *121*, 261–270. [CrossRef]
9. Królczyk, J.B. Metrological changes in the surface morphology of cereal grains in the mixing process. *Int. Agrophys.* **2016**, *30*, 193–202. [CrossRef]
10. Gogolin, A.; Wasilewski, M.; Ligus, G.; Wojciechowski, S.; Gapinski, B.; Krolczyk, J.B.; Zajac, D.; Krolczyk, G.M. Influence of geometry and surface morphology of the U-tube on the fluid flow in the range of various velocities. *Measurement* **2020**, *164*, 108094. [CrossRef]
11. Pluta, K.; Florkiewicz, W.; Malina, D.; Rudnicka, K.; Michlewska, S.; Królczyk, J.B.; Sobczak-Kupiec, A. Measurement methods for the mechanical testing and biocompatibility assessment of polymer-ceramic connective tissue replacements. *Measurement* **2021**, *171*, 108733. [CrossRef]
12. Mathia, T.G.; Pawlus, P.; Wieczorowski, M. Recent trends in surface metrology. *Wear* **2011**, *271*, 494–508. [CrossRef]
13. Kacalak, W.; Lipiński, D.; Szafraniec, F.; Zawada-Tomkiewicz, A.; Tandecka, K.; Królczyk, G. Metrological basis for assessing the state of the active surface of abrasive tools based on parameters characterizing their machining potential. *Measurement* **2020**, *165*, 108068. [CrossRef]
14. Kacalak, W.; Lipiński, D.; Różański, R.; Królczyk, G.M. Assessment of the classification ability of parameters characterizing surface topography formed in manufacturing and operation processes. *Measurement* **2021**, *170*, 108715. [CrossRef]
15. Pawlus, P.; Reizer, R.; Zelasko, W. Prediction of Parameters of Equivalent Sum Rough Surfaces. *Materials* **2020**, *13*, 4898. [CrossRef]
16. Pawlus, P.; Reizer, R.; Wieczorowski, M. Reverse Problem in Surface Texture Analysis—One-Process Profile Modeling on the Basis of Measured Two-Process Profile after Machining or Wear. *Materials* **2019**, *12*, 4169. [CrossRef] [PubMed]
17. Pawlus, P.; Reizer, R.; Wieczorowski, M. Conditions of the Presence of Bimodal Amplitude Distribution of Two-Process Surfaces. *Materials* **2020**, *13*, 4037. [CrossRef] [PubMed]
18. Bartkowiak, T.; Berglund, J.; Brown, C.A. Multiscale Characterizations of Surface Anisotropies. *Materials* **2020**, *13*, 3028. [CrossRef] [PubMed]

19. Marteau, J.; Deltombe, R.; Bigerelle, M. Quantification of the Morphological Signature of Roping Based on Multiscale Analysis and Autocorrelation Function Description. *Materials* **2020**, *13*, 3040. [CrossRef] [PubMed]
20. Eseholi, T.; Coudoux, F.X.; Corlay, P.; Sadli, R.; Bigerelle, M. A Multiscale Topographical Analysis Based on Morphological Information: The HEVC Multiscale Decomposition. *Materials* **2020**, *13*, 5582. [CrossRef]
21. Xiao, H.; Han, W.; Tang, W.; Duan, Y. An Efficient and Adaptable Path Planning Algorithm for Automated Fiber Placement Based on Meshing and Multi Guidelines. *Materials* **2020**, *13*, 4209. [CrossRef] [PubMed]
22. Kaplonek, W.; Mikolajczyk, T.; Pimenov, D.Y.; Gupta, M.K.; Mia, M.; Sharma, S.; Patra, K.; Sutowska, M. High-Accuracy 3D Optical Profilometry for Analysis of Surface Condition of Modern Circulated Coins. *Materials* **2020**, *13*, 5371. [CrossRef] [PubMed]
23. Grochalski, K.; Wieczorowski, M.; Pawlus, P.; H'Roura, J. Thermal Sources of Errors in Surface Texture Imaging. *Materials* **2020**, *13*, 2337. [CrossRef]
24. Aslantas, K.; Danish, M.; Hasçelik, A.; Mia, M.; Gupta, M.; Ginta, T.; Ijaz, H. Investigations on Surface Roughness and Tool Wear Characteristics in Micro-Turning of Ti-6Al-4V Alloy. *Materials* **2020**, *13*, 2998. [CrossRef] [PubMed]
25. Struzikiewicz, G.; Sioma, A. Evaluation of Surface Roughness and Defect Formation after The Machining of Sintered Aluminum Alloy AlSi10Mg. *Materials* **2020**, *13*, 1662. [CrossRef]
26. Bartkowiak, T.; Mendak, M.; Mrozek, K.; Wieczorowski, M. Analysis of Surface Microgeometry Created by Electric Discharge Machining. *Materials* **2020**, *13*, 3830. [CrossRef] [PubMed]
27. Sutowska, M.; Kapłonek, W.; Pimenov, D.Y.; Gupta, M.K.; Mia, M.; Sharma, S. Influence of Variable Radius of Cutting Head Trajectory on Quality of Cutting Kerf in the Abrasive Water Jet Process for Soda–Lime Glass. *Materials* **2020**, *13*, 4277. [CrossRef] [PubMed]
28. Singh, S.; Prakash, C.; Pramanik, A.; Basak, A.; Shabadi, R.; Królczyk, G.; Bogdan-Chudy, M.; Babbar, A. Magneto-Rheological Fluid Assisted Abrasive Nanofinishing of β-Phase Ti-Nb-Ta-Zr Alloy: Parametric Appraisal and Corrosion Analysis. *Materials* **2020**, *13*, 5156. [CrossRef] [PubMed]
29. Singh, T.P.; Singla, A.K.; Singh, J.; Singh, K.; Gupta, M.K.; Ji, H.; Song, Q.; Liu, Z.; Pruncu, C.I. Abrasive Wear Behavior of Cryogenically Treated Boron Steel (30MnCrB4) Used for Rotavator Blades. *Materials* **2020**, *13*, 436. [CrossRef]
30. Bazan, A.; Kawalec, A.; Rydzak, T.; Kubik, P.; Olko, A. Determination of Selected Texture Features on a Single-Layer Grinding Wheel Active Surface for Tracking Their Changes as a Result of Wear. *Materials* **2021**, *14*, 6. [CrossRef]
31. Krawiec, P.; Rozanski, L.; Czarnecka-Komorowska, D.; Warguła, Ł. Evaluation of the Thermal Stability and Surface Characteristics of Thermoplastic Polyurethane V-Belt. *Materials* **2020**, *13*, 1502. [CrossRef] [PubMed]

materials

Article

Determination of Selected Texture Features on a Single-Layer Grinding Wheel Active Surface for Tracking Their Changes as a Result of Wear

Anna Bazan [1,*], Andrzej Kawalec [1], Tomasz Rydzak [1], Paweł Kubik [1] and Adam Olko [2]

[1]   Faculty of Mechanical Engineering and Areonautics, Rzeszow University of Technology, Powstancow Warszawy 12, 35-959 Rzeszow, Poland; ak@prz.edu.pl (A.K.); t.rydzak@prz.edu.pl (T.R.); p.kubik@prz.edu.pl (P.K.)
[2]   Pratt&Whitney Rzeszow, Hetmańska 120, 35-001 Rzeszow, Poland; adam.olko@prattwhitney.com
*   Correspondence: abazan@prz.edu.pl; Tel.: +48-17-865-1371

Abstract: Measurements of the active surface microgeometry of the grinding wheel by contact and optical methods are commonly used to obtain a cloud of points representing the surface of the examined tool. Parameters that can be determined on the basis of the above-mentioned measurements can be universal parameters, which are commonly used to assess the geometric structure of a surface or parameters taking into account specific properties of the grinding wheel active surface (GWAS) structure. This article proposes a methodology for determining the average level of binder, which allows the definition the cut-off level required to separate from the measurement data: (i) the areas representing grains, (ii) the areas of gumming up of the grinding wheel, and (iii) deep cavities in approximately the same places on the investigated grinding wheel, regardless of the degree of its wear. This, in turn, allows one to track changes in characteristic parameters computed from measurements of texture in the above-mentioned areas due to different GWAS wear processes. The research was based on the analysis of data obtained from measurements of single-layer grinding wheels using the replica technique. The adopted measurement methodology enables measurement of approximately the same (94% coverage) areas of the GWAS at four stages of grinding wheel operation. Errors that were computed related to the determination of the volume of abrasive on the GWAS at various stages of wear using the developed methodology were lower, on average, by 48% compared to the automatic recognition of islands made with a commercial software.

Keywords: electroplated grinding wheel; grinding wheel wear; grinding wheel surface texture

check for **updates**

Citation: Bazan, A.; Kawalec, A.; Rydzak, T.; Kubik, P.; Olko, A. Determination of Selected Texture Features on a Single-Layer Grinding Wheel Active Surface for Tracking Their Changes as a Result of Wear. *Materials* **2021**, *14*, 6. https://dx.doi.org/10.3390/ma14010006

Received: 31 October 2020
Accepted: 17 December 2020
Published: 22 December 2020

Publisher's Note: MDPI stays neutral with regard to jurisdictional claims in published maps and institutional affiliations.

1. Introduction

A grinding wheel's surface microgeometry belongs to the most important features influencing interactions between any grinding wheel and a work-piece during grinding. It decides, among other things, the magnitude of such parameters associated with the grinding process as grinding force, energy, and grinding temperature [1–3]. Thereby, a grinding wheel's geometry influences the grinding process flow and its effects, as well as the quality of the manufactured surface.

Microgeometry, apart from the types and properties of abrasive grain materials and binders and the features of the grinding wheel structure, is one of the important factors determining the cutting ability of this tool, i.e., the ability of the grinding wheel to remove machining allowance [4,5]. Microgeometry, in turn, depends on the topography of the grinding wheel's active surface (GWAS) and geometric features of the abrasive grains, e.g., their height, slope steepness, or characteristic angles [6–9]. To describe each of these features quantitatively, several parameters can be used, which are determined in various ways and characterized by a better or worse ability to characterize the considered feature. For example, the height of the grains on the GWAS can be represented by the parameter Sq

(root mean square height) [7,8], the mean height of the elevations above the cut-off point determined to the highest elevation [10], mean or maximum height of elevations above the designated mean area [11], and medium or maximum height of the motifs determined using motif analysis [12]. Information about the GWAS microgeometry is important during the entire time of grinding wheel operation.

The cutting potential of single-layer grinding wheels (SLGWs) is associated with a very limited amount of abrasive. Their cutting properties cannot be restored by dressing. However, they are often used for grinding products that are subject to high dimensional and shape requirements and made of difficult-to-machine materials, e.g., integrally bladed rotors made of nickel superalloy [13,14] or high-hardness steel gears [15–17]. Therefore, any instability in a manufacturing process implementing such grinding wheels can result in significant financial losses. This is one of the main reasons for insightful testing of SLGWs, including their topography.

Quantitative information on the active surface of the grinding wheel is most often obtained by analyzing 2D images, obtained with, e.g., a scanning electron microscope (SEM) [7,18–24], atomic force microscope (AFM) [25,26], or optical microscope [18,27–34], and from 3D measurements of surface topography using, e.g., contact profilometers [7,8,35,36], confocal microscopes [12,37,38], and interferometers [37,39–41]. Indirect measurement methods, such as measuring the weight of the grinding wheel before and after grinding, allow one to specify the volume of the grinding tool consumed during the process [24,42,43] and to determine the grinding ratio G. Unlike the indirect methods, direct observation and measurement methods of the GWAS topography allow one to obtain data with a much wider range of applications. However, the interpretation of collected and computed data is still difficult and relevant. Visual assessment of the microscopic GWAS images enables one to distinguish between static and active grains and to detect various forms of grinding wheel wear. Thanks to this, some researchers determine some quantitative parameters directly on the basis of the above-mentioned 2D images. These parameters are, e.g., the number of active grains per unit area and the number of grains torn out of the binder per unit area [6,18,32,44,45].

The developed methods of image analysis applied to 2D GWAS views allow one to determine, among other things, the distance between the grains, the surface area and the percentage of the surface area of the grains and areas of gumming up of grinding wheels [46], and the maximum and minimum diameter of the Feret grains [47]. Compared to a microscopic image analysis performed directly by a researcher, computer image analysis methods allow larger areas of the GWAS to be analyzed in less time. On the other hand, direct analysis by a researcher enables precise determination of the number of grains and their boundaries. In addition, the places where some grains have been torn out from SLGWs and cavities in the bond have formed can be relatively easily recognized by a researcher. It becomes very difficult or even impossible to determine them at all using known image analysis methods [48].

The determination of the height and volume parameters of texture is possible from the results of 3D measurement of GWAS topography. Several surface texture (ST) parameters can be calculated according to the ISO standards [49,50] or the European Union report [51]. They require, however, insightful analysis. Nguyen and Butler et al. [7,8] interpreted the parameter *Sds* (summit density) as the density of cutting edges. The parameter *Ssc* (mean summit curvature) was, however, associated by them with the radius of the rounding of the grain, and thus with the sharpness of the grains. The same interpretation of the parameters *Sds* and *Ssc* was used by Yan et al. [40]. Moreover, the parameter *Sdq* (root mean square gradient) was supposed to indicate the angles of the slopes of the grains. The authors of Ref. [40] also associated the sum of the parameters *Vvc* (core void volume) and *Vvv* (dale void volume) with the volume of space on the GWAS where chips can collect. Wang et al. [52] observed the aforementioned parameters *Ssc* and *Sdq* as well as the peak–peak height (*Sz*) while examining the wear of grains on the abrasive belt. Kapłonek and Nadolny [53,54] indicated the suitability of grinding wheels for the assessment of the surface of grinding wheels in terms of their sticking and abrasion of such ST parameters as *Sdr* (surface development factor) and *Sk* (surface core height). Vidal et al. [38] investigated several selected surface texture parameters, of which *Svk* showed the greatest changes due to the GWAS wear, and in the case of dressing, the parameter *Sa* changed the most.

The above-mentioned ST parameters are related to different GWAS features only to a limited extent. The analysis of elevations above the cut-off level (particles) and pits or cavities below the cut-off level (pores) can provide important information about the condition of the GWAS. With the appropriate cut-off level, the particles can represent areas associated with abrasive grains or areas of sticking, while pores can represent voids or pores in the binder. Therefore, the determined parameters of particles and pores have a direct reference to grinding wheel design.

The distribution and shape of abrasive grain areas above a certain level may indicate the activity of abrasive grains. They may be used to evaluate the wear processes of the GWAS and to select appropriate machining parameters. For example, in order to evaluate the cutting potential of a GWAS, Kacalak et al. developed, on the basis of the Teager–Kaisear energy operator [55], a parameter depending on the height and sharpness of the slopes of the tops of grains above the level dependent on the area of the average measured topography [56]. On the other hand, the parameters of the pores are factors that influence, among other things, the lubrication and cooling conditions during grinding. Hence, they can be used to model the phenomena related to the flow of cutting fluid through the grinding zone [57,58].

The natural separation boundary for particles and pores in the case of SLGWs is the binder, and that approach was used by Setti et al. [59]. However, it is difficult to automatically determine the level of separation between the grain and the binder on measured GWAS topographies [37,60]. Ismail et al. presented a method for determining a reference surface that would represent the level of binder using the so-called inverse method (reversal method), which consists of measuring the same surface in several rotational positions [12]. The authors assessed that this method proved to be effective for 66% of the investigated areas. In the remaining cases, it was not possible to determine the reference area due to the low repeatability of the ST measurements performed.

In their investigations, Kacalak et al. applied [9,10] watershed segmentation and the related analysis of motifs to determine the parameters of abrasive grains. The separation of motifs can be done in two ways—to recognize motifs related to either valleys or hills. In the first case, a single motif consists of the valleys and the surrounding area. In the second case, the motif covers the hill with its immediate surroundings (Figure 1). This means that in both cases, the motifs do not only contain data on grinding wheel grains or deep cavities, but also some more information on surface texture. The authors of Ref. [61] also used the analysis of motifs in the study of grinding wheel topography.

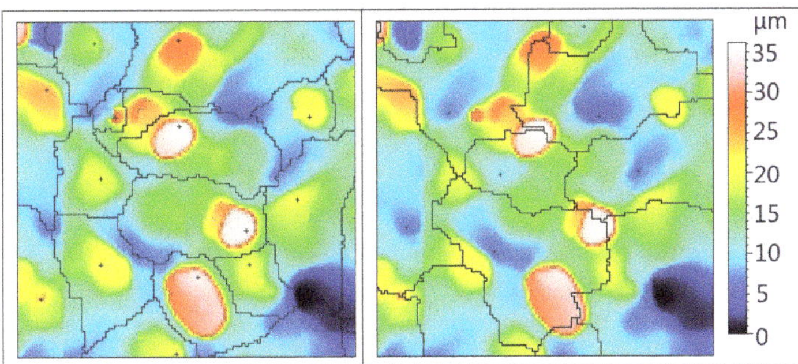

Figure 1. Segmentation of motifs associated with hills (**left**) and valleys (**right**).

Various segmentation methods from three-dimensional data for different types of surfaces were analyzed in [62–65]. The problem of determining the critical points necessary to conduct spatial segmentation is discussed in [66]. It should be remembered that the issue of segmentation does not only refer to surface macrogeometry. It is a technique with a wider application and is also used to analyze macrogeometry [67].

In the Image Metrology's commercial Scanning Probe Image Processor (SPIP^{TM}) software for ST analysis, watershed segmentation can isolate entire motifs or only elevations and pits without their surroundings. The main disadvantage of watershed segmentation for ST assessment of GWASs is the difference of the cut-off levels of individual extracted elements in both the analysis of motifs and the recognition of hills and cavities. Consequently, the height and volume parameters are not determined from a common reference surface. Therefore, it is not possible to compare the volume or the maximum height of two selected abrasive grains.

Commercial programs for the analysis of surface texture, such as SPIP by Image Metrology or MountainsMap by DigitalSurf, allow one to analyze particles and pores segmented with a single surface. However, it is important that particles and pores on a GWAS are determined with the cut-off level independent from the degree of GWAS wear. In order to always cut off and analyze the same parts of the considered ST features, the cut-off level should be selected in a way that keeps the same or approximately the same place for a given grain independently from sticking or recess processes.

The cut-off level can be determined within SPIP or DigitalSurf at a user-defined distance from the reference element, which may be, e.g., the average area or the highest or lowest measured point. Self-determination of the reference level by the user allows full control over it. The cut-off level can be guided by visual inspection of the measured topography. Moreover, the result is not dependent on algorithms that are unknown to the user in the computational (mathematical) layer. On the other hand, the independent determination of the reference level by the user is more labor-intensive and time-consuming in analysis of a large number of measured surfaces.

To reduce such effects, the extraction of the considered features from the measured topographies should be automated. Of the two above-mentioned software packages, only SPIP has an automatic cut-off function. The conducted research described in Section 3 showed, however, that the use of this function did not lead to satisfactory results related to the separation of areas corresponding to the abrasive grains. For this reason, a proprietary algorithm for calculating cut-off levels for particles and pores was developed in the paper. It was implemented in the SPIP software as a plug-in. It enables determination of the cut-off level automatically and processing of many measured examples of GWASs, as well as the determination of the required ST parameters of investigated GWASs in batch processing.

In Section 7, the results of tests related to the use of the developed algorithm for determining the cut-off level and separation of areas associated with abrasive grains, areas of sticking, and cavities in the binder from the measured texture of GWASs are shown. The analyses were done for one single-layer electroplated cubic boron nitride (cBN) grinding wheel at various stages of grinding wheel wear. The paper ends with conclusions drawn from the research and bibliography analysis.

2. Research Methodology

The surface topography studies concerned the active surfaces of 17 SLGWs with a nickel bond applied by electroplating with a cBN abrasive with grain number B35 (average grain diameter of $d_g = 35$ μm). The grinding wheels had a conical shape with a maximum diameter of $d_s = 100$ mm and a cone angle of 140°.

The ground items were made of high-alloy Pyrowear 53 steel after thermo-chemical treatment, with a surface layer hardness of 81 HRA. The surface grinding processes were carried out on a Fortis grinder by Michael Deckel in the presence of grinding oil. Each grinding wheel was operated with a different set of adjustable parameters until it was completely worn (manifested by an intensive increase in grinding force) or until the specific volume of material was removed ($V' = 2652$ mm^3/mm). The adjustable parameters of the grinding process were changed in the following ranges:

- Grinding speed (for diameter $d_s = 100$ mm): $v_s = 20$–40 m/s, which corresponds to the rotational speed range of the grinder spindle: $n = 4000$–8000 rev/min;
- Feed speed: $v_w = 1000$–7500 mm/min;
- Grinding depth: $a_e = 7$–30 μm (0.2–$0.86 \cdot d_g$).

The active surface topography of each of the tested grinding wheels was measured several times at different stages of grinding wheel wear. The GWAS tests were planned after the removal of the following specific volume of material V' [mm^3/mm]: 0 (new grinding wheel), 204 or 272, 408 or 476, 1224 or 1360, 2652, or when the grinding wheel was worn out. Replicas were used to map the GWAS topography. The use of replicas was due to two reasons. First, by using replicas, it was not necessary to remove the grinding wheel from the machine and tool holder to measure GWAS topography. This significantly shortened the testing time and also improved the stability of the grinding wheel's operating conditions throughout its lifetime. The second reason was related to difficulties with direct measurement of the GWAS on the InfiniteFocus microscope. The areas of bonding near the grains were of complex shape and intractable to measure. The problems of performing direct measurements on the grinding wheel were probably caused by big differences in color between grains and the bond. It was not possible to set such measurement conditions, i.e., brightness and contrast, to lighten up grains and the bond properly at the same time. The replicas were made using the Struers RepliSet system. According to information provided by the manufacturer, the material of the replicas was a black silicone rubber with the ability to reproduce details above 0.1 μm. In order to visualize the precision of the replication method, the views of some grains obtained directly from the grinding wheel with the use of the InfiniteFocus microscope were compared to the related views from the replicas. It can be seen in Figure 2 that only the smallest details located on the investigated grains, such as some lines from cleavage planes, were not mapped.

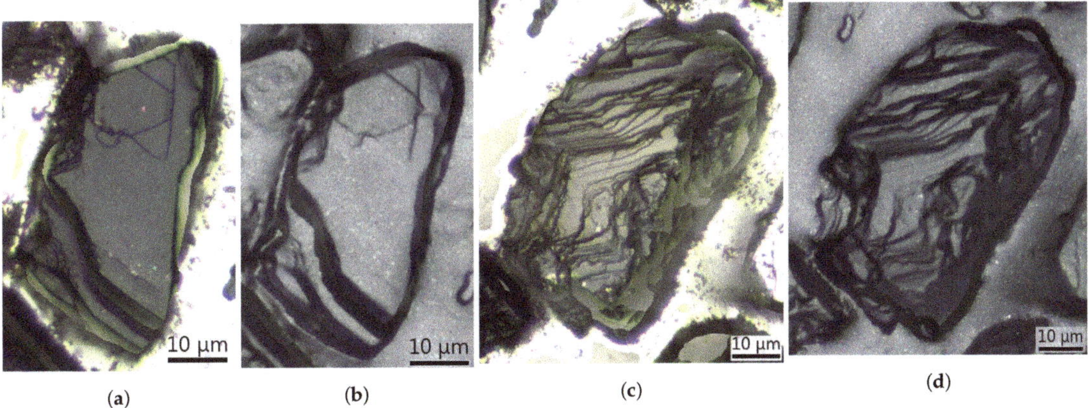

(a)	(b)	(c)	(d)

Figure 2. Views of the grains observed directly on the InfiniteFocus microscope (**a**,**c**) and the same grains mapped by the replica (**b**,**d**).

Thanks to the use of replicas, in order to test the microgeometry of the grinding wheel, it was not necessary to remove the tool (grinding wheel) from the machine tool. This reduced the work and time consumption of the research. It is worth mentioning that replicas can also be helpful in the case of measurements of large grinding wheels that may not fit on the measuring device; e.g., a profilometer or microscope.

An Alicona InfiniteFocus microscope with a ×20 lens was used to measure the topography of the GWAS replicas. At each of the tested grinding stages, six areas that were approximately the same with dimensions of 2.35 mm × 2.59 mm were measured. The measured surfaces were spaced at 120° around the axis of rotation of the grinding wheel (Figure 3). Two surfaces with different radial positions were measured at each angular position. The measurement parameters of the GWAS replicas are presented in Table 1.

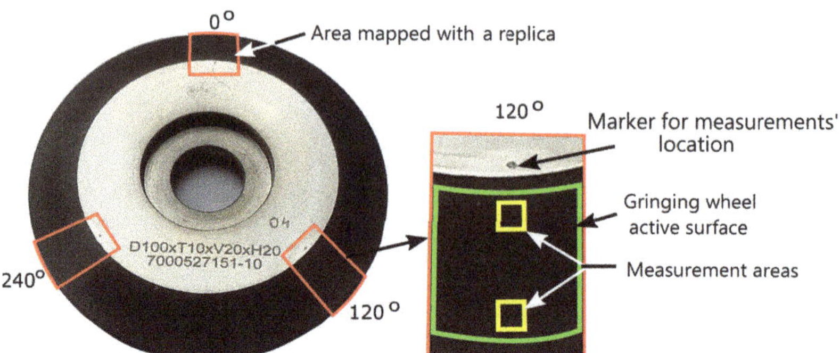

Figure 3. Replicas' production sites and measuring surfaces on the grinding wheel active surface (GWAS).

Table 1. Parameters for measuring topography on the Alicona Focus Variation microscope.

Lens	×20
Single imaging field	0.71 mm × 0.54 mm
Number of imaging fields in the X and Y axes	4 × 5
Measurement area	2.35 mm × 2.59 mm
Analysis area	2.25 mm × 2.50 mm
Horizontal resolution	5 μm
Vertical resolution	100 nm
Sampling step	0.44 μm × 0.44 μm

Positioning of Measuring Surfaces

As mentioned in the introduction, the GWAS topography analysis can be very useful in tests related to wear, and then in the supervision of the cutting properties of grinding wheels. In order to obtain the necessary data, the GWAS measurement should be carried out at various stages of the grinding wheel wear, i.e., after removing a different volume of material. Wear information of the best quality was obtained by continuously measuring the same places on the grinding wheel. Changes in the calculated GWAS parameters at various stages of wear resulted not only from the wear itself, but also from the variance of the values of these parameters within the GWAS. The analysis of the grinding wheel wear based on the measurement of various areas of the GWAS after the removal of subsequent volumes of material only gave the possibility of tracing the generalized changes in the topography parameters within the GWAS. When constantly comparing the same GWAS areas, the changes observed in the GWAS parameters resulted primarily from the wear of the grinding wheel. The conducted analyses were not burdened with an error resulting from possible changes of these parameters within the GWAS. In addition, by measuring the same places on the grinding wheel, it was possible to determine how specific abrasive grains were worn out.

In order to measure the GWAS topography in approximately the same places at different stages of wear, marks were made on the face of the grinding wheel with a vibrating pen in the form of small pits. These characteristic points were spaced around the axis of rotation of the grinding wheel every 120° (Figure 3). They indicated the places where replicas were to be made. At the stage of measuring the GWAS topography mapped with replicas on the InfiniteFocus microscope, the markers were used for the initial orientation of the measurement surfaces. The precise orientation of the measurement surfaces was based on a selected characteristic detail of the GWAS, such as an abrasive grain with a distinctive shape.

The reproducibility of the positioning of the measurement surfaces—to ensure the measurement of the GWAS topography in approximately the same places at different stages of the study—was checked for one randomly selected measurement spot on the grinding wheel, which was considered the most worn after visual inspection of the topography maps. The research on the appropriate positioning of the measuring surfaces is illustrated in Figure 4. Four topography maps (marked as V1, V2, V3, and V4) obtained as a result of measurements carried out on four replicas made at different stages of grinding wheel wear were analyzed. From each of the topography maps with dimensions of 2.25 mm × 2.50 mm, two areas (A and B) of approximately 0.22 mm × 0.24 mm were located in opposite corners of maps. Then, four views of the A areas (coming from four replicas) and four views of the B areas were compared with each other. To make the comparison easier, in views A and B, the characteristic fragments of the topography that were found on the corresponding views were selected. In Figure 4, these fragments are outlined. The comparison of the opposite corners of the four topography maps allowed the identification of the area common to all analyzed maps.

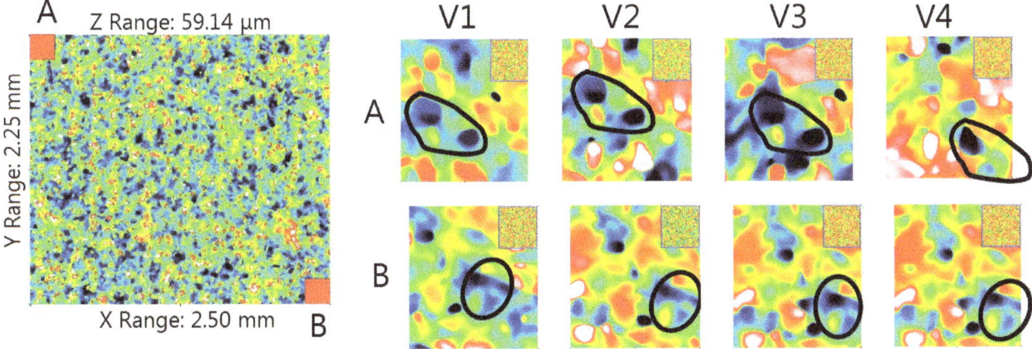

Figure 4. Views of the opposite corners of the topography maps, marked as A and B (left), obtained as a result of measuring corresponding areas on four replicas of $V1 \div V4$ made at different stages of grinding wheel wear (right).

In the four analyzed topography maps showing fragments of the GWAS with area 5.625 mm^2, the surface area of the repeating area (the area common to all four maps) was approximately 5.29 mm^2. It accounted for approximately 94% of the measuring area.

3. Automatic Determination of the Cut-Off Level in the SPIP 6.4.2 Software

In order to obtain data on particles and pores separated from the measured GWAS topography, which would be most useful during the analysis of the grinding wheel wear, two conditions were met: At different stages of wear, the same places on the grinding wheel were constantly measured, and the cut-off levels for particles and pores were determined in such a way that they cut off the same grain fragments, areas of gumming up of the GWAS, and cavities. In this part of the article, the results of the study showing the separation of particles using the automatically selected cut-off level in the SPIP 6.4.2 software will be presented.

Figure 5 shows a fragment of the measured active surface topography of the new grinding wheel (the actual material loss $V' = 0$ mm^3/mm), that found after removing the specific material volumes equal to $V' = 204$ mm^3/mm, and that found after complete wheel wear ($V' = 680$ mm^3/mm). The total wear of the grinding wheel was associated with an intensive increase in the grinding force. On the measured surface of the new grinding wheel, the cut-off level was clearly "higher" than on the measured topographies of the used grinding wheel. In other words, in the case of a used wheel topography, more areas belonging to the bond were above the cut-off level than was the case with the wheel before its operation. The results for the automatically determined (AT) cut-off level are also presented in Section 6.

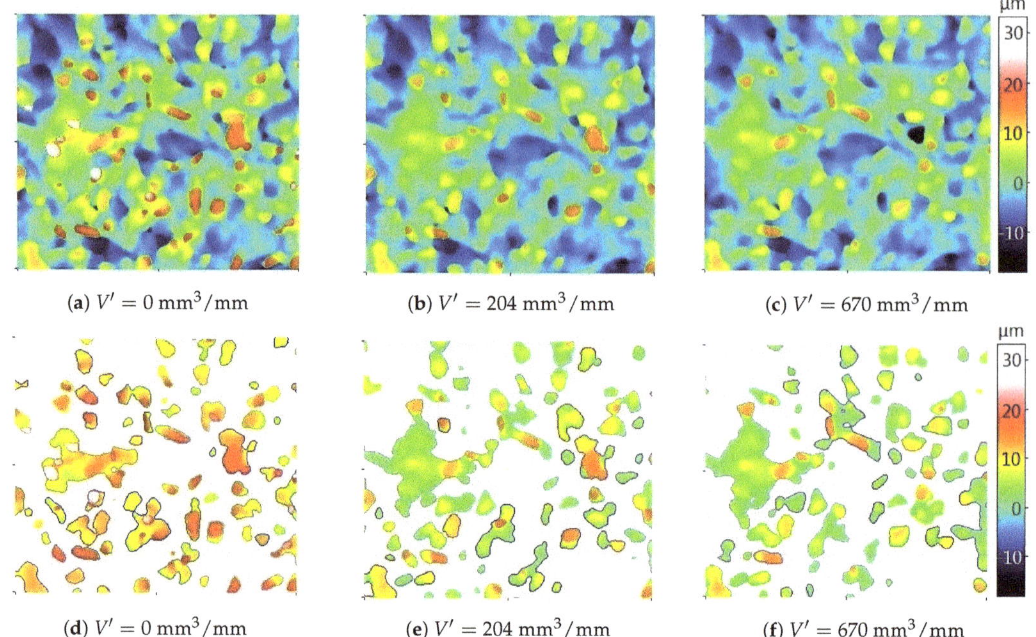

(**a**) $V' = 0$ mm³/mm   (**b**) $V' = 204$ mm³/mm   (**c**) $V' = 670$ mm³/mm

(**d**) $V' = 0$ mm³/mm   (**e**) $V' = 204$ mm³/mm   (**f**) $V' = 670$ mm³/mm

Figure 5. Maps of the new (**a**) and used (**b,c**) grinding wheel topography and the corresponding areas of particles above the automatically determined cut-off level (**d–f**). The color palette is the same for all six images. Grinding parameters: $v_s = 30$ m/s (for maximum grinding wheel diameter $d_s = 100$ mm), tangential feed speed $v_w = 4250$ mm/min, grinding depth $a_e = 20$ μm.

The percentage of the surface particles presented in Figure 5 was $A\% = 22.19\%$ for a new grinding wheel, $A\% = 22.81\%$ for the specific material loss $V' = 204$ mm³/mm, and $A\% = 22.20\%$ for a completely worn wheel. The analysis of the replica photos showed that there was no chip build-up on the GWAS. Therefore, increasing the proportion of particles (corresponding to abrasive grains) on the grinding wheel in the used condition compared to the new grinding wheel is illogical. The analysis of the percentage share of particles confirms the visual assessment of the areas above the cut-off level and the conclusion that the automatically selected cut-off level changes significantly depending on the degree of wear of the grinding wheel. A quantitative measure of the error in determining the cut-off level for abrasive grains at different stages of wear using the automatic threshold (AT) function is presented in Section 5.

4. Calculation of the Average Level of the Binder

The function of automatic determination of the cut-off level for pores and particles available in the SPIP 6.4.2 program for the analyzed topographies of the tested grinding wheels did not meet the expectations. For this reason, a proprietary algorithm for determining the cut-off level was developed, which depended on the level of binder on the grinding wheels. The main idea of the algorithm is based on the analysis of the bearing area curve (BAC), also called the Abbott–Firestone curve, and was already presented in Ref. [68]. It is also briefly illustrated in Figures 6 and 7.

The measurement points on the BAC related to the binder correspond particularly to the core area, i.e., the central part of the curve. Abrasive grains and deep pits after grain extraction on the BAC correspond, respectively, to the areas on the left side of the curve with the lowest values of the material share and the right side of the curve with the highest values of the material share (Figure 7). Compared to the grains and grain pits, the point height differences within the binder are small. The developed concept assumes that in order to determine the average level of the binder (average level, because the binder is related to a certain range of ordinate values, and not only a single value), the tangent line to the BAC with the smallest slope at a specific span of the search window should be determined. An analogous secant with a window span of 40% is used to calculate the topography parameters *Sk*, *Spk*, and *Svk* according to the [49] standard. The algorithm for determining the average level of the binder is shown in Figure 6.

Adoption of the span of the secant search window
with the smallest slope

⇓

Determining the secant with the smallest slope

⇓

Determining the values of the ordinates
at the boundaries of the search window
with the smallest inclination of the secant

⇓

Computing of the average height of the measurement points
within the prescribed limits
and accepting it as the average level of the binder

Figure 6. Algorithm for determining the average level of the binder.

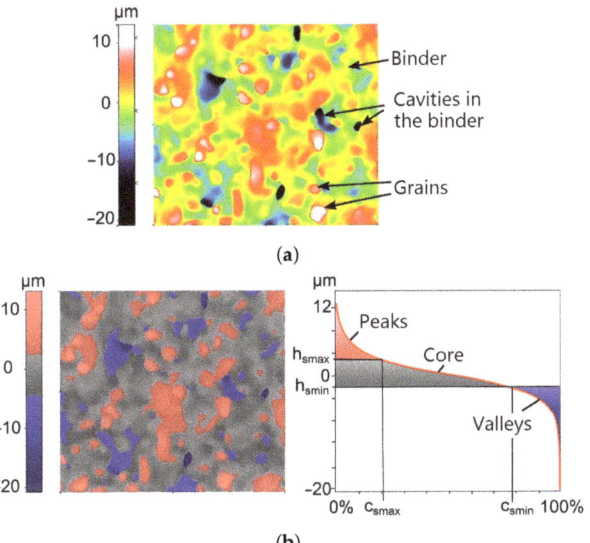

Figure 7. The areas of grains, binder, and cavities in the binder (**a**), as well as their corresponding areas on the bearing area curve (**b**) [68].

Based on the analysis of the surface share of the binder on the maps of the topography of the tested grinding wheels, in Ref. [68], the span of the search window for the smallest inclination was assumed for 40% of the material share. In the described research, it was checked how the width of the search window affects the determined value of the average level of the binder. For this purpose, 10 measured GWAS topographies with different degrees of grinding wheel wear were randomly selected. On each of these topographies, the average level of the binder was determined with a window span of 20% to 70% of the material content in 10% steps. Group differences were tested using the Wilcoxon signed-rank test with a significance level of $\alpha = 0.05$.

It was noticed that the seven study groups could be divided into two sets. The first contained groups related to the window spans of 20%, 30%, and 40% (set 20–40), and the second contained groups related to the window spans of 50%, 60%, and 70% (set 50–70). Within each set, the groups showed no statistically significant differences. On the other hand, significant differences were observed when comparing groups from different collections.

On average, in the set 50–70, the determined binder level was 0.17 μm higher than in the set 20–40. This was probably due to the fact that with the larger slope search windows, with the smallest inclination, those windows also contained points belonging to the grains. For this reason, a search window equal to 40% of material share was adopted for further research.

5. Comparison of Results Obtained Using Automatically Determined (AT) Cut-Off Level and the Developed Algorithm (OA)

In order to compare the effects of extracting grain areas using the automatically determined (AT) cut-off level and our own developed algorithm (OA), 10 pairs of measurement data were used. Each pair was related to the same measurement area, but at a different stage of wear. For each pair, the level of particle cut-off was manually determined, so that for a given pair, it ran in the same place and cut off analogous grain fragments.

Manual determination of the cut-off level included simultaneous analysis of 2–3 surface maps established at the same location on the grinding wheel and at different stages of its wear. One of the analyzed 2D maps was always a new grinding wheel surface map. On each map, a characteristic arrangement of details was selected—a pattern of several grains or a cavity with a characteristic shape. Views of the maps were enlarged to dimensions of about 0.5 mm × 0.5 mm because individual grains and cavities were clearly visible in such a window. The color palette on each map was always chosen to be similar to other analyzed maps (Figure 8). Thanks to that, it was easy to find and compare individual grains on different maps. After such preparations, the cut-off level from the cavities to the hills was manually "moved in space" and set at the level ensuring that most of the binder was below it. The process of moving the cut-off level can be compared to flooding the surface with water. For determining the cut-off level, the map of a new grinding wheel was used as the reference map. Given that all maps showed distinctive grains, attempts were made to determine the cut-off level in such a way that the grains were separated from their surroundings at the same level in all cases.

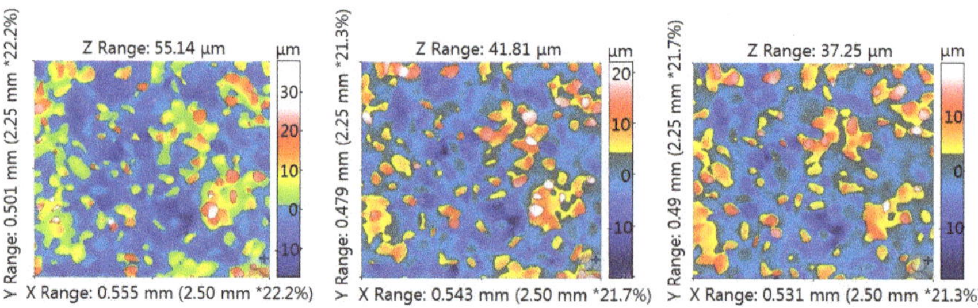

Figure 8. View of three maps analyzed at the same time to manually determine the cut-off level at different stages of wear; the area below the cut-off level is marked as blue.

For all data, the volume of particles above the cut-off level determined manually (V_m), automatically (V_{AT}), and with the use of the developed algorithm (V_{OA}) was calculated. Then, for pairs of surfaces and a given method of determining the cut-off level, the difference between the volume of particles at the earlier and later stages of wear (ΔV) was calculated. The quality of the effects of the automated computation of the threshold AT and OA cut-off methods was determined by comparing ΔV_{AT} and ΔV_{AT} to ΔV_m. The error (err) of the method was assumed to be the difference between the value obtained when determining the manual cut-off level (ΔV_m) and the value (ΔV) related to this method. The absolute error ($err_\%$) of the method was determined using the formula: $err_\% = err / \Delta V_m$.

For all 10 pairs of surfaces, the error in determining the cut-off level was lower with the use of the developed algorithm than with the automatic method. The quality of the cut-off level determination improved, on average, by 48% (Table 2). The maximum improvement was over 125%. It is also worth noting that the third error quartile in the OA method is smaller than the minimum error determined for AT. Therefore, it can be concluded that the developed algorithm allows the separation of the same grain areas at different stages of grinding wheel wear to a much better extent.

The average error in determining the cut-off level using the developed algorithm was over 9%. It is therefore appropriate to continue to search for and refine methods for determining the cut-off level to separate grain regions and pits on the measured GWAS topographies. It should be remembered that the manual method is also not ideal and is burdened with an error resulting from the subjective determination of the cut-off level, depending on the researcher.

Figures 9 and 10 show areas of particles above the cut-off level determined automatically and on the basis of the average binder level (the methodology for calculating the cut-off levels for particles and pores based on the average level of the binder is presented later in the article).

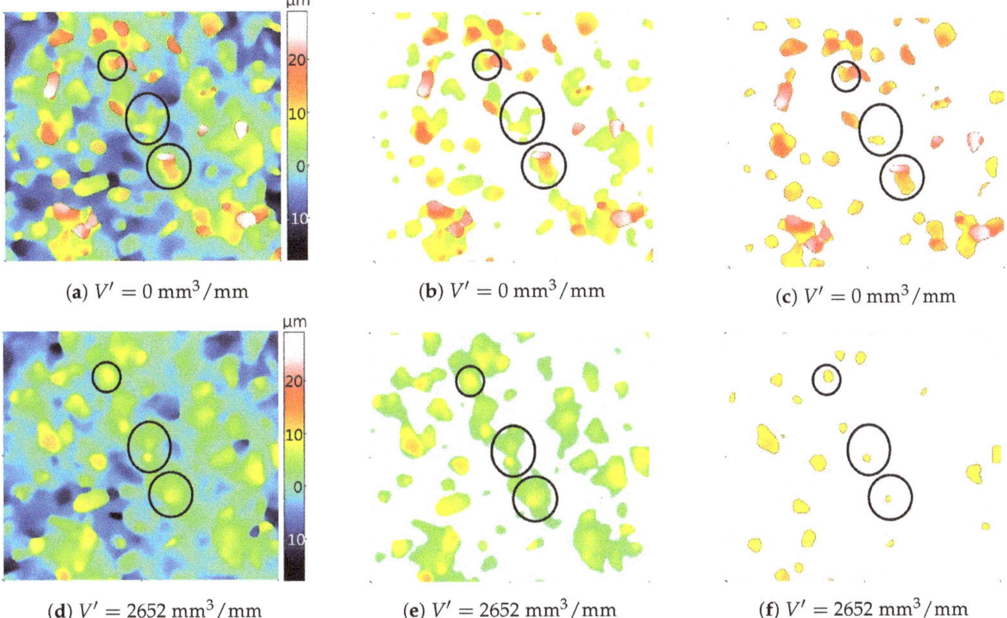

(**a**) $V' = 0\ \mathrm{mm}^3/\mathrm{mm}$ (**b**) $V' = 0\ \mathrm{mm}^3/\mathrm{mm}$ (**c**) $V' = 0\ \mathrm{mm}^3/\mathrm{mm}$

(**d**) $V' = 2652\ \mathrm{mm}^3/\mathrm{mm}$ (**e**) $V' = 2652\ \mathrm{mm}^3/\mathrm{mm}$ (**f**) $V' = 2652\ \mathrm{mm}^3/\mathrm{mm}$

Figure 9. Maps of the new (**a**) and used (**d**) grinding wheel topography and the corresponding areas of particles above the automatically determined cut-off level (**b,e**) and the cut-off level determined by the developed algorithm (**c,f**). The color palette is the same for all six images. Grinding parameters: $v_s = 40\ \mathrm{m/s}$ (for the maximum grinding wheel diameter $d_s = 100\ \mathrm{mm}$), tangential feed speed $v_w = 7500\ \mathrm{mm/min}$, grinding depth $a_e = 20\ \mathrm{\mu m}$.

Table 2. Mean value, standard deviation, and quartiles Q_1, Q_2, and Q_3 computed for the analyzed GWAS areas.

	OA Error [%]	AT Error [%]	Improvement [%]
mean	9.35	57.64	48.29
std	5.6	46.17	42.87
min	0.67	14.76	7.61
Q_1	5.34	25.85	14.94
Q_2	9.17	43.92	34.75
Q_3	13.95	80.36	71.24
max	17.65	141.00	125.74

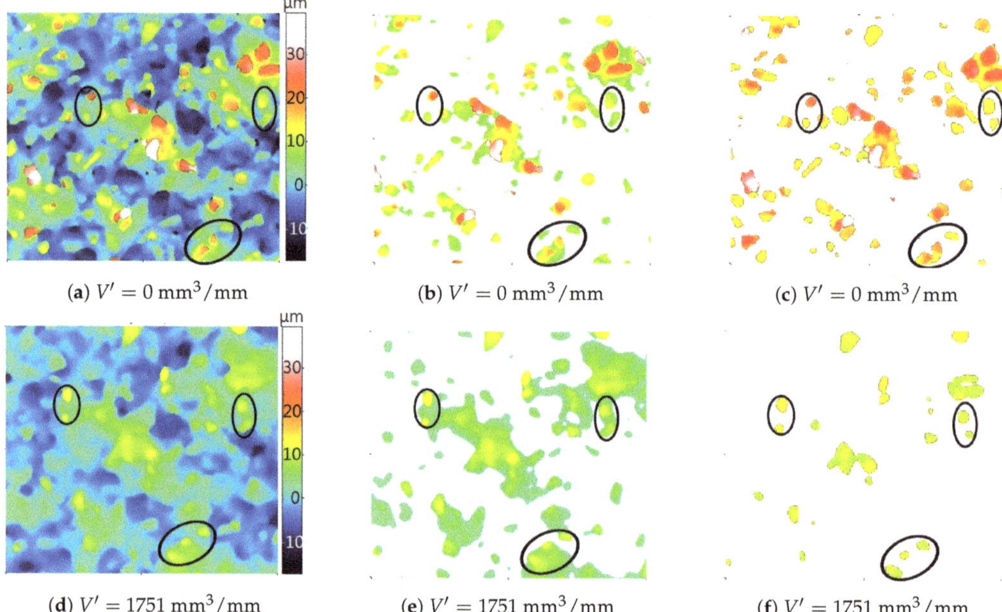

(**a**) $V' = 0 \text{ mm}^3/\text{mm}$　　(**b**) $V' = 0 \text{ mm}^3/\text{mm}$　　(**c**) $V' = 0 \text{ mm}^3/\text{mm}$

(**d**) $V' = 1751 \text{ mm}^3/\text{mm}$　　(**e**) $V' = 1751 \text{ mm}^3/\text{mm}$　　(**f**) $V' = 1751 \text{ mm}^3/\text{mm}$

Figure 10. Maps of the new (**a**) and used (**d**) grinding wheel topography and the corresponding areas of particles above the automatically determined cut-off level (**b,e**) and the cut-off level determined by the developed algorithm (**c,f**). The color palette is the same for all six images. Grinding parameters: $v_s = 20 \text{ m/s}$ (for the maximum grinding wheel diameter $d_s = 100 \text{ mm}$), tangential feed speed $v_w = 4250 \text{ mm/min}$, grinding depth $a_e = 10 \text{ μm}$.

For two grinding wheels, their topography was analyzed before starting the grinding process and after removing a certain volume of material. In order to facilitate the comparison of the results obtained with the two analyzed methods of determining the cut-off level, three groups of grains were distinguished (circled) for each of the grinding wheels. In both presented cases, the cut-off level determined automatically on the new grinding wheel (the specific material loss $V' = 0 \text{ mm}^3/\text{mm}$) was clearly "higher" than on the used grinding wheel. The determination of the cut-off level in relation to the average bond level resulted in the fact that the cut-off level at different stages of the grinding wheel wear varied to a lesser extent. As a result, it was possible to more reliably determine changes in height and volume parameters of particles due to wear.

6. Segmentation of Grains, Areas of Sticking, and Pores on GWAS Topographies

Segmentation of grains into the GWAS topography consisted of separating the measured point cloud particles located above the appropriately selected cut-off level. The particles on the measured object corresponded to the abrasive grains or their highest frag ments and the areas of sticking, which were associated with gumming up of grinding wheels. In turn, the segmentation of the pores on the measured topographies was associated with the emergence of areas from these topographies that constitute cavities in the binder due to their unevenness or resulting from grain breaking.

The most detailed information about the GWAS can be obtained when one particle corresponds to only one grain or stagnation area. Likewise, the pores observed should represent single cavities. To follow the aim of segmentation, the cut-off levels for particles and pores were not carried out at the designated average level of the binder. In the case of particles, the value of the mean level of the binder increased by 10 μm was taken as the value that cuts off some of the points for further analysis. The "elevation" of the cut-off level compared to the average level of the binder served to remove more measuring points corresponding to the binder and low-lying portions of the grains, which improved the quality of grain recognition (Figure 11).

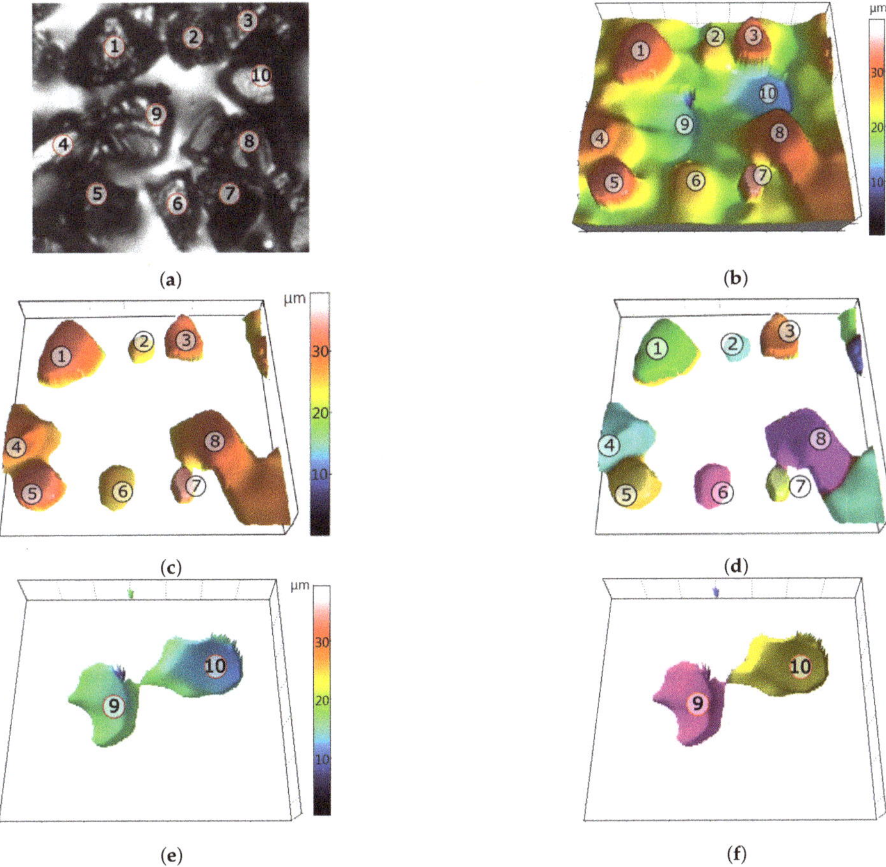

Figure 11. An example of particle and pore segmentation from the measured topography: (**a**) an image of a replica fragment with dimensions of 140 μm × 140 μm, (**b**) the corresponding map of the topography, (**c**) areas of particles above the cut-off level, (**d**) separated particles, (**e**) areas of pores below the cut-off level, and (**f**) isolated pores.

The purpose of isolating deep depressions on the measured topographies and determining their parameters was to provide information on the processes of grain extraction and the collection of chips and material derived from grains on the GWAS. The cut-off level below which the recesses were analyzed was set at the mean level of the binder reduced by 5 μm. The "lowering" of the cut-off level compared to the average level of the adhesive served to improve pore separation by removing relatively shallow depressions resulting from uneven surfaces of the binder.

After the cut-off levels for particles and pores had been established, the *Threshold* command was used to detect these elements. During the analysis of particles, they were divided into two categories (Figure 12):

- Particles of type "particles": particles with an area in the range $[100, 2500]$ μm^2,
- Particles of type "sticking": particles with an area bigger than 2500 μm^2.

The limit values of the particles' surface area for the above-mentioned categories were determined on the basis of the topography analysis of new grinding wheels. Limiting the size of the "grains" above 100 μm was to eliminate any residual binder, peaks resulting from the optical measurement system used, and peaks of low grains with low cutting potential. The value distinguishing "grains" from "sticking" was due to the fact that on each of the examined fragments of the topography of new grinding wheels, at least 99% of grains had a surface area smaller than 2500 μm.

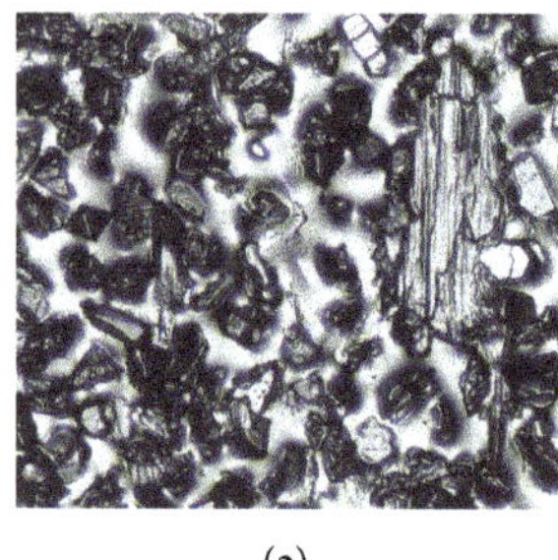 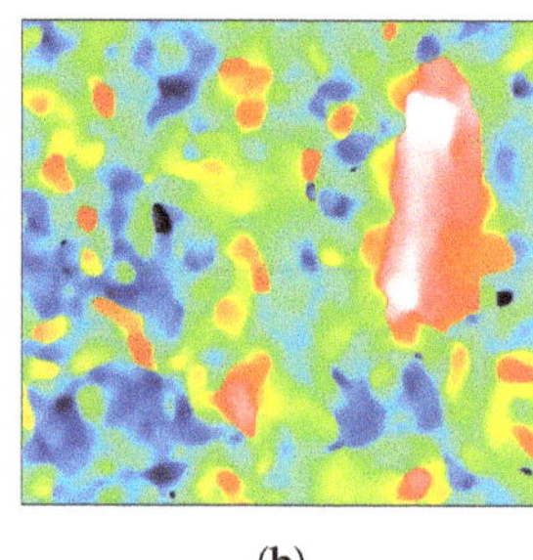 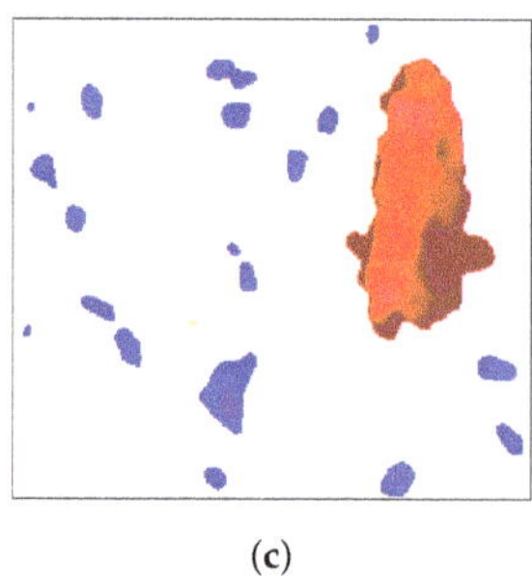

(a) (b) (c)

Figure 12. Example of grain-type and stick-type particles: replica image (**a**), topography map (**b**), grain-type particles (blue), and stick-type particles (red) (**c**).

7. The Results of the Grinding Wheel Topography Tests During the Service Life

In the previous sections, the methodology of segmentation of the areas of abrasive grains, sticking, and pores was presented, including the determination of the average level of the binder using the proposed method. The results of the research on the active surface topography of the grinding wheel throughout its life are presented for a grinding wheel operating with the following adjustable parameters: $v_s = 20$ m/s (for the maximum wheel diameter, $d_s = 100$ mm), tangential feed speed $v_w = 4250$ mm/min, and grinding depth $a_e = 10$ μm. The moment when an intensive increase in the grinding force was observed was assumed as the end of the grinding wheel's life. The tested grinding wheel was completely worn after achieving a specific material loss equal to $V' = 1751$ mm^3/mm. The GWAS topography tests were carried out five times—at different stages of wear. Each time, six measurement areas were measured on the GWAS replicas.

Figure 13 shows the fragments of the measured surface (0.5 mm $\times$ 0.5 mm) after removing various volumes of material. The height of the grains above the binder and the number of grains above the cut-off are visible as the volume of material removed increases. The changes in the depression areas in the binder are more difficult to notice. This is due to the fact that these changes occur mainly due to the removal of grains from the binder, which took place relatively rarely. The analysis of topography maps and microscopic images of GWAS replicas showed that the dominant type of wear was grain breaking. Hence, the differences in images and parameters of particles are more visible.

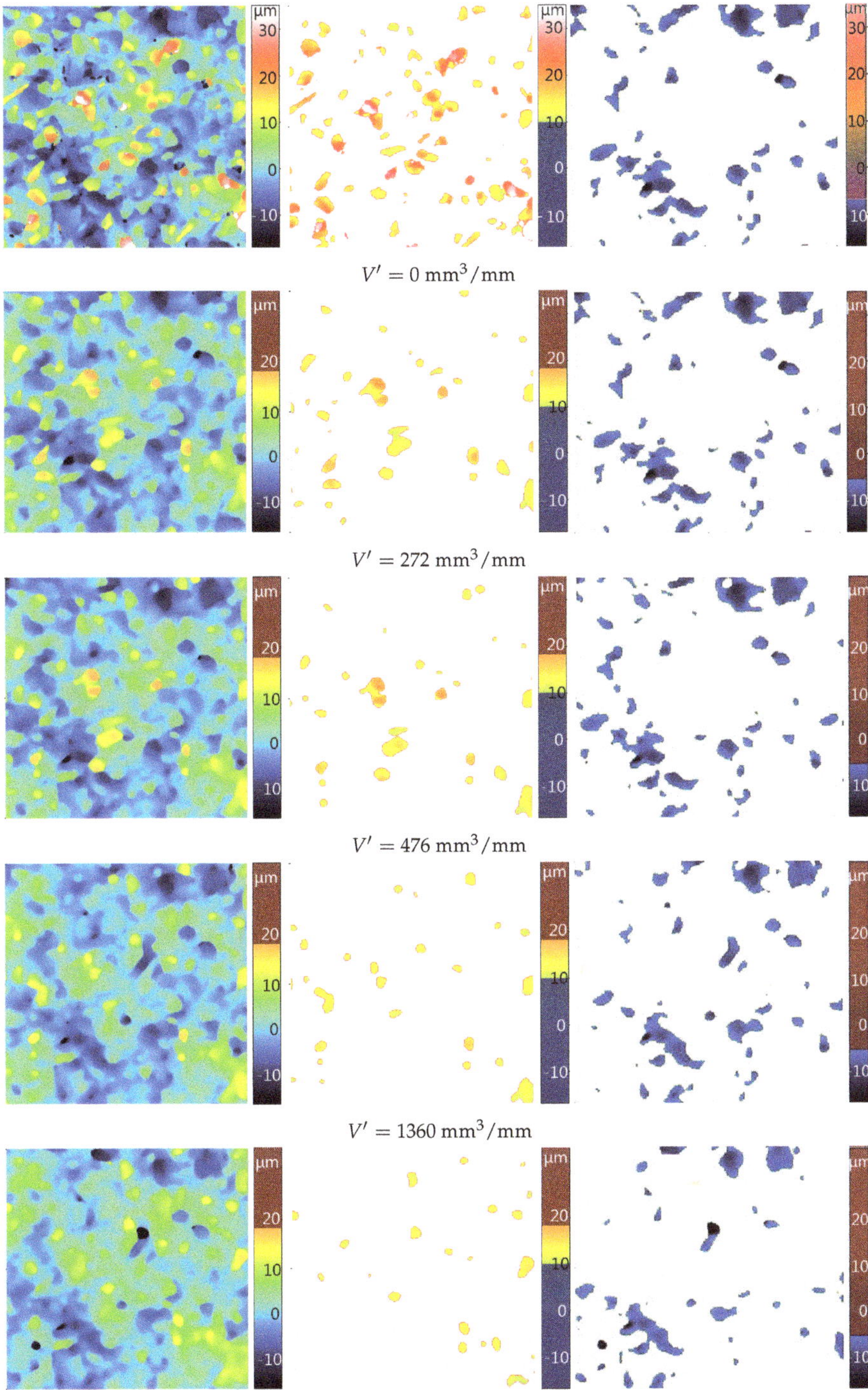

$$V' = 0 \text{ mm}^3/\text{mm}$$

$$V' = 272 \text{ mm}^3/\text{mm}$$

$$V' = 476 \text{ mm}^3/\text{mm}$$

$$V' = 1360 \text{ mm}^3/\text{mm}$$

$$V' = 1751 \text{ mm}^3/\text{mm} \text{ (after the GWAS was worn out)}$$

Figure 13. GWAS topography maps (**left**), corresponding particles maps above the cut-off level (**center**), and pore maps (**right**) for different values of the actual material loss V'.

The segmentation of particles and pores allowed for the determination of quantitative parameters of these elements. An example of a quantitative analysis of the GWAS at various stages of consumption is presented on the basis of volumetric parameters, i.e., grain volume ($Vsum(g)$), pore volume ($Vsum(p)$), and sticking area volume ($Vsum(s)$) per unit surface (Figure 14).

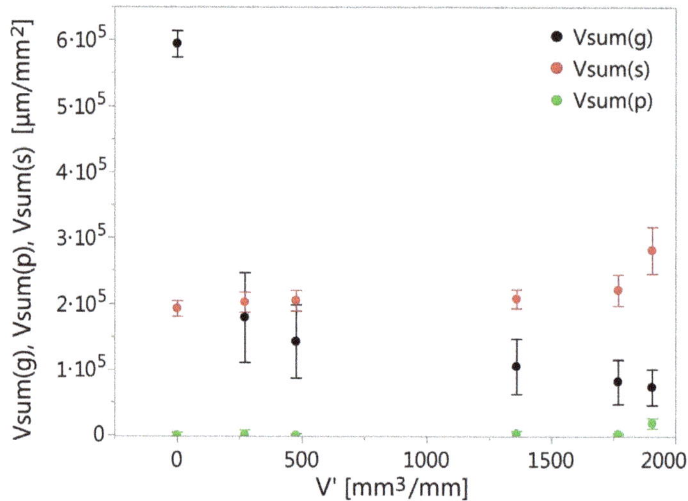

Figure 14. Grain volume ($Vsum(g)$), pore volume ($Vsum(p)$), and sticking area volume ($Vsum(s)$) per unit surface after removing different volumes of material.

At the beginning of the grinding wheel's operation, the grain volume decreased very intensively. The wear of the grains occurred mainly due to their chipping, as the pore volume increased slightly during this time. The largest pore volume gradient was observed at the end of the grinding wheel's life. Therefore, it can be concluded that, at that stage of grinding wheel's operation, the extraction of the grains took place with the greatest intensity. During the life of the wheel, $552 \cdot 10^3$ $\mu m^3/mm^2$ of the abrasive was lost. The pore volume increased by about $79 \cdot 10^3$ $\mu m^3/mm^2$. After the wheel was completely worn out, there was $35.5 \cdot 10^3$ $\mu m^3/mm^2$ of the material that acted as a sealing material in the process of gumming up of the investigated GWAS.

More examples concerning the application of the proposed methodology of determining the cut-off level and segmentation of grains, areas of sticking, and pores on a GWAS in the context of grinding wheel performance are presented in [69]. That paper describes the results of research related to determination of GWAS parameters that are particularly sensitive to wear, including the parameters of grains, areas of sticking, and pores segmented using the developed methodology. It also presents the established mathematical relationships between the grinding process parameters and the specific material loss as well as a few selected GWAS parameters.

8. Conclusions

Based on the research on the active surface topography of a single-layer grinding wheel with a binder applied with the galvanic method and with cBN coating, the following conclusions can be drawn:

- Based on the measurement of the GWAS topography, it is possible to observe changes in the microgeometry of the grinding wheel occurring as a result of wear, and to perform a qualitative and quantitative analysis based on horizontal, height, and volume characteristics.

- The developed methodology for GWAS measurement allowed for the measurement of a GWAS in approximately the same places on the grinding wheel at different stages of wear. For the four surveyed areas, the common area accounted for 94% of their area.
- The analysis of particles above the specified cut-off level and pores below the cut-off level provides quantitative information, including height and volume information, on such characteristic GWAS elements as: abrasive grains, cavities in the binder caused by grain breakout, and areas of sticking.
- When analyzing changes in the GWAS topography areas related to grains, deep cavities, and sticking areas associated with gumming up of GWAS, which occur as a result of wear, it is important that the same fragments of these elements are constantly separated at different stages of wear. The cut-off levels for particles and pores should be at the same position on the wheel.
- The developed algorithm for determining the average level of the binder, against which the cut-off levels for particles and pores were determined, allows one to obtain more useful information about particles and pores analyzed in terms of grinding wheel wear than the available algorithm of automatic determination of the cut-off level in commercial software.
- Among the particles, there are areas corresponding to abrasive grains and sticking areas corresponding to gumming up of the GWAS. The categorization of particles into "grains" and "sticking" was established based on the surface area of the particle. The limit value for grain differentiation and sticking in the case of a grinding wheel with grain number B35 was set at 2500 µm.
- For a grinding wheel with grinding speed $v_s = 20$ m/s (for a maximum grinding wheel diameter of $d_s = 100$ mm), feed rate of $v_w = 4250$ mm/min, and grinding depth of $a_e = 10$ µm, the largest volume of grains were broken in the initial period of the grinding wheel's operation. The extraction of grains from the bond was most intense at the end of the grinding wheel's life.

Author Contributions: Conceptualization, A.K., A.B., T.R., P.K., and A.O.; methodology, A.K. and A.B.; software, A.B., P.K., and A.O.; validation, A.B. and A.K.; investigation, A.K. and A.B.; writing—original draft preparation, A.B.; writing—review and editing, A.K.; visualization, A.B. and T.R.; supervision, A.K. All authors have read and agreed to the published version of the manuscript.

Funding: This research received no external funding.

Conflicts of Interest: The authors declare no conflict of interest.

Abbreviations

The following abbreviations are used in this manuscript:

AT	Automatically determined
BAC	Bearing area curve
cBN	Cubic boron nitride
GWAS	Grinding wheel active surface
ISO	International Standards Organization
OA	Developed algorithm
ST	Surface texture
SLGW	Single-layer grinding wheel

References

1. Qi, H.; Rowe, W.; Mills, B. Experimental investigation of contact behaviour in grinding. *Tribol. Int.* **1997**, *30*, 283–294. [CrossRef]
2. Pal, B.; Chattopadhyay, A.; Chattopadhyay, A. Development and performance evaluation of monolayer brazed cBN grinding wheel on bearing steel. *Int. J. Adv. Manuf. Technol.* **2010**, *48*, 935–944. [CrossRef]
3. Bhaduri, D.; Kumar, R.; Jain, A.; Chattopadhyay, A. On tribological behaviour and application of TiN and MoS2-Ti composite coating for enhancing performance of monolayer cBN grinding wheel. *Wear* **2010**, *268*, 1053–1065. [CrossRef]

4. Klocke, F.; Soo, S.L.; Karpuschewski, B.; Webster, J.A.; Novovic, D.; Elfizy, A.; Axinte, D.A.; Tonissen, S. Abrasive machining of advanced aerospace alloys and composites. *CIRP Ann. Manuf. Technol.* **2015**, *64*, 581–604. [CrossRef]

5. Malkin, S.; Guo, C. *Grinding Technology: Theory and Application of Machining with Abrasives*; Industrial Press: New Tork, NY, USA, 2008.

6. Shi, Z.; Malkin, S. Wear of Electroplated CBN Grinding Wheels. *J. Manuf. Sci. Eng.* **2005**, *128*, 110–118. [CrossRef]

7. Nguyen, A.T.; Butler, D.L. Correlation of grinding wheel topography and grinding performance: A study from a viewpoint of three-dimensional surface characterisation. *J. Mater. Process. Technol.* **2008**, *208*, 14–23. [CrossRef]

8. Butler, D.; Blunt, L.; See, B.; Webster, J.; Stout, K. The characterisation of grinding wheels using 3D surface measurement techniques. *J. Mater. Process. Technol.* **2002**, *127*, 234–237. [CrossRef]

9. Lipiński, D.; Kacalak, W. Metrological Aspects of Abrasive Tool Active Surface Topography Evaluation. *Metrol. Meas. Syst.* **2016**, *23*, 567–577. [CrossRef]

10. Kacalak, W.; Tandecka, K. A method and new parameters for assessing the active surface topography of diamond abrasive films. *J. Mach. Eng.* **2016**, *16*, 95–108.

11. Xie, J.; Xu, J.; Tang, Y.; Tamaki, J. 3D graphical evaluation of micron-scale protrusion topography of diamond grinding wheel. *Int. J. Mach. Tools Manuf.* **2008**, *48*, 1254–1260. [CrossRef]

12. Ismail, M.F.; Yanagi, K.; Isobe, H. Geometrical transcription of diamond electroplated tool in ultrasonic vibration assisted grinding of steel. *Int. J. Mach. Tools Manuf.* **2012**, *62*, 24–31. [CrossRef]

13. Guo, C.; Ranganath, S.; McIntosh, D.; Elfizy, A. Virtual high performance grinding with CBN wheels. *CIRP Ann. Manuf. Technol.* **2008**, *57*, 325–328. [CrossRef]

14. Xun, L.; Fanjun, M.; Wei, C.; Shuang, M. The CNC grinding of integrated impeller with electroplated CBN wheel. *Int. J. Adv. Manuf. Technol.* **2015**, *79*, 1353–1361. [CrossRef]

15. You, H.Y.; Ye, P.Q.; Wang, J.; Deng, X.Y. Design and application of CBN shape grinding wheel for gears. *Int. J. Mach. Tools Manuf.* **2003**, *43*, 1269–1277. [CrossRef]

16. Lv, M.; Zhang, M.D.; Zhao, H. The Deformation Analysis on Tooth Profiles with Electroplated CBN Hard Gear-Honing-Tools. *Adv. Mater. Res.* **2012**, *426*, 159–162. [CrossRef]

17. Kohler, J.; Schindler, A.; Woiwode, S. Continuous generating grinding. Tooth root machining and use of CBN-tools. *CIRP Ann. Manuf. Technol.* **2012**, *61*, 291–294. [CrossRef]

18. Naik, D.N.; Mathew, N.T.; Vijayaraghavan, L. Wear of Electroplated Super Abrasive CBN Wheel during Grinding of Inconel 718 Super Alloy. *J. Manuf. Process.* **2019**, *43*, 1–8. [CrossRef]

19. Rao, X.; Zhang, F.; Lu, Y.; Luo, X.; Ding, F.; Li, C. Analysis of diamond wheel wear and surface integrity in laser-assisted grinding of RB-SiC ceramics. *Ceram. Int.* **2019**, *45*, 24355–24364. [CrossRef]

20. Kang, M.; Zhang, L.; Tang, W. Study on three-dimensional topography modeling of the grinding wheel with image processing techniques. *Int. J. Mech. Sci.* **2020**, *167*, 105241. [CrossRef]

21. Reddy, P.P.; Ghosh, A. Some critical issues in cryo-grinding by a vitrified bonded alumina wheel using liquid nitrogen jet. *J. Mater. Process. Technol.* **2016**, *229*, 329–337. [CrossRef]

22. Ichida, Y. Fractal Analysis of Micro Self-Sharpening Phenomenon in Grinding with Cubic Boron Nitride (cBN) Wheels. In *Scanning Electron Microscopy*; Kazmiruk, V., Ed.; Intech: Rijeka, Croatia, 2012.

23. Capela, P.; Carvalho, S.; Guedes, A.; Pereira, M.; Carvalho, L.; Correia, J.; Soares, D.; Gomes, J. Effect of sintering temperature on mechanical and wear behaviour of a ceramic composite. *Tribol. Int.* **2018**, *120*, 502–509. [CrossRef]

24. Ahmed, A.; Bahadur, S.; Russell, A.; Cook, B. Belt abrasion resistance and cutting tool studies on new ultra-hard boride materials. *Tribol. Int.* **2009**, *42*, 706–713. [CrossRef]

25. Dai, C.; Ding, W.; Xu, J.; Ding, C.; Huang, G. Investigation on size effect of grain wear behavior during grinding nickel-based superalloy Inconel 718. *Int. J. Adv. Manuf. Technol.* **2017**, 1–11. [CrossRef]

26. Dai, C.; Ding, W.; Xu, J.; Fu, Y.; Yu, T. Influence of grain wear on material removal behavior during grinding nickel-based superalloy with a single diamond grain. *Int. J. Mach. Tools Manuf.* **2017**, *113*, 49–58. [CrossRef]

27. Choudhary, A.; Naskar, A.; Paul, S. *Evaluation of Surface Morphology of Yttria-Stabilized Zirconia with the Progress of Wheel Wear in High-Speed Grinding*; Springer: Singapore, 2019; pp. 315–323.

28. Liu, W.; Deng, Z.; Shang, Y.; Wan, L. Parametric evaluation and three-dimensional modelling for surface topography of grinding wheel. *Int. J. Mech. Sci.* **2019**, *155*, 334–342. [CrossRef]

29. Shen, J.; Wang, J.; Jiang, B.; Xu, X. Study on wear of diamond wheel in ultrasonic vibration-assisted grinding ceramic. *Wear* **2015**, *332–333*, 788–793. [CrossRef]

30. Hood, R.; Cooper, P.; Aspinwall, D.; Soo, S.; Lee, D. Creep feed grinding of gamma-TiAl using single layer electroplated diamond superabrasive wheels. *CIRP J. Manuf. Sci. Technol.* **2015**. [CrossRef]

31. Liang, Z.; Wang, X.; Wu, Y.; Xie, L.; Liu, Z.; Zhao, W. An investigation on wear mechanism of resin-bonded diamond wheel in Elliptical Ultrasonic Assisted Grinding EUAG of monocrystal sapphire. *J. Mater. Process. Technol.* **2012**, *212*, 868–876. [CrossRef]

32. Hwang, T.W.; Malkin, S.; Evans, C.J. High Speed Grinding of Silicon Nitride With Electroplated Diamond Wheels, Part 2: Wheel Topography and Grinding Mechanisms. *J. Manuf. Sci. Eng.* **1999**, *122*, 42–50. [CrossRef]

33. Liu, Q.; Chen, X.; Gindy, N. Assessment of Al2O3 and superabrasive wheels in nickel-based alloy grinding. *Int. J. Adv. Manuf. Technol.* **2006**, *33*, 940–951. [CrossRef]

44. Pellegrin, D.V.D.; Corbin, N.D.; Baldoni, G.; Torrance, A.A. Diamond particle shape: Its measurement and influence in abrasive wear. *Tribol. Int.* **2009**, *42*, 160–168. [CrossRef]

45. Blunt, L.; Ebdon, S. The application of three-dimensional surface measurement techniques to characterizing grinding wheel topography. *Int. J. Mach. Tools Manuf.* **1996**, *36*, 1207–1226. [CrossRef]

46. Jourani, A.; Hagège, B.; Bouvier, S.; Bigerelle, M.; Zahouani, H. Influence of abrasive grain geometry on friction coefficient and wear rate in belt finishing. *Tribol. Int.* **2013**, *59*, 30–37. [CrossRef]

47. Ismail, M.F.; Yanagi, K.; Isobe, H. Characterization of geometrical properties of electroplated diamond tools and estimation of its grinding performance. *Wear* **2011**, *271*, 559–564. [CrossRef]

48. Vidal, G.; Ortega, N.; Bravo, H.; Dubar, M.; González, H. An Analysis of Electroplated cBN Grinding Wheel Wear and Conditioning during Creep Feed Grinding of Aeronautical Alloys. *Metals* **2018**, *8*, 350. [CrossRef]

49. Cai, R.; Rowe, W.B. Assessment of vitrified CBN wheels for precision grinding. *Int. J. Mach. Tools Manuf.* **2004**, *44*, 1391–1402. [CrossRef]

50. Yan, L.; Rong, Y.; Jiang, F.; Zhou, Z. Three-dimension surface characterization of grinding wheel using white light interferometer. *Int. J. Adv. Manuf. Technol.* **2011**, *55*, 133–141. [CrossRef]

51. Zhao, Q.; Guo, B. Ultra-precision grinding of optical glasses using mono-layer nickel electroplated coarse-grained diamond wheels. Part 1: {ELID} assisted precision conditioning of grinding wheels. *Precis. Eng.* **2015**, *39*, 56–66. [CrossRef]

52. Wang, Z.; Zhang, Z.; Sun, Y.; Gao, K.; Liang, Y.; Li, X.; Ren, L. Wear behavior of bionic impregnated diamond bits. *Tribol. Int.* **2016**, *94*, 217–222. [CrossRef]

53. Tan, S.; Zhang, W.; Duan, L.; Pan, B.; Rabiei, M.; Li, C. Effects of MoS2 and WS2 on the matrix performance of WC based impregnated diamond bit. *Tribol. Int.* **2019**, *131*, 174–183. [CrossRef]

54. Shi, Z.; Malkin, S. An Investigation of Grinding with Electroplated CBN Wheels. *CIRP Ann. Manuf. Technol.* **2003**, *52*, 267–270. [CrossRef]

55. Schoenhagen, Y.; Vasquez, J. *Process Simulation of Mono-Layer Super Abrasive Grinding Wheels*; WPI: Shanghai, China, 2012.

56. Kapłonek, W.; Nadolny, K. Assessment of the grinding wheel active surface condition using SEM and image analysis techniques. *J. Braz. Soc. Mech. Sci. Eng.* **2013**, *35*, 207–215. [CrossRef]

57. Kaplonek, W.; Nadolny, K. SEM-based Morphological Analysis of the New Generation AlON-based Abrasive Grains (Abral[®]) with Reference to Al2O3/SiC/cBN Abrasives. *Acta Microsc.* **2015**, *24*, 64–78.

58. Hwang, T.W.; Evans, C.; Whitenton, E.P.; Malkin, S. High Speed Grinding of Silicon Nitride With Electroplated Diamond Wheels, Part 1: Wear and Wheel Life. *J. Manuf. Sci. Eng.* **1999**, *122*, 32–41. [CrossRef]

59. ISO. 25178-2:2012 Geometrical Product Specifications (GPS). In *Surface Texture: Areal—Part 2: Terms, Definitions and Surface Texture Parameters*; ISO: Geneva, Switzerland, 2012.

60. ASME. B46:1 Surface Texture. In *Surface Roughness, Waviness, and Lay*; ASME: New York, NY, USA, 2019.

61. Stout, K.J.; Sullivan, P.J.; Dong, W.P.; Mainsah, E.; Luo, N.; Mathia, T.; Zahouani, H. *The Development of Methods for the Characterization of Roughness on three Dimensions*; Commission of the European Communities: Luxembourg, 1994.

62. Wang, W.; Salvatore, F.; Rech, J.; Li, J. Comprehensive investigation on mechanisms of dry belt grinding on AISI52100 hardened steel. *Tribol. Int.* **2018**, *121*, 310–320. [CrossRef]

63. Nadolny, K. The method of assessment of the grinding wheel cutting ability in the plunge grinding. *Open Eng.* **2012**, *2*, 399–409. [CrossRef]

64. Nadolny, K. Wear phenomena of grinding wheels with sol-gel alumina abrasive grains and glass-ceramic vitrified bond during internal cylindrical traverse grinding of 100Cr6 steel. *Int. J. Adv. Manuf. Technol.* **2015**, *77*, 83–98. [CrossRef]

65. Kaiser, J.F. On a simple algorithm to calculate the 'energy' of a signal. In Proceedings of the International Conference on Acoustics, Speech, and Signal Processing, Albuquerque, NM, USA, 3–6 April 1990; Volume 1, pp. 381–384. [CrossRef]

66. Kacalak, W.; Lipiński, D.; Szafraniec, F.; Tandecka, K. The methodology of the grinding wheel active surface evaluation in the aspect of their machining potential. *Mechanik* **2018**, *91*, 690–697. [CrossRef]

67. Chong-Ching, C. An application of lubrication theory to predict useful flow-rate of coolants on grinding porous media. *Tribol. Int.* **1997**, *30*, 575–581. [CrossRef]

68. Chang, C.C.; Wang, S.H.; Szeri, A.Z. On the Mechanism of Fluid Transport Across the Grinding Zone. *ASME J. Manuf. Sci. Eng.* **1996**, *3*, 332–338. [CrossRef]

69. Setti, D.; Kirsch, B.; Aurich, J.C. Characterization of micro grinding tools using optical profilometry. *Optics Lasers Eng.* **2019**, *121*, 150–155. [CrossRef]

70. Kapłonek, W.; Nadolny, K.; Tomkowski, R.; Valicek, J. High-accuracy surface topography measurements of abrasive tools using a 3D optical profiling system. *Pomiary Autom. Kontrola* **2012**, *58*, 443–447.

71. Ye, R.; Jiang, X.; Blunt, L.; Cui, C.; Yu, Q. The application of 3D-motif analysis to characterize diamond grinding wheel topography. *Measurement* **2016**, *77*, 73–79. [CrossRef]

72. Senin, N.; Blunt, L.A.; Leach, R.K.; Pini, S. Morphologic segmentation algorithms for extracting individual surface features from areal surface topography maps. *Surf. Topogr. Metrol. Prop.* **2013**, *1*, 015005. [CrossRef]

73. Lou, S.; Jiang, X.; Scott, P.J. Application of the morphological alpha shape method to the extraction of topographical features from engineering surfaces. *Measurement* **2013**, *46*, 1002–1008. [CrossRef]

64. Lou, S.; Pagani, L.; Zeng, W.; Jiang, X.; Scott, P. Watershed segmentation of topographical features on freeform surfaces and it application to additively manufactured surfaces. *Precis. Eng.* **2020**, *63*, 177–186. [CrossRef]
65. Li, D.; Wang, B.; Qiao, Z.; Jiang, X. Ultraprecision machining of microlens arrays with integrated on-machine surface metrolog. *Opt. Express* **2019**, *27*, 212–224. [CrossRef]
66. Scott, P.J. An algorithm to extract critical points from lattice height data. *Int. J. Mach. Tools Manuf.* **2001**, *41*, 1889–1897. [CrossRef]
67. Zhao, H.; Anwer, N.; Bourdet, P. Curvature-based Registration and Segmentation for Multisensor Coordinate Metrology. *Procedi. CIRP* **2013**, *10*, 112–118. [CrossRef]
68. Bazan, A.; Kawalec, A. Methods of grain separation from single-layer grinding wheel topography. *Mechanik* **2018**, *91*, 926–928. doi:10.17814/mechanik.2018.10.164. [CrossRef]
69. Bazan, A.; Kawalec, A.; Rydzak, T.; Kubik, P. Variation of Grain Height Characteristics of Electroplated cBN Grinding-Wheel Activ. Surfaces Associated with Their Wear. *Metals* **2020**, *10*, 1479. [CrossRef]

materials

Article

A Multiscale Topographical Analysis Based on Morphological Information: The HEVC Multiscale Decomposition

Tarek Eseholi [1], François-Xavier Coudoux [1], Patrick Corlay [1], Rahmad Sadli [1] and Maxence Bigerelle [2,*]

1   Opto-Acousto-Electronics Department, Institute of Electronics,
    Microelectronics and Nanotechnology (IEMN), UMR-CNRS 8520,
    Polytechnic University of Hauts-de-France, Le Mont Houy, 59313 Valenciennes, France;
    tareksaad.eseholi@gmail.com (T.E.); francois-xavier.coudoux@uphf.fr (F.-X.C.); Patrick.Corlay@uphf.fr (P.C.);
    rahmadsadli@gmail.com (R.S.)
2   Laboratory of Industrial and Human Automation Control, Mechanical Engineering and Computer
    Science, (LAMIH) UMR-CNRS 8201, Polytechnic University of Hauts-de-France, Le Mont Houy,
    59313 Valenciennes, France
*   Correspondence: maxence.bigerelle@uphf.fr

Received: 25 October 2020; Accepted: 29 November 2020; Published: 7 December 2020

Abstract: In this paper, we evaluate the effect of scale analysis as well as the filtering process on the performances of an original compressed-domain classifier in the field of material surface topographies classification. Each surface profile is multiscale analyzed by using a Gaussian Filter analyzing method to be decomposed into three multiscale filtered image types: Low-pass (LP), Band-pass (BP), and High-pass (HP) filtered versions, respectively. The complete set of filtered image data constitutes the collected database. First, the images are lossless compressed using the state-of-the art High-efficiency video coding (HEVC) video coding standard. Then, the Intra-Prediction Modes Histogram (IPHM) feature descriptor is computed directly in the compressed domain from each HEVC compressed image. Finally, we apply the IPHM feature descriptors as an input of a Support Vector Machine (SVM) classifier. SVM is introduced here to strengthen the performances of the proposed classification system thanks to the powerful properties of machine learning tools. We evaluate the proposed solution we called "HEVC Multiscale Decomposition" (HEVC-MD) on a huge database of nearly 42,000 multiscale topographic images. A simple preliminary version of the algorithm reaches an accuracy of 52%. We increase this accuracy to 70% by using the multiscale analysis of the high-frequency range HP filtered image data sets. Finally, we verify that considering only the highest-scale analysis of low-frequency range LP was more appropriate for classifying our six surface topographies with an accuracy of up to 81%. To compare these new topographical descriptors to those conventionally used, SVM is applied on a set of 34 roughness parameters defined on the International Standard GPS ISO 25178 (Geometrical Product Specification), and one obtains accuracies of 38%, 52%, 65%, and 57% respectively for Sa, multiscale Sa, 34 roughness parameters, and multiscale ones. Compared to conventional roughness descriptors, the HEVC-MD descriptors increase surfaces discrimination from 65% to 81%.

Keywords: mechanical engineering; surface roughness; roughness analysis; high-efficiency video coding (HEVC); texture feature descriptors; texture image classification; support vector machine (SVM)

---

1. Introduction

Topographic characterization of rough surfaces plays a major role in the field of surface science. It covers various fields such as tribology, corrosion, electrical or thermal contact, biocompatibility, adhesion, gloss, etc. There are two categories of topographical surface analyses: the first one consists in understanding the mechanism of surface creation that can be voluntary (tooling, surface finishing, etc.) or fortuitous (wear, corrosion, etc.). The second one consists in understanding how surface roughness influences the surface functionality to optimize the surface topography by appropriate surface texturing. For example, Min et al. [1] modified the texture of dental zirconia ceramics by a picosecond laser to enhance hydrophilicity. Typically, the surface texture is characterized by local pixel variations repeated in regular or random patterns in the spatial domain, which provide useful information about spatial distribution. In particular, the surface profile represents the roughness, the primary form, and the waviness with three different frequency scales. Brown et al. [2] discussed the different methods used to determine these scales such as wavelet, fractal, modal, or Fourier analyses. Whatever the recording systems (laser scanning microscopy, interferometry and confocal microscopy, Atomic force microscopy, 3D profilometer, etc.), a discretized map is obtained and can be seen as a grayscale image encoded on 32 bits (float), representing the height of roughness amplitude. However, the topographical image analysis is a challenging task because of the significant change in the material surface texture appearance, depending on process parameters such as length scales and local physical properties. The information depends on the scale, and it is introduced in the multifractal concept as the information dimension. Ghosh and Pandey. [3] used an Atomic Force Microscopy and showed that In-doped ZnO thin films deposited on glass exhibited multifractal behavior such that entropy of the surface topography depends on the scale. By simulation of random deposition, Hosseinabadi. [4] observed a multi-affinity of surface topography due to the diffusion of particles.

This physical information described by entropy can be linked to the information theory firstly described by Shannon. The Shannon entropy is intensively used in image analyses and classification: in surface engineering, Huaian et al. [5] measured grinding surface roughness and showed that singular value entropy is strongly correlated with actual roughness, with the monotonicity of the entropy decreasing more significantly as the roughness increases. Pahuja and Ramulu. [6] analyzed roughness of machining polymer matrix composites and proposed an indicator (ratio of wavelet packet energy to entropy) to characterize the surface and to better predict the surface quality as a function of machine tool, material, and process variables. As entropy is linked to data compression, Bigerelle and Iost. [7] analyzed, for the first time in material sciences, the compression ratio of images of an interface during a diffusion process and demonstrated that the compression ratio (lossless compression) is linked thermodynamically to the entropy of the system. Recently, Zhang and Wang. [8] proposed a lossless algorithm to compress 3D surfaces and showed that the compression ratio can be even higher than 100 for smooth 3D surfaces.

Topographic measuring devices are becoming more and more efficient, and stitching techniques allow a significant increase in the range of measurements. In 2019, Elkhuizen et al. [9] compared three 3D measurements techniques to capture the entire surface topology of the Girl with a Pearl Earring by Johannes Vermeer and obtained, by an original 3D scan based on fringe-encoded stereo imaging, a resolution of $55,714 \times 63,571$ pixels with 5 μm precision depth. Such large resolutions require multiscale analyses: Le Goïc et al. [10] proposed a multiscale analysis method and showed that the filtering surface in the Fourier transform domain is well adapted for fractal surfaces and allows, thanks to ANOVA analyses, for differentiation of the effect of pressure. In this paper, we propose to analyze if a digital image compression algorithm can be used to characterize and classify 3D topographical surfaces. We refer here to the compaction of a known type of image, where the decompressed image is used as an input of an image-processing-based materials science engineering process. In this case, the decompressed image is not viewed by a human and does not need to look close to the original image. Rather, this application involves decoded images containing as minimum data as needed to guarantee that material analysis results are of good quality from a mechanical engineering point of view. Generally, image feature descriptors can be extracted directly from the visual information (color, texture, and shape) either in the pixel domain or in the frequency domain after image transform like Fast Fourier Transform, Discrete cosine transform, Gabor wavelets. Mistry et al. [11] proposed a content-based image retrieval system using such hybrid image features. Moreover, when considering digital image compression, one can benefit from relevant compressed-domain information pertaining to the visual content feature extraction techniques. Such information includes distribution of transform coefficients [12], motion vectors [13], and block-based segmentation [14]. Recently, Zargari et al. [15–18] showed that the intra-prediction modes used by intra-coding in the H.264/AVC and High-efficiency video coding (HEVC) video coding standards can be considered possible efficient image feature descriptors.

In this paper, we present an original method for multiscale decomposed surface classification in the compressed domain and we determine on which filtering range and scale length the surface category should be optimally analyzed to be classified efficiently. In the present case, we are interested in topographical multiscale analysis. The surface profiles are classified at three different frequency ranges (e.g., high-pass, low-pass, and band-pass). These three frequency ranges are separated by the Gaussian multiscale analysis method. Le Goïc et al. [10] showed that this constitutes the most efficient multiscale analysis method for characterizing highly complex topographies. The proposed algorithm applies in the HEVC compressed domain to extract texture feature descriptors. Simultaneously, we need to keep the visual quality good for visual analysis of the mechanical image by experts. To do both, lossless HEVC is used, which guarantees the preservation of the original material parameters to be analyzed with moderate compression ratios depending on image complexity. Descriptors consist of the so-called Intra-Prediction Modes Histogram (IPHM) of each lossless HEVC compressed image. Zargari et al. [17] introduced firstly the IPHM descriptors; these descriptors are computed directly from compressed image data without full decoding of the entire image, which is of great interest in terms of computational complexity. Then, the compressed-domain texture feature extraction is combined by machine learning to strengthen the performances of mechanical material surface classification, which constitutes the main originality of the proposed work. Support Vector Machine (SVM) classification is used to discriminate between the highly similar IPHM descriptors taken from either same or different material surfaces. Simulation results obtained on our topographical images database show that the proposed SVM-based classifier in the HEVC compressed domain gives very high-quality classification performances with accuracy of 81% when applied on the highest length scale of a separated Low-pass (LP) topographical image data set.

The paper is organized as follows: Section 2 presents the material set and methodology. Section 3 describes the proposed algorithm and starts with a brief description of the high-efficiency video coding intra-prediction coding technique. The Intra-Prediction Modes Histogram (IPMH) feature is then detailed as well as its powerful exploitation by SVM for image texture classification in the compressed domain. Simulation results are detailed and discussed in Section 4, showing the effectiveness of the proposed method. Finally, Section 5 gives the conclusions and the perspectives.

2. Materials and Methods

Digital image compression is a key point for reducing the computational complexity, where compression will simultaneously reduce the bit rate and offer an efficient image feature descriptor. In this section, we first give a brief overview on the collected data base characteristics.

2.1. Surface Processing

2.1.1. Surface Texturing

To create a set of topographical maps to analyze the morphological features by our new methodology and to obtain surface topography, the tube/wire blasting process will be used. Tube shot blasting machine is designed for processing of tubes/wires, round bars, and other cylindrical elements external surfaces. The shot blasting process allows for obtaining finished products with clean surface and for increasing the durability of surface-protective applications (coating, painting, etc.), cells adhesion, interface adhesion of fiber-reinforced polymer rods and concrete in structural strengthening, and producing hydrophilic conductor. High efficiency of the shot blasting machine is provided by the continuous abrasive handling. Workpieces move through the tube shot blasting machine on a conical cylindrical conveyor or on a conveyor with skew rollers, providing simultaneous rotation and transition movement of the product through the blasting machine, which is a condition for evenly blasted surfaces.

In the present study, five texturing conditions are applied on initial rods of pure aluminum (99.99%) of 1 m long and 3 mm diameter each (Figure 1b). The different conditions are indexed from $i = 1$ to $i = 6$. The aluminum rods are sandblasted with corundum media (Al_2O_3) with 5 different increasing pressures P_i ($P_1 = 5$ bars and $P_5 = 6$ bars). Sample 6 is the rod before sandblasting treatment. To practice morphological measurements, rods are then cut every 10 cm to obtain a sample of 1 cm (ten samples by rod). To evaluate the texturing process repeatability, 2 rods are investigated for each process condition, leading to $6 \times 2 \times 10 = 120$ surface topographies for future investigations.

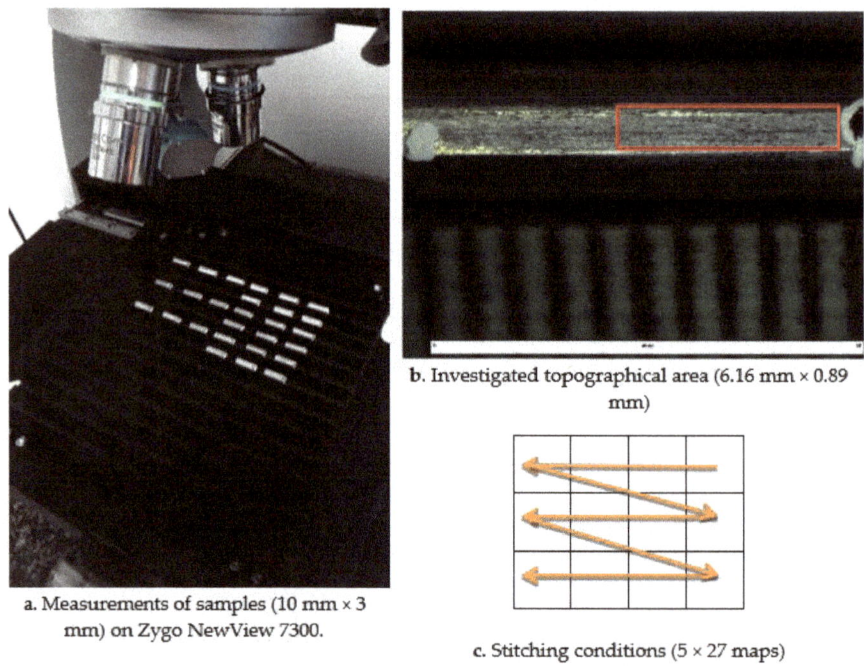

b. Investigated topographical area (6.16 mm × 0.89 mm)

a. Measurements of samples (10 mm × 3 mm) on Zygo NewView 7300.

c. Stitching conditions (5 × 27 maps)

Figure 1. Measurements of rods on Zygo NewView 7300 (**a**) with stitching method (**c**) on a 6.16 mm × 0.89 mm area (**b**).

2.1.2. Topographical Measurements

Topography was measured on each sample using a white light interferometer (NewView 7300, Zygo™, Middlefield, OH, USA) with magnification 50× (Figure 1a). The idea of light interferometer is based on using the wave properties of light to generate the 3D topography precisely [19]. It uses a scanning white light interferometry for producing surface row image and measuring the microstructure of surfaces in three dimensions: it measures the height (Z-axis) over an area with X and Y length and width [20]. To obtain a representative surface area, the stitching method (Figure 1c) processes with 20% overlap (135 topographical maps with 640 × 480-pixel resolution of each individual map; see (Table 1)). Finally, for each of the 120 investigated surfaces, a 13,952 × 2014 approximately 30 mega pixels map is analyzed on an area of 6.16 mm × 0.89 mm (Figure 1b) with a lateral resolution of 0.44 µm. A primary study based on topographical map segmentation has shown that these conditions allow for the detection of 2000 fine craters on the investigated surface due to the sandblasting process.

Table 1. Measurement conditions (Zygo NewView 7300) for each sample.

Lens Magnification	50×
Map resolution (pixel)	640 × 480
Number of stitches	5 × 27, 20% overlapping
Final investigated area (mm)	6.16 × 0.89
Lateral resolution (µm)	0.44
Final resolution (pixel)	13,952 × 2014

2.1.3. Surface Pretreatment

From the initial investigated area (Figure 2a), surface topographies are recorded according to the conditions described in (Figure 2b). As it can be visually observed, cylindrical forms of the rod and reference measurement plane are obtained that can be modelled by a third-degree polynomial equation (Figure 2c) and removed to obtain final topographical maps to investigate (Figure 2d).

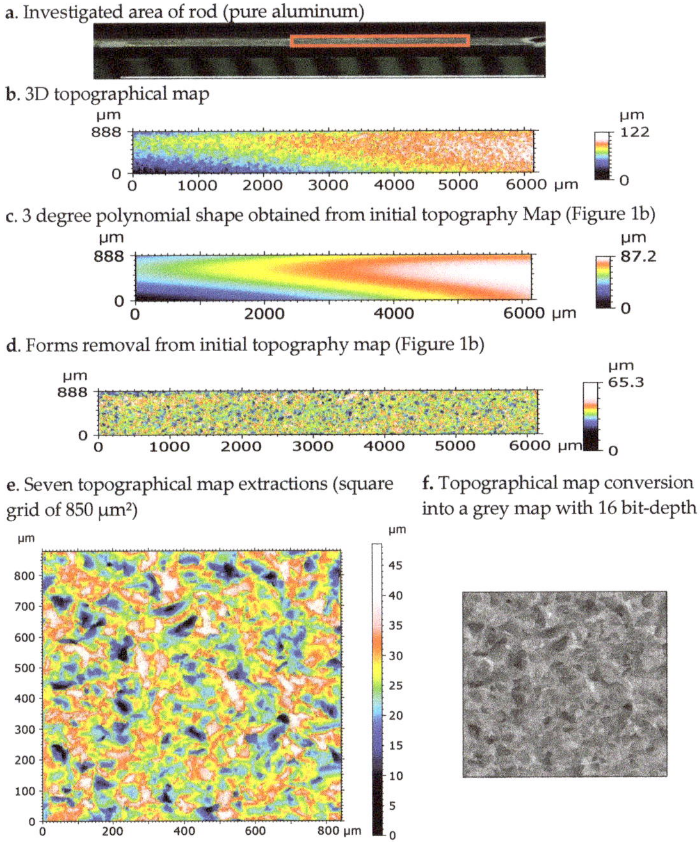

Figure 2. The seven states of treatments of topography measurements to obtain a final 16-bit image.

The 3D topographies corresponding to the six process conditions are shown in Figure 3. To visualize the crater impact, the motif method is applied (defined in the Geometrical Product Specification, ISO 25178-2 standard) and a histogram of the height amplitudes (in µm) is plotted.

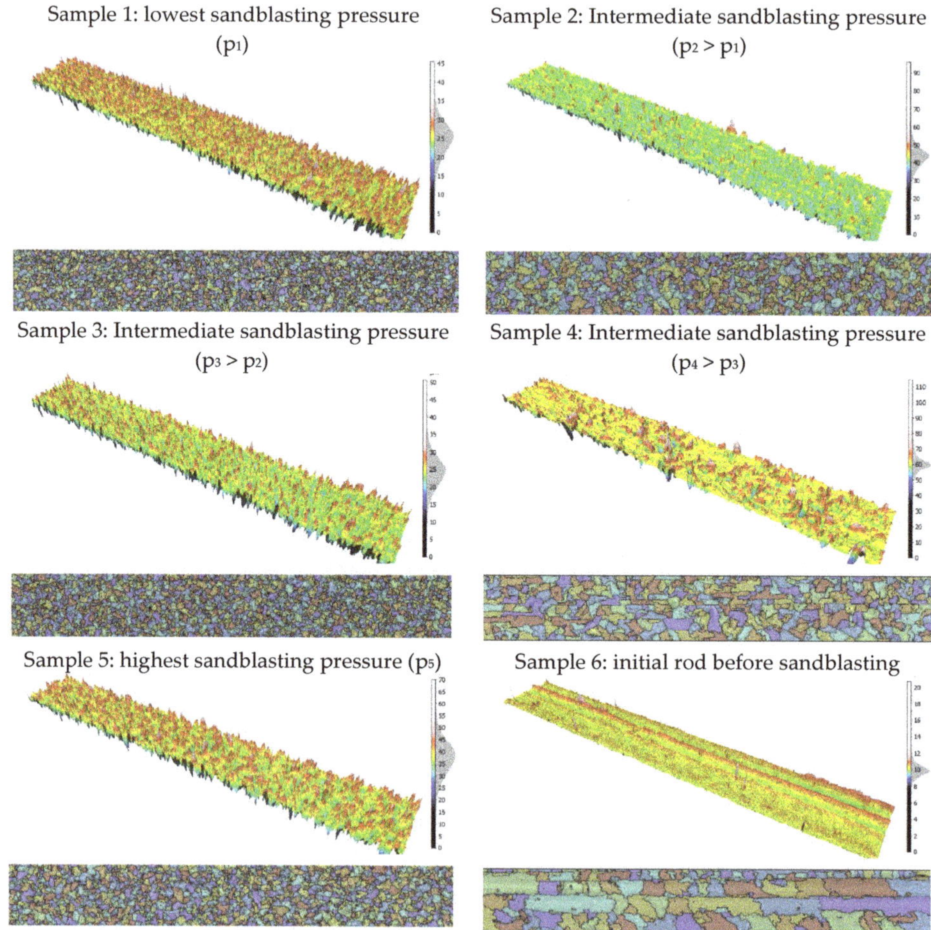

Figure 3. Three-dimensional topography and motif maps with histograms of the height amplitudes (in µm) corresponding of the six process configurations.

Then, seven topographies are extracted from this surface (Figure 2e). Surfaces are resampled using spline interpolation to obtain a 1024×1024 squared topography map corresponding to a 878×844 µm^2 area (these extractions allow for the division of each rectangular map into 7 squared maps that will allow for quantification of the uncertainties of the original map). After that, the topographical map is converted into a grey map with 16 bit-depth (Figure 2f). To obtain this transformation, the amplitude is normalized by the ratio of Sz (roughness parameter which represents the maximum amplitude of the surface topography). This transformation makes it possible to free oneself from the amplitude of the roughness in order to consider only the information contained in the topography. The amplitude parameter Rz can then be introduced later in the classification analysis, thus decorrelating the spatial information from the roughness amplitude.

The decomposition steps are applied for each set of surfaces (5 sets, Figure 4(1)–(5)) and to initial surfaces before treatment (Figure 4(6)); 110 surfaces are investigated that lead to a databank of $110 \times 6 = 660$ 16-bit-depth images.

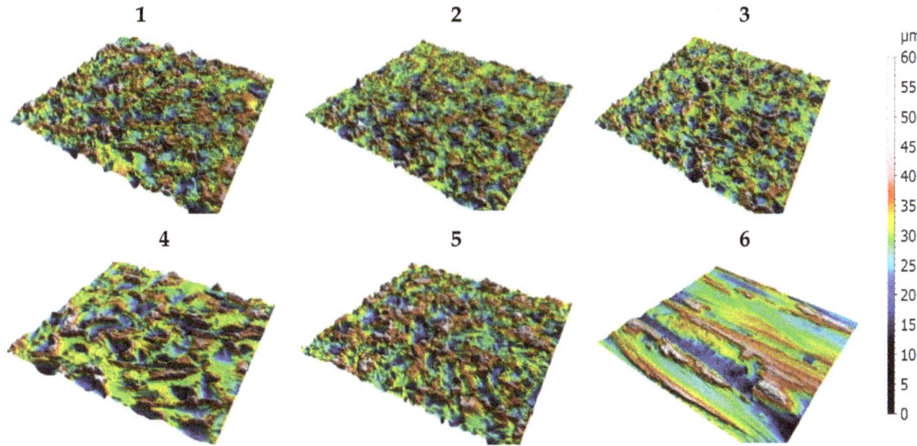

Figure 4. The five topographies (**1–5**) obtained with different mechanical treatments applied on initial surface (**6**).

2.1.4. Multiscale Roughness Analysis

The multiscale surface filtering decomposition techniques have proven their efficiency in roughness functional analysis [21]. Each topography map is multiscale analyzed by using the Gaussian filter recommended by ISO 11562-1996 and ASME B46.1-1995 standards to discover at which scale it extended. Each procedure parameter influences the morphology of the surface. This filter was adapted in order to filter the 3D surfaces with a given frequency cutoff value. Le Goïc et al. [10] described the low-pass, band-pass, and high-pass filters used in this study. Our system will then filter all surfaces with different cutoff values in order to obtain a multiscale decomposition.

The 30 consecutive steps are used in this decomposition, with a cutoff varying from 2 µm to 360 µm. The set of cutoff values is selected to cover the spectrum of the topographical map taking account the transmission characteristics at 50% of transmission centered on a bandwidth of ±40%. The high pass filtering with two cutoff values (8 and 78 µm) is applied on the surfaces with two mechanical treatments, 1 and 4, as well as the initial surface, 6 (Figure 5). This filtering allows to see the different scales of surface deformation. As it can be observed, surface one presents a homogeneous deformation at the two scales due to impact of the mechanical treatments. The mechanical treatment is based on shot penning that creates craters on the initial surfaces. Surface four is less homogeneous due to a lack of recovery of the shoot penning process. The initial surface presents grooves due to the drawing process during metal forming manufacturing.

The low pass filtering with cutoff of 78 µm is also applied on surfaces with two mechanical treatments, 1 and 4, as well as the initial surface, 6 (Figure 6). This filtering allows to describe waviness of the surface that quantifies principal metal deformations (main craters and grooves). Finally, pass band filtering allows for quantification of some particular scales of deformations (Figure 7).

High Pass Filter
Mechanical Surface Treatment

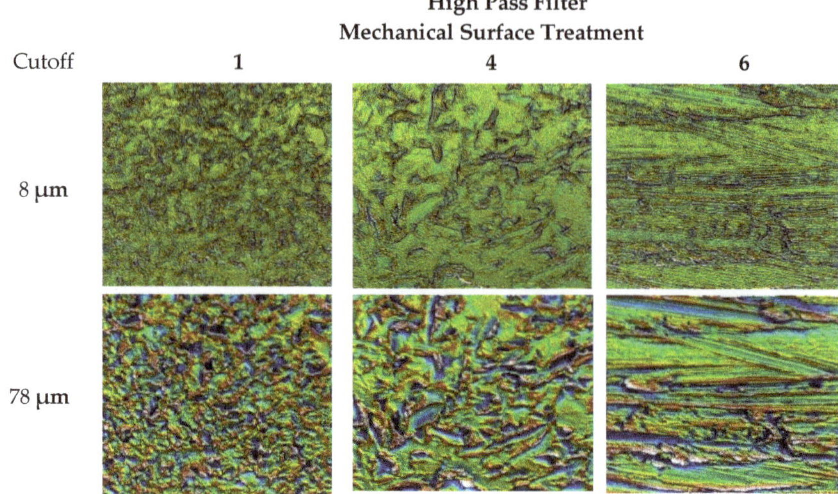

Figure 5. High pass filtering with two cutoffs (8 and 78 µm) applied on the surfaces with two mechanical treatments, 1 and 4, and the initial surface, 6.

Low Pass Filter
Mechanical Surface Treatment

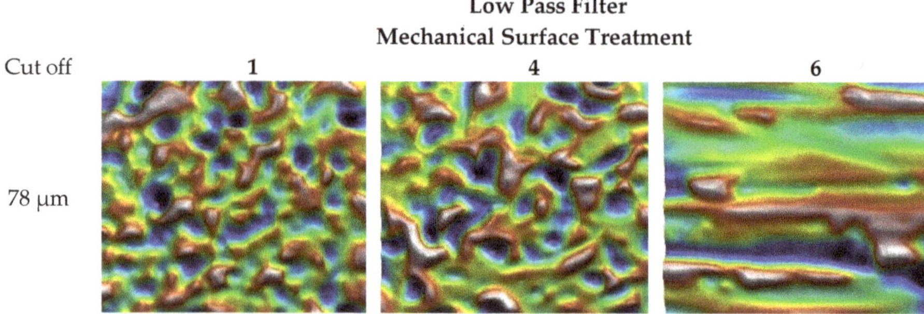

Figure 6. Low pass filtering with a cutoff of 78 µm applied on the surfaces with two mechanical treatments, 1 and 4, and the initial surface, 6.

Band Pass Filter
Mechanical Surface Treatment

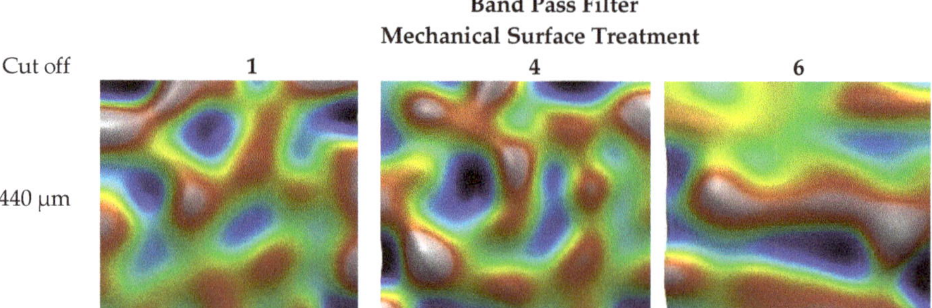

Figure 7. Band pass filtering with cutoff of 440 µm applied on the surfaces with two mechanical treatments, 1 and 4, and the initial surface, 6.

2.2. Topographical Materials Texture Image Data Set

The collected mechanical topographic image data set consists of nearly 42,000 images that represent six mechanical material categories with a resolution of 1024 × 1024 pixels and two internal bit-depths: 8 and 16 bits, respectively. Here, each surface topography includes seven surface regions. Each surface region profile is decomposed into three different types of filtered images: Low-pass (LP), Band-pass (BP), and High-pass (HP) filtered image. Each filtered image represents the roughness, the primary form, and the waviness of the surface, respectively. Finally, each filtered image type decomposes into 18 different spatial length-scales to result totally in 42,000 images.

It can be noticed visually that there is a high similarity between any two surfaces topography images from different mechanical surface categories. The 36 decomposed surface images from different six materials regions are presented in Figure 8, where each material surface topography was decomposed into three filtering techniques: LP, BP, and HP at two length-scales. For example, the first column presents six different materials' surface decomposed images after LP filtering at first zooming scale. Similarly, the third column presents the six different materials' surface decomposed images filtered with HP filtering techniques at first zooming scale, and the fourth column presents the different six materials' surface decomposed images filtered with LP surface filtering techniques at second zooming scale, which have high similarity to the third column.

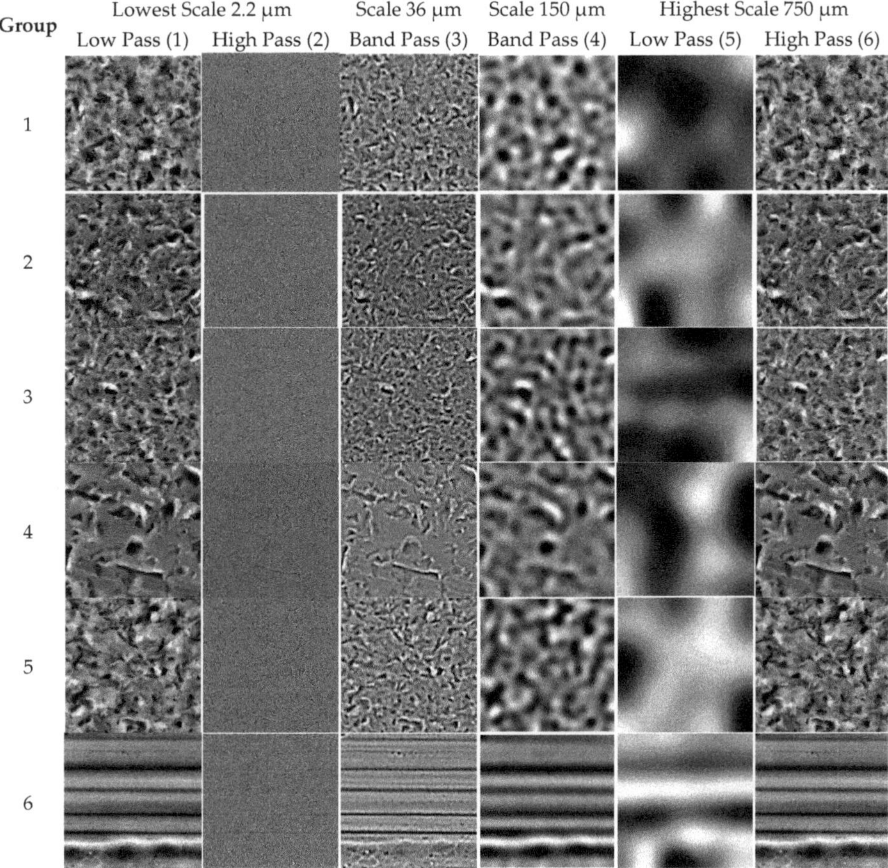

Figure 8. One image (resolution of 1024 × 1024 pixels) from six mechanical material categories (1 to 6) with four different length scales and three filtering methods.

2.3. Topographical Analysis from the GPS ISO 25178 Standard Using SVM Decomposition

The purpose of this paper is to propose a new multiscale analysis based only on information contained in a topographical map. To compare these new topographical descriptors to those conventionally used, a common tool of relevancy quantification must be used. The questions to answer can be sum up in the term: "a relevant method is the method allowing classified with accuracy the surfaces tooled with different process conditions". The classification method used in this paper is a well-known deep learning tools called the Support Vector Machine (SVM), which is close to the discriminant analysis that we have proven to be relevant to discriminate topographical maps. The SVM method needs a set of parameters to differentiate surfaces maps. In our cases, four set of parameters are used (see Appendix A for details).

Set 1. a single parameter, Sa, the most used parameters in surface topography;

Set 2. a set of Sa parameters computed at 30 different cutoff filters for LP, HP, and BP Gaussian filters;

Set 3. 34 roughness parameters defined by the International Standard GPS ISO 25178 (Geometrical Product Specification); and

Set 4. 34 roughness parameters defined by the International Standard GPS ISO 25178 (Geometrical Product Specification) computed at 30 different cutoff filters for LP, HP, and BP Gaussian filters.

After SVM computation, one obtains percentages of 38%, 52%, 65%, and 57% of good classification respectively for the four sets.

2.4. Information, Lossless Compression, and Topographical Caracterisation

Bigerelle et al. [22] have shown that the compressibility of an image can characterize a physical mechanism and can be quantified by the compression ratio using lossless algorithms (run length encoding and Lempel-Ziv-Welch) or a combination of such algorithms (RLE + LZW and LZW + RLE). The compressibility of the information is in fact related to the entropy contained in the topographic surface. Bigerelle et al. [23] showed that the compression ratio of simulated images based on diffusion mechanisms described the scaling laws of statistical physics with concept of entropy. Bigerelle et al. [24] confirmed that compression ratio of images characterizes the kinetics of nanostructures patterns obtained by Monte Carlo methods and can be used to find physical parameters in inverse method. Dalla-Costa et al. [25] formulated the compression ratio in the multi fractal formalism using Legendre transform and showed that the compression ratio can be used to characterize the mechanism of abrasion in tribology. In order to verify that the lossless compressed information allows characterization of topographical maps, all the images of the six surface categories (with their associated filtering) contained in the database (see Section 2.2) are compressed by the LZW algorithm and the compressed image size is computed. We can then perform an analysis of variance (single-factor ANOVA) where the factor is the surface number. The F-test is then computed (F = variance between 6 surfaces/variance in a surface) at all scales with the three filtering methods (high pass, band pass, and low pass). Statistical significance is given for $F > 1$: the higher F, the more discrimination is obtained by the compression ratio. To find the most relevant scale, F values are plotted versus the scale (in μm) for the three filtering methods (Figure 9).

For the band pass, high pass, and low pass filters, one obtains respectively maximal F values of 558, 414, and 548 corresponding to the scales of respectively 78.2, 78.2, and 29.6 μm (Table 2). At theses scales of maximal relevance, histograms of compressed image sizes are plotted (Figure 10) and one can visually observe the efficient discrimination of the different categories of surfaces.

This clearly means that the compression ratio well discriminates the different surface topographies without computing any roughness parameters. It can be noticed that lossy compression algorithms can have an interest with topographical data recorded by photo goniometers due to the amount of data [26].

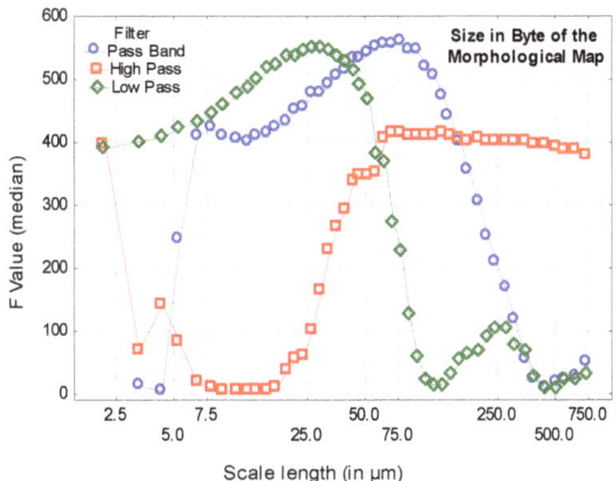

Figure 9. F median plots of the one factor ANOVA (factor: 6 surface categories) versus the scale (in µm) for the 3 filtering methods (band pass, high pass, and low pass).

Table 2. Statistics of maximal F values for the three filtering methods (see Figure 9).

Filter	Scale	Fmean	$F_5{}^{th}$	$F_{50}{}^{th}$	$F_{95}{}^{th}$
Band pass	78.2	566	441	558	721
High pass	78.2	418	334	414	515
Low pass	29.6	552	454	548	666

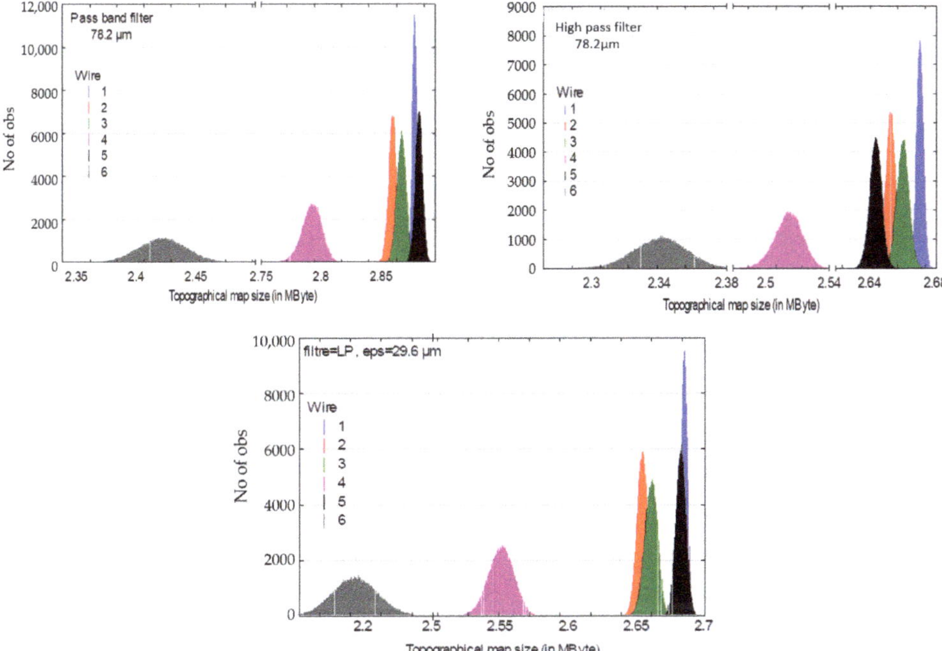

Figure 10. Histograms of size (mean) of the compressed image at the maximal scales of relevance.

3. Description of the Proposed Algorithm

Here, we describe the proposed Machine Learning-based classification algorithm, based on SVM classifier applied in the compressed image domain. As mentioned earlier, digital compression is applied to the image database aiming to optimize storage capacities. However, the used image compression technique must not affect the structural image properties, which are further exploited during mechanical analysis of the materials.

3.1. HEVC Intra-Prediction Coding

HEVC is the current state-of-the-art digital video compression standard, with bit rate savings of about 50% compared to its predecessor H.264/AVC for the same perceptual quality. Such performances are made possible thanks to the introduction of new coding tools as well as the optimization of existing ones [27]. In particular, intra-prediction coding has been significantly improved. The block partition is more flexible, ranging from 4×4 up to 32×32 blocks, and the number of intra-prediction modes has been extended to 35 modes compared to 11 modes in H.264/AVC [28].

Intra-prediction allows exploitation in a very efficient way: spatial redundancy inherent to image contents. It is done by extrapolating sample values from the reconstructed reference samples positioned at the left and upper boundaries of the block to be predicted, depending on the 33 directional angles as shown in Figure 11.

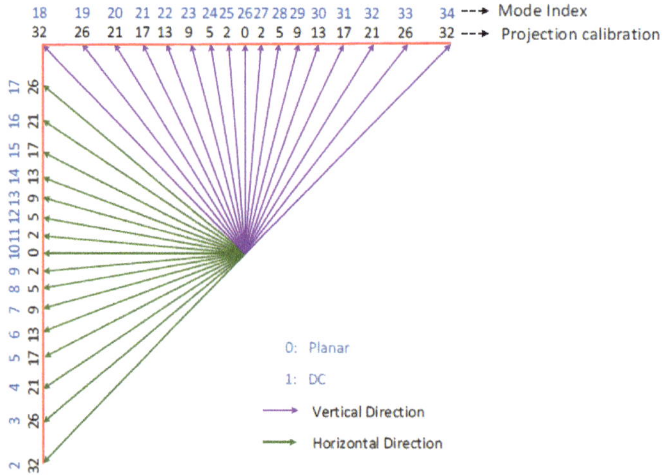

Figure 11. Intra-prediction modes in High-efficiency video coding (HEVC) [29].

These 33 HEVC directional angles are used to model different image blocks' directional structures. Additionally, the so-called DC mode is used for predicting smoothed areas by using the mean of reconstructed neighboring samples, and the planar mode is used for predicting complex texture blocks by performing two-dimensional linear interpolation from block reference neighbor samples [28]. Once all prediction modes have been computed, the Sum of Absolute Errors (SAE) is evaluated between the original block and each of the predicted ones. The predicted block which minimizes the SAE is selected as the best candidate.

Flynn et al. [30] presented the range extensions of HEVC version 2 that define three profiles for high bit-depth image coding to cover a broad range of video requirements:

- HEVC Main 4:4:4 16 Still Picture (MSP) profile only considers intra-coding;
- Main-RExt (main_444_16_intra) and High Throughput 4:4:4 16 Intra apply both intra- and inter-coding.

In our case, we consider the MSP profile for HEVC still-image lossless intra-compression.

This profile supports up to 16-bit depth still-image compression. We implement the MSP profile using the HEVC reference software HM 16.12 version; the lossless coding parameters are enabled, causing bypass of the transformation, quantization, and all the in-loop filtering operations (Table 3). For well characterized texture in a localized image area, we fix the Prediction Unit (PU) size to 4×4 blocks to have the finest analysis size.

Table 3. Summary of the HM 16.12 reference software encoder configuration.

Coding Options	Chosen Parameter
Encoder version	16.12
Profile	Main-still-picture
Internal bit depth	8
Frames to be encoded	1
Max CU width	16
Max CU height	16
GOP	1
Search range	64
Quantization parameter	0
Transform skip	Disabled
Transform skip Fast	Disabled
Deblocking filter	0
Sample adaptive offset	Disabled
Trans quant bypass ena	0
CU Trans quant bypass	0

The resulting compression is lossless from a mechanical point-of-view, with lossless compression ratios ranging from 2:1 to 6:1, depending on the image complexity.

3.2. HEVC IPHM-Based Classification

The ability of the HEVC intra-prediction process to efficiently predict texture image contents is illustrated in Figure 12.

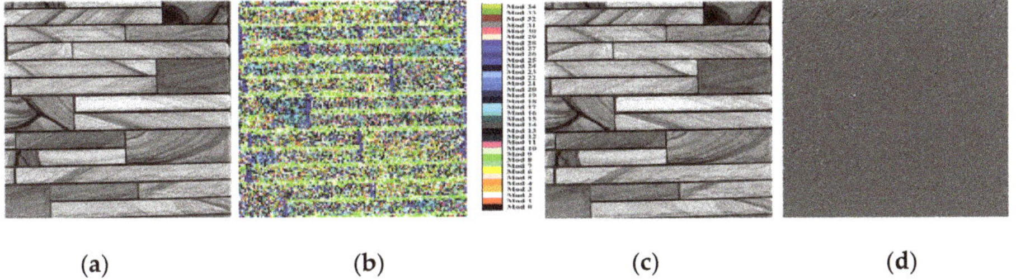

Figure 12. Illustrative example of HEVC intra-prediction efficiency: (**a**) original image of 1024 × 1024 pixels; (**b**) selected modes to predict the original image (each prediction mode is represented here by one among 35 different colors); (**c**) intra-predicted image; and (**d**) the residual image.

It is clear that the predicted version of the image inherits most of the main textural characteristics of the original image leading to a significantly low residual signal. Hence, the HEVC intra-prediction process is very well suited to capture the texture features which represent one of the more important visual descriptors in the field of image classification, pattern recognition, or computer vision. Traditionally, several methods have been studied in the literature to extract and characterize the texture feature descriptors. Humeau. [31] categorized the texture feature extraction methods into seven classes: statistical approaches, structural approaches, transform-based approaches, model-based approaches, graph-based approaches, learning-based approaches, and entropy- based approaches. He also gave drawbacks and presented examples of applications for each method. Texture analysis is widely used for various applications like medical imaging [32], remote sensing [33], or industrial automation [34]. Hence, intra-prediction results should constitute a good candidate for texture feature extraction. Recently, Zargari et al. [15–18] developed a compressed-domain texture feature descriptor based on the occurrence of prediction modes used for intra-coding. The so-called Intra-Prediction Modes Histogram (IPMH) descriptor consists in counting the number of blocks predicted by each of the 35 available intra prediction modes. IPMHs are calculated directly from the compressed image data without the need to decode the whole image, hence reducing the computational complexity. Zargari et al. [17] presented the different steps to extract the IPMHs listed below:

- Compress the entire topographical image database with HEVC lossless intra-prediction coding by computing the 35 intra-prediction modes for Prediction Units (PU) of size 4×4 pixels.
- Search for the best prediction mode that minimizes the Sum of Absolute Difference (SAD). The selected mode indicates the relation between the pixels inside the Prediction Unit (PU) and the boundary neighbor pixels.
- Count the frequently utilized prediction modes to arrange each mode in one histogram bin as given by the following equation:

$$H'_i = \{h_i \, 0 \leq i \leq 34\} \tag{1}$$

where H'_i is the bin of the histogram for the mode (i). h_i indicates the number of blocks in the coded picture which are predicted by mode (i).

The normalized IPMH is generated as follows:

$$H_i = \frac{H'_i}{X} \tag{2}$$

where X represents the total number of 4×4 blocks in the image (65,536 blocks in the case of a 1024×1024 image).

Finally, Zargari et al. [15–17] proposed to measure the similarity between every two images based on the intersection between their corresponding normalized IPMH defined as follows:

$$\text{Sim}_{a,b} = \sum_{i=0}^{34} \min((H_i, a, H_i, b)) \tag{3}$$

where (a) is the first image and (b) is the second image.

Zargari et al. [17] validated this method firstly in the H.264/AVC compressed domain and then in the HEVC one, using VisTex conventional image databases of natural scenes.

Unfortunately, the similarity measurements performed on our image database indicate high correlation between many pairs of IPMHs whether they belong to the same or different categories.

In order to illustrate this major drawback, we consider six images from each surface category that presented in Figure 8 to evaluate the proposed similarity measurement method on surface categories classification. One query sample image was used from each category, and other images were used for testing. The first five retrieved images from each query images category are ranked in descending order based on the similarity value. This leads to classification of the corresponding 36 surface images with poor accuracy not up to 20%, as illustrated in (Figure 13).

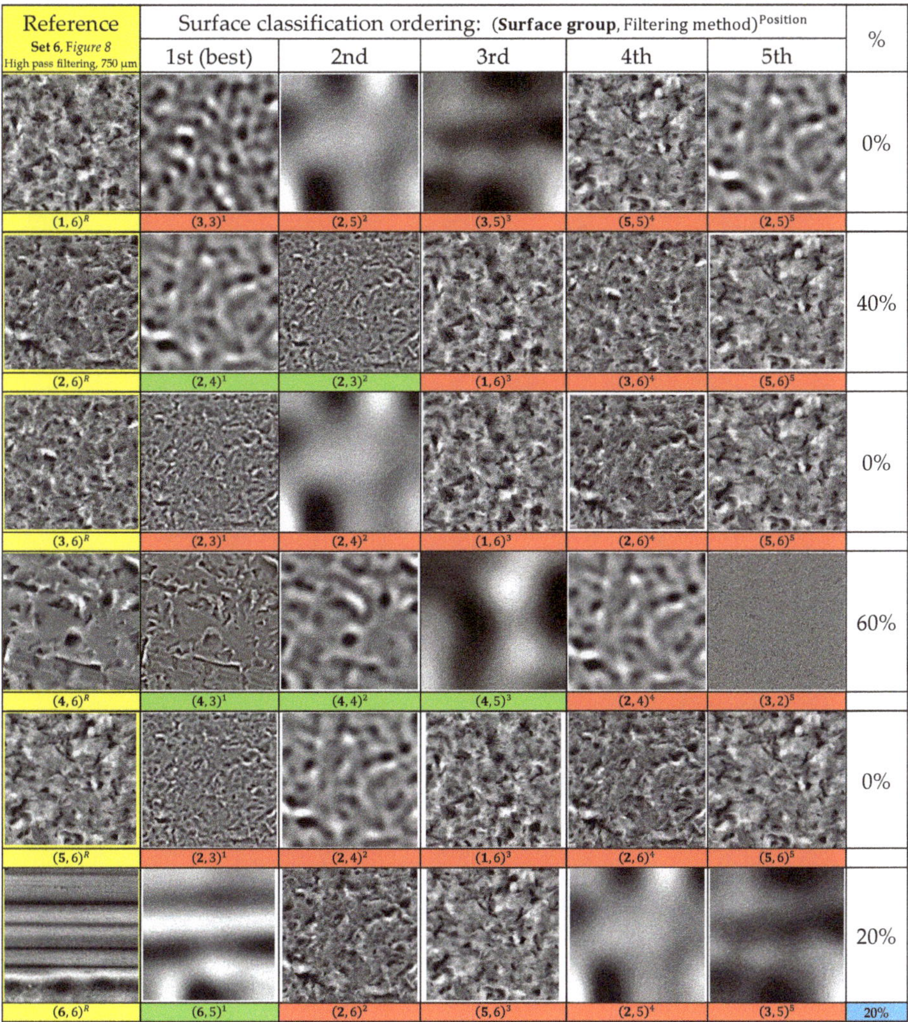

Reference Set 6, *Figure 8* High pass filtering, 750 μm	Surface classification ordering: (**Surface group**, Filtering method)^Position					%
	1st (best)	2nd	3rd	4th	5th	
$(1,6)^R$	$(3,3)^1$	$(2,5)^2$	$(3,5)^3$	$(5,5)^4$	$(2,5)^5$	0%
$(2,6)^R$	$(2,4)^1$	$(2,3)^2$	$(1,6)^3$	$(3,6)^4$	$(5,6)^5$	40%
$(3,6)^R$	$(2,3)^1$	$(2,4)^2$	$(1,6)^3$	$(2,6)^4$	$(5,6)^5$	0%
$(4,6)^R$	$(4,3)^1$	$(4,4)^2$	$(4,5)^3$	$(2,4)^4$	$(3,2)^5$	60%
$(5,6)^R$	$(2,3)^1$	$(2,4)^2$	$(1,6)^3$	$(2,6)^4$	$(5,6)^5$	0%
$(6,6)^R$	$(6,5)^1$	$(2,6)^2$	$(5,6)^3$	$(2,5)^4$	$(3,5)^5$	20%

Figure 13. First five retrieved images for six images tests (categories 1 to 6) using Intra-Prediction Modes Histogram (IPMH) which indicate a classification accuracy of 20%.

In particular, we can see the false classification for the first, third, and fifth surface image categories (see Figure 13, first, third, and fifth rows). Also, we can verify poor classification for the second surface image category, with three false classifications out of five in total (see Figure 13, second row).

This leads us to develop an original robust classification algorithm by combining IPMH with SVM machine learning tools to find the optimal separator between nonlinear surface image categories.

3.3. The Proposed Method

In order to strengthen the classification process, we propose to combine the IPMH solution described in the previous section with the nonlinear SVM model. Several studies have been already proposed in the literature for image classification based on the combination of machine learning tools with the texture descriptors such as locally binary pattern (LBP) features [35], filter bank features [36], or cooccurrence matrix-based features [37]. However, these solutions are often applied in the pixel domain. The complete block diagram of the proposed algorithm is presented in Figure 14.

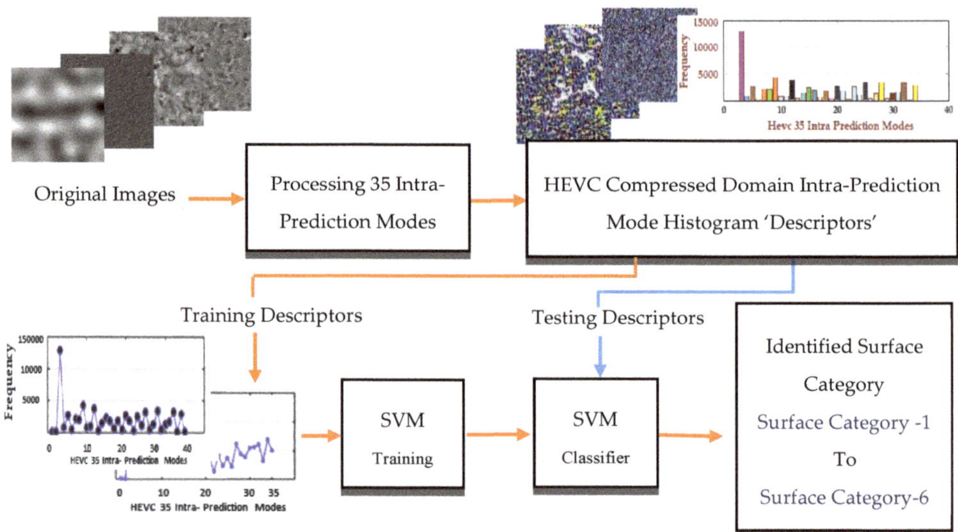

Figure 14. Block diagram of the proposed model integrates the HEVC lossless intra prediction model with nonlinear support vector machine model.

First, the IPMHs are computed in the compressed domain from the HEVC lossless intra-predicted images. Each histogram is a vector of dimension equal to the total number of intra-prediction modes, i.e., 35 in the HEVC case. These histograms are then used as input features for training the nonlinear SVM. SVM is one of the essential supervised machine learning tools; it has been proposed in many scientific classification fields, such as bioinformatics [38], medical diagnosis [39], environment monitoring [40], and material scientific classification [41]. Designed initially to solve two-class binary classification, SVM has been extended to multiclass classification with two different approaches: One vs. Rest and One vs. One [41,42]. SVM uses training data (features) to give the computers acknowledgement without previous programming based on recent advances in statistical learning theory, aiming to maximize the distance between the hyperplane and the support vectors (the samples that effect on the hyperplane) [42]. SVM solves the nonlinear classification problem by increasing the dimensionality to find the optimal hyperplane in kernel space. Its complexity depends on the number of training samples and does not depend on the kernel space dimensionality [35,42,43].

4. Simulation Results

In this section, we evaluate the performances of the proposed SVM-based classification algorithm in the HEVC compressed domain, using the topographic image database described in Section 2.2.

Firstly, we will present the achieved compression ratio for each surface filtered image type. Secondly, we will present the effectiveness of the proposed image texture descriptor to characterize the surface topography with different analyzing conditions. Then, we will present the impact of multiscale surface filtering types on the model classification performance. Finally, the effect of scale analysis on the model performance will be also evaluated.

4.1. The Impact of Surface Topography Filtering Types on Achieved Compression Ratios

In general, the achieved lossless compression ratios depend on image complexity. The compression ratio is high at the lowest scale of analysis, except for high pass filtered images where there is no difference between compression ratios achieved at any scale value as illustrated in the following figures (Figures 15–17). The HEVC lossless compression ratios for the six multiscale low-pass filtered surfaces image categories are presented in Figure 15. The average of the compression ratio is also given.

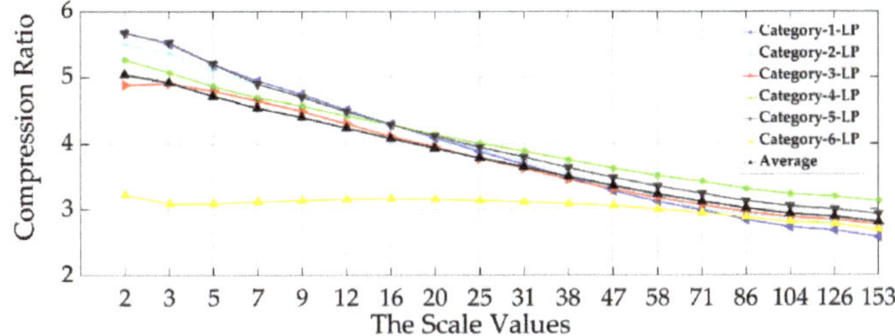

Figure 15. Relationship between the scale of analysis and the six surface categories compression performance by using the multiscale low-pass (LP) data sets.

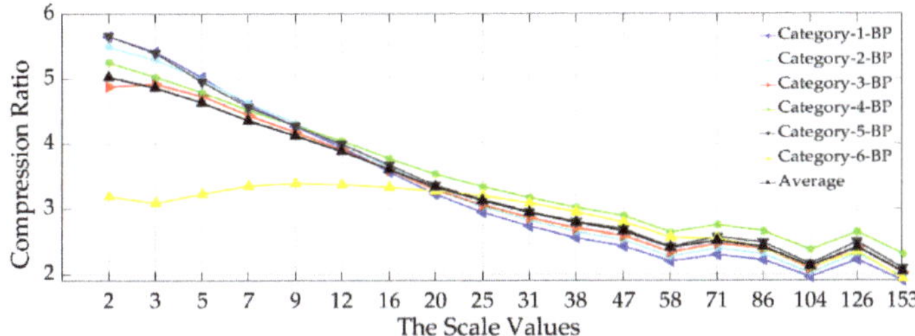

Figure 16. Relationship between the scale of analysis and the six surface categories' compression performance by using the multiscale band-pass (BP) data sets.

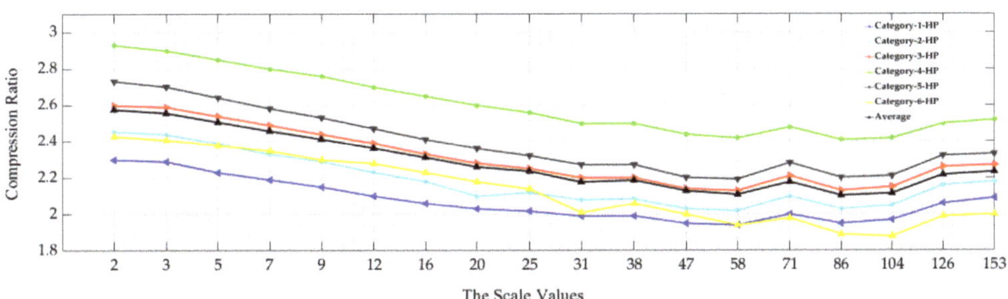

Figure 17. Relationship between the scale of analysis and the six surface categories compression performance by using the multiscale high-pass (HP) data sets.

The compression ratios vary between 2.7:1 and 5.7:1 depending on the scale value. The first scale value (2) indicates variance between the six achieved compression ratios. However, for the next eleven analysis scale values (3 to 47), the compression ratio values are much closer between the six LP multiscale surface categories except for category_6. Globally, the scale of analysis and the achieved Compression Ratio (CR) are inversely proportional, where CR increases as the scale of analysis decreases.

The compression ratios also are inversely proportional relative to the scale of analysis in the case of the six BP multiscale surface categories as presented in Figure 16.

The best $CR_{Average}$ (= 5:1) was achieved at the lowest analysis scale, while the $CR_{Average}$ (= 2:1) was obtained at the highest length-scale of analysis (Figure 16).

There is no significant difference between the compression ratio for the six high-pass multiscale surface categories at different scales of analysis compared to the average of the computed CR averages at all available analysis scales ($CR_{Average}$ = 2.3:1) as shown in (Figure 17).

4.2. Evaluating IPMH as Texture Feature Descriptor

As already mentioned in Section 3.2, the proposed texture feature descriptor is highly related to the specific pattern of the predicted blocks of pixels. The 33 angular prediction modes can predict all frequency components for specific predicted 4×4 directional blocks in a topography image with a residual signal nearly null, as illustrated in Figure 18.

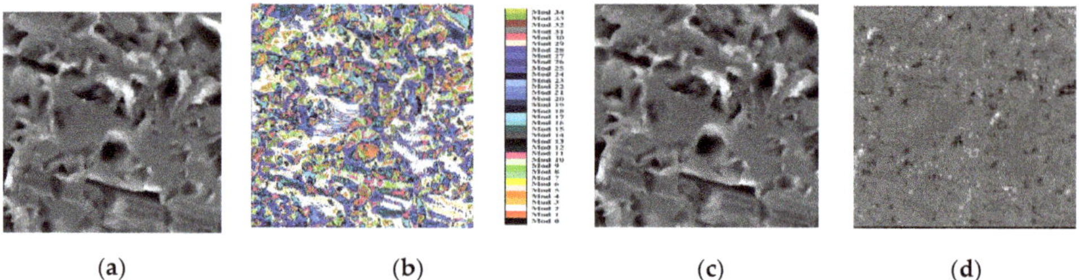

(a)   (b)   (c)   (d)

Figure 18. Illustrative example of HEVC intra-prediction efficiency for characterizing topographical image: (**a**) original image of 1024×1024 pixels; (**b**) selected modes to predict the original image (each prediction mode is represented here by one among 35 different colors); (**c**) intra-predicted image; and (**d**) the residual image (an anamorphic transformation is done on the whole gray scale to see morphological details).

The first three subfigures in Figure 19 compare the IPMH averages for the six categories at three different multiscale filtered image types: LP, BP, and HP filtered image data sets. The last subfigure presents the IPMH averages for the three different multiscale filtered image data sets.

a. Comparison between the LP-IPMHs averages for six multiscale surfaces categories.

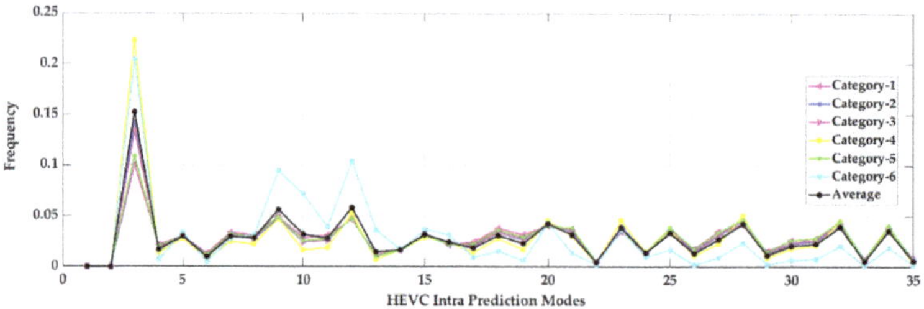

b. Comparison between the BP-IPMHs averages for six multiscale surfaces categories.

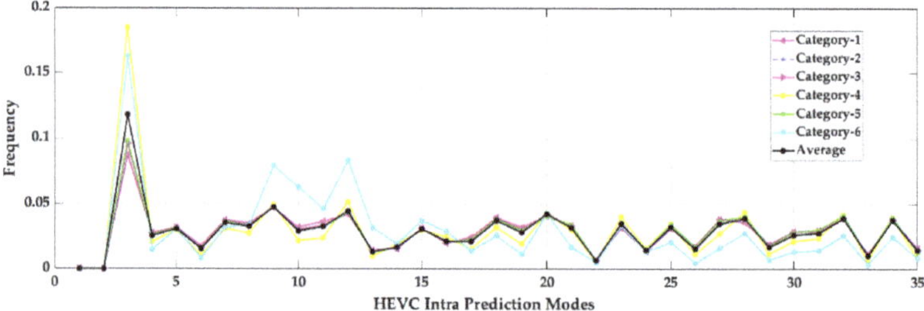

Figure 19. Cont.

c. Comparison between the **HP-IPMHs** averages for six multiscale surfaces categories.

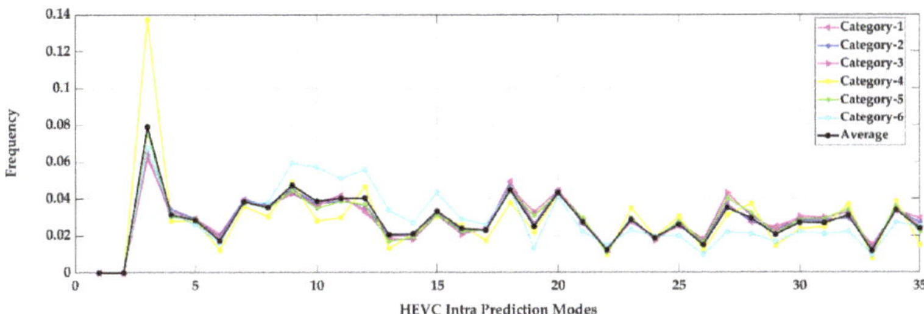

d. Comparison between the averages of six multiscale surfaces categories **LP-IPMHs** average,

BP-IPMHs average, and **HP-IPMHs** average.

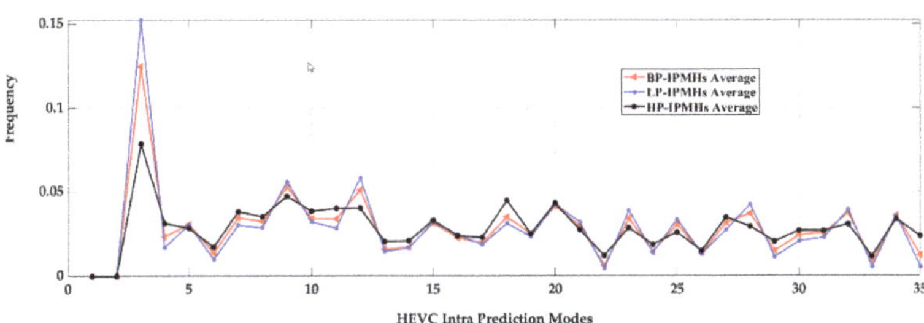

Figure 19. Comparison between the IPMH averages for three different filtered image data sets: LP, BP, and HP data sets.

The comparison between the average IPMHs was nearly similar for the first five material categories at different prediction modes, while the sixth category has a small IPMH difference compared to the others.

4.3. The Impact of Surface Topography Filtering Types on Topographical Images Classification Accuracy

As we previously illustrated in Section 2.2., the surface topography profile has decomposed into three different filtering methods with eighteen different length-scales. We propose to use SVM to find the optimal separation between these three multiscales filtered image data sets to evaluate the following:

- Case-1: the impact of considering the three-filtered image data sets together on the six surfaces categories' classification performances.
- Case-2: the impact of each filter separately on the six surfaces categories' classification performances.
- Case-3: the impact of each scale of analysis on the six surfaces categories' classification performances.

To perform that, firstly, the data set is separated into two partitions: a training data set in order to build the classifier and a testing data set to evaluate the classifier. Different data set sizes are considered in order to evaluate the impact of training data set size on the proposed model performance.

Secondly, for model training, we use a variable number of specific training data sets in each simulation case while the rest of the data set is used for testing. For example, in case-1 where the three data set images are considered together (41,580 images), the training data sets are 7% (by using just the first region from each surface), 14% (using the first and last regions from each surface), 21% (using the first, fifth, and last regions from each surface), 28% (using the first, third, fifth, and last regions from each surface), 35% (using the first, third, fourth, fifth, and last regions from each surface), 42% (using the first, third, fourth, fifth, sixth, and last regions from each surface), and 50% (using all the seven regions from each surface) IPMHs from each surface category respectively, while the rest of the data set is used for testing. That for case-2 is the same (when considering the three separated data sets with all available scale of analysis separately). The same is used for case-3 except each image data set is divided into two equal partitions: one used as the training data set and the second used for evaluating the proposed compressed-domain topographies classifier.

Thirdly, to evaluate linear, Poly, and RBF (LIBSVM_MODELS) learning algorithms, we perform the 5-k Cross-Validation using the training data set to select the kernel model and to tune the model parameters in each simulation case. The procedure for learning and testing the nonlinear SVM model is illustrated in Figure 20, in the case that the total data set was split into $\alpha\%$ for learning ($0 \leq \alpha \leq 1$) and the remaining was used for model validation.

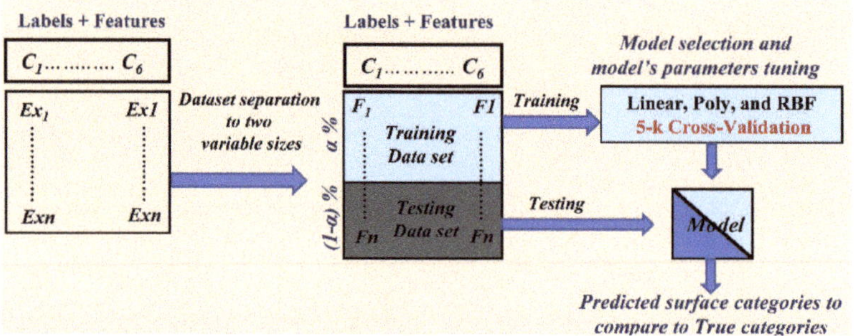

Figure 20. Block diagram depicting the procedure for learning and testing the support vector machine (SVM) model.

During SVM evaluation, the polynomial function kernel gave better classification performance in the three cases, with different optimized kernel parameters (C and *gamma*) for each simulation.

Finally, we trained the SVM models for case-1 and case-2 with a varied number of randomized training data sets to evaluate the impact of increasing the number of training data set on the classification performance.

The IPMH feature descriptors are not able to classify a mix of three multiscale surface filtered image data sets. The classification accuracy reaches 52% by using 21% of the total data set as the training data set (8732 IPMHs) while using the rest of the data set (33,848 IPMHs) for testing the proposed model. The classification accuracy does not have a proportional relationship with the size of the training data set, as it is clearly noticed in Figure 21a by plotting the average of the achieved accuracies while classifying the six surfaces categories.

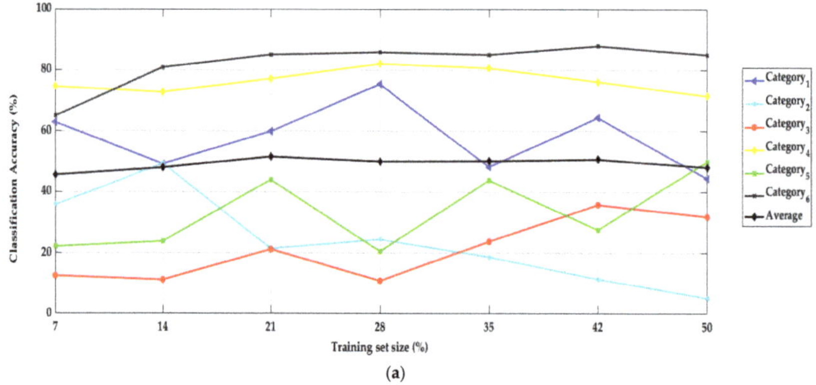

Figure 21. *Cont.*

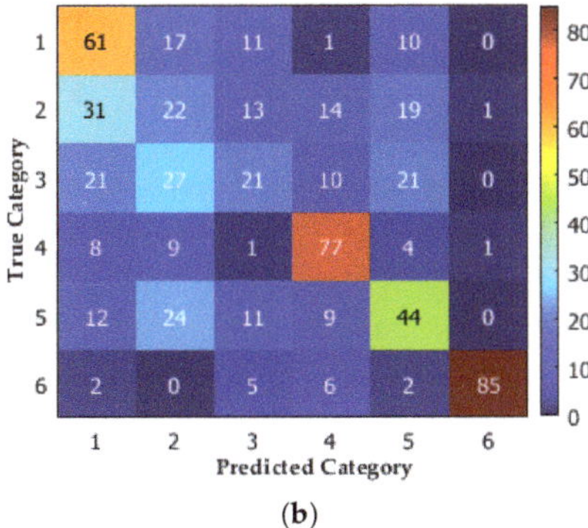

Figure 21. The topography classification performance of mixed three multiscale surface filtered image data sets: (**a**) the effect of increasing the training set size on the classification accuracy and (**b**) a confusion matrix for six surface category classifications by using 21% of the mixed data set for training.

The classification accuracy is reported in the confusion matrix for the six topography categories while considering 21% of the data set for training (Figure 21b), where the columns and the rows represent the predicted and the actual classes, respectively. The values located at the diagonal of the matrix indicate the exact prediction percentage. For example, the prediction percentage for category 1 is equal to 61%.

In the case of LP filtered images data set, we considered seven different percentages (from 7% to 50%) of the total data set (13,860 IPMHs) for the training data set, as illustrated in Figure 22a.

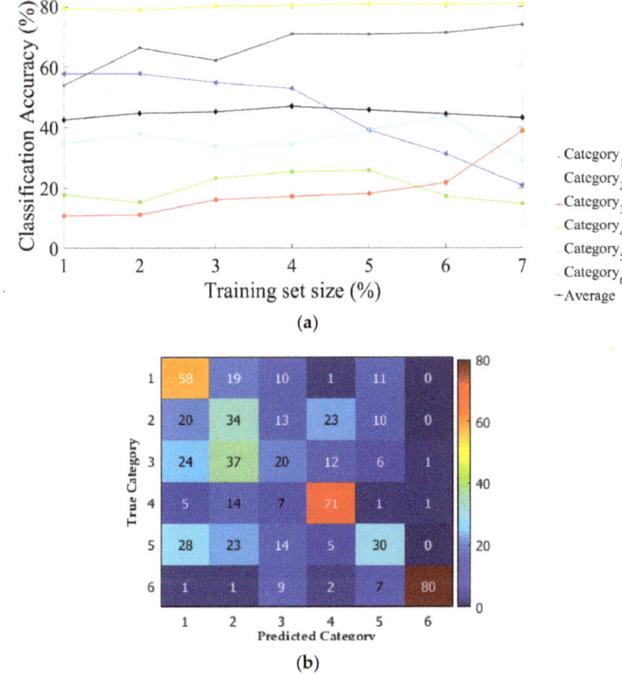

Figure 22. The topographies classification performance of multiscale LP data set: (**a**) the effect of increasing the training set size on the classification accuracy and (**b**) a confusion matrix for six surface category classifications by using 42% of the LP data set for training.

The classification accuracy can reach 49% by using 42% of the LP data set (5821 IPMHs) for training.

The confusion matrix (Figure 22b) presents the classification accuracy for the six categories by using 42% of the total LP filtered image data set.

In the case of BP filtered image data sets, we also considered seven different percentages (from 7% to 50%) of the total data set (13,860 IPMHs) for the training data set, as illustrated in Figure 23a. The classification accuracy can reach 47% by using 28% of the BP data set (3881 IPMHs) for training.

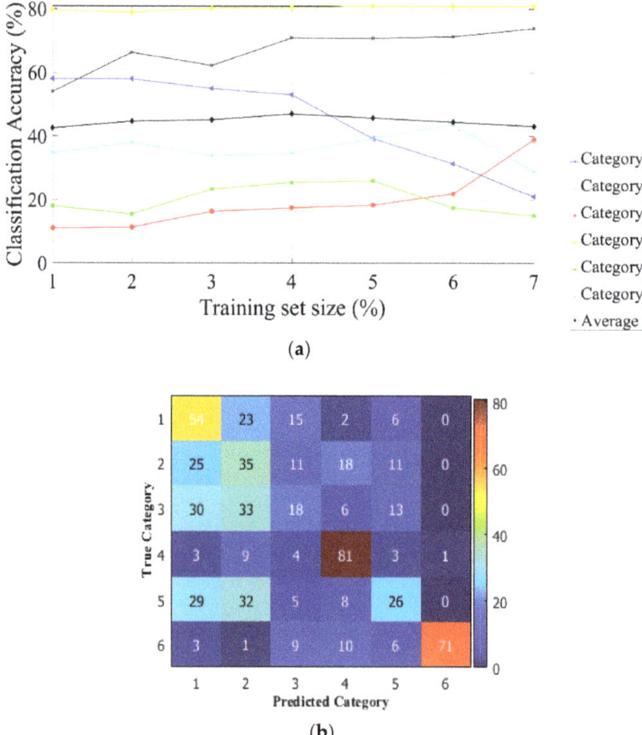

Figure 23. The topographies classification performance of the multiscale BP data set: (**a**) the effect of increasing the training set size on the classification accuracy and (**b**) a confusion matrix for six surface category classifications by using 28% of the BP data set for training.

The confusion matrix for the six categories for the BP filtered image data set is given in Figure 23b. We can deduce from Figure 23a that the obtained classification accuracy by using the BP filtered image data set is less good than that obtained while using the LP filtered image data set (Figure 22a).

In a similar way, in the case of the HP filtered image data set, we considered seven different percentages (from 7% to 50%) of the total data set (13,860 IPMHs) for the training data set, as illustrated in Figure 24a, which demonstrated the strength of compressed-domain classifier with no directly proportional relationship between the size of the training data set and the classification accuracy.

The classification accuracy can reach 70% by using just of 35% from HP data set (4851 IPMHs) for training.

Moreover, the classification accuracy decreased significantly when considering a large-size training data set. For example, the classifier reached 64% accuracy by using 9702 (50%) IPMHs for training. From the confusion matrix (Figure 24b) for the six categories for the HP filtered image data set, we can notice an 18% average increase in the classification accuracy when considering only the HP filtered data set. The prediction percentage for category 1 is equal to 71%.

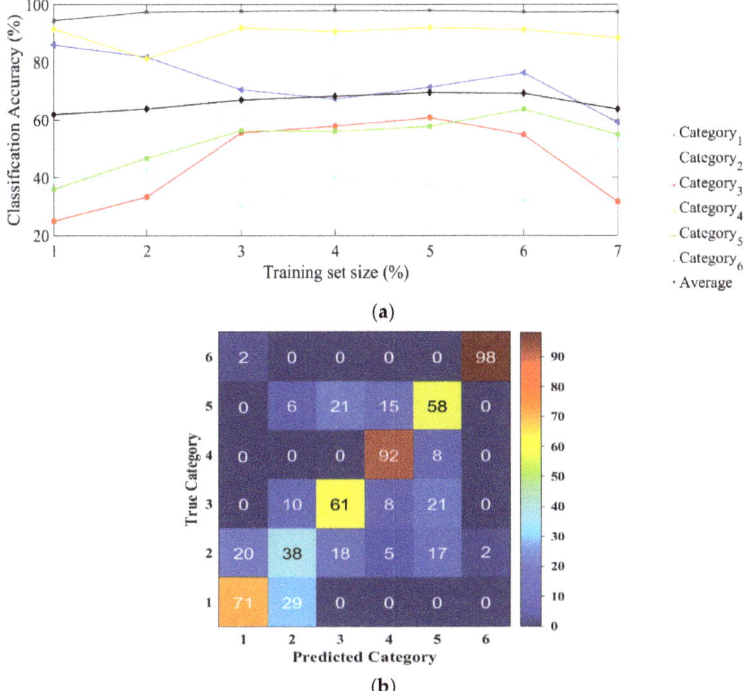

Figure 24. The topographies classification performance of the multiscale HP data set: (**a**) the effect of increasing the training set size on the classification accuracy and (**b**) a confusion matrix for six surface category classifications by using 35% of the HP data set for training.

4.4. *The Impact of Scale of Analysis on Topographical Images Classification Accuracy*

As we previously illustrated, the surface topography profile decomposes into three different filtering methods (low-pass, band-pass, and high-pass filters) with eighteen different length-scales.

In this section, we aim to evaluate the effect of each length-scale on system classification accuracy for three different cases: LP filtered data set, BP filtered data set, and HP filtered data set. The comparison between the achieved accuracies from these three separated data sets at different scales of analysis is shown in Figure 25. We can note the significant improvement for six topography categories' classification accuracies by using a single analysis scale of each separated data set. Therefore, the single scale analysis was more appropriate than multiscale analysis in the case of classifying the LP, BP, and HP data set separately.

The results have indicated a significant improvement for classification accuracy in the case of the LP filtered image data set at the highest scale of analysis. The average accuracy reached 81% by using 50% of the total highest scale of LP data sets for training, where the average accuracy was enhanced by 32% compared to the case of the multiscale LP data set. For the BP separated data set, the best-achieved classification accuracy of 68% was obtained from the fiftieth scale of analysis, where the average accuracy was enhanced by 21% compared to the case of the multiscale BP data set.

In addition, the ninth scale of analysis of the separated HP data sets gives a better classification accuracy of 73%. The robust performance achieved by using the separated scales of LP data sets for classifying six multiscale surface categories at different compression ratios and scales of analysis, as shown in Figure 26a.

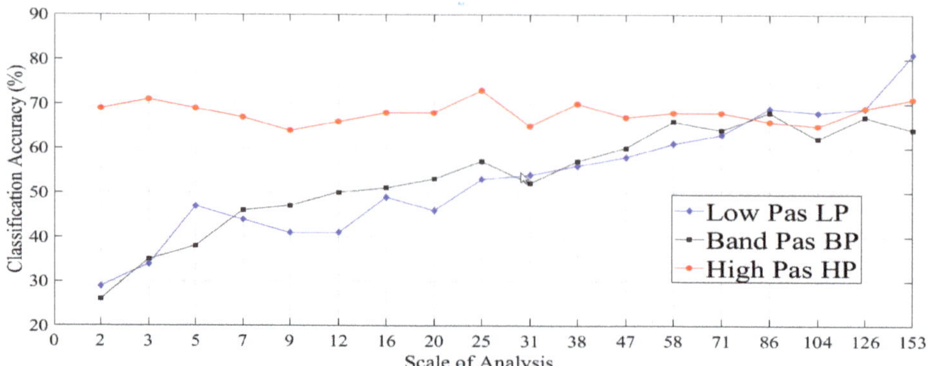

Figure 25. Comparison between the achieved accuracy averages for three different filtered image data sets: LP, BP, and HP data set at all available scales of analysis.

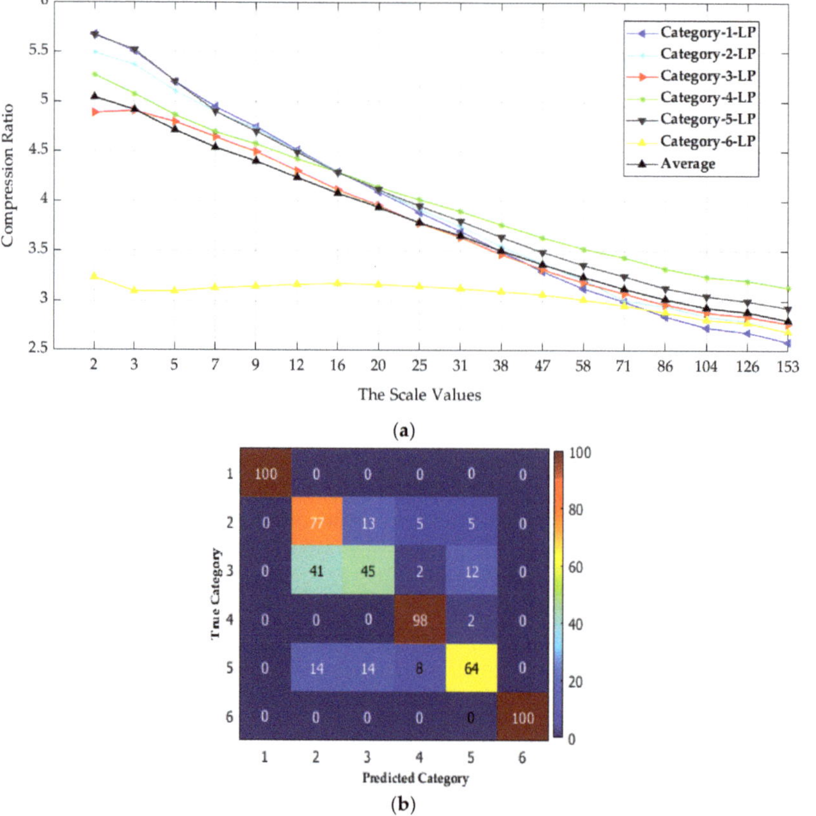

Figure 26. The impact of scale of analysis on the performances of a compressed-domain classifier: (**a**) the relation between the scale of analysis and the six surface categories' compression and classification performances by using the multiscale LP data sets and (**b**) a confusion matrix for six surface category. classifications by using 50% of the highest-scale LP data set for training.

In order to increase the classification accuracy, we could use higher-scale analysis at lower compression ratios. For example, we can obtain a classification accuracy of 81% with an average compression ratio CR = 2.5:1 at the highest length-scale 153. The six highest-scale low-pass surface categories robust performances were reported in the confusion matrix shown in Figure 26b. The prediction percentages for category 1 and category 6 are equal to 100%.

For the six multiscale band-pass surface categories case, we selected scale of analysis = 86 and average CR = 2.16:1 to obtain a classification accuracy of 68%, as illustrated in Figure 27a. The fiftieth scale band-pass surface category performances were reported in the confusion matrix shown in Figure 27b.

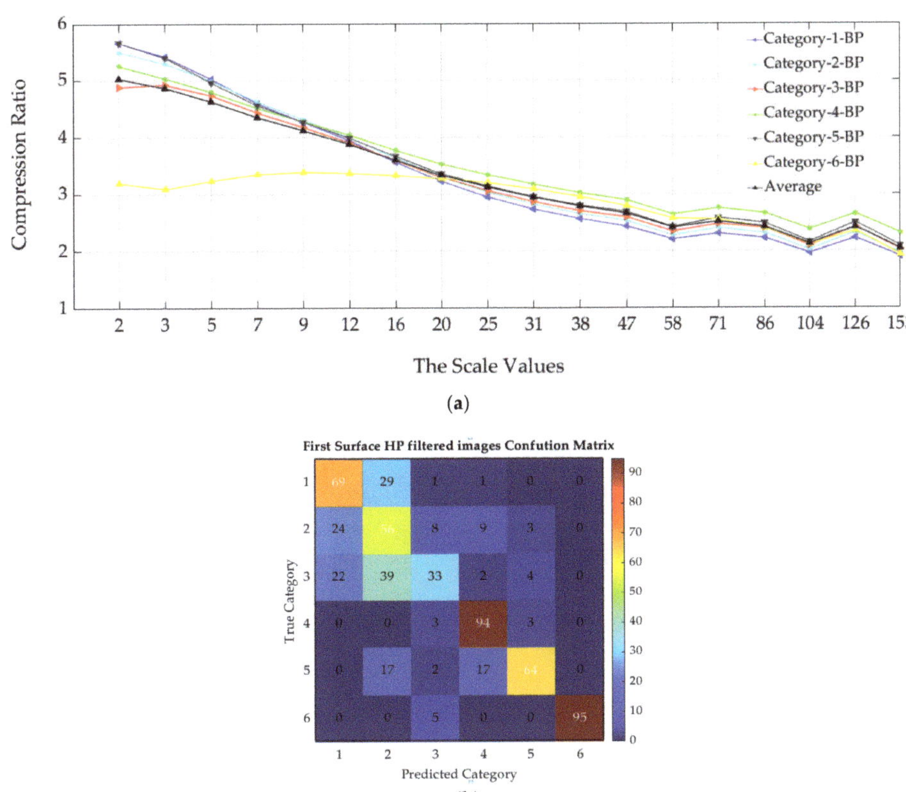

Figure 27. The impact of scale of analysis on the performances of a compressed-domain classifier: (**a**) the relation between the scale of analysis and the six surface categories' compression and classification performances by using the multiscale BP data sets and (**b**) a confusion matrix for six surface categories classification by using 50% of the highest-scale BP data set for training.

Finally, the following (Figure 28) represents the performances achieved by using separated scales of the HP data sets, where the scale of analysis does not have a big impact on the classification accuracy or the compression ratio.

There is no significant difference between the obtained classification accuracies and compression ratios at the highest and the lowest scales of analysis with Acc = 69% and $CR_{Average}$ = 2.5:1 and Acc = 71% and $CR_{Average}$ = 2.2:1 respectively as shown in Figure 28a.

Consequently, the six highest-scale high-pass surface category performances were reported in the confusion matrix shown in Figure 28b.

We have also compared our compressed-domain classifier for multiscale topographical images with four conventional methods based on a set of roughness parameters based on Eur 15178N and ISO

25178. We found that the third methods that include the whole scale analysis with all parameters lead to an accuracy 65% for classifying these six different mechanical surfaces. All these obtained results are provided in Appendix A.

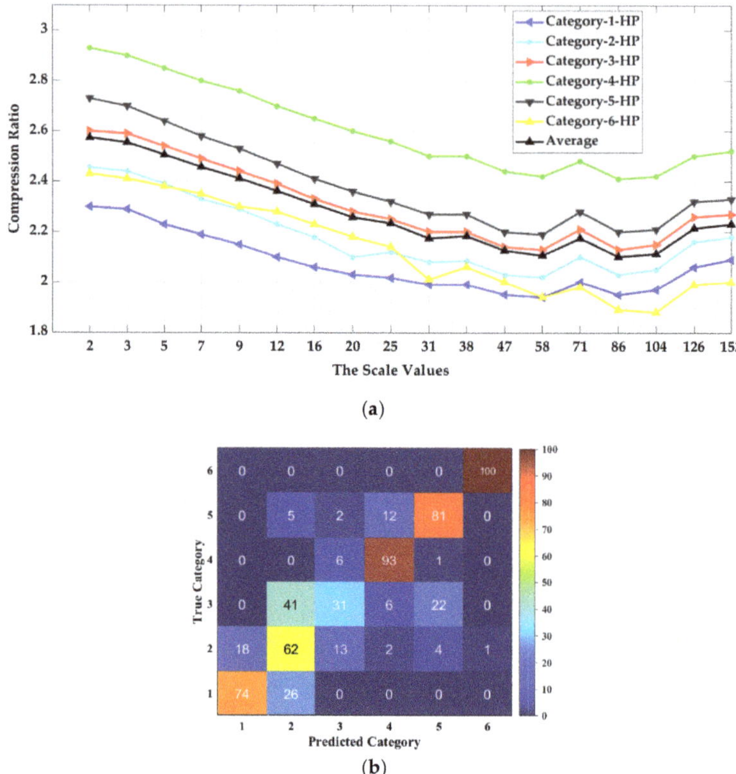

(a)

(b)

Figure 28. The impact of scale of analysis on the performances of a compressed-domain classifier: (**a**) the relation between the scale of analysis and the six surface categories' compression and classification performances by using the multiscale HP data sets and (**b**) a confusion matrix for six surface categories classification by using 50% of the highest-scale HP data set for training.

5. Conclusions

This paper has evaluated the effects of surface filtering types and the scale of analysis on the performance of six mechanical multiscale decomposed surface classification. The surface profile was analyzed by using the Gaussian filter multiscale analyzing technique by different filters, LP, BP, and HP filters at all available analysis scales, and finally, we collect three different multiscale images data sets. The collected 42,000 multiscale topographic images were compressed using the HEVC lossless compression technique which guaranteed to preserve the original material parameters. Also, the proposed texture feature descriptor was extracted from the HEVC compressed domain aiming to reduce the computation complexity. Finally, these extracted feature descriptors are fed into SVM for enhancing the system classification accuracy. The results demonstrated that the robust compressed-domain topographies classifier is based on multiscale analysis methodologies. The low-frequency components (LP data set) of the surface profile were more appropriate for characterizing our surface topographies. The best accuracy for the LP image data set was 81% in the case of the highest-scale classification with a moderate compression ratio average = 2.8:1. In a further study, we will investigate the impact of a lossy compression in mechanical surface topography classification performance.

Author Contributions: Conceptualization, M.B., T.E. and F.-X.C.; methodology, P.C., F.-X.C., and T.E.; software: M.B., T.E., and R.S.; validation, M.B., T.E., and R.S.; formal analysis, M.B., P.C., and F.-X.C.; writing—original draft preparation, T.E.; writing—review and editing, T.E. and M.B.; administration, M.B. and P.C.; funding acquisition, M.B. All authors have read and agreed to the published version of the manuscript.

Funding: This research received no external funding.

Conflicts of Interest: The authors declare no conflict of interest.

Appendix A. Analysis by Conventional Methods

In order to evaluate the performances of the proposed classification method, it is necessary to compare with more conventional methods. Conventional methods are based on a set of roughness parameters based on Eur 15178N and ISO 25178. Table A1 shows the set of parameters calculated for the original cable 5 image (Figure A1) (i.e., unfiltered).

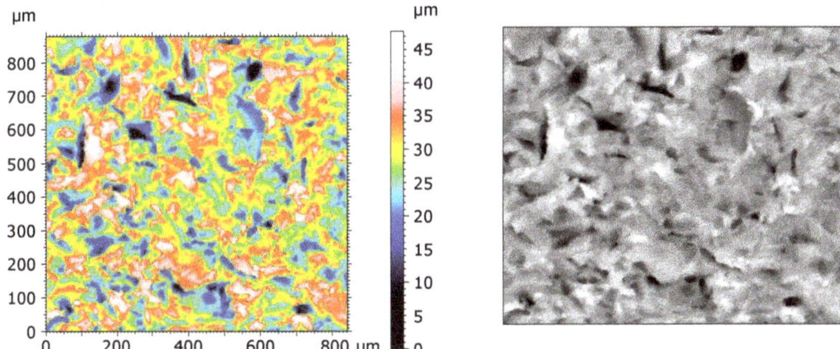

Figure A1. Original cable 5 topography (left) and it is corresponding grey level image (right).

Table A1. List of the 34 roughness parameters and their numerical values corresponding to the map of cable 5 (Figure A1).

ISO 25178			
Height Parameters			
Sq	6.42	µm	Root-mean-square height
Ssk	−0.468		Skewness
Sku	3.48		Kurtosis
Sp	18.8	µm	Maximum peak height
Sv	29.0	µm	Maximum pit height
Sz	47.8	µm	Maximum height
Sa	5.04	µm	Arithmetic mean height
Functional Parameters (Volume)			
Vm	0.243	$\mu m^3/\mu m^2$	Material volume
Vv	8.02	$\mu m^3/\mu m^2$	Void volume
Vmp	0.243	$\mu m^3/\mu m^2$	Peak material volume
Vmc	5.68	$\mu m^3/\mu m^2$	Core material volume
Vvc	7.13	$\mu m^3/\mu m^2$	Core void volume
Vvv	0.889	$\mu m^3/\mu m^2$	Pit void volume

Table A1. *Cont.*

ISO 25178			
Functional Parameters (Stratified surfaces)			
Sk	15.5	μm	Core roughness depth
Spk	4.78	μm	Reduced summit height
Svk	8.12	μm	Reduced valley depth
Smr1	8.62	%	Upper bearing area
Smr2	87.1	%	Lower bearing area
Spatial Parameters			
Sal	26.6	μm	Auto correlation length
Str	0.819		Texture-aspect ratio
Feature Parameters			
Spd	0.000312	$1/\mu m^2$	Density of peaks
Spc	0.306	$1/\mu m$	Arithmetic mean peak curvature
S10z	38.2	μm	Ten-point height
S5p	15.4	μm	Five-point peak height
S5v	22.7	μm	Five-point pit height
Sda	2333	μm	Mean dale area
Sha	3135	μm	Mean hill area
Sdv	1745	μm	Mean dale volume
Shv	1663	μm	Mean hill volume
EUR 15178N			
Hybrid Parameters			
Sdq	0.981		Root-mean-square slope
Sds	0.00402	$1/\mu m^2$	Density of summits
Ssc	0.248	$1/\mu m$	Arithmetic mean summit curvature
Sdr	38.4	%	Developed interfacial area
Sfd	2.51		Fractal dimension of the surface

In order to compare with the proposed new method, we will similarly formulate the approach proposed in this article with the same tools of discrimination but use several types of characterization used in the bibliography.

In a generic way, the same technique of machine learning (support vector machine) will be used but with predictors that are roughness parameters which depend on the method used.

There exist mainly four methods of roughness analysis:

- **Method 1. Sa Analyses.** Sa is the arithmetic average value of roughness determined from deviations about the center plane. Sa is by far the most common roughness parameter, though this is often for historical reasons and not for particular merit, as the early roughness meters could only measure it. Whitehouse discusses the advantages of this parameter (robust and easy to understand) and the inconvenience (unable to characterize the skewness of the surface amplitude, i.e., difference of peaks and valleys, unable to characterize the size of peaks and valleys) [44]. The Sa is computed without filtering, i.e., at the whole scale. SVM classification is performed with this unique roughness parameter.

- **Method 2. Sa Multiscale Analysis.** As presented in Section 2.1, multiscale can be used to practice a multiscale decomposition and Sa roughness parameters are computed for all scales for the three Gaussian filters (pass band, low pass, and high pass). Giljean et al. [45] have shown that this multiscale analysis allows for the Sa roughness parameters to detect the size of the peaks and valleys, avoiding the main critic claimed by Whitehouse [44]. One obtains a set of parameters Sa (F,ε), where F is the filter and ε the scale length (cut off filter). From this set, SVM classification is processed.

- **Method 3. Whole-scale analysis by a set of roughness parameters.** Thirty-four R_i roughness parameters (see Table A1 for their descriptions) with $i = \{1, 2 \ldots ,34\}$ are computed without filtering, i.e., at the whole scale. Najjar et al. [46] has shown that the measure of functionality of a surface must be analyzed with the amplitude, spatial, and hybrid parameters to find the best one that characterizes the effect of roughness. They proposed a relevance function to classify the efficiency of roughness parameters based on variance analysis. One obtains a set of parameters R_i from which SVM classification is processed.

- **Method 4. Multiscale analysis by a set of roughness parameters.** Thirty-four R_i roughness parameters (see Table A1 for their descriptions) with $i = \{1, 2 \ldots ,34\}$ are computed for all scales for the three filters (pass band, low pass, and high pass). By analyzing all the roughness parameters of the GPS standard, Le Goic et. [10] showed that, with different types of filtering at different scales, ANOVA discriminates a wide range of tribological mechanisms by classification indexes based on databank of F values created from ANOVA [10]. One obtains a set of parameters Ri (F, ε), where F is the filter and ε is the scale length. From this set, SVM classification is processed.

The results of SVM classification with their associated confusion matrix for the four methods of roughness characterization are presented in Figure A2). For method 1, only 38% is well classified. It is well known that Sa only characterizes the amplitude of the roughness and is unable to quantify difference between peaks and valleys. However, filtering allows us to define some particularity of roughness surface and SVM allows to characterize the small amplitude and high amplitude, form, and waviness that increase the quality of classification by Sa (52%).

Method 1 Sa Roughness Parameter Whole Scale							Method 2 Sa Roughness Parameter Multi-Scale							Method 3 All 34 Roughness Parameters Whole Scale							Method 4 All 34 Roughness Parameters Multi-scale					
Confusion Matrix: Predicted Categories (horizontal) Versus True Categories (vertical)																										
	1	2	3	4	5	6		1	2	3	4	5	6		1	2	3	4	5	6		1	2	3	4	5	6
1	0	96	0	0	4	0	1	0	100	0	0	0	0	1	50	50	0	0	0	0	1	9	90	0	0	0	0
2	0	97	0	0	3	0	2	0	59	34	0	7	0	2	5	64	28	0	3	0	2	0	61	31	0	8	0
3	0	100	0	0	0	0	3	0	53	43	4	0	0	3	0	58	42	0	0	0	3	0	49	49	0	2	0
4	0	38	0	0	60	2	4	0	0	0	57	42	1	4	0	0	0	52	48	0	4	0	0	0	53	47	0
5	0	69	0	0	31	0	5	0	31	0	14	54	0	5	0	17	0	0	83	0	5	0	29	0	0	71	0
6	0	0	0	0	0	100	6	0	0	0	0	0	100	6	0	0	0	0	0	100	6	0	0	0	0	0	100
38 %							52 %							65 %							57 %						

Figure A2. The results of SVM classification for the four methods of roughness characterization.

However, taking into account all roughness parameters will introduce frequency parameters that define peaks and valleys that increase classification with 65% of well-classified data. Finally, including the multiscale analyses with all parameters allows for definition of all multiscale signatures of the topography and will lead to 57% of well-classified data.

References

1. Ji, M.; Xu, J.; Chen, M.; EL Mansori, M. Enhanced hydrophilicity and tribology behavior of dental zirconia ceramics based on picosecond laser surface texturing. *Ceram. Int.* **2019**, *46*. [CrossRef]
2. Brown, C.A.; Hansen, H.N.; Jiang, X.J.; Blateyron, F.; Berglund, J.; Senin, N.; Bartkowiak, T.; Dixon, B.; Le Goïc, G.; Quinsat, Y.; et al. Multiscale analyses and characterizations of surface topographies. *CIRP Ann.* **2018**, *67*, 839–862. [CrossRef]
3. Ghosh, K.; Pandey, R. Fractal and multifractal analysis of in-doped ZnO thin films deposited on glass, ITO, and silicon substrates. *Appl. Phys. A* **2019**, *125*. [CrossRef]
4. Hosseinabadi, S.; Karimi, Z.; Masoudi, A.A. Random deposition with surface relaxation model accompanied by long-range correlated noise. *Phys. A Stat. Mech. Appl.* **2020**, *560*, 125130. [CrossRef]

5. Huaian, Y.; Xinjia, Z.; Le, T.; Yonglun, C.; Jie, Y. measuring grinding surface roughness based on singular value entropy of quaternion. *Meas. Sci. Technol.* **2020**, *31*, 115006. [CrossRef]

6. Pahuja, R.; Ramulu, M. Characterization of surfaces generated in milling and abrasive water jet of CFRP using wavelet packet transform. *IOP Conf. Ser. Mater. Sci. Eng.* **2020**, *842*, 12001. [CrossRef]

7. Bigerelle, M.; Iost, A. Characterisation of the diffusion states by data compression. *Comput. Mater. Sci.* **2002**, *24*, 133–138. [CrossRef]

8. Zhang, A.; Wang, K.C.P. The fast prefix coding algorithm (FPCA) for 3D pavement surface data compression. *Comput. Civ. Infrastruct. Eng.* **2017**, *32*, 173–190. [CrossRef]

9. Elkhuizen, W.S.; Callewaert, T.W.J.; Leonhardt, E.; Vandivere, A.; Song, Y.; Pont, S.C.; Geraedts, J.M.P.; Dik, J. Comparison of Three 3D scanning techniques for paintings, as applied to vermeer's 'girl with a pearl earring. *Herit. Sci.* **2019**, *7*, 89. [CrossRef]

10. Le Goïc, G.; Bigerelle, M.; Samper, S.; Favrelière, H.; Pillet, M. multiscale roughness analysis of engineering surfaces: A comparison of methods for the investigation of functional correlations. *Mech. Syst. Signal Process.* **2016**, *66–67*, 437–457. [CrossRef]

11. Mistry, Y.; Ingole, D.T.; Ingole, M.D. Content based image retrieval using hybrid features and various distance metric. *J. Electr. Syst. Inf. Technol.* **2017**. [CrossRef]

12. Mehrabi, M.; Zargari, F.; Ghanbari, M. compressed domain content based retrieval using H.264 DC-pictures. *Multimed. Tools Appl.* **2012**, *60*, 443–453. [CrossRef]

13. Rahmani, F.; Zargari, F. Temporal feature vector for video analysis and retrieval in high efficiency video coding compressed domain. *Electron. Lett.* **2018**, *54*, 294–295. [CrossRef]

14. Zargari, F.; Rahmani, F. Visual information retrieval in HEVC compressed domain. In Proceedings of the 23rd Iranian Conference on Electrical Engineering, Tehran, Iran, 10–14 May 2015; pp. 793–798. [CrossRef]

15. Rahmani, F.; Zargari, F. Compressed domain visual information retrieval based on I-Frames in HEVC. *Multimed. Tools Appl.* **2017**, *76*, 7283–7300. [CrossRef]

16. Yamaghani, M.; Zargari, F. Classification and retrieval of radiology images in H.264/AVC compressed domain. *Signal Image Video Process.* **2017**, *11*, 573–580. [CrossRef]

17. Zargari, F.; Mehrabi, M.; Ghanbari, M. Compressed domain texture based visual information retrieval method for I-Frame coded pictures. *IEEE Trans. Consum. Electron.* **2010**, *56*, 728–736. [CrossRef]

18. Yamghani, A.R.; Zargari, F. Compressed domain video abstraction based on I-Frame of HEVC coded videos. *Circuits Syst. Signal Process.* **2019**, *38*, 1695–1716. [CrossRef]

19. Zygo Corporation. *NewView 7200 & 7300 Operating Manual*, OMP-0536, Rev. E, ed.; Zygo Corporation: Middlefield, CT, USA, 2011.

20. Yoshizawa, T. *Handbook of Optical Metrology Principles and Applications*, 2nd ed.; CRC Press: Boca Raton, FL, USA, 2015. [CrossRef]

21. Bigerelle, M.; Guillemot, G.; Khawaja, Z.; El Mansori, M.; Antoni, J. Relevance of wavelet shape selection in a complex signal. *Mech. Syst. Signal Process.* **2013**, *41*, 14–33. [CrossRef]

22. Bigerelle, M.; Abdel-Aal, H.A.; Iost, A. Relation between entropy, free energy and computational energy. *Int. J. Mater. Prod. Technol.* **2010**, *38*, 35–43. [CrossRef]

23. Bigerelle, M.; Haidara, H.; Van Gorp, A. Monte carlo simulation of gold nano-colloids aggregation morphologies on a heterogeneous surface. *Mater. Sci. Eng. C* **2006**, *26*, 1111–1116. [CrossRef]

24. Bigerelle, M.; Dalla-Costa, M.; Najjar, D. Multiscale similarity characterization of abraded surfaces. *Proc. Inst. Mech. Eng. Part B J. Eng. Manuf.* **2007**, *221*, 1473–1482. [CrossRef]

25. Dalla Costa, M.; Bigerelle, M.; Najjar, D. A new methodology for quantifying the multi-scale similarity of images. *Microelectron Eng.* **2007**, *84*, 424–430. [CrossRef]

26. Lemesle, J.; Robache, F.; Le Goic, G.; Mansouri, A.; Brown, C.A.; Bigerelle, M. Surface reflectance: An optical method for multiscale curvature characterization of wear on ceramic-metal composites. *Materials* **2020**, *13*, 1024. [CrossRef] [PubMed]

27. Sullivan, G.J.; Ohm, J.; Han, W.; Wiegand, T. Overview of the high efficiency video coding (HEVC) standard. *IEEE Trans. Circuits Syst. Video Technol.* **2012**, *22*, 1649–1668. [CrossRef]

28. Sze, V.; Budagavi, M.; Sullivan, G.J. *High Efficiency Video Coding (HEVC): Algorithms and Architectures*, 1st ed.; Springer Publishing Company, Incorporated: Cham, Switzerland, 2014. [CrossRef]

29. Li, J.; Li, B.; Xu, J.; Xiong, R.; Gao, W. Fully connected network-based intra prediction for image coding. *IEEE Trans. Image Process.* **2018**, *27*, 3236–3247. [CrossRef]

30. Flynn, D.; Marpe, D.; Naccari, M.; Nguyen, T.; Rosewarne, C.; Sharman, K.; Sole, J.; Xu, J. Overview of the range extensions for the HEVC Standard: Tools, profiles, and performance. *IEEE Trans. Circuits Syst. Video Technol.* **2016**, *26*, 4–19. [CrossRef]

31. Humeau-heurtier, A. Texture feature extraction methods: A survey. *IEEE Access* **2019**, *7*, 8975–9000. [CrossRef]

32. Faust, O.; Acharya, U.R.; Meiburger, K.M.; Molinari, F.; Koh, J.E.W.; Yeong, C.H.; Kongmebhol, P.; Ng, K.H. Comparative assessment of texture features for the identification of cancer in ultrasound images: A review. *Biocybern. Biomed. Eng.* **2018**, *38*, 275–296. [CrossRef]

33. Lan, Z. Study on multi-scale window determination for GLCM texture description in high-resolution remote sensing image geo-analysis supported by GIS and domain knowledge. *Int. J. Geo Inf.* **2018**, *7*, 175. [CrossRef]

34. Pérez-Barnuevo, L. Automated recognition of drill core textures: A geometallurgical tool for mineral processing prediction. *Miner. Eng.* **2017**, *118*. [CrossRef]

35. García-ordás, M.T.; Alegre-gutiérrez, E.; Alaiz-rodríguez, R.; González-castro, V. Tool wear monitoring using an online, automatic and low cost system based on local texture. *Mech. Syst. Signal Process.* **2018**, *112*, 98–112. [CrossRef]

36. Binias, B.; Myszor, D. A Machine learning approach to the detection of pilot' s reaction to unexpected events based on EEG signals. *Comput. Intell. Neurosci.* **2018**, *14*, 1–9. [CrossRef]

37. Zhang, X.; Cui, J.; Wang, W.; Lin, C. A study for texture feature extraction of high-resolution satellite images based on a direction measure and gray level co-occurrence matrix fusion algorithm. *Sensors* **2017**, *17*, 1474. [CrossRef] [PubMed]

38. Gao, L.; Ye, M.; Wu, C. Cancer classification based on support vector machine optimized by particle swarm optimization. *Molecules* **2017**, *22*, 2086. [CrossRef] [PubMed]

39. Zeng, N.; Qiu, H.; Wang, Z.; Liu, W.; Zhang, H.; Li, Y. A new switching-delayed-PSO-based optimized SVM algorithm for diagnosis of alzheimer's disease. *Neurocomputing* **2018**, *320*, 195–202. [CrossRef]

40. Machine, V. A novel method for the recognition of air visibility level based on the optimal binary tree support vector machine. *Atmosphere* **2018**, *9*, 481. [CrossRef]

41. Ortegon, J.; Ledesma-Alonso, R.; Barbosa, R.; Castillo, J.; Castillo Atoche, A. Material phase classification by means of support vector machines. *Comput. Mater. Sci.* **2017**, *148*, 336–342. [CrossRef]

42. Ben-david, S.; Shalev-Shwartz, S. *Understanding Machine Learning: From Theory to Algorithms*; Cambridge University Press: Cambridge, UK, 2014.

43. Zhang, Z.; Chen, H.; Xu, Y.; Zhong, J.; Lv, N.; Chen, S. Multisensor-based real-time quality monitoring by means of feature extraction, selection and modeling for Al alloy in arc welding. *Mech. Syst. Signal Process.* **2015**, *61*, 151–165. [CrossRef]

44. Whitehouse, D.J. *Handbook of Surface Metrology*, 1st ed.; Institute of Physics Publishing for Rank Taylor Hobson Co., Bristol: London, UK, 1994.

45. Giljean, S.; Najjar, D.; Bigerelle, M.; Alain, I. Multiscale analysis of abrasion damage on stainless steel. *Surf. Eng.* **2008**, *24*, 8–17. [CrossRef]

46. Najjar, D.; Bigerelle, M.; Iost, A. The Computer-based bootstrap method as a tool to select a relevant surface roughness parameter. *Wear* **2003**, *254*, 450–460. [CrossRef]

Publisher's Note: MDPI stays neutral with regard to jurisdictional claims in published maps and institutional affiliations.

 *materials*

Article

High-Accuracy 3D Optical Profilometry for Analysis of Surface Condition of Modern Circulated Coins

Wojciech Kapłonek [1], Tadeusz Mikolajczyk [2], Danil Yurievich Pimenov [3], Munish Kumar Gupta [3,4], Mozammel Mia [5,*], Shubham Sharma [6], Karali Patra [3,7] and Marzena Sutowska [1]

[1] Department of Production Engineering, Faculty of Mechanical Engineering, Koszalin University of Technology, Racławicka 15-17, 75-620 Koszalin, Poland; wojciech.kaplonek@tu.koszalin.pl (W.K.); marzena.sutowska@tu.koszalin.pl (M.S.)

[2] Department of Production Engineering, UTP University of Science and Technology, Al. prof. S. Kaliskiego 7, 85-796 Bydgoszcz, Poland; tami@utp.edu.pl

[3] Department of Automated Mechanical Engineering, South Ural State University, Lenin Prosp. 76, 454080 Chelyabinsk, Russia; danil_u@rambler.ru (D.Y.P.); munishguptanit@gmail.com (M.K.G.); kpatra@iitp.ac.in (K.P.)

[4] Key Laboratory of High Efficiency and Clean Mechanical Manufacture, Ministry of Education, School of Mechanical Engineering, Shandong University, Jinan 250061, China

[5] Department of Mechanical Engineering, Imperial College London, Exhibition Rd., South Kensington, London SW7 2AZ, UK

[6] Department of Mechanical Engineering, IK Gujral Punjab Technical University, Jalandhar-Kapurthala Road, Kapurthala 144603, India; shubham543sharma@gmail.com

[7] Department of Mechanical Engineering, Indian Institute of Technology Patna, Patna 800013, India

[*] Correspondence: m.mia19@imperial.ac.uk

Received: 25 October 2020; Accepted: 23 November 2020; Published: 26 November 2020

Abstract: The article shows that noncontact measurement techniques can be an important support to X-ray-based methods when examining the surface condition of modern circulated coins. The forms and degrees of wear of such coins, affecting their utility values, qualifying them as a legal tender in a given country, can be measured and analyzed, among other things, using advanced high-accuracy optical profilometry methods. The authors presented four analyses carried out for reverses and obverses of round coins (1 zloty, 1 franc, 50 bani, 5 pens) characterized by different degrees of surface wear. All of the coins were measured using 3D optical profilometers (Talysurf CLI 2000 and S neox) representing two generations of these types of systems. The obtained results confirm the validity of the applied high-accuracy measurement systems in conjunction with dedicated software in the presented applications. Examples of the analyses carried out can be a significant source of information on the condition of coins in the context of maintaining their functional properties (selection of appropriate wear–resistant alloys and correctness of the production process).

Keywords: circulated coins; surface condition; optical methods; measurements and analysis

---

1. Introduction

Coins, as cash marks bearing the issuers' marks and used as legal tenders in countries, have long been an interesting and intriguing object of scientific research. They exist, similar to banknotes, in many world monetary systems. Coin production has not changed significantly for centuries. Today, their production process has been greatly improved—production time has been shortened, and the use of new, promising materials has positively influenced their quality and performance characteristics, as reported by de Toit et al. and Mendoza-López et al. in their works [1,2], but the very idea of production has remained the same—coins were and are embossed.

Ancient coins, widely described in the works of Dow [3] and Metcalf [4], obtained by archaeological methods are extremely rich sources of historical information. Such information is usually received based on analyses of the chemical composition, corrosion products, surface morphology, microstructure, and physical properties, and then correlated with known manufacturing processes, sources of raw materials, and finally, geographical distribution of ancient mints. The above analyses (depending on the characteristics of the coin, complexity of the process of preparing the coin for research, and intensity of the research program) are carried out using many advanced observation measurement X-ray-based methods. Among them, the most significant methods are those using diffractometry, spectroscopy, electron microscopy, and computed tomography. A multitude of these methods and their variations, metrological characteristics, and specificity of application (not every technique is appropriate for the assessment of a given coin or group of coins) allow for the perception of the complexity of carried-out analyses and the real difficulty of assessing ancient coins. This is confirmed by numerous scientific papers from this research area published over the last several years. A representative review of the above X-ray-based methods and their application with adequate references are given in Table 1, and in Figure 1, the selected X-ray- and optical-based experimental setups carrying out the ancient coin analyses are presented.

Table 1. Selected X-ray-based methods used for the analysis of ancient coins.

Application	Method	References
Analysis of chemical composition and microstructure	EDS (EDX)	[5] [1], [6] [2]
	EDXRF	[7] [4], [8–11]
	GRT	[12]
	LAMQS	[8]
	LIBS	[13,14]
	ND	[15,16]
	PIXE	[6] [2], [17] [3], [18], [6] [2], [19], [20] [3], [21–26]
	RBS	[25]
	XPS	[27] [4], [28]
	XRD	[5] [1], [7] [4], [16,28]
	XRF	[12,13,29,30], [31] [3]
	SRXRF	[6] [2], [19] [4]
	SEM	[32] [3], [33,34]
	TEM	[33,35]
Analysis of the corrosion products	EDS (EDX)	[36] [1,5]
	PIXE	[37,38] [3]
	XPS	[36] [1,5]
	XRD	[36] [1,5], [38] [3], [39] [4]
	XRF	[38] [3]
Reconstruction	CT	[40–42]

[1] Authors also used a combination of scanning electron microscopy (SEM) and optical microscopy (OM); [2] authors also used SEM; [3] based on chemical composition/corrosion products, authors carried out an analysis of the authenticity of the coins; [4] for a study of the homogeneity/heterogeneity/ presence of surface enrichments, authors also used a combination of SEM and energy-dispersive X-ray spectrometry (EDS/EDX); [5] authors also used electron microprobe analysis (EMPA).

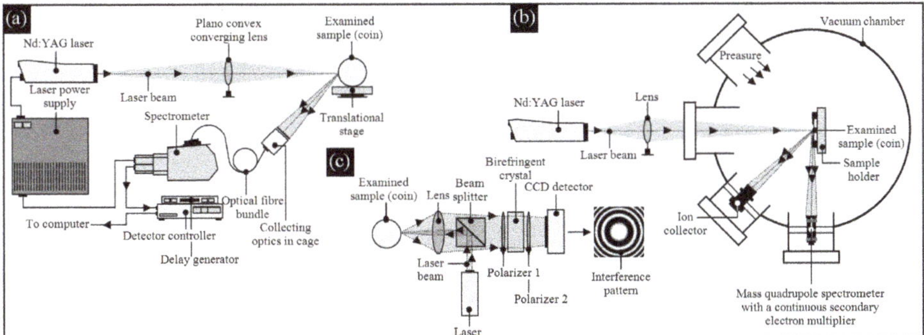

Figure 1. Selected experimental setups used for analyzing of the ancient coins: (**a**) laser-induced breakdown spectroscopy (LIMB), presented in the work of Awasthi et al. [14] (coins collected from the G.R. Sharma Memorial Museum, University of Allahabad, India, dating back to VI Cent. A.D.); (**b**) laser ablation coupled with mass quadrupole spectrometry (LAMQS), presented in the work of Torrisi et al. [8] (Mediterranean basin bronze coins, dating back to the II–X Cent. A.D.); (**c**) high-resolution laser microprofilometry using conoscopic holography (CH), presented in the work of Spagnolo et al. [43] (ancient Roman coins, dating back to the I Cent. A.D.).

Coins in their modern form are mainly circulation coins. Similarly, coins such as antique ones produced of metals and their alloys are subject to the wear phenomenon. It is a natural consequence of circulation. Wear occurs with varying intensities over the entire surface of a given coin in a whole or fragmentary manner as a result of a range of physical and chemical interactions. Everyday handling of coins (e.g., tossing and/or removing them from cash registers; storing them in wallets, pockets, and bags; accidentally dropping them on a hard surface; etc.) causes small losses of material. Therefore, apart from their purely mint (uncirculated) conditions, from the time they are put into circulation, coins are subject to wear. This process can be divided into several phases:

- Phase I: Slight wear (the coin entering the circulation begins to show signs of manipulation, abrasion, or slight wear, which can be stated after a visual analysis of changes in fragments of the surface texture and differences in the sharpness of fine elements (mint mark, motto, etc.), as well as slight differences in their color).
- Phase II: Average wear (the coin in circulation, after a period of time, begins to show signs of medium wear. It can be stated after a visual analysis of the design's highest points, which begin to lose their sharpness and slightly round or flatten, while fine elements close together begin to connect).
- Phase III: Extensive wear (the coin in circulation, after a long time, begins to show signs of extensive wear. It can be stated after a visual analysis of the design sharpness. High points begin to connect with the next lower elements of the design. After flattening the protective rim, the entire surface begins to flatten—most of the details connect with itself or partially with the surface).
- Phase IV: Critical wear (the coin in circulation, after a long time, begins to show signs of critical wear. This phase causes the completely flattening of the protective rim, design, and all other elements of the coin. Both the obverse and reverse are difficult to identify. The coin cannot fulfill the tender function and must be withdrawn from circulation).

Similar to ancient coins, modern coins are subjected to numerous studies that allow the determination of their chemical composition, as presented in the work of Roumie et al. [44], for high accuracy and surface micromorphology characterization, as reported in the work of Papp and Kovacs [45], and allow the localization of surface defects (stains), as presented in the work of Corregidor et al. [46]. The above studies are carried out using advanced X-ray-based observation measurement methods [47], which are often supported by other noncontact techniques. A significant number of such techniques are optical profilometry methods [48]. These methods and their many variants allow for comprehensive studies of coins with the use of advanced geometric (dimensional shape) analyses and surface texture measurements (mainly roughness). Modern 3D optical profilometers are excellent platforms for this type of study. These platforms offer, among others, the possibility of integration of noncontact and high-accuracy measurement methods in one instrument, wide measurement range, short time of surface scanning, wide range of analysis, and surface visualization modes available in specialized software cooperating with the measurement instrument. A representative review of the above methods and their applications with adequate references are given in Table 2, whereas in Figure 2, selected optical-based experimental setups for analyses of modern coins are presented.

Table 2. Selected optical profilometry-based methods used for the analysis of modern coins.

Application	Method	References
Measurements of shape and surface profile (2D)	FP	[49]
	FP + SP	[50]
	FP + PS	[51]
	FP + PS + TPU + DIA	[52]
	MI	[53]
Measurements of shape (3D)	SL	[54]
Measurements of surface profile and topography (3D)	FP	[54–57]
	FVM	[58]
	CLSM	[59]
	CS	[60,61]
	SL	[62]

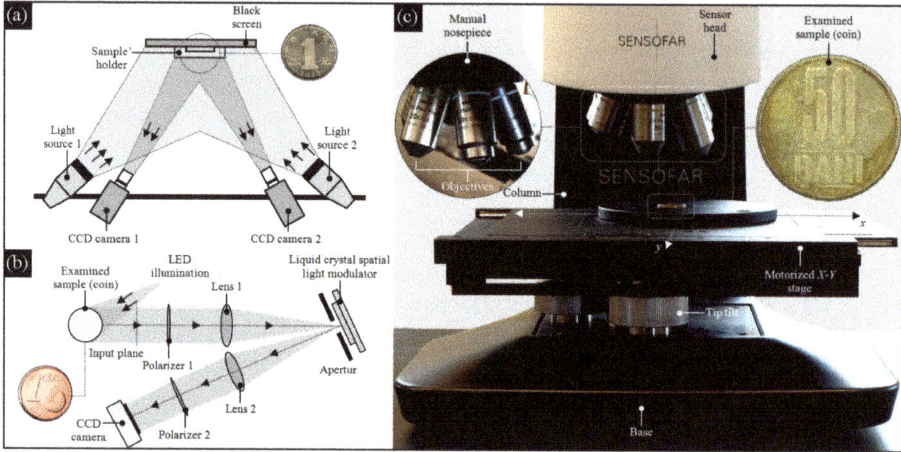

Figure 2. Selected experimental setups used for analyzing modern coins: (**a**) three-dimensional digital image correlation (3D-DIC), presented in the work Yan et al. [63] (1 yuan coin); (**b**) computational shear interferometry (CSI), shown in the work of Falldorf et al. [64] (1 euro cent coin); (**c**) confocal laser scanning microscopy (CLSM) (50 bani coin).

To familiarize readers with selected issues related to the assessment of the microtopography of circulated coins (Section 2.1) with various degrees of surface wear using optical profilometry methods, the authors of this article carried out experimental studies. During these experiments, measurement systems (Section 2.2) representing two generations of high-accuracy 3D optical profilometers were used. The main part of the work focused on the presentation of selected analyses of circulated coins representing various degrees of surface wear. In Section 3, four of such analyses were discussed. Each of them consisted of information on the type of measurement system used, parameters of the measurement process, and selected results of the analyses carried out, along with a detailed description and interpretation.

2. Experimental Studies

2.1. Selection and Characteristics of the Coins

For the experimental studies, a set of six round-shaped coins in uncirculated (one coin used as reference) and circulated (five coins used in main studies) conditions was selected. Each of the coins had individual physical (composition, geometrical dimensions) and visual (relief) features. A main criterion for selecting the coins was their visually observed surface condition (forms and wear degrees). The general characteristics of the analyzed modern coins are presented in Table 3.

Table 3. Characteristics of the modern coins used in experimental studies.

Coin Value	Country	Type	Years	Composition (Alloy)	Diameter (mm)	Thickness (mm)	Weight (g)
1 złoty	Poland	Standard circulation coin	1990–2016	$Cu_{75}Ni_{25}$	23.00	1.70	5.00
1 franc	France		1959–2001	Ni	24.00	1.79	6.00
50 bani	Romania		2005–2017	$Cu_{80}Zn_{15}Ni_5$	23.75	1.90	6.10
5 pence	Great Britain		2011–2015	Ni-plated steel	18.00	1.89	3.25

Before starting the measurements, the surfaces of the coins were cleaned with undiluted acetone, then washed in water with the addition of mild detergent (soap), and carefully dried with compressed air.

2.2. Characteristics of Observation Measurement Systems

To carry out accurate measurements and imaging of the surfaces of small-size objects (coins), the authors selected appropriate metrological systems. Two types of profilometric-based measurement systems using optical methods were used in the experimental studies.

The first of the systems was the multisensory 3D optical profilometer Talyscan CLI 2000 (Taylor- Hobson, Leicester, UK). The system was characterized by the following properties: capacity: $200 \times 200 \times 200$ mm; traverse length: 200 mm at resolution 0.5 µm; measuring speed: 0.5, 1.5, 10, 15, and 30 mm/s; and positioning: 30 mm/s. During the experiments, the measuring head of the instrument was equipped with the confocal chromatic (CLA) sensor RB-800 (scanning frequency: 5000 Hz; measuring range: 0.8 mm; resolution (vertical): 0.025 µm; speed: 30 mm/s). The Talyscan CLI 2000 2.6 software was used for realizing the measurement process, whereas the TalyMap Silver 4.1 software using Mountains Technology® (Digital Surf, Besançon, France) was utilized for the analysis and visualization of the surface microtopography. A detailed description of this older-generation instrument (2003–2004) was presented in the work of Kapłonek et al. [60], whereas its exemplary applications were given by Fan et al. [65], Saremi-Yarahmadi et al. [66], Beamud et al. [67], and Genna et al. [68].

The second of the systems was the 3D optical profilometer S neox (Sensofar Metrology, Terrassa, Spain). This new-generation instrument used three main optical methods (confocal, interferometry, and focus variation) and was characterized by the following properties: capacity: $700 \times 600 \times 40$ mm; maximum vertical scanning range: 20, 100, 10 (interferometry), and 37 mm (confocal, focus variation); resolution and linearity (z-axis): 2 nm; and <0.5 µm/mm linear stage. The motorized revolving nosepiece of the S neox was equipped with five Brightfield type TU Plan Fluor EPI (semiapochromat) 5×, 10×, 20×, and 50× and TU Plan Apo EPI (apochromat) 150× microscopic lenses (Nikon Corp., Tokyo, Japan). The SensoSCAN 2.0 software provided the correct course of the measurement process, whereas SensoMAP Premium software using Mountains Technology® (Digital Surf, Besançon, France) was used for advanced analysis and visualization of the surface microtopography. A comprehensive description of this measurement system was given in the work of Artigas et al. [69], whereas a review of its applications can be found in the works of Ding et al. [70], Leksycki and Królczyk [71], Leksycki et al. [72], and Tato et al. [73].

The abovementioned measurement systems were supported by the 3D laser microscope LEXT OLS4000 (Olympus Corp., Shinjuku, Tokyo, Japan), bench-type multisensory coordinate measuring machine VideoCheck® IP 250 (Werth Messtechnik, Gießen, Germany), and digital microscope Omni Core (Ash Technologies Ltd., Kildare, Ireland).

3. Results and Discussion

3.1. Comparative Analysis of Uncirculated/Circulated 1 Złoty (Obverse) Coin's Surface Condition

In many cases, circulated coins' wear is determined by advanced instrumental observations (microscopy) or accurate measurements (profilometry) of their obverse or reverse, often using uncirculated coins as reference. In this subsection, such type of comparative analysis was presented.

Figure 3 depicts a modern Polish 1 złoty coin (alloy: $Cu_{75}Ni_{25}$; edge: alternately smooth and serrated; diameter: 23 mm; thickness: 1.70 mm; weight: 5 g). Figure 3a presents an image obtained by the bench-type multisensory coordinate measuring machine (CMM) VideoCheck® IP 250 for a fragment (4.278 × 1.290 mm) of an uncirculated coin issued in 1994. Uncirculation, in this case, caused the coin's surface to present an excellent condition. Obverse (engraver: S. Wątróbska-Frindt) side elements—a fragment of relief, mint mark (enlarged miniature at bottom right), and legend—are clear and sharp. Additionally, the spatial nature of the above elements (they are convex) is visible.

An image from Figure 3a corresponding to the 2D height map (indexed colors) of a fragment (3.823 × 1.290 mm) of the same coin obtained by the 3D optical profilometer Talysurf CLI 2000 is presented in Figure 3c. The height of the field is in the range from ~0.06 to 0.08 mm, whereas the height of the elements on it is ~0.08 to ~0.16 mm. This shows the relatively high differentiation of the height observed on a surface of such uncirculated coin. The circulated 1 złoty coin issued in 1990 is presented in Figure 3b. The differences in geometric dimensions (from ~0.04 to 0.1 mm) of the selected elements of this coin as compared with the uncirculated coin (Figure 3a) indicate a significant average wear (Phase II). The relief is clearly flattened, as are the mint mark (enlarged miniature at bottom right) and legend. The height change of the elements can also be analyzed on the 2D height map (Figure 3d), where a slight surface wear, worn areas of the coin (marked as 1–8), is presented. Using measurement data obtained by the 3D optical profilometer Talysurf CLI 2000 and 3D laser microscope LEXT OLS4000, the image fusion presented in Figure 3e,f was generated. A vast fragment (10 × 10 mm) of the uncirculated (Figure 3e) and circulated (Figure 3f) coins was presented in the form of a 2D height map (indexed colors) with a fragment (3.844 × 1.135 mm) of a 2D pseudo-color height map. This combination perfectly reflects the differences in the heights of the depicted relief elements (eagle claws) of the uncirculated (clearly convex) and circulated (clearly flattened) coins and strongly corresponds to previous analyses (Figure 3c,d). Additionally, Figure 3e,f provides the values of selected amplitude (surface) parameters, Sa, Sq, Sp, Sv, and St, included in the ISO 25178-2:2012 standard [74] and EUR 15178 EN report [75]. These parameters are correlated and have a significant impact on the operating properties. The essential parameters from this group—Sa (arithmetic mean deviation of the surface) and Sq (root-mean-square deviation of the surface)—represent an overall measure of the surface texture. The Sq parameter is usually used to characterize optical surfaces (more smooth) and Sa machined surfaces (more irregular). The use of these parameters was dictated by the similar characteristics of the analyzed surfaces—uncirculated coin (more smooth) and circulated coin (more irregular). The values of these parameters showed slight differences in the range from 0.008 mm (Sa) to 0.010 mm (Sq). The additional parameters Sp (maximum height of summits) and Sv (maximum depth of valleys) showed slightly larger differences in the range from 0.016 mm (Sp) to 0.030 mm (Sv), whereas the value of the parameter St (total height of the surface) for both surfaces was the same and amounted to 0.280 mm. The analysis of the values of selected amplitude (surface) parameters allows for the conclusion that, generally, the values obtained for the circulated coin were ~7% to ~27% lower than the parameters obtained for the uncirculated coin. The surface of the circulated coin was, therefore, distinguished by the average wear characteristic described in Introduction Phase II.

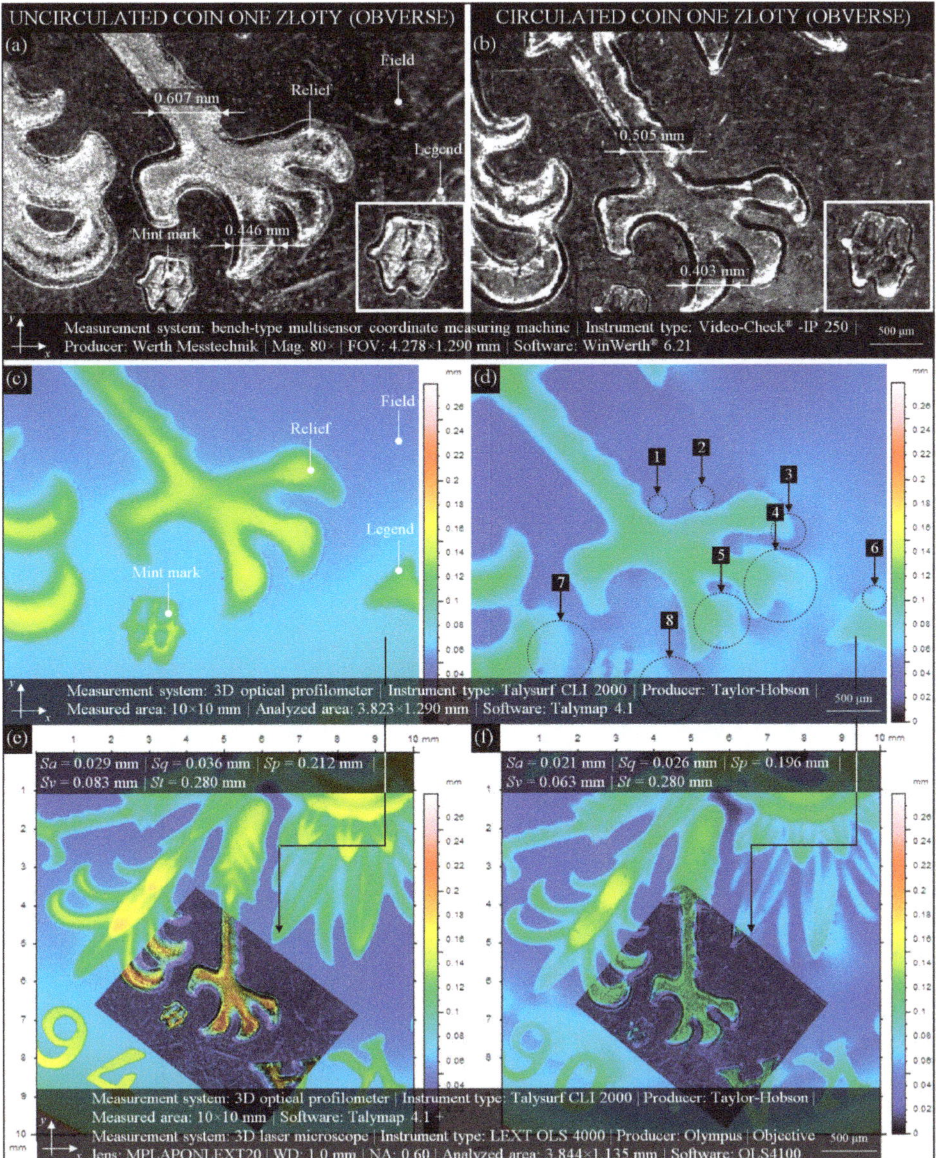

Figure 3. Selected results of observation and analysis of the surface condition of modern Polish uncirculated and circulated 1 złoty coin: images of the obverse of (**a**) uncirculated and (**b**) circulated 1 złoty coin issued in 1994 and 1990, respectively, acquired by bench-type multisensory coordinate measuring machine VideoCheck® IP 250 (Werth Messtechnik, Gießen, Germany) with characteristic elements and dimensions corresponding to images from Figure 1a,b, the 2D height map (indexed colors) obtained by the 3D optical profilometer Talysurf CLI 2000 (Taylor-Hobson, Leicester, Great Britain) for the obverse of (**c**) uncirculated and (**d**) circulated 1 złoty coin with the visible worn surface of the field as well as relief, mint mark, and legend; image fusion of a vast fragment (10 × 10 mm) of (**e**) uncirculated and (**f**) circulated coin presented in the form of a 2D height map (indexed colors) with a fragment (3.844 × 1.135 mm) obtained by the 3D laser microscope LEXT OLS4000 (Olympus Corp., Shinjuku, Tokyo, Japan) in the form of a 2D pseudo-color height map.

3.2. *Analysis of the Circulated 1 Franc and 50 Bani Coins' Edge Surface Condition*

It seems that the obverse and reverse of the coin are the elements most often exposed to wear. Another equally important element that is also subject to the wear process is the edge. There are different types of edges, most often plain (smooth) or patterned (mostly reeded and lettered). Edges usually have a decorative function, but also prevent coin clipping and counterfeiting.

In this subsection, studies carried out on the edges of two modern circulation coins—French 1 franc and Romanian 50 bani—were presented. Each of the analyses included the following elements: general view of the coin, a fragment of the coin's edge, and based on marks on the edge of the area of interest (AOI), 2D height map, surface microtopography, and close-up view with calculated values of selected surface texture parameters and extracted single surface profiles.

The image of the obverse (engraver: L. O. Roty) of a modern French 1 franc coin (alloy: Ni; edge: reeded; diameter: 24 mm; thickness: 1.79 mm; weight: 6 g) acquired by the digital microscope Omni Core (lens: +5; WD: 200 mm; magnification: 45×) is presented in Figure 4a with characteristic elements (the *Sower*, designed by O. Roty in 1900, a national emblem of the French Republic).

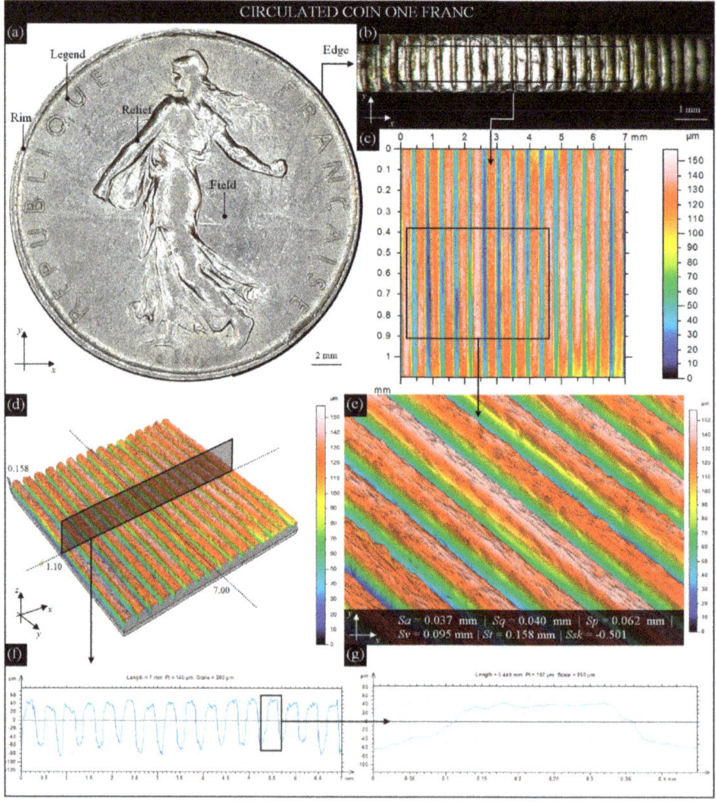

Figure 4. Selected results of observation and analysis of the surface condition of modern circulated French 1 franc coin obtained by the digital microscope Omni Core (Ash Technologies Ltd., Kildare, Ireland) and 3D optical profilometer S neox (Sensofar Metrology, Terrassa, Spain): (**a**) general view of coin obverse; (**b**) an image (10.93 × 1.79 mm) of a fragment of the coin's reeded edge; (c) a 2D height map (indexed colors) of the AOI (7.11 × 1.10 mm) from Figure 4b; (**d**) surface microtopography (7.11 × 1.10 × 0.15 mm); (**e**) close-up view (4.49 × 0.51 mm) of the reeded edge with calculated amplitude (surface) parameters; (**f**) extracted from Figure 4d, a single surface profile with visible grooves; (**g**) extracted from Figure 4f and enlarged image of a single groove.

For a fragment of the right side of the coin, mounted vertically in a holder, an image (10.93 × 1.79 mm) of its reeded edge (Figure 4b) was acquired. The condition of the edge surface showed minimal wear on the upper and lower parts, whereas relatively higher wear occurred on the grooves. They were slightly flattened, and some were vertically deformed. Despite this, the wear was considered to be average (Phase II) as convex of the grooves was visible, which were usually strongly flattened for extensively worn coins (Phase III).

On the image of the reeded edge (Figure 4b), the AOI (7.11 × 1.10 mm) was marked, for which measurements by the 3D optical profilometer S neox were carried out. The obtained results in the form of a 2D height map (indexed colors) and surface microtopography (area of the topography (axes x, y, z): 7.00 × 1.10 × 0.15 mm; number of profile points (axis x): 7001; distance between profile points (axis x): 1 μm; number of profiles (axis y): 158; distance between profile points (axis y): 7 μm) are presented in Figure 4c,d. On the 2D height map (indexed colors), an AOI (4.49 × 0.51 mm) was marked.

For this AOI, a close-up view of the reeded edge was extracted (Figure 4e). Additionally, the values of the selected amplitude (surface) parameters, *Sa*, *Sq*, *Sp*, *Sv*, *St*, and *Ssk*, were added. From surface microtopography (Figure 4d), a single surface profile (type: west–east; surface size: 7.00 × 1.00 mm; profile size: 7.00 mm (7001 points)) was extracted (Figure 4f). The profile consisted of 16 slightly worn grooves on the edge, and one of them was additionally enlarged and is shown in Figure 4g. Analysis of data from the 3D optical profilometer revealed apart from the wear on the outer surfaces of the grooves, the wear also occurred in the spaces between the subsequent grooves (Figure 4c). With an average depth of ~0.050 mm, there were numerous valleys even below 0.060 mm (Figure 4f). The *Ssk* (skewness of the height distribution) parameter, defined as the degree of symmetry of the surface heights to the mean plane, was used in the presented analysis. In this case, the value of *Ssk* is > 0, which indicates the predominance of peaks composing the surface.

The studies prepared for the second coin was identical and included elements listed at the beginning of this subsection. The image in Figure 5a presents a reverse of a modern Romanian 50 bani coin (alloy: $Cu_{80}Zn_{15}Ni_5$, edge: smooth and lettered; diameter: 23.75 mm; thickness: 1.90 mm; weight: 6.10 g) acquired by the digital microscope Omni Core (lens: +5; WD: 200 mm; magnification: 45×) with the characteristic element (lettering 50 BANI). For a fragment of the left side of the coin, mounted vertically in a holder, an image (11.42 × 1.90 mm) of its smooth and lettered edge (Figure 5b) was acquired. The condition of the edge shows average wear on the smooth surface (numerous scratches and slightly lost material) and around the letters R O M. The strongest deformation was observed in the lower part of the letter O. Circulation of this coin caused a visible average/extensive (locally) wear of the surface. Its intensity allows the conclusion that the coin surface wear process is between Phase II and Phase III. On the image of the smooth and lettered edge (Figure 5b), the AOI (7.01 × 1.30 mm) was marked, for which the optical measurements by the 3D optical profilometer S neox were carried out. This high-accuracy system allowed for obtaining a set of measurement data (Figure 5c,d) in the same form as previously presented for the 1 franc coin—the 2D height map (indexed colors) and surface microtopography (area of topography (axes x, y, z): 7.01 × 1.130 × 0.10 mm; number of profile points (axis x): 7001; distance between profile points (axis x): 1 μm; number of profiles (axis y): 187; distance between profile points (axis y): 7 μm). From the marked AOI in Figure 5c (5.10 × 1.00 mm), a fragment of the surface in an area of letter R and O was extracted. This close-up view of the smooth surface of the bani coin with calculated amplitude (surface) parameters *Sa*, *Sq*, *Sp*, *Sv*, *St*, and *Ssk* is presented in Figure 5e. In a similar way as in previous analysis, from surface microtopography (Figure 5d), a single surface profile (type: west–east, surface size: 7.01 × 1.30 mm; profile size: 7.01 mm (7001 points)) was extracted (Figure 5f). The profile passes through two letters of the edge—R and O. The deformation of the letter O (Figure 5g) is visible in the form of its depression being more than twice as large (~0.072 mm) on the left side relative to the right side (~0.039 mm). The values of the amplitude (surface) parameters *Sa*, *Sq*, *Sp*, *Sv*, *St*, and *Ssk*, calculated for circulated 50 bani coins, were on average more than 50% lower than for the circulated 1 franc coin. In the case of the skewness of the height distribution value determined for the 1 franc coin, *Ssk* > 0 indicates the predominance of peaks composing the surface.

Visual analysis and values above the parameters confirmed the generally worse surface condition of this coin, with the process being between Phase II and Phase III.

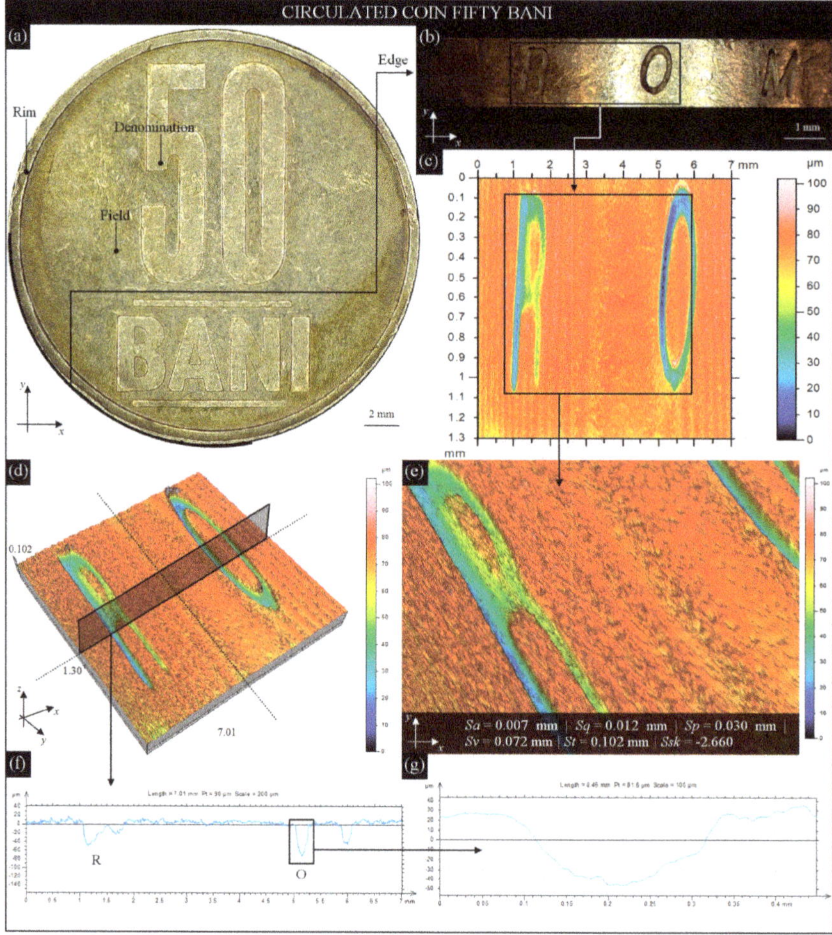

Figure 5. Selected results of observation and analysis of the surface condition of modern Romanian circulated 50 bani coin obtained by the digital microscope Omni Core (Ash Technologies Ltd., Kildare, Ireland) and 3D optical profilometer S neox (Sensofar Metrology, Terrassa, Spain): (**a**) general view of coin reverse; (**b**) image (11.42 × 1.90 mm) of a fragment of the coin's smooth and lettered edge; (**c**) 2D height map (indexed colors) of the AOI (7.01 × 1.30 mm) from Figure 4b; (**d**) surface microtopography (7.01 × 1.30 × 0.10 mm); (**e**) close-up view (5.10 × 1.00 mm) of the smooth and lettered edge with calculated amplitude (surface) parameters; (**f**) extracted from Figure 4d, a single surface profile with deformed left side of the letter O; (**g**) extracted from Figure 5f and enlarged letter O.

3.3. Comparative Analysis of the Circulated 50 Bani (Reverse) Coin's Surface Condition

In Section 3.1, a comparative analysis of the obverses of two 1 złoty Polish coins was presented. The circulated coin representing average degree of wear (Phase II) was compared with the reference uncirculated coin. A similar analysis is presented in this section, although two of the same face value circulated coins, but with totally different surface conditions, were compared in this case. The results, showing the differences in the denomination state in the central part of the field, were presented in the form of individual surface profiles extracted from 2D height maps (indexed colors).

For analysis, two modern Romanian 50 bani coins (alloy: $Cu_{80}Zn_{15}Ni_5$; edge: smooth and lettered; diameter: 23.75 mm; thickness: 1.90 mm; weight: 6.10 g) were prepared. The condition of the coins was selected for analysis in such a way as to obtain visible differences during the visual observation. The first coin (Figure 6a) represented a condition between average (Phase II) and extensive wear (Phase III), whereas the second (Figure 6d) represented critical wear (Phase IV).

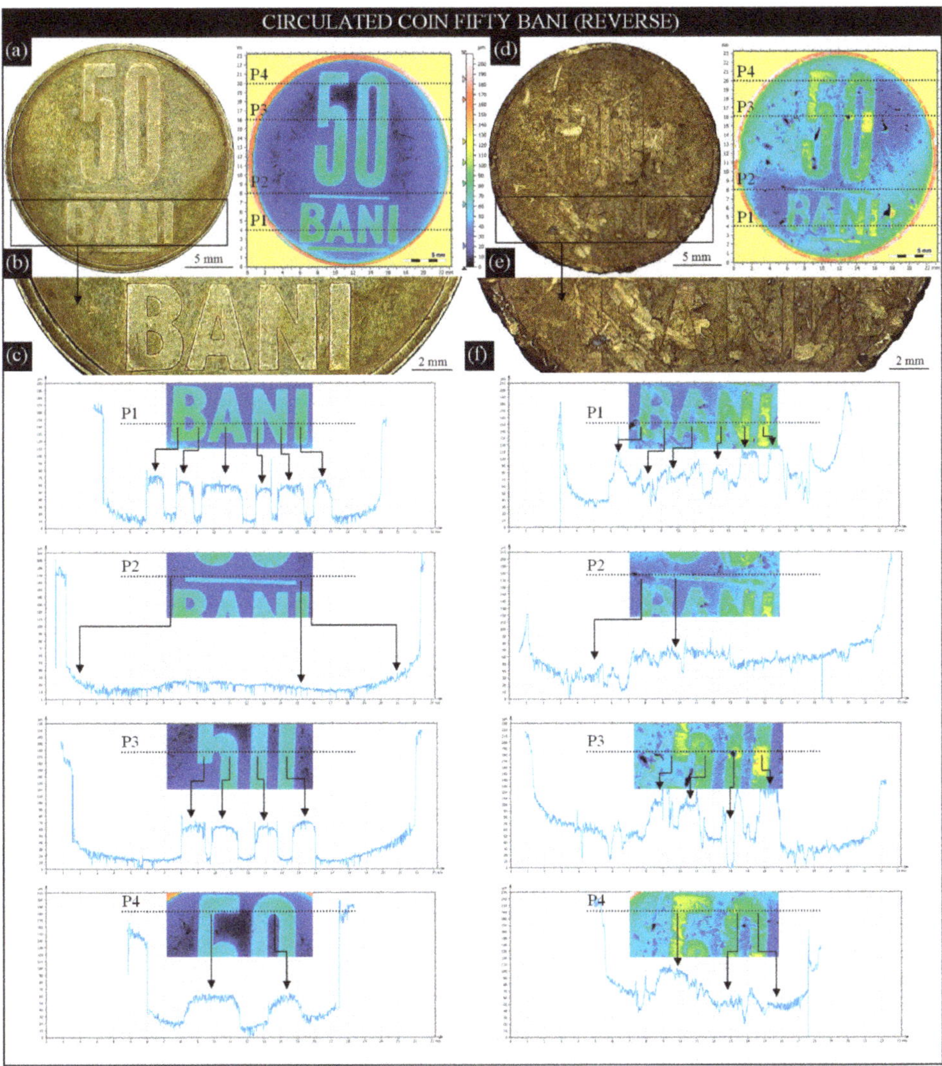

Figure 6. Selected results of observation and analysis of the surface condition of modern Romanian circu- lated 50 bani coins obtained by the digital microscope Omni Core (Ash Technologies Ltd., Kildare, Irela- nd) and 3D optical profilometer S neox (Sensofar Metrology, Terrassa, Spain): (**a–d**) general view of the coins (left: in condition between average and extensive wear (Phase II/III); right: in condition representing critical wear (Phase IV)) with a corresponding 2D height map (indexed colors); (**b–e**) AOIs (22.70 × 4.91 mm) extracted from Figure 6a–d presenting enlarged lettering B A N I for various conditions of surface wear; (**c–f**) surface profiles (P1–4 mm, P2–8, P3–16, and P4–20 mm) extracted from each of the coins.

With these two images acquired by the digital microscope Omni Core (lens: +5; WD: 200 mm; magnification: 45×), 2D height maps (indexed colors) corresponded. The maps were obtained for the entire surface of both coins (area (axes *x*, *y*): 30.30 × 27.30 mm; number of profile points (axis *x*): 11,755; distance between profile points (axis *x*): 2.58 μm; number of profiles (axis *y*): 10,587; distance between profile points (axis *y*): 2.58 μm by the 3D optical profilometer S neox. Figure 6b shows an AOI (22.70 × 4.91 mm) extracted from Figure 6a presenting an enlarged lettering of B A N I. The letters in the field of this coin are slightly distorted at the ends, and there are also visible slight scratches on their surface. Despite these, the lettering is legible. On the abovementioned 2D height map (Figure 6a), a set of four surface profiles (type: west–east; surface size: 23.00 × ~0.22 mm) was marked and presented in Figure 6c. Each of the profiles were determined for the characteristic element of the denomination: P1–for lettering B A N I, P2—for the area of the field above the line, P3—for the lower part of the number 50, P4—for the upper part of the number 50. All the described elements were well recognizable on the profiles (they were additionally marked with arrows), which confirms the relatively good (in this context) condition of this part of the coin. Figure 6d–f presents the same analyses as above, prepared for the seconds of the analyzed coins. They include a general view of the coin with a 2D height map (indexed colors) (Figure 6d); AOI (22.70 × 4.91 mm) extracted from Figure 6d presenting enlarged lettering B A N I (Figure 6e); and a set of four surface profiles (type: west–east, surface size: 23.00 × ~0.22 mm) (Figure 6f). Visual analysis of the second circulation coin allows us to state a high degree of its wear corresponding to the critical wear (Phase IV). The protective rim was flattened also, and numerous material losses were visible. The edge was strongly deformed. The lettering B A N I was highly flattened and difficult to identify. The field was characterized by numerous scratches, abrasions, and material losses. The 2D height map (indexed colors) revealed details of the above defects. A clearly lower surface height was visible in the central region of the coin, while numerous losses of the material dominated around the central part. The study of the profiles (P1–P4) showed a large surface degradation in terms of height. The elements of the denomination were highly flattened and difficult to identify, which qualifies the coin for immediate withdrawal from circulation.

3.4. Analysis of the Circulated 5 Pence (Reverse) Coin's Surface Condition

The last analysis presented in this section aims to show the possibility of visualizing surfaces using the confocal fusion processing algorithm, combining measurement data from confocal and focus-variation images of the surface. This algorithm was widely described in the works by Artigas et al. [76] and Bermudez et al. [77], whereas selected examples of its application are given in the works of Flys et al. [78], Hatami et al. [79], and Maruda et al. [80].

The reverse image (engraver: M. Dent) of a modern British 5 pence coin (alloy: Ni-plated steel, edge: reeded; diameter: 18 mm; thickness: 1.89 mm; weight: 3.25 g) acquired by the digital microscope Omni Core (lens: +5; WD: 200 mm; magnification: 45×) is presented in Figure 7a with characteristic elements (royal shield of arms and the lettering FIVE PENCE in the central position). Using the 3D optical profilometer Talysurf CLI 2000, the entire coin was measured and is presented in Figure 7b in the form of a 2D height map (indexed colors). In the marked AOI (13.78 × 13.78 mm), typical for everyday handling of the coin, the relative slight wear (in terms of surface height) was clearly visible.

Based on the obtained measurement data, selected amplitude (surface) parameters *Sa*, *Sq*, *Sp*, *Sv*, *St*, *Ssk*, and *Sku* were calculated and presented below this figure. Analysis of their values shows that the overall wear of the coin's surface is between slight (Phase I) and average (Phase II). Visualizations of selected fragments of the 5 pence coin using the confocal fusion processing algorithm based on mea-surement data obtained by the 3D optical profilometer S neox are presented in Figure 7c–e. Extracted from Figure 7a, the AOI (3.32 × 2.92 mm) depicting a lion claw (front foot) is shown in Figure 7c.

Figure 7. Selected results of observation and analysis of the surface condition of a modern British circulated 5 pence coin obtained by the digital microscope Omni Core (Ash Technologies Ltd., Kildare, Ireland), 3D optical profilometer Talysurf CLI 2000 (Taylor-Hobson, Leicester, Great Britain), and 3D optical profilometer S neox (Sensofar Metrology, Terrassa, Spain): (**a**) general view of coin reverse; (**b**) 2D height map (indexed colors) with calculated amplitude (surface) parameters; (**c**) extracted from Figure 1a, AOI (3.32 × 2.92 mm) depicting a lion claw (front foot); (**d**) extracted from Figure 7a, AOI (5.52 × 4.42 mm) depicting a lion head; (**e**) extracted from Figure 7a, AOI (3.32 × 2.92 mm) depicting a lion claw (rear foot).

This high-resolution visualization (3665 × 2757 pixels) shows the structure of the coin in an area of relief and a fragment of the characteristic texture of the field. The next AOI (5.52 × 4.42 mm) represents a lion head (Figure 7d), and another AOI (3.32 × 2.92 mm) depicts a lion claw (rear foot) (Figure 7e). The above high-resolution visualizations were obtained by precise point-by-point scanning of the coin's surface, and processing obtained in this way a single scan by smart confocal fusion algorithm. As a result, a set of surface topographies was created, which were stitched together (image stitching procedure), allowing to obtain an output large area topography. The great advantage of such topography was the fact that it retained a high-resolution and represented the same quality of the detail. This, in turn, allowed for the precise observation of various forms and intensities of wear in selected areas of the coin. For all of the AOIs, an individual set of selected amplitude (surface) parameters, *Sa*, *Sq*, *St*, *Ssk*, and *Sku*, was added. The differences between the parameters were relatively small and resulted mainly from the height of the elements of a given relief as well as local, more intensive wear of the surface. The values of the *Ssk* parameter showed the predominance of peaks, whereas the values of the *Sku* parameter indicated lack of inordinately high peaks or deep valleys in the measured areas.

4. Conclusions

This article was an attempt to familiarize the readers with issues related to the analysis of circulation coins in the context of observation, visualization, and measurement of various forms of wear occurring on their surfaces. Additionally, the authors' intention was to present the use of advanced methods based on optical profilometry in this specific type of applications. The obtained results of measurements and studies allowed the authors to draw the following detailed conclusions:

1. Most of the analyses of modern circulation coins focus on the precise determination of their chemical composition. In this case, modern varieties of X-ray-based observation measurement methods, such as EDS (EDX), EDXRF, LAMQS, and, PIXE, were used (Section 1). To extend the results of the spectroscopic examinations, 2D/3D dimensional-shape measurements (macroscale) and surface texture measurements (micro scale) were carried out. Additionally, the authors located and recognized the surface defects. Modern measurement methods, such as CLSM, FVM, and advanced variants of interferometry, using optical profilometry, prove to be helpful in these activities.

2. The authors of the article showed that optical profilometry could be successfully used in the analysis of the surface condition of modern circulation coins representing Phases I–IV of surface wear. The measurement capabilities of the systems used (Section 2.1) turned out to be sufficient for the needs of the carried-out experimental tests. The undoubted advantages of the 3D optical profilometers used were relatively quick measurement time, noncontact method of assessment, high resolution and measurement range, integration of (optical) measurement methods in one instrument, and advanced processing and visualization of measurement data. In more complex cases, this type of tests may be carried out with the use of additional measuring instruments or more specialized computer software—it depends on the application.

3. The analyses presented in the article (Section 3) are general and illustrative—they are shown against the background of hardware and software capabilities. Analyses of the surface condition of coins can be much more advanced (comprehensive) or strictly focused on a specific feature of the assessed surface. For example, it can relate with mint-made errors (errors from a three fundamental groups, including blank planchet, fundamental die-setting, and broadstrike) generated during the coin manufacturing process. Such errors occurring during the minting practice must be located and adequately analyzed. Another interesting area is concerning the analyses carried out to establish the authenticity of a given coin or group of coins. Such activities usually refer to the antique coin(s), but in justified cases, they may also refer to coins illegally released into circulation.

4. Special computer software (especially based on Mountains Technology®) provided significant support in the carried-out analyses. Its universal character and the number of implemented functions it performed were useful in characterizing coin surfaces. In case of a need to generate the output large area topography, the image stitching procedure was conducive.

5. In the authors' opinion, the subject discussed in this article is extremely interesting and has a chance to be further developed. There is a plan to carry out a more detailed study (e.g., analysis of surface wear in the context of changes of coin relief height). Such changes are extremely important because critical wear makes it impossible to use a given coin as a legal tender and, as a result, causes its withdrawal from circulation. The authors also plan to carry out a wider research program using advanced X-ray-based observation measurement methods supported by noncontact techniques (optical profilometry), where issues regarding the influence of elemental composition on the wear process will be considered.

Author Contributions: Supervision, W.K.; conceptualization, W.K., T.M., M.K.G. and S.S.; research methodology, T.M., D.Y.P., M.M., K.P. and M.S.; investigation, W.K., M.K.G., S.S. and M.S.; formal analysis, D.Y.P., M.K.G., M.M. and M.S.; writing—original draft preparation, W.K., D.Y.P.; writing—review and editing, W.K. and T.M., D.Y.P., M.K.G., M.M., S.S., K.P. and M.S. All authors have read and agreed to the published version of the manuscript.

Funding: This research received no external funding.

Acknowledgments: The authors express their gratitude to the peoples who actively supported the studies presented in this paper: Krzysztof Maciejewski (Department of Production Engineering, Faculty of Mechanical Engineering, Koszalin University of Technology, Koszalin, Poland) for his technical support during optical measurements using the bench-type multisensory coordinate measuring machine VideoCheck® IP 250, Mariusz Władowski (Optotom, Warszawa, Poland) for his technical support during optical measurements using the 3D optical profilometer Sneox, and Marcin Surman, for providing Polish coins for the tests and for his insightful consultations regarding the circulation coins.

Conflicts of Interest: The authors declare no conflict of interest.

Nomenclature

3D-DIC	Three-dimensional digital image correlation (3D-DIC)
AOI	Area of interest
CH	Conoscopic holography
CS	Confocal sensor
CSI	Computational shear interferometry
CLA	Chromatic light aberration phenomenon
CLSM	Confocal laser scanning microscopy
CMM	Coordinate measuring machine
CT	X-ray computed tomography
DIA	Digital image acquisition
EDS	Energy-dispersive X-ray spectrometry
EDXRF	Energy-dispersive X-ray fluorescence spectrometry
EMA	Electron microprobe analysis
FP	Fringe projection
FVM	Focus variation microscopy
GRT	Gamma-ray transmission
LAMQS	Laser ablation coupled to mass quadrupole spectrometry
LIBS	Laser-induced breakdown spectroscopy
MI	Moiré interferometry
ND	Neutron diffraction
OM	Optical microscopy
PIXE	Particle (proton)-induced X-ray emission
PS	Phase stepping
RBS	Rutherford backscattering spectrometry
XPS	X-ray photoelectron spectroscopy
XRD	X-ray diffraction
SEM	Scanning electron microscopy
SL	Structured light
SP	Speckle projection
SRXRF	Synchrotron radiation X-ray fluorescence spectrometry
TEM	Transmission electron microscopy
TPU	Temporal phase unwrapping
WD	Working distance, mm
Sa	Arithmetic mean deviation of the surface, μm
Sku	Kurtosis of the height distribution, -
Sp	Maximum height of summits, μm
Sq	Root-mean-square deviation of the surface, μm
Ssk	Skewness of the height distribution, -
St	Total height of the surface, μm
Sv	Maximum depth of valleys, μm

Glossary

Circulated—a term used to refer to a coin characterized by various conditions of surface wear (from slight wear to critical wear).

Circulation—a term used to refer to a coin that is currently or was in the past in circulation as a means of payment (money).

Denomination—a value assigned by a government to a given coin.

Design—motif, pattern, or emblem used in the coin.

Edge—the third side of a coin, containing reeds, lettering, or other ornaments. The edge can also be plain.

Field—a flat (or slightly curved) area of a coin with no emblem or inscription.

Motto—word, sentence, or phrase inscribed on a coin to express a guiding national principle.

Obverse—front or head side of the coin.

Relief—any element of a coin's design that is raised above the field. The opposite of relief is incuse.

Reverse—back or tail side of the coin.

Rim—raised portion of the design along the edge that protects the coin from wear.

Uncirculated—theoretically, a coin that has never circulated and thus retains all of its original mint conditions (a coin without wear with excellent surface condition).

References

1. Du Toit, M.; van der Lingen, E.; Glaner, L.; Süss, R. The development of a novel gold alloy with 995 fineness and increased hardness. *Gold Bull.* **2002**, *35*, 46–52. [CrossRef]
2. Mendoza-López, M.L.; Pérez-Bueno, J.J.; Rodríguez-García, M.E. Characterizations of silver alloys used in modern Mexican coins. *Mater. Charact.* **2009**, *60*, 1041–1048. [CrossRef]
3. Dow, J.A. *Ancient Coins through the Bible*; Tate Publishing & Enterprises: Mustang, OK, USA, 2011.
4. Metcalf, W.E. (Ed.) *The Oxford Handbook of Greek and Roman Coinage*; Oxford University Press: Oxford, UK, 2012.
5. He, L.; Liang, J.; Zhao, X.; Jiang, B. Corrosion behavior and morphological features of archeological bronze coins from ancient China. *Microchem. J.* **2012**, *99*, 203–212. [CrossRef]
6. Rodrigues, M.; Schreiner, M.; Melcher, M.; Guerra, M.; Salomon, J.; Radtke, M.; Alram, M.; Schindel, N. Characterization of the silver coins of the Hoard of Beçin by X-ray based methods. *Nucl. Instrum. Methods Phys. Res. B* **2011**, *269*, 3041–3045. [CrossRef]
7. Mousser, H.; Amri, R.; Madani, A.; Darchen, A.; Mousser, A. Microchemical surface analysis of two Numidian coins. *Appl. Surf. Sci.* **2011**, *257*, 5961–5965. [CrossRef]
8. Torrisi, L.; Italiano, A.; Torrisi, A. Ancient bronze coins from Mediterranean basin: LAMQS potentiality for lead isotopes comparative analysis with former mineral. *Appl. Surf. Sci.* **2016**, *387*, 529–538. [CrossRef]
9. Pitarch, A.; Queralt, I. Energy dispersive X-ray fluorescence analysis of ancient coins: The case of Greek silver drachmae from the Emporion site in Spain. *Nucl. Instrum. Methods Phys. Res. B* **2010**, *268*, 1682–1685. [CrossRef]
10. Pitarch, A.; Queralt, I.; Alvarez-Perez, A. Analysis of Catalonian silver coins from the Spanish war of independence period (1808–1814) by energy dispersive X-ray fluorescence. *Nucl. Instrum. Methods Phys. Res. B* **2011**, *269*, 308–312. [CrossRef]
11. Gore, D.B.; Davis, G. Suitability of transportable EDXRF for the on-site assessment of ancient silver coins and other silver artifacts. *Appl. Spectrosc.* **2016**, *70*, 840–851. [CrossRef]
12. Ager, F.J.; Gómez-Tubío, B.; Paúl, A.; Gómez-Morón, A.; Scrivano, S.; Ortega-Feliu, I.; Respaldiza, M.A. Combining XRF and GRT for the analysis of ancient silver coins. *Microchem. J.* **2016**, *126*, 149–154. [CrossRef]
13. Pardini, L.; El Hassan, A.; Ferretti, M.; Foresta, A.; Legnaioli, S.; Lorenzetti, G.; Nebbia, E.; Catalli, F.; Harith, M.A.; Diaz Pace, D.; et al. X-ray fluorescence and laser-induced breakdown spectroscopy analysis of Roman silver denarii. *Spectrochim. Acta B* **2012**, *74*, 156–161. [CrossRef]
14. Awasthi, S.; Kumar, R.; Rai, G.K.; Rai, A.K. Study of archaeological coins of different dynasties using LIBS coupled with multivariate analysis. *Opt. Laser. Eng.* **2016**, *79*, 29–38. [CrossRef]
15. Corsi, J.; Grazzi, F.; Lo Giudice, A.; Re, A.; Scherillo, A.; Angelici, D.; Allegretti, S.; Barello, F. Compositional and microstructural characterization of Celtic silver coins from northern Italy using neutron diffraction analysis. *Microchem. J.* **2016**, *126*, 501–508. [CrossRef]
16. Griesser, M.; Kockelmann, W.; Hradil, K.; Traum, R. New insights into the manufacturing technique and corrosion of high leaded antique bronze coins. *Microchem. J.* **2016**, *126*, 181–193. [CrossRef]
17. Flament, C.; Marchetti, P. Analysis of ancient silver coins. *Nucl. Instrum. Methods Phys. Res. B* **2004**, *226*, 179–184. [CrossRef]

18. Hajivaliei, M.; Mohammadifar, Y.; Ghiyasi, K.; Jaleh, B.; Lamehi-Rachti, M.; Oliaiy, P. Application of PIXE to study ancient Iranian silver coins. *Nucl. Instrum. Methods Phys. Res. B* **2008**, *266*, 1578–1582. [CrossRef]

19. Hajivaliei, M.; Nadooshan, F.K. Compositional study of Parthian silver coins using PIXE technique. *Nucl. Instrum. Methods Phys. Res. B* **2012**, *289*, 56–58. [CrossRef]

20. Šmit, Ž.; Šemrov, A. Analysis of Greek small coinage from the classic period. *Nucl. Instrum. Methods Phys. Res. B* **2018**, *417*, 100–104. [CrossRef]

21. Tripathy, B.B.; Rautray, T.R.; Rautray, A.C.; Vijayan, V. Elemental analysis of silver coins by PIXE technique. *Appl. Radiat. Isot.* **2010**, *68*, 454–458. [CrossRef]

22. Rautray, T.R.; Nayak, S.S.; Tripathy, B.B.; Das, S.; Das, M.R.; Das, S.R.; Chattopadhyay, P.K. Analysis of ancient Indian silver punch-marked coins by external PIXE. *Appl. Radiat. Isot.* **2011**, *69*, 1385–1389. [CrossRef]

23. Abdelouahed, H.B.; Gharbi, F.; Roumié, M.; Baccouche, S.; Romdhane, K.B.; Nsouli, B.; Trabelsi, A. PIXE analysis of medieval silver coins. *Mater. Charact.* **2010**, *61*, 59–64. [CrossRef]

24. Beck, L.; Bosonnet, S.; Réveillon, S.; Eliot, D.; Pilon, F. Silver surface enrichment of silver–copper alloys: A limitation for the analysis of ancient silver coins by surface techniques. *Nucl. Instrum. Methods Phys. Res. B* **2004**, *226*, 153–162. [CrossRef]

25. Beck, L.; Alloin, E.; Berthier, C.; Réveillon, S.; Costa, V. Silver surface enrichment controlled by simultaneous RBS for reliable PIXE analysis of ancient coins. *Nucl. Instrum. Methods Phys. Res. B* **2008**, *266*, 2320–2324. [CrossRef]

26. Denker, A.; Opitz-Coutureau, J.; Griesser, M.; Denk, R.; Winter, H. Non-destructive analysis of coins using high-energy PIXE. *Nucl. Instrum. Methods Phys. Res. B* **2004**, *226*, 163–171. [CrossRef]

27. Ingo, G.M.; Riccucci, C.; Faraldi, F.; Pascucci, M.; Messina, E.; Fierro, G.; Di Carlo, G. Roman sophisticated surface modification methods to manufacture silver counterfeited coins. *Appl. Surf. Sci.* **2017**, *421*, 109–119. [CrossRef]

28. Caridi, F.; Torrisi, L.; Cutroneo, M.; Barreca, F.; Gentile, C.; Serafino, T.; Castrizio, D. XPS and XRF depth patina profiles of ancient silver coins. *Appl. Surf. Sci.* **2013**, *272*, 82–87. [CrossRef]

29. Gorghinian, A.; Esposito, A.; Ferretti, M.; Catalli, F. XRF analysis of Roman Imperial coins. *Nucl. Instrum. Methods Phys. Res. B* **2013**, *309*, 268–271. [CrossRef]

30. Del Hoyo-Meléndez, J.M.; Świt, P.; Matosz, M.; Woźniak, M.; Klisińska-Kopacz, A.; Bratasz, Ł. Micro-XRF analysis of silver coins from medieval Poland. *Nucl. Instrum. Methods Phys. Res. B* **2015**, *349*, 6–16. [CrossRef]

31. Hložek, M.; Trojek, T. Silver and tin plating as medieval techniques of producing counterfeit coins and their identification by means of micro-XRF. *Radiat. Phys. Chem.* **2017**, *137*, 234–237. [CrossRef]

32. Marjo, C.E.; Davis, G.; Gong, B.; Gore, D.B. Spatial variability of elements in ancient Greek (ca. 600–250 BC) silver coins using scanning electron microscopy with energy dispersive spectrometry (SEM-EDS) and time of flight-secondary ion mass spectrometry (ToF-SIMS). *Powder Diffr.* **2017**, *32*, 95–100. [CrossRef]

33. Vasiliev, A.L.; Kovalchuk, M.V.; Yatsishina, E.B. Electron microscopy methods in studies of cultural heritage sites. *Crystallogr. Rep.* **2016**, *61*, 873–885. [CrossRef]

34. Khramchenkova, R.; Safina, I.; Drobyshev, S.; Batasheva, S.; Nuzhdin, E.; Fakhrullin, R. Advanced microscopy techniques for nanoscale diagnostic of cultural heritage: Scanning electron microscopy for investigation of medieval coins and frescos from the Republic of Tatarstan. In *Nanotechnologies and Nanomaterials for Diagnostic, Conservation and Restoration of Cultural Heritage*; Lazzara, G., Fakhrullin, R., Eds.; Elsevier: Amsterdam, The Netherlands, 2019; pp. 1–23.

35. Pistofidis, N.; Vourlias, G.; Pavlidou, E.; Dilo, T.; Civici, N.; Stamati, F.; Gjongecaj, S.; Prifti, I.; Bilam, O.; Stergioudis, G.; et al. Microscopical examination of ancient silver coins. *AIP Conf. Proc.* **2004**, *899*, 798.

36. Liang, C.; Yang, C.; Huang, N. Investigating the tarnish and corrosion mechanisms of Chinese gold coins. *Surf. Interface Anal.* **2011**, *43*, 763–769. [CrossRef]

37. Cruz, J.; Corregidor, V.; Alves, L.C. Simultaneous use and self-consistent analyses of μ-PIXE and μ-EBS for the characterization of corrosion layers grown on ancient coins. *Nucl. Instrum. Methods Phys. Res. B* **2017**, *406*, 324–328. [CrossRef]

38. Fierascu, R.C.; Fierascu, I.; Ortan, A.; Constantin, F.; Mirea, D.A.; Tatescu, M. Complex archaeometallurgical investigation of silver coins from the XVIth-XVIIIth century. *Nucl. Instrum. Methods Phys. Res. B* **2017**, *401*, 318–324. [CrossRef]

39. Abdel-Kareem, O.; Al-Zahrani, A.; Arbach, M. Authentication and conservation of corroded archaeological Qatabanian and Himyarite silver coins. *J. Archaeol. Sci. Rep.* **2016**, *9*, 565–576. [CrossRef]

40. Miles, J.; Mavrogordato, M.; Sinclair, I.; Hinton, D.; Boardman, R.; Earl, G. The use of computed tomography for the study of archaeological coins. *J. Archaeol. Sci. Rep.* **2016**, *6*, 35–41. [CrossRef]

41. Bozzini, B.; Gianoncelli, A.; Mele, C.; Siciliano, A.; Mancini, L. Electrochemical reconstruction of a heavily corroded Tarentum hemiobolus silver coin: A study based on microfocus X-ray computed microtomography. *J. Archaeol. Sci.* **2014**, *52*, 24–30. [CrossRef]

42. Van Loon, L.L.; Nelson, A.J.; Kagan, U.W.; Barron, K.; Banerjee, N.R. Bubbles in the bullion: Micro-CT imaging of the internal structure of ancient coins. *Microsc. Microanal.* **2019**, *25*, 420–421. [CrossRef]

43. Spagnolo, G.S.; Majo, R.; Carli, M.; Ambrosini, D.; Paoletti, D. Virtual gallery of ancient coins through conoscopic holography. *Proc. SPIE* **2003**, *5146*, 202–209.

44. Roumie, M.; Nsouli, B.; Chalhoub, G.; Hamdan, M. Quality control of coins mint using PIXE and RBS analysis. *Nucl. Instrum. Methods Phys. Res. B* **2010**, *268*, 1916–1919. [CrossRef]

45. Papp, Z.; Kovacs, I. Surface analysis of a modern silver coin: SEM/EDS measurements. *Rev. Roum. Chim.* **2013**, *58*, 65–67.

46. Corregidor, V.; Alves, L.C.; Cruz, J. Analysis of surface stains on modern gold coins. *Nucl. Instrum. Methods Phys. Res B* **2013**, *306*, 232–235. [CrossRef]

47. Kapłonek, W.; Nadolny, K.; Ungureanu, M.; Pimenov, D.Y.; Zieliński, B. SEM-based observations and analysis of the green silicon carbide grinding wheel active surfaces after the graphite and silicone impregnation process. *Int. J. Surf. Sci. Eng.* **2019**, *13*, 181–200. [CrossRef]

48. Bustillo, A.; Pimenov, D.Y.; Matuszewski, M.; Mikolajczyk, T. Using artificial intelligence models for the prediction of surface wear based on surface isotropy levels. *Robot. Comput. Integr. Manuf.* **2018**, *53*, 215–217. [CrossRef]

49. Peng, J.; Wang, M.; Deng, D.; Liu, X.; Yin, Y.; Peng, X. Distortion correction for microscopic fringe projection system with Scheimpflug telecentric lens. *Appl. Opt.* **2015**, *54*, 10055–10062. [CrossRef]

50. Liu, C.; Chen, L.; He, X.; Thang, V.D.; Kofidis, T. Coaxial projection profilometry based on speckle and fringe projection. *Opt. Commun.* **2015**, *341*, 228–236. [CrossRef]

51. Quan, C.; He, X.Y.; Wang, C.F.; Tay, C.J.; Shang, H.M. Shape measurement of small objects using LCD fringe projection with phase shifting. *Opt. Commun.* **2001**, *189*, 21–29. [CrossRef]

52. Spagnolo, G.S.; Cozzella, L.; Leccese, F. Projected fringes profilometry for cultural heritage studies. In Proceedings of the 2019 IMEKO TC-4 International Conference on Metrology for Archaeology and Cultural Heritage, Florence, Italy, 4–6 December 2019; pp. 435–438.

53. Steckenrider, J.J.; Steckenrider, J.S. High-resolution moiré interferometry for quantitative low-cost, real-time surface profilometry. *Appl. Opt.* **2015**, *54*, 8298–8305. [CrossRef]

54. Liu, S.; Feng, W.; Zhang, Q.; Liu, Y. Three-dimensional shape measurement of small object based on tri-frequency heterodyne method. *Proc. SPIE* **2015**, *9623*, 96231C-1–96231C-5.

55. Quan, C.; He, X.; Tay, C.J.; Shang, H.M. 3D surface profile measurement using LCD fringe projection. *Proc. SPIE* **2001**, *4317*, 511–516.

56. Quan, C.; Tay, C.J.; Kang, X.; He, X.Y.; Shang, H.M. Shape measurement by use of liquid-crystal display fringe projection with two-step phase shifting. *Appl. Opt.* **2003**, *42*, 2329–2335. [CrossRef] [PubMed]

57. Chen, J.; Guo, T.; Wang, L.; Wu, Z.; Fu, X.; Hu, X. Microscopic fringe projection system and measuring method. *Proc. SPIE* **2013**, *8759*, 87594U.

58. Deininger, L.; Francese, S.; Clench, M.R.; Langenburg, G.; Sears, V.; Sammon, C. Investigation of infinite focus microscopy for the determination of the association of blood with fingermarks. *Sci. Justice* **2018**, *58*, 397–404. [CrossRef] [PubMed]

59. Fabich, M. Advancing confocal laser scanning microscopy: The advantage of optical metrology. *Opt. Photonik* **2009**, *4*, 40–43. [CrossRef]

60. Kapłonek, W.; Sutowska, M.; Ungureanu, M.; Çetinkaya, K. Optical profilometer with confocal chromatic sensor for high-accuracy 3D measurements of the uncirculated and circulated coins. *J. Mech. Energy Eng.* **2018**, *2*, 181–192. [CrossRef]

61. Song, Z.; Chung, R.; Zhang, X.T. An accurate and robust strip-edge-based structured light means for shiny surface micromeasurement in 3-D. *IEEE Trans. Ind. Electron.* **2013**, *60*, 1023–1032. [CrossRef]

62. Carcagni, P.; Daffara, C.; Fontana, R.; Gambino, M.C.; Mastroianni, M.; Mazzotta, C.; Pampaloni, E.; Pezzati, L. Optical micro-profilometry for archaeology. *Proc. SPIE* **2005**, *5857*, 58570F.

63. Yan, T.; Su, Y.; Zhang, Q. Precise 3D shape measurement of three-dimensional digital image correlation for complex surfaces. *Sci. China Technol. Sci.* **2018**, *61*, 68–73. [CrossRef]

64. Falldorf, C.; Agour, M.; Bergmann, R.B. Digital holography and quantitative phase contrast imaging using computational shear interferometry. *Opt. Eng.* **2015**, *54*, 024110. [CrossRef]

65. Fan, W.C.; Cao, P.; Long, L. Degradation of joint surface morphology, shear behavior and closure characteristics during cyclic loading. *J. Cent. South Univ.* **2018**, *25*, 653–661. [CrossRef]

66. Saremi-Yarahmadi, S.; Binner, J.; Vaidhyanathan, B. Erosion and mechanical properties of hydrothermally-resistant nanostructured zirconia components. *Ceram. Int.* **2018**, *44*, 10539–10544. [CrossRef]

67. Beamud, E.M.; Núñez, P.J.; García-Plaza, E.; Amaro, F.R. Characterization of surface texture in electropolishing processes using 3D surface topography parameters. *Procedia Manuf.* **2019**, *41*, 114–120. [CrossRef]

68. Genna, S.; Giannini, O.; Guarino, S.; Ponticelli, G.S.; Tagliaferri, F. Laser texturing of AISI 304 stainless steel: Experimental analysis and genetic algorithm optimisation to control the surface wettability. *Int. J. Adv. Manuf. Technol.* **2020**, *110*, 3005–3022. [CrossRef]

69. Artigas, R.; Pinto, A.; Laguarta, F. Three-dimensional micromeasurements on smooth and rough surfaces with a new confocal optical profiler. *Proc. SPIE* **1999**, *3824*, 93–104.

70. Ding, W.; Dai, C.; Yu, T.; Xu, J.; Fu, Y. Grinding performance of textured monolayer CBN wheels: Undeformed chip thickness nonuniformity modeling and ground surface topography prediction. *Int. J. Mach. Tools Manuf.* **2017**, *122*, 66–80. [CrossRef]

71. Leksycki, K.; Królczyk, J.B. Comparative assessment of the surface topography for different optical profilometry techniques after dry turning of Ti6Al4V titanium alloy. *Measurement* **2020**, *169*, 108378. [CrossRef]

72. Leksycki, K.; Feldshtein, E.; Królczyk, G.M.; Legutko, S. On the chip shaping and surface topography when finish cutting 17-4 PH precipitation-hardening stainless steel under near-dry cutting conditions. *Materials* **2020**, *13*, 2188. [CrossRef]

73. Tato, W.; Blunt, L.; Llavori, I.; Aginagalde, A.; Townsend, A.; Zabala, A. Surface integrity of additive manufacturing parts: A comparison between optical topography measuring techniques. *Procedia CIRP* **2020**, *87*, 403–408. [CrossRef]

74. International Organization for Standardization. *Geometrical Product Specification (GPS)—Surface Texture: Areal—Part. 2: Terms, Definitions and Surface Texture Parameters*; Technical Report No. 25178-2:2012; International Organization for Standardization: Geneva, Switzerland, 2012.

75. Stout, K.J.; Liam, B.; Dong, W.; Mainsah, E.; Luo, N.; Mathia, T.; Sullivan, P.; Zahouani, H. *The Development of Methods for the Characterization of Roughness in Three Dimensions*; Technical Report, No. EUR 15178 EN; BCR: Brussels, Belgium, 1993.

76. Artigas, R.; Matilla, A.; Mariné, J.; Pérez, J.; Cadevall, C. Three-dimensional measurements with a novel technique combination of confocal and focus variation with a simultaneous scan. *Proc. SPIE* **2016**, *9890*, 98900B.

77. Bermudez, C.; Matilla, A.; Aguerri, A. Confocal fusion: Towards the universal optical 3D metrology technology. In Proceedings of the 12th LAMDAMAP, Renishaw Innovation Center, Wotton-Under Edge, UK, 15–16 March 2017.

78. Flys, O.; Berglund, J.; Rosen, B.G. Using confocal fusion for measurement of metal AM surface texture. *Surf. Topogr.* **2020**, *8*, 024003. [CrossRef]

79. Hatami, S.; Ma, T.; Vuoristo, T.; Bertilsson, J.; Lyckfeldt, O. Fatigue strength of 316 L stainless steel manufactured by selective laser melting. *J. Mater. Eng. Perform.* **2020**, *29*, 3183–3194. [CrossRef]

80. Maruda, R.W.; Krolczyk, G.M.; Wojciechowski, S.; Powalka, B.; Klos, S.; Szczotkarz, N.; Matuszak, M.; Khanna, N. Evaluation of turning with different cooling-lubricating techniques in terms of surface integrity and tribologic properties. *Tribol. Int.* **2020**, *148*, 106334. [CrossRef]

Publisher's Note: MDPI stays neutral with regard to jurisdictional claims in published maps and institutional affiliations.

Article

Magneto-Rheological Fluid Assisted Abrasive Nanofinishing of β-Phase Ti-Nb-Ta-Zr Alloy: Parametric Appraisal and Corrosion Analysis

Sunpreet Singh [1], Chander Prakash [2,*], Alokesh Pramanik [3], Animesh Basak [4], Rajasekhara Shabadi [5], Grzegorz Królczyk [6,*], Marta Bogdan-Chudy [6] and Atul Babbar [7]

[1] Department of Mechanical Engineering, National University of Singapore, Singapore 119077, Singapore; snprt.singh@gmail.com

[2] School of Mechanical Engineering, Lovely Professional University, Phagwara, Punjab 144411, India

[3] Department of Mechanical Engineering, Curtin University Australia, Perth 6102, Australia; alokesh.pramanik@curtin.edu.au

[4] Adelaide Microscopy, The University of Adelaide, Adelaide, SA 5005, Australia; animesh.basak@adelaide.edu.au

[5] Unité Matériaux et Transformations CNRS UMR 8207, Université de Lille, 59000 Lille, France; rajashekhara.shabadi@univ-lille.fr

[6] Department of Mechanical engineering, Opole University of Technology, 45-758 Opole, Poland; m.bogdan-chudy@po.edu.pl

[7] Mechanical Engineering Department, Shree Guru Gobind Singh Tricentenary University, Gurugram 122505, India; atulbabbar123@gmail.com

* Correspondence: chander.mechengg@gmail.com (C.P.); grzegorz.krolczyk@wp.pl (G.K.)

Received: 17 October 2020; Accepted: 13 November 2020; Published: 16 November 2020

Abstract: The present work explores the potential of magneto-rheological fluid assisted abrasive finishing (MRF-AF) for obtaining precise surface topography of an in-house developed β-phase Ti-Nb-Ta-Zr (TNTZ) alloy for orthopedic applications. Investigations have been made to study the influence of the concentration of carbonyl iron particles (CIP), rotational speed (Nt), and working gap (Gp) in response to material removal (MR) and surface roughness (Ra) of the finished sample using a design of experimental technique. Further, the corrosion performance of the finished samples has also been analyzed through simulated body fluid (SBF) testing. It has been found that the selected input process parameters significantly influenced the observed MR and Ra values at 95% confidence level. Apart from this, it has been found that Gp and Nt exhibited the maximum contribution in the optimized values of the MR and Ra, respectively. Further, the corrosion analysis of the finished samples specified that the resistance against corrosion is a direct function of the surface finish. The morphological analysis of the corroded morphologies indicated that the rough sites of the implant surface have provided the nuclei for corrosion mechanics that ultimately resulted in the shredding of the appetite layer. Overall results highlighted that the MRF-AF is a potential technique for obtaining nano-scale finishing of the high-strength β-phase Ti-Nb-Ta-Zr alloy.

Keywords: β-phase TNTZ alloy; nano-finishing; magnetic abrasive finishing; surface roughness; material removal; optimization; parametric appraisal

---

1. Introduction

Roughly 80% of biomedical implants are developed using metallic materials, including stainless steel, cobalt-chromium, Nitinol, and titanium alloys. This is mainly due to the fact that the metallic biomaterials play a remarkable role in the recovery of dysfunctional organs and improving the life of human beings [1]. Further, the need of the Ti-based biomaterials is consistently growing due to the rapid increase in the population of the elderly population, road accidents, and sports injuries. Ti-alloy based biomaterials have been used for the development of organs due to their excellent bio-mechanical performance [2]. It has been reported that the most popular class of Ti alloys, Ti6Al4V, suffers from poor tribological properties and is mainly used for the restricted non-tribological applications [3]. Further, in [4], it has been reported that the surface flaws can lead to the implant failure due to propagation of the cracks. The intrinsic characteristics of this alloy tend to release aluminum and vanadium ions, which results in their accumulations on the host tissues and causes toxic reactions [5,6]. Underlining such facts, many research interests are focused on the development of an effective alternative, β-phase Ti-Nb-Ta-Zr (β-TNTZ) biomedical alloy, for orthopedic implants, especially actabular cup, shoulder joint, and knee joint assemblies [7–16].

However, the poor finishability of the β-TNTZ alloy is one of critical barriers that limit the performance quality by attracting bacterial infection [17] and vulnerability to attract plaques [18], which entail inflammation around the implant surface [19–21]. Indeed, a wide range of finishing processes, including grinding [22], honing [23], ball burnishing [24], flexible abrasive tools [25], etc., have been developed for processing the free-form surfaces of the developed implants. As per [26], the manual finishing of the implant surfaces is non-effective, imprecise, and takes more time. Apart from the conventional finishing processes, the chemical mechanical polishing is effective in polishing Ti implants [27] and to obtain mirror-polished surfaces without any contamination and reacted layers [28]. Furthermore, polishing techniques such as electro-polishing, magneto-electro-polishing [29], and electron beam radiation [30] are useful to surface finishing in nano-scale.

The magneto-rheological fluid assisted abrasive finishing (MRF-AF) process has been successful in producing nano-finished precise components [26]. Further, the finishing of β-phase Ti-Nb-Ta-Zr alloy is difficult because of low surface hardness as compared to other biomaterial [31]. Therefore, in the present study, the in-house developed β-phase Ti-Nb-Ta-Zr alloy has been heat-treated to increase the surface hardness of the alloy and to make it suitable for abrasive finishing. The magneto-rheological (MR) fluid used as a polishing medium consisted of carbonyl-iron-particles (CIP) and hard-abrasive-powder particles in base medium of synthetic mineral oil and grease. The rheological characteristics and the yield strength of the developed MR fluid affects by the externally applied magnetic field [32]. Ultimately, the CIP in the MR fluid develops an interlinked chain along the direction of magnetic field, resulting in a semisolid abrasive tool to process hard surfaces [33]. Barman et al. studied the effect of magnetorheological polishing fluid compositions on the surface finish of Ti-alloy. Ultra-fine surface roughness, ranging 10–70 nm, has been achieved using different types of rheological fluids [34]. Barman et al. studied the effect of tool paths such as spiral and raster on the surface finishing of Ti-based bio-medical alloy. It has been observed that at tool rotational speed of 1200 rpm, working gap of 1 mm, and finishing time of 6.30 h, using a raster path provided the best surface finish and surface topography [35]. Parameswari et al. studied the effect of abrasive particle concentration on surface finishing. The finishing rate has been significantly affected by initial roughness and concentration of abrasive particles [36]. Nagdeve et al. developed a rotational-magnetorheological abrasive flow finishing (R-MRAFF) process based special tool for nano-finishing of femoral component of knee joint and surface finish in the range of 78–89 nm was attained, by considering the effect of various input process parameters [37]. Barman et al. studied the influence of magnetic field-assisted finishing (MFAF) process on the various surface finishing and the average surface roughness obtained was 11.32 nm. The roughness parameters have obtained the values in the range of nano-meters and rendered better surface topography [38].

From the available literature, the nano-finishing of β-phase Ti-Nb-Ta-Zr biomedical alloy has not been reported yet. The novelty of research work is that the high-strength β-phase Ti-Nb-Ta-Zr is very tough to process using convectional finishing processes. The MR-fluid based abrasive-finishing set was developed in-house and the capability of nano-finishing on heat-treated β-TNTZ substrate has been investigated using single and multi-objective optimization. The material removal (MR) and percentage change in surface roughness (%ΔRa) of the implant surface have been studied in response to input process parameters, such as carbonyl iron particles (CIP) concentration, rotational speed (Nt), and working gap (Gp). Further, the surface morphology and rendered image analysis have been performed to obtain the characteristics of the processed surfaces. The simulated body fluid (SBF) test has also been carried out to identify the corrosion resistivity of the MRF-AF finished β-TNTZ substrate specimens. Further, the as-corroded surfaces have been characterized to observe the effect of surface roughness on the achieved corrosion characteristics.

2. Materials and Methods

High-strength β-phase Ti-Nb-Ta-Zr alloy has been developed using vacuum-arc melting process. The samples of size 10 × 5 mm for the finishing process were cut from the as-developed ingot through a wire-cut electric discharge machining process (Model Ecocut, Electronica, India). After that, the prepared specimens were subjected to a heat-treatment process to improve the mechanical properties of β-phase Ti-Nb-Ta-Zr as reported in previous study [39]. The microstructure of the samples before and after heat-treatment were examined by field emission scanning electron micrograph (FE-SEM; JEOL 7600F; JEOL Inc., Peabody, MA, USA) and associated energy dispersive spectroscopy (EDS, FE-SEM; JEOL 7600F; JEOL Inc., Peabody, MA, USA). From the microstructure analysis of untreated samples, it has been observed that the material comprised majorly β-type phases with grain size 250 μm, as can be seen in Figure 1a. The related EDS spectrum conform to the elemental composition and wt.% of each elements present in the material; refer to Figure 1b. After heat treatment, microstructure is refined and grain size becomes finer in the range of 100–150 μm; refer to Figure 1c. The heat-treated microstructure comprised α-type and ω-type phases, which further improved the mechanical properties of alloy. As a result, the ultimate compressive-strength and surface-hardness of the developed alloy was enhanced to 1195 MPa and 515 HV, respectively, as suitable for load-bearing implants requisites. Figure 1d shows the EDS spectrum and elemental composition of alloy after heat-treatment. The observations are close with the previous research studies [39–42].

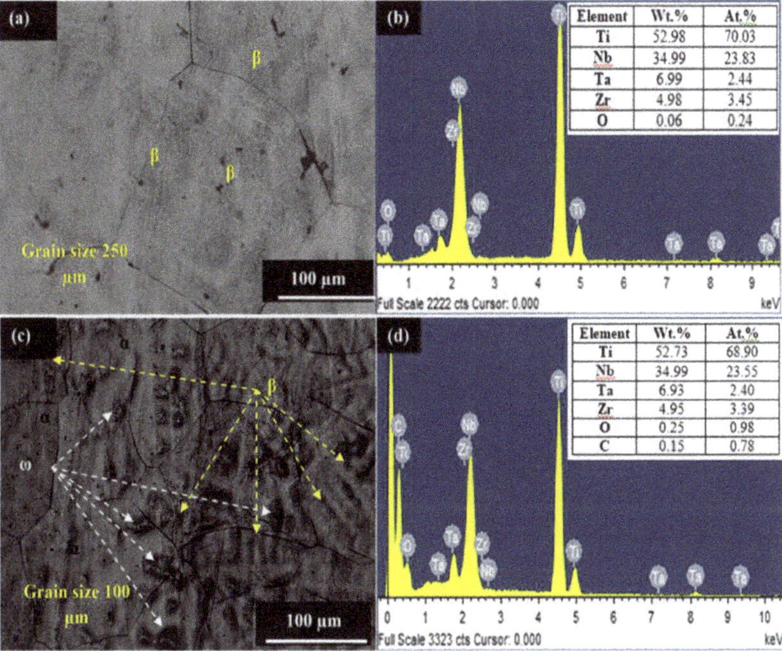

Element	Wt.%	At.%
Ti	52.98	70.03
Nb	34.99	23.83
Ta	6.99	2.44
Zr	4.98	3.45
O	0.06	0.24

Element	Wt.%	At.%
Ti	52.73	68.90
Nb	34.99	23.55
Ta	6.93	2.40
Zr	4.95	3.39
O	0.25	0.98
C	0.15	0.78

Figure 1. Microstructure and EDS spectrum: (**a,b**) β-TNTZ and (**c,d**) β-TNTZ after heat treatment.

The heat-treated β-phase Ti-Nb-Ta-Zr alloy specimens were then finished using an in-house developed MRF-AF setup; refer to Figure 2. The MRF-AF processing consisted of three stages, such as development of magnetorheological-fluid, preparation of customized finishing magnetic assisted tool, and finishing of the work surfaces. A permanent magnet tool of material neodymium-iron (Nd-Fe-B) with magnetic flux intensity ~0.45 Taxella Gauss was used as tool for experimentation to provide the required magnetic field in the finishing zone. Generally, the working gap between the abrasive tool and β-phase Ti-Nb-Ta-Zr alloy workpiece was filled with the abrasive media that acted as a ball-end polishing brush. The speed at which the tool rotates plays a crucial role in attaining the required cutting forces to chip out the small amount of material from the work surface. Table 1 shows the process parameters and their levels.

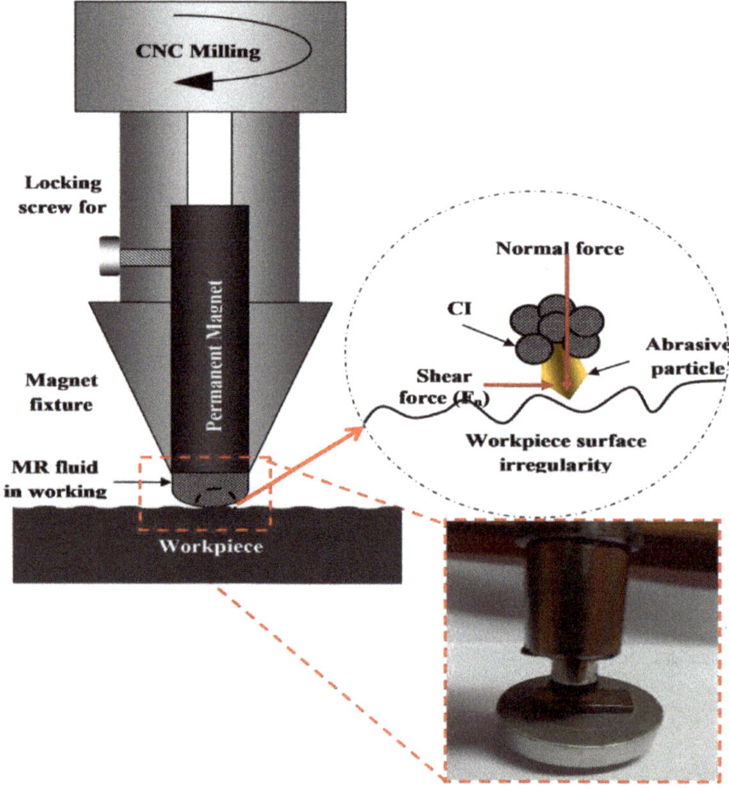

Figure 2. Pictorial representation of MRF-AF setup.

Table 1. MRF-AF process parameters and their working levels.

Symbol	Process Parameters	Unit	Range
CIP	Iron (Fe); size~25 μm	% by vol.	30, 35, 40
Nt	Rotational speed of tool	rpm	600, 900, 1200
Gp	Work-gap	Mm	1, 1.5, 2

Presently, the three most crucial input process parameters of MRF-AF process have been selected (such as CIP, Nt, and Gp) to identify their impact on achieved MR and Ra. The MR from the work surface has been calculated by using Equation (1):

$$MR = \rho_{workpice} \times V_{total\ material\ removed} \tag{1}$$

where, $\rho_{workpice}$ is the density of the workpiece and $V_{total\ material\ removed}$ is the total volume of material removed. Digital weighing balance (Scientech, Delhi, India) of accuracy 0.01 mg was used for the MR calculations. Further, the Ra values of the as-finished work surfaces have been calculated by using a non-contact three-dimensional (3D) Surface Profilometer (Talysurf CCI Lite, Leicester, UK) that uses a white light interferometer equipped with the TalyMap Platinum 6.0. The measurement of Ra was taken at three different locations. The design of experimentation technique, based on Taguchi L9 orthogonal array, was used to perform the statistical analysis on the observed output responses (such as MR and Ra), and to identify the statistical importance of the selected input process parametric levels on the observed responses using analysis of variance (ANOVA) [43]. Table 2 illustrates the control log of experimentation. Furthermore, corrosion performance parameter, corrosion-current, of the as-finished β-phase Ti-Nb-Ta-Zr alloy specimens has been studied in SBF medium using potentiodynamic polarization-based electrochemical system-1000E (make: Gamry Instruments, Warminster, PA, USA). The concentration of the SBF medium (pH 7.2) consisted of 9, 0.24, 0.43, and 0.2 g/L of NaCl, CaCl$_2$, KCl, and NaHCO$_3$ [44]. For this, potential rate and scan range has been selected as 1 mV/s and −250OCP to +250OCP mV, respectively. Tafel extrapolation technique was used to calculate the corrosion-current (ICOR). Before evaluating ICOR, the specimens were dipped in the SBF solution for about 24 h and the test was conducted at 37 ± 0.1 °C.

Table 2. Control log of experimentation.

Experiment No.	CIP	Nt	Gp
1	35	600	1.0
2	35	900	1.5
3	35	1200	2.0
4	40	600	1.5
5	40	900	2.0
6	40	1200	1.0
7	45	600	2.0
8	45	900	1.0
9	45	1200	1.5

3. Results and Discussion

3.1. Parametric Optimization

In the present research work, to understand the effect of the input process parameters on the material removal (MR) and surface roughness (Ra), has been studied through the use of design of experimentation. The concentration of the diamond abrasive particles has been selected as 3.5%vol. on the basis of pilot experimentation. As noticed, the 3.5%vol. of the diamond abrasive particles corresponded to the higher MR and lower Ra. The output responses observed after performing the set of experimentations, following Table 3, have been given in Table 3. The signal/noise (S/N) ratio has been calculated using Minitab-17 statistical software package. In the case of MR, the S/N ratio has been optimized at "larger-the-better" options, whereas in the case of Ra, the S/N ratio has been optimized at "smaller-the-better" option.

Table 3. Control log of experimentation.

Experiment No.	MR					Ra				
	Raw Value (g)			Mean	S/N	Raw Value (nm)			Mean	S/N
1	60	50	70	60	35.56	16	15	11	14	−22.92
2	50	55	75	60	35.56	8	10	6	8	−18.06
3	28	35	30	31	29.83	60	55	65	60	−35.56
4	68	55	57	60	35.56	10	12	8	10	−20.00
5	65	50	65	60	35.56	4	2	6	4	−12.04
6	68	67	60	65	36.26	45	50	40	45	−33.06
7	23	27	25	25	27.96	21	20	22	21	−26.44
8	66	62	67	65	36.26	14	15	10	13	−22.28
9	30	28	32	30	29.54	75	50	55	60	−35.56
Overall mean S/N ratio (mo)					33.56					25.10

Note: The unit of S/N ratio is decibel (dB).

Figure 3 shows the S/N ratio plot for the MR of the abrasive finished β-phase Ti-Nb-Ta-Zr alloy. It can be seen in the case of process parameter 'CIP' that the MR increased by increasing the concentration of the iron particles in the MR fluid from 35 to 40%vol. This is mainly due to the fact that, as the concentration of iron particles increased, the abrasive tool became stronger and more efficient to process the hard work surfaces. The iron particles are the prime source of producing a magnetically held semi-solid tool; hence, the CIP proportion of 40%vol. was favoured in obtaining stronger abrasive tools capable of withstanding higher cutting forces executed while processing the surface. However, as the CIP proportion was further increased to 45%vol., the MR has reduced drastically. This is mainly due to the fact that the higher proportion of iron particles has dominated the existence of abrasive diamond particles, as a result of the abrasive particles trapped within the rich iron particles. Owing to this, the cutting action of the resulting abrasive tool has been sacrificed. Further, in the case of "Nt", it can be seen that the MR has first increased by increasing the rotational speed from 600 to 900 rpm; however, with a further increase in the rotational speed to 1200 rpm, the MR has been dropped, significantly. Noticeably, at 900 rpm, the abrasive tool exerted greater cutting forces on the work material and therefore resulted in greater material removal. Further increase in the rotational speed to 1200 rpm has widened the magnetic flux density area and impeded the strength of the abrasive cutting tool. As regards to "Gp", it has been seen that when the working gap has been increased from 1 to 2 mm, the MR of the MRF-AF processed β-phase Ti-Nb-Ta-Zr alloy reduced. This can be attributed towards the reason that due to an increase in the working gap, the normal tangential force exerted by the abrasive cutting tool on work surface has reduced, resulting in weak cutting action.

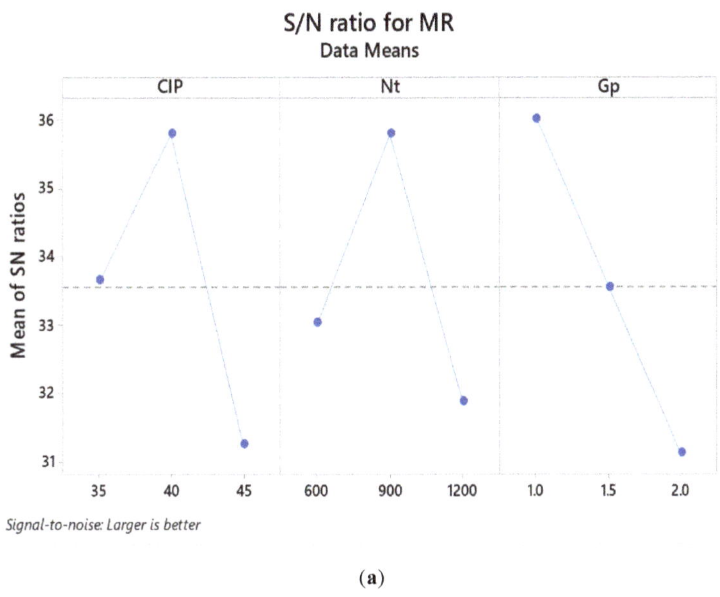

(**a**)

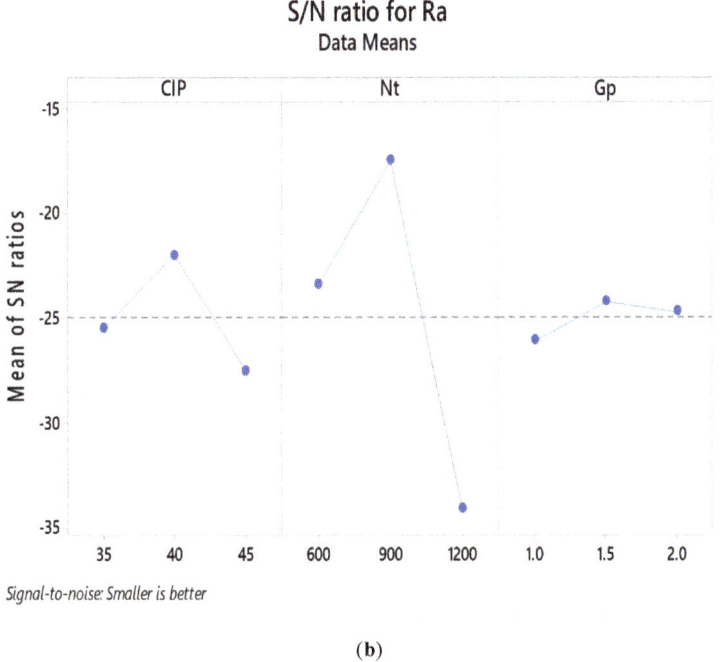

(**b**)

Figure 3. S/N ratio response of the input process parameters on MR (**a**) and Ra (**b**).

Further, in the case of 'Ra', it has been found that the surface roughness of the MRF-AF processed β-phase Ti-Nb-Ta-Zr alloy reduced by increasing the CIP process parametric level from 35 to 40%vol. This is due to the fact that, with an increase in the CIP level to 40%vol., the magnetic flux density of the abrasive tool has increased, resulting in an increase in the cutting strength of the abrasive finishing tool. However, by further increasing the CIP to 45%vol., the cutting action of the diamond abrasive particles has been dropped due to undesirable increase in the content of iron particles in the abrasive tool. Further, in the case of "Nt", it can be seen that the surface finishing of the processed alloy increases by increasing the rotational speed from 600 to 900 rpm. The reason behind this observation is the same as discussed for the MR. At 900 rpm, the cutting thrust force exerted by the developed abrasive tool has increased to result in the removal of the peak from the surface of the alloy. With the increase in the rotational speed the cutting action of the abrasive particles reduced [45,46]. At 1200 rpm, the brush of abrasive particles flared due to that magnetic flux density area increased as a result the materials removal reduced. The surface morphology and atomic force microscopy (AFM) analysis of the MRF-AF processed β-phase alloy—refer to Figure 4—also indicate that the presence of rough surface textures on the metallic implant surface are high. There are severe surface scratches on the metallic surface at 600 and 1200 rpm. However, comparatively, the surface is quite finished in the case of 900 rpm. Furthermore, it can be seen that, in the case of 900 rpm, the maximum heights of the peaks on the processed surfaces ranged from 150–160 nm. As regards the "Gp", it can be seen that the surface finishing of the MRF-AF processed β-phase alloy increased by increasing the work gap from 1 to 1.5 mm and further underwent a slight drop after increasing the work gap to 2.0 mm. This is due to the fact that, when the processing work gap has been increased, the tangential cutting force on the work surface reduced that contained the cutting action of the abrasive tool only limited to the removal of the surface peaks.

Table 4 shows the analysis of variance (ANOVA) analysis of S/N ration for MR and Ra. It can be seen that, in the case of MR, all the input process parameters are significant at 95% confidence level as their respective p-value is less than 5%. Further, the percentage contribution of input process parameters, such as CIP, Nt, and Gp is 33.64, 26.44, 39.28%, respectively. In the case of Ra, it has been found that only Nt ($p < 0.05$) has obtained statistically significant effect on the surface roughness of the processed metallic alloy. Further, process parameters such as CIP, Nt, and Gp has obtained 9.39, 85.57, and 1.12% contribution for Ra. The optimized levels of the input process parameters in response of MR are: CIP—45%vol., Nt—900 rpm, and Gp—1 mm. Whereas, in the case of Ra, the optimized input process parametric levels are: CIP—45%vol., Nt—900 rpm, and Gp—1.5 mm. Further, Table 5 shows the response table of S/N ratio for various levels of selected input process parameters. These given values have been used for the prediction of optimized response of the output parameters as per the suggested optimum levels of the input processing parameters.

Table 4. ANOVA for S/N ratio for MR and Ra.

Source	Degree of Freedom	Sum of Square	Variance	F-Test	p-Value	Contribution (%)
			MR			
CIP	2	30.9716	15.4858	53.19	0.018 *	33.64
Nt	2	24.3382	12.1691	41.80	0.023 *	26.44
Gp	2	36.1651	18.0825	62.11	0.016 *	39.28
Residual Error	2	0.5823	0.2912			0.632
Total	8	92.0573				100
			Ra			
CIP	2	47.086	23.543	2.40	0.294	9.39
Nt	2	429.047	214.523	21.85	0.044 *	85.57
Gp	2	5.630	2.815	0.29	0.777	1.12
Residual Error	2	19.639	9.819			3.91
Total	8	501.401				100

* Indicates significant parameters, F-test is the Fisher's test, and p-value is the probability.

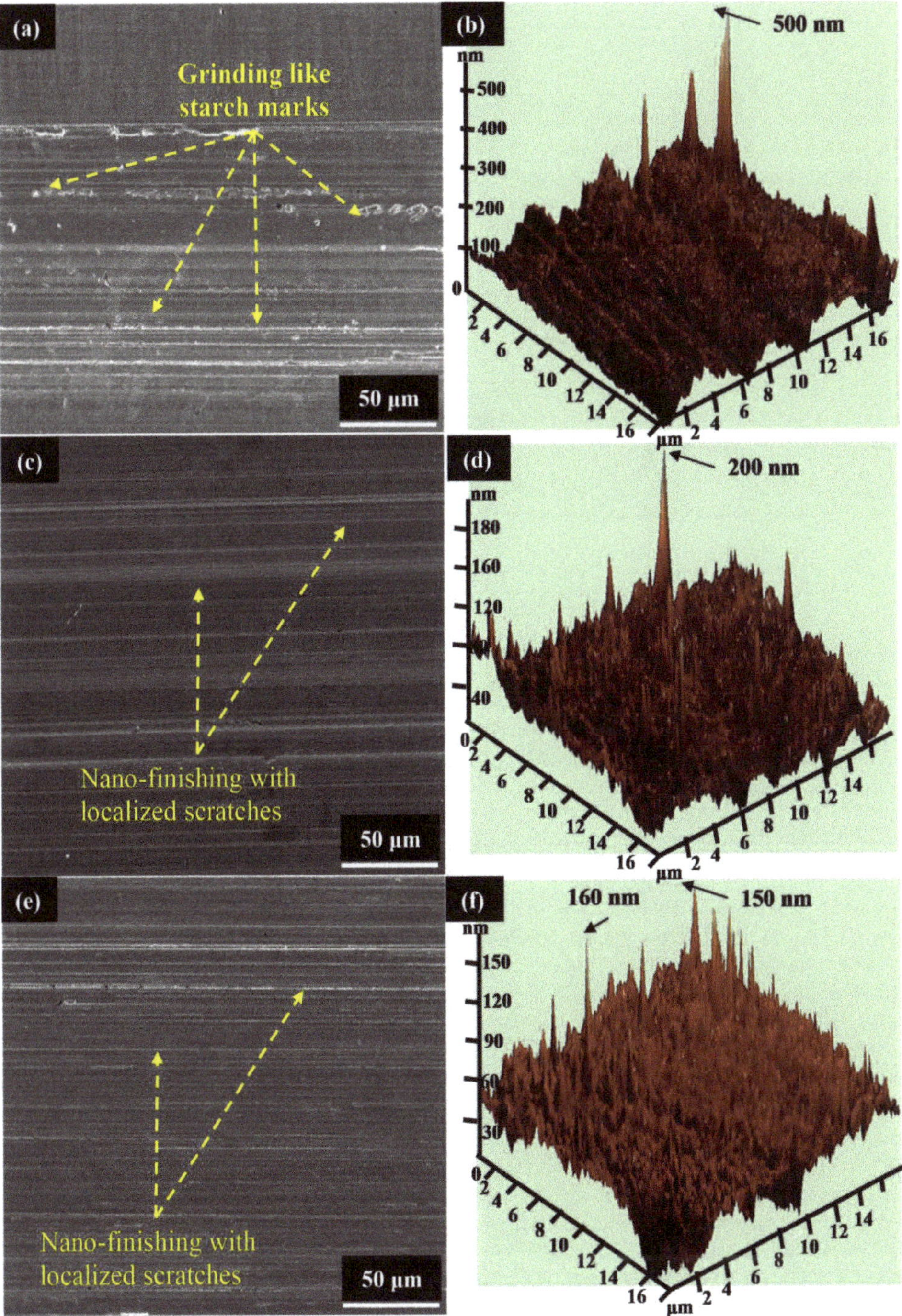

Figure 4. Surface morphology and AFM imaging of MRF-AF processed alloy surface (**a**,**b**) at 600 rpm, (**c**,**d**) at 1200 rpm, and (**e**,**f**) at 900 rpm.

Table 5. S/N response of input process parameters.

Level	CIP	Nt	Gp
		MR	
1	33.65	33.03	36.03 *
2	35.79 *	35.79 *	33.56
3	31.25	31.88	31.12
Delta	4.54	3.92	4.91
Rank	2	3	1
		Ra	
1	−25.52	−23.40	−26.09
2	−21.98 *	−17.46 *	−24.23 *
3	−27.51	−34.14	−24.68
Delta	5.53	16.68	1.86
Rank	2	1	3

* Indicates the optimum parameters.

In order to confirm the accuracy of the predicted results, a set of five confirmatory experiments have been conducted on the suggested parametric levels of input parameters. Table 6 shows the average values of responses on confirmatory experimental results.

Table 6. Predicted and confirmatory experimentation results at optimized levels of input parameters.

Responses	Predicted Values	Confirmatory Values	Difference (±)
MR (g)	105.8	103.7	2.1
Ra (nm)	4.63	4.67	0.04

It can be seen that the confirmatory experimentation results are in good agreement with the predicted results, highlighting the accuracy of the applied design of experimentation approach in obtaining the desirable output responses.

3.2. Corrosion Performance

Reportedly, the resulting surface topographical features of biomedical implant plays a significant role in the osteoblast adhesion, differentiation, extracellular matrix secret, and corrosion resistance [47–49]. Further, the nanostructured implant surface, being conductive to the body fluid, accelerates the osseointegration. There are numerous reports indicating that the synergistic effects of microstructure and nanostructure are often desirable to mediate the mechanism of cell attachment, growth, differentiation, and appetite formation [50]. Underlining these facts, the present study investigates the effect of the MRF-AF processing of β-phase Ti-Nb-Ta-Zr alloy to obtain nano-scale surface finish to enhance the biological performance of the resulting implant surface, especially to identify the obvious differences of corrosion resistance in various obtained surface. As the corrosion performance of the finished implants mainly depends on the quality of surfaces (Ra value) produced after MRF-AF processing, the statistical analysis (as presented in Table 5) has been used for performing the SBF test runs. Three different categories of test runs have been conducted to visualize the impact of selected input processing parameters (CIP, Nt, and Gp) and on ICOR.

The set #1 consisted of three samples with variable parameter 'CIP' at three selected levels (35, 40, and 45%vol.) and considering the optimum level the others (Nt, and Gp) as per Figure 3b. Further, the set #2 contains the three samples at three different levels of Nt (600, 900, and 1200 rpm) and considering the optimum levels of CIP and Gp. Lastly, the set #3 contains the three samples corresponding to three levels of Gp (1.0, 1.5, and 2.0 mm), while taking the levels of CIP and Nt as optimum. The corresponding plot for the ICOR is presented in Figure 5. From Figure 5, it can be seen that the highest corrosion resistance in the case of set #1 (ICOR~−8 μA/cm^2) has been corresponded to CIP of 40%vol. The observed trend in-line with the observation made while the optimization of Ra response parameter. Further in the case of set #2, it has been found the parametric levels of Nt has maximum impact on the obtained values of the ICOR. As observed, at 900 rpm minimum corrosion-current value has been obtained, signifying it as the best parametric level for obtaining the highest possible corrosion resistance. This particular parameter has been previously identified as statistically significant; therefore, to elaborate further the significance of Nt, a morphological study on the set #3 samples has been performed.

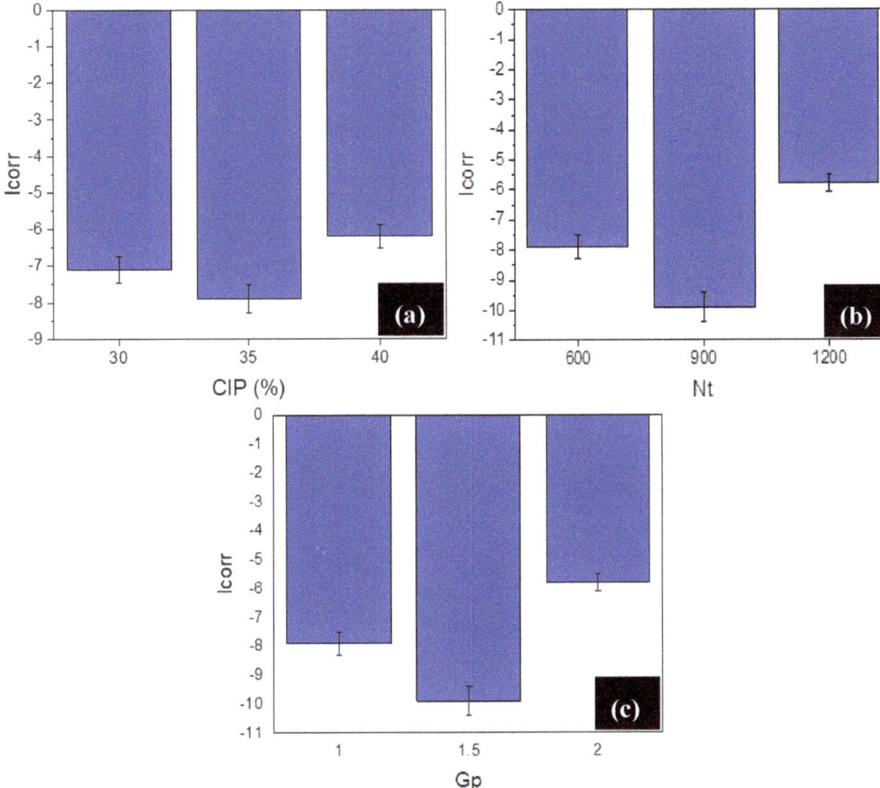

Figure 5. Results of ICOR for set #1 (**a**), set #2 (**b**), and set #3 (**c**). (Note: The unit of ICOR is $\mu A/cm^2$).

Figure 6 shows the corroded surface morphology and associated EDS spectrum of the set #2 samples. It can be seen from the Figure 6a that the surface has undergone severe shredding of the obtained apatite layer owing to the rough nano-texture. The rough sites existing on the surface have acted as nuclei sites to originate the corrosion mechanism. The corresponding higher ICOR values have subsisted the applied potential resulting in pitting and galvanic corrosion. The corresponding EDS spectrum (refer to Figure 6b) indicated the formation of 'O', 'P', and 'Ca' elements representing the existence of non-uniformly formed apatite layer. Further, in the case of 1200 rpm, it can be seen from Figure 6c that, although apatite layer has formed on the finished implant surface, due to higher ICOR value the brittle fracture and erosion of the apatite layer has been witnessed. The apatite layer acts as a barrier to the corrosion but severe shredding and higher ICOR value forced the formed apatite layer to undergo brittle fracture and its removal from the implant's surface. Indeed, such phenomenon is never desirable for biomedical implant due to the possibilities of biological complications (such as toxicity and hypersensitivity) owing to the release of metal ions. However, as regards 900 rpm, no shredding, pit formation, and galvanic corrosion has been identified in the case of Figure 6e. This is primarily due to the fact that, at 900 rpm, a very fine finished surface has been obtained that encouraged the formation of uniform apatite layer to act as a corrosion barrier. Further the observed ICOR value in this case is minimum (~ -10 μA/cm^2) that diminished the formation of electric potential.

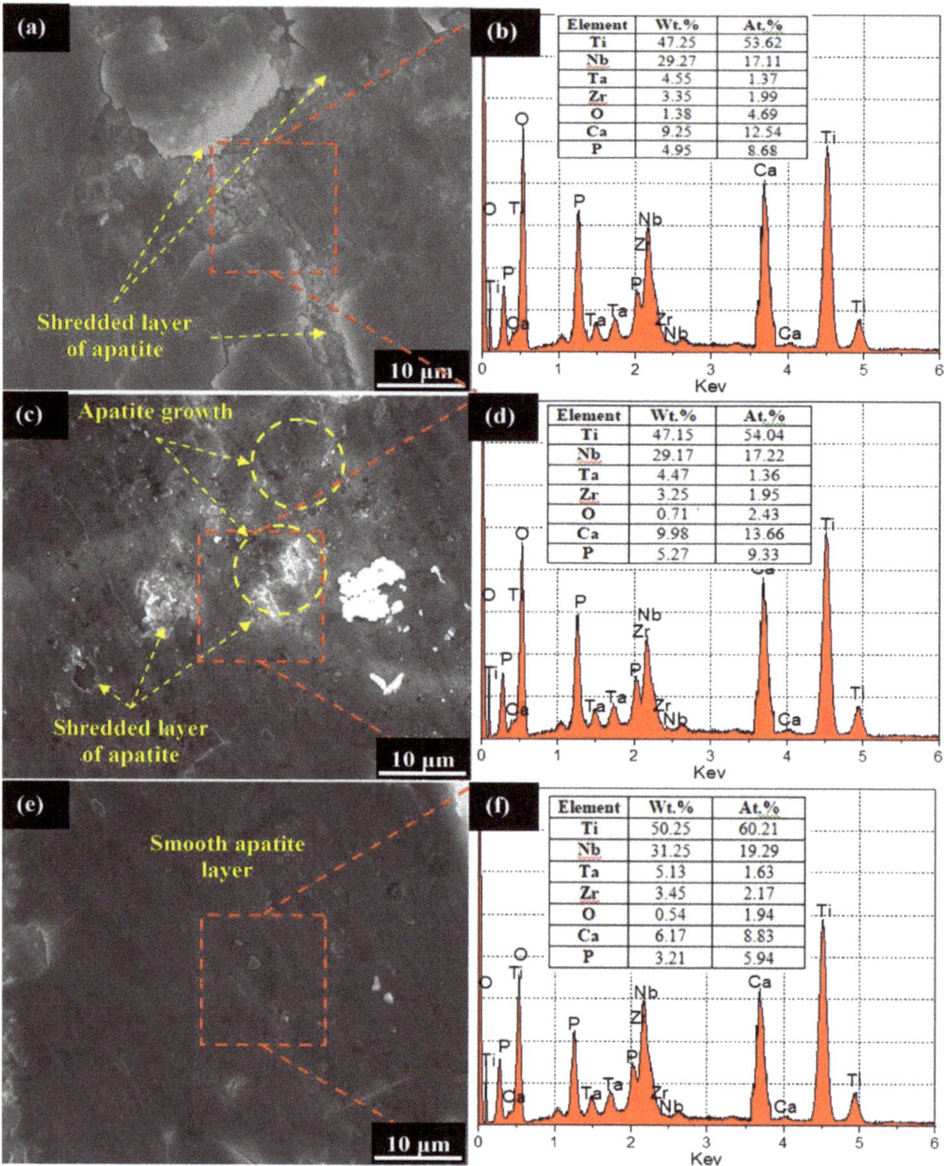

Element	Wt.%	At.%
Ti	47.25	53.62
Nb	29.27	17.11
Ta	4.55	1.37
Zr	3.35	1.99
O	1.38	4.69
Ca	9.25	12.54
P	4.95	8.68

Element	Wt.%	At.%
Ti	47.15	54.04
Nb	29.17	17.22
Ta	4.47	1.36
Zr	3.25	1.95
O	0.71	2.43
Ca	9.98	13.66
P	5.27	9.33

Element	Wt.%	At.%
Ti	50.25	60.21
Nb	31.25	19.29
Ta	5.13	1.63
Zr	3.45	2.17
O	0.54	1.94
Ca	6.17	8.83
P	3.21	5.94

Figure 6. Corroded morphology and EDS spectrum of finished surface: (**a**,**b**) 600 rpm, (**c**,**d**) 1200 rpm, and (**e**,**f**) 900 rpm.

4. Conclusions

In the present study, an investigation has been made to analyze the influence of the input process parameters of MRF-AF process on the MR and Ra of the finished β-phase Ti-Nb-Ta-Zr alloy. Besides this, the influence of the input process parameters on corrosion-resistance of the finished samples has also been studied. Based on the key findings, the following conclusions can be drawn:

It has been found that the MR of the processed alloy specimens has been significantly affected by all the selected input process parameters of MRF-AF. Furthermore, the optimized parametric levels as regards to MR are: CIP—40%vol., Nt—900 rpm, and Gp—1.0 mm.

However, in the case of Ra, it has been found that except Nt, none of the input process parameters are statistically significant. In this case, the optimized parametric levels identified are: CIP—40%vol., Nt—900 rpm, and Gp—1.5 mm. The confirmatory experimentation results have been found in good correlation with the predicted responses.

The results of the corrosion analysis of the developed samples highlighted that the corrosion resistance of the finished samples depends on their surface topography. It has been found that the samples possessed high surface finish developed a uniform layer of apatite in SBF medium that performed as a corrosion barrier. On the other side, the rough sites on the implant surface acted as the nuclei to propagate the corrosion mechanics that later resulted in shredding, pitting, and galvanic corrosion.

Overall, the results highlight that the MRF-AF process is highly suitable for producing nano-scale finishing of the biomedical implants made of high-strength β-phase Ti-Nb-Ta-Zr alloy.

Author Contributions: Conceptualization, S.S. and C.P.; methodology, C.P.; software, G.K. and A.P.; validation, A.B.(Animesh Bassak), R.S. and M.B.-C.; formal analysis, S.S., C.P, and A.B.(Atual babbar); investigation, C.P, R.S., A.B.(Animesh Bassak) and A.P.; resources, C.P and S.S; data curation, G.K and M.B.-C.; writing—C.P. and S.S.; writing—review and editing, A.P., A.B. (Animesh Bassak), A.B.(Atual babbar), C.P and S.S; visualization, A.B.(Animesh Bassak), A.B. (Atual babbar); supervision, C.P.; project administration, C.P. and S.S; funding acquisition, G.K and M.B.-C. All authors have read and agreed to the published version of the manuscript.

Funding: This research received no external funding.

Acknowledgments: There exists no conflict of interest.

Conflicts of Interest: The authors declare no conflict of interest

References

1. Niinomi, M.; Nakai, M.; Hieda, J. Development of new metallic alloys for biomedical applications. *Acta Biomater.* **2012**, *8*, 3888–3903. [CrossRef] [PubMed]
2. Niinomi, M.; Nakai, M. Titanium-based biomaterials for preventing stress shielding between implant devices and bone. *Int. J. Biomater.* **2011**, *2011*, 836587. [CrossRef] [PubMed]
3. Hu, T.; Hu, L.; Ding, Q. Effective solution for the tribological problems of Ti-6Al-4V: Combination of laser surface texturing and solid lubricant film. *Surf. Coat. Technol.* **2012**, *206*, 5060–5066. [CrossRef]
4. Collipriest, J.E., Jr. An experimentalist's view of the surface flaw problem. In *The Surface Crack—Physical Problems and Computational Solutions*; ASME: New York, NY, USA, 1972; pp. 43–61.
5. Lee, T.M.; Chang, E.; Yang, C.Y. A comparison of the surface characteristics and ion release of Ti6Al4V and heat-treated Ti6Al4V. *J. Biomed. Mater. Res.* **2000**, *50*, 499–511. [CrossRef]
6. Zaffe, D.; Bertoldi, C.; Consolo, U. Accumulation of aluminium in lamellar bone after implantation of titanium plates, Ti-6Al-4V screws, hydroxyapatite granules. *Biomaterials* **2004**, *25*, 3837–3844. [CrossRef]
7. Wei, Q.; Wang, L.; Fu, Y.; Qin, J.; Lu, W.; Zhang, D. Influence of oxygen content on microstructure and mechanical properties of Ti-Nb-Ta-Zr alloy. *Mater. Des.* **2011**, *32*, 2934–2939. [CrossRef]
8. Stráský, J.; Harcuba, P.; Václavová, K.; Horváth, K.; Landa, M.; Srba, O.; Janeček, M. Increasing strength of a biomedical Ti-Nb-Ta-Zr alloy by alloying with Fe, Si and O. *J. Mech. Behav. Biomed. Mater.* **2017**, *71*, 329–336. [CrossRef]
9. Prakash, C.; Kansal, H.K.; Pabla, B.S.; Puri, S. Experimental investigations in powder mixed electric discharge machining of Ti–35Nb–7Ta–5Zrβ-titanium alloy. *Mater. Manuf. Process.* **2017**, *32*, 274–285. [CrossRef]

10. Prakash, C.; Kansal, H.K.; Pabla, B.S.; Puri, S. Processing and characterization of novel biomimetic nanoporous bioceramic surface on β-Ti implant by powder mixed electric discharge machining. *J. Mater. Eng. Perform.* **2015**, *24*, 3622–3633. [CrossRef]

11. Prakash, C.; Kansal, H.K.; Pabla, B.S.; Puri, S. Multi-objective optimization of powder mixed electric discharge machining parameters for fabrication of biocompatible layer on β-Ti alloy using NSGA-II coupled with Taguchi based response surface methodology. *J. Mech. Sci. Technol.* **2016**, *30*, 4195–4204. [CrossRef]

12. Prakash, C.; Uddin, M.S. Surface modification of β-phase Ti implant by hydroaxyapatite mixed electric discharge machining to enhance the corrosion resistance and in-vitro bioactivity. *Surf. Coat. Technol.* **2017**, *326*, 134–145. [CrossRef]

13. Prakash, C.; Kansal, H.K.; Pabla, B.S.; Puri, S. Powder mixed electric discharge machining: An innovative surface modification technique to enhance fatigue performance and bioactivity of β-Ti implant for orthopedics application. *J. Comput. Inf. Sci. Eng.* **2016**, *16*, 041006. [CrossRef]

14. Prakash, C.; Kansal, H.K.; Pabla, B.S.; Puri, S. Effect of surface nano-porosities fabricated by powder mixed electric discharge machining on bone-implant interface: An experimental and finite element study. *Nanosci. Nanotechnol. Lett.* **2016**, *8*, 815–826. [CrossRef]

15. Singh, S.; Prakash, C.; Singh, H. Deposition of HA-TiO2 by plasma spray on β-phase Ti-35Nb-7Ta-5Zr alloy for hip stem: Characterization, mechanical properties, corrosion, and in-vitro bioactivity. *Surf. Coat. Technol.* **2020**, *398*, 126072. [CrossRef]

16. Singh, H.; Prakash, C.; Singh, S. Plasma Spray Deposition of HA-TiO2 on β-phase Ti-35Nb-7Ta-5Zr Alloy for Hip Stem: Characterization of Bio-mechanical Properties, Wettability, and Wear Resistance. *J. Bionic Eng.* **2020**, *17*, 1029–1044. [CrossRef]

17. Teughels, W.; Van Assche, N.; Sliepen, I.; Quirynen, M. Effect of material characteristics and/or surface topography on biofilm development. *Clin. Oral Implant. Res.* **2006**, *17*, 68–81. [CrossRef]

18. Bain, C.A.; Moy, P.K. The association between the failure of dental implants and cigarette smoking. *Int. J. Oral Maxillofac. Implant.* **1993**, *8*, 609–615.

19. Brocard, D.; Barthet, P.; Baysse, E.; Duffort, J.F.; Eller, P.; Justumus, P. A multicenter report on 1.022 consecutively placed ITI implants: A 7-year longitudinal study. *Int. J. Oral Maxillofac. Implant.* **2000**, *15*, 691–700.

20. Karoussis, I.K.; Kotsovilis, S.; Fourmousis, I. A comprehensive and critical review of dental implant prognosis in periodontally compromised partially edentulous patients. *Clin. Oral Implant. Res.* **2007**, *18*, 669–679. [CrossRef]

21. Balshe, A.A.; Eckert, S.E.; Koka, S.; Assad, D.A.; Weaver, A.L. The effects of smoking on the survival of smooth- and rough-surface dental implants. *Int. J. Oral Maxillofac. Implant.* **2008**, *23*, 1117–1122.

22. Aaraj Khodaii, S.J.; Barazandeh, F.; Adibi, H.; Sarhan, A. Optimization of Grinding partially stabilized zirconia (PSZ) for dental Implant application. *Modares Mech. Eng.* **2018**, *18*, 187–194.

23. Rao, P.S.; Jain, P.K.; Dwivedi, D.K. Optimization of key process parameters on electro chemical honing (ECH) of external cylindrical surfaces of titanium alloy Ti 6Al 4V. *Mater. Today Proc.* **2017**, *4*, 2279–2289. [CrossRef]

24. Affatato, S.; Ruggiero, A. Surface analysis on revised hip implants with stem taper for wear and failure incidence evaluation: A first investigation. *Measurement* **2019**, *145*, 38–44. [CrossRef]

25. Tian, Y.; Shi, C.; Fan, Z.; Zhou, Q. Experimental investigations on magnetic abrasive finishing of Ti-6Al-4V using a multiple pole-tip finishing tool. *Int. J. Adv. Manuf. Technol.* **2020**, *106*, 3071–3080. [CrossRef]

26. Singh, H.; Singh, S.; Prakash, C. *Current Trends in Biomaterials and Bio-manufacturing*; Springer: Berlin, Germany, 2019.

27. Basim, G.B.; Ozdemir, Z.; Mutlu, O. Biomaterials applications of chemical mechanical polishing. In Proceedings of the Planarization/CMP Technology (ICPT 2012), International Conference VDE 2012, Grenoble, France, 15–17 October 2012; pp. 1–5.

28. Okawa, S.; Watanabe, K. Chemical mechanical polishing of titanium with colloidal silica containing hydrogen peroxide—Mirror polishing and surface properties. *Dent. Mater. J.* **2009**, *28*, 68–74. [CrossRef] [PubMed]

29. Zur Rahman Pompa, L.; Haider, W. Influence of electropolishing and magnetoelectropolishing on corrosion and biocompatibility of titanium implants. *J. Mater. Eng. Perform.* **2014**, *23*, 3907–3915.

30. Okada, A.; Uno, Y.; Yabushita, N.; Uemura, K.; Raharjo, P. High efficient surface finishing of bio-titanium alloy by large-area electron beam irradiation. *J. Mater. Process. Technol.* **2004**, *149*, 506–511. [CrossRef]

31. Sidpara, A.; Jain, V.K. Analysis of forces on the freeform surface in magnetorheological fluid based finishing process. *Int. J. Mach. Tools Manuf.* **2013**, *69*, 1–10. [CrossRef]
32. Kumar, S.; Jain, V.K.; Sidpara, A. Nanofinishing of freeform surfaces (knee joint implant) by rotational-magnetorheological abrasive flow finishing (R-MRAFF) process. *Precis. Eng.* **2015**, *42*, 165–178. [CrossRef]
33. Barman, A.; Das, M. Force analysis during spot finishing of titanium alloy using novel tool in magnetic field assisted finishing process. *Int. J. Precis. Technol.* **2019**, *8*, 190–200.
34. Barman, A.; Das, M. Nano-finishing of bio-titanium alloy to generate different surface morphologies by changing magnetorheological polishing fluid compositions. *Precis. Eng.* **2018**, *51*, 145–152. [CrossRef]
35. Barman, A.; Das, M. Toolpath generation and finishing of bio-titanium alloy using novel polishing tool in MFAF process. *Int. J. Adv. Manuf. Technol.* **2019**, *100*, 1123–1135. [CrossRef]
36. Parameswari, G.; Jain, V.K.; Ramkumar, J.; Nagdeve, L. Experimental investigations into nanofinishing of Ti6Al4V flat disc using magnetorheological finishing process. *Int. J. Adv. Manuf. Technol.* **2019**, *100*, 1055–1065. [CrossRef]
37. Nagdeve, L.; Jain, V.K.; Ramkumar, J. Preliminary investigations into nano-finishing of freeform surface (femoral) using inverse replica fixture. *Int. J. Adv. Manuf. Technol.* **2019**, *100*, 1081–1092. [CrossRef]
38. Barman, A.; Das, M. Magnetic field assisted finishing process for super-finished Ti alloy implant and its 3D surface characterization. *J. Micromanufacturing* **2018**, *1*, 154–169. [CrossRef]
39. Singh, S.; Prakash, C. Effect of cryogenic treatment on the microstructure, mechanical properties and finishability of β-TNTZ alloy for orthopedic applications. *Mater. Lett.* **2020**, *278*, 128461. [CrossRef]
40. Málek, J.; Hnilica, F.; Veselý, J.; Smola, B.; Bartáková, S.; Vaněk, J. The influence of chemical composition and thermo-mechanical treatment on Ti–Nb–Ta–Zr alloys. *Mater. Des.* **2012**, *35*, 731–740. [CrossRef]
41. Majumdar, P. Microstructural Evaluation of Boron Free and Boron Containing Heat-Treated Ti-35Nb-7.2 Zr-5.7 Ta Alloy. *Microsc. Microanal.* **2012**, *18*, 295. [CrossRef]
42. Tang, X.; Ahmed, T.; Rack, H.J. Phase transformations in Ti-Nb-Ta and Ti-Nb-Ta-Zr alloys. *J. Mater. Sci.* **2000**, *35*, 1805–1811. [CrossRef]
43. Quinn, G.P.; Keough, M.J. *Experimental Design and Data aAnalysis for Biologists*; Cambridge University Press: Cambridge, UK, 2002.
44. Rosalbino, F.; De Negri, S.; Scavino, G.; Saccone, A. Electrochemical corrosion behaviour of binary magnesium-heavy rare earth alloys. *Metall. Ital.* **2018**, *2*, 34–43.
45. Babbar, A.; Prakash, C.; Singh, S.; Gupta, M.K.; Mia, M.; Pruncu, C.I. Application of hybrid nature-inspired algorithm: Single and bi-objective constrained optimization of magnetic abrasive finishing process parameters. *J. Mater. Res. Technol.* **2020**, *4*, 7961–7974. [CrossRef]
46. Nguyen, D.N.; Dao, T.P.; Prakash, C.; Singh, S.; Pramanik, A.; Krolczyk, G.; Pruncu, C.I. Machining parameter optimization in shear thickening polishing of gear surfaces. *J. Mater. Res. Technol.* **2020**, *9*, 5112–5126. [CrossRef]
47. Abron, A.; Hopfensperger, M.; Thompson, J.; Cooper, L.F. Evaluation of a predictive model for implant surface topography effects on early osseointegration in the rat tibia model. *J. Prosthet. Dent.* **2001**, *85*, 40–46. [CrossRef]
48. Ogawa, T.; Nishimura, I. Genes differentially expressed in titanium implant healing. *J. Dent. Res.* **2006**, *85*, 566–570. [CrossRef]
49. Wang, G.; Wan, Y.; Wang, T.; Liu, Z. Corrosion behavior of titanium implant with different surface morphologies. *Procedia Manuf.* **2017**, *10*, 363–370. [CrossRef]
50. Kubo, K.; Tsukimura, N.; Iwasa, F.; Ueno, T.; Saruwatari, L.; Aita, H.; Chiou, W.A.; Ogawa, T. Cellular behavior on TiO_2 nanonodular structures in a micro-to-nanoscale hierarchy model. *Biomaterials* **2009**, *30*, 5319–5329. [CrossRef] [PubMed]

Publisher's Note: MDPI stays neutral with regard to jurisdictional claims in published maps and institutional affiliations.

 *materials* 

Article

Prediction of Parameters of Equivalent Sum Rough Surfaces

Pawel Pawlus [1],*, Rafal Reizer [2]  and Wieslaw Zelasko [1]

[1] Faculty of Mechanical Engineering and Aeronautics, Rzeszow University of Technology, Powstancow Warszawy 12 Street, 35-959 Rzeszow, Poland; w.zelasko@prz.edu.pl

[2] College of Natural Sciences, University of Rzeszow, Rejtana 16C, 35-959 Rzeszow, Poland; rreizer@ur.edu.pl

* Correspondence: ppawlus@prz.edu.pl; Tel.: +48-17-865-1183

Received: 28 September 2020; Accepted: 30 October 2020; Published: 31 October 2020

Abstract: In statistical models, the contact of two surfaces is typically replaced by the contact of a smooth, flat, and an equivalent rough sum surface. For the sum surface, the zeroth, second, and fourth moments of the power spectral density m_0, m_2, and m_4 respectively, are the sum of spectral moments of two contacted surfaces. In this work, the selected parameters of the sum surfaces were predicted when the parameters of individual surfaces are known. During parameters selection, it was found that the pair of parameters: Sp/Sz (the emptiness coefficient) and Sq/Sa, better described the shape of the probability ordinate distribution of the analyzed textures than the frequently applied pair: the skewness Ssk and the kurtosis Sku. It was found that the RMS height Sq and the RMS slope Sdq were predicted with very high accuracy. The accuracy of prediction of the average summit curvature Ssc, the areal density of summits Sds, and parameters characterizing the shape of the ordinate distribution Sp/Sz and Sq/Sa was also good (the maximum relative errors were typically smaller than 10%).

Keywords: contact mechanics; equivalent sum rough surface; surface topography; parameters

1. Introduction

All surfaces are rough. Surface topography characterization is important during studies of various phenomena such as friction, wear, and contact resistance. Surface topography is of prime importance in problems of contact mechanics [1]. The contact of random isotropic Gaussian surfaces was typically studied. From among various models of random surface description, the model proposed by Nayak [2] was frequently applied. This model is based on earlier works of Longuet-Higgins [3,4], who described ocean surfaces. However, many surfaces are anisotropic. Nayak's model was extended to anisotropic surfaces [5].

According to Nayak, the statistical parameters characterizing an isotropic Gaussian surface can be expressed in terms of spectral moments of profiles. The m_0 moment is the profile variance, the m_2 moment is the profile mean square slope, and the m_4 moment is the mean square curvature of the surface profile. Greenwood and Tripp [6] found that the contact of two surfaces can be replaced by the contact of the equivalent sum rough surface and the smooth flat surface. The elastic modulus of the equivalent surface can be obtained from the following equation:

$$E' = \left(\frac{1 - v_{12}}{E_1} + \frac{1 - v_{22}}{E_2} \right)^{-1} \tag{1}$$

where E_i and v_i ($i = 1, 2$) are Young's moduli and Poisson's ratios of the two contacting elements [6].

Figure 1 shows a scheme of the contact of two rough surfaces. A separation of surfaces measured from the summits mean plane is called d, while h means a separation based on surface heights. R_1, R_2, and R_3 are radii of the summits in contact. One can see that radius of summits, density, and height of summits are important parameters in contact mechanics.

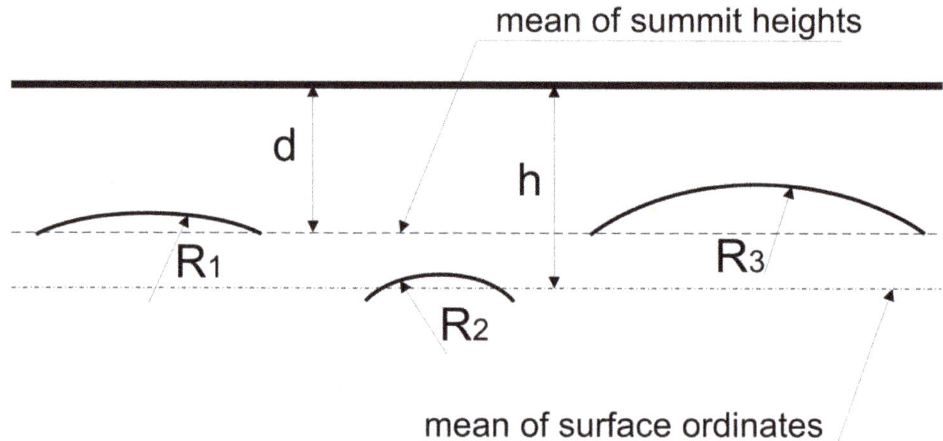

Figure 1. Scheme of the contact of rough surfaces.

To simplify the problem of the contact of two rough surfaces, researchers typically considered an equivalent surface in the contact with a smooth plane [7–9]. They found that the contact of two surfaces was negligibly different from the contact of a smooth flat and the equivalent sum rough surface. The m_0 moment is the profile variance, the m_2 moment is the profile mean square slope, and the m_4 moment is the mean square curvature of the surface profile. These moments are substantial in contact of two surfaces. This simplification can be used not only in statistical models of the elastic contact [7,9], but also in statistical models of the elastic–plastic contact [10–13]. The contact parameters such as the number of contacting asperities, the real area of contact, and the contact load, for any given separation between the equivalent sum rough surface and a smooth flat, can be also calculated by summing the contributions of all the contacting asperities using the summit identification model [14–16]. Combining two rough surfaces onto one equivalent rough surface and a smooth plane can also be helpful in deterministic contact models considering the Boussinesq problem [17–19]. In all cases, the ordinates of the equivalent sum rough surface should be the sum of the ordinates of two contacted surfaces. For the equivalent rough surface, the spectral moments are the sum of the spectral moments of two individual surfaces [20]. The parameters important in rough contact mechanics, such as the average curvature of summits, Ssc, and the areal summit density, can be computed from the spectral moments [2,21]:

$$Ssc = \frac{8m_4^{\frac{1}{2}}}{3\pi^{\frac{1}{2}}} \tag{2}$$

$$Sds = \frac{1}{6\pi\sqrt{3}}\left(\frac{m_4}{m_2}\right) \tag{3}$$

These equations were proved only for surfaces of Gaussian ordinate distribution.

However, the distribution of surface height is often different from Gaussian [20,22,23]. Specifically, two-process surfaces characterized by negative values of the skewness, Ssk, and high values of the kurtosis, Sku, are of high functional importance [16,24–26]. Two-process textures have traces of two processes (machining or wear). The contact of deterministic surfaces is also possible.

In this work, the selected parameters of the equivalent sum rough surfaces will be predicted when the parameters of each surface are known.

2. Analyzed Textures

Surface textures were measured using a white light interferometer Talysurf CCI Lite (produced by Taylor Hobson Ltd., Leicester, UK). A vertical resolution was 0.01 nm. The measured area 3.3×3.3 mm^2 and contained 1024×1024 points. Spikes and isolated deep and narrow valleys were eliminated by truncation of the height corresponding to material ratios of 0.01–99.99%. Before calculations of parameters, flat surfaces were leveled, while forms of the curved surfaces were removed using the polynomials of the second degree. Digital filtration was not used. We tried to analyze surfaces of various types: isotropic, anisotropic, random, deterministic, or mixed, of symmetric and non-symmetric ordinate distribution, one-process, and two-process. Therefore, surfaces after vapor blasting, polishing, lapping, milling, one- and two-process honing (plateau honing), and vapor blasting followed by lapping were measured. In most cases, the measured surfaces were machined using typical techniques.

Examples of analyzed textures are shown in Figures 2–5. They present isometric views, ordinate distributions, and directionality plots of selected surfaces. Surface after vapor blasting shown in Figure 2 is an isotropic texture of comparatively big height and Gaussian ordinate distribution. Surface presented in Figure 3 is a smooth random mixed Gaussian texture after polishing. Figure 4 shows deterministic texture after milling of an asymmetric ordinate distribution. These surfaces represent one-process textures. Figure 5 presents anisotropic cross-hatched random two-process texture after plateau honing of asymmetric ordinate distribution.

Forty textures were measured. In order to obtain equivalent rough surfaces, the ordinate distributions of measured surfaces were summed using TalyMap software: 50 equivalent sum surfaces were obtained. Parameters of individual surfaces and parameters of the sum surface were also calculated using the TalyMap software.

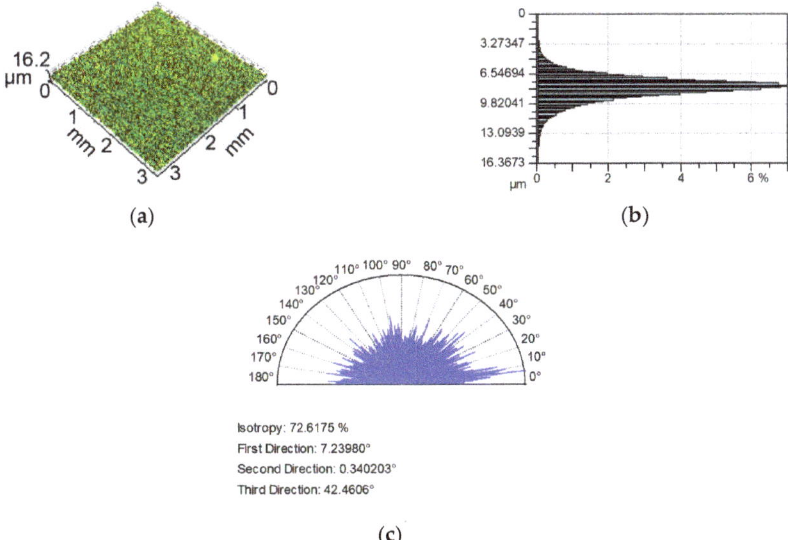

(a)

(b)

(c)

Figure 2. Isometric view (**a**), ordinate distribution (**b**), and directionality plot (**c**) of surface texture after vapor blasting.

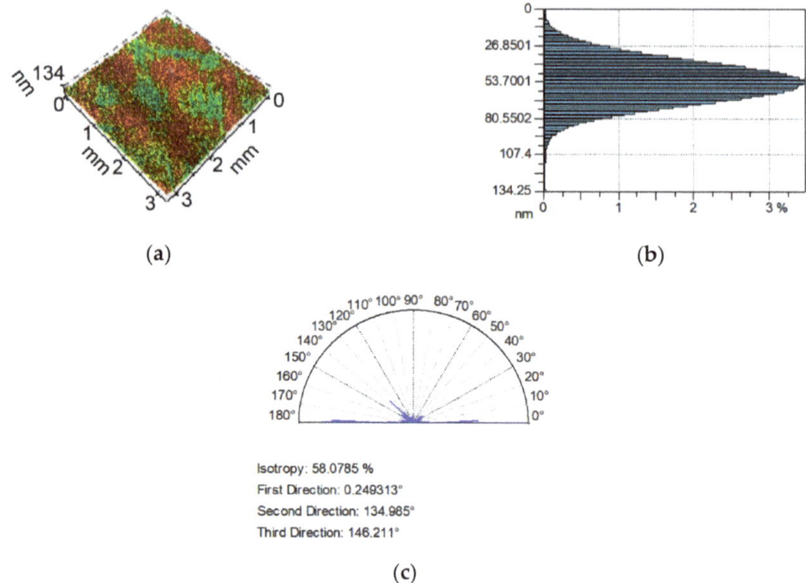

Figure 3. Isometric view (**a**), ordinate distribution (**b**), and directionality plot (**c**) of surface texture after polishing.

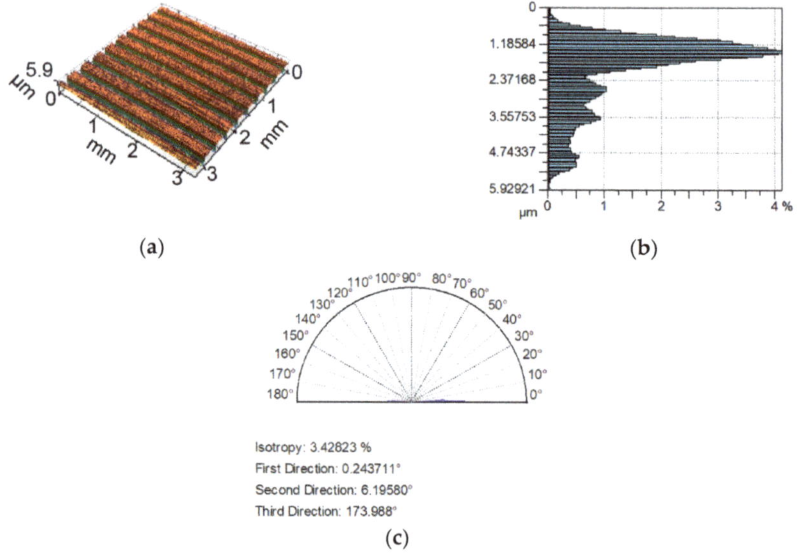

Figure 4. Isometric view (**a**), ordinate distribution (**b**), and directionality plot (**c**) of surface texture after milling.

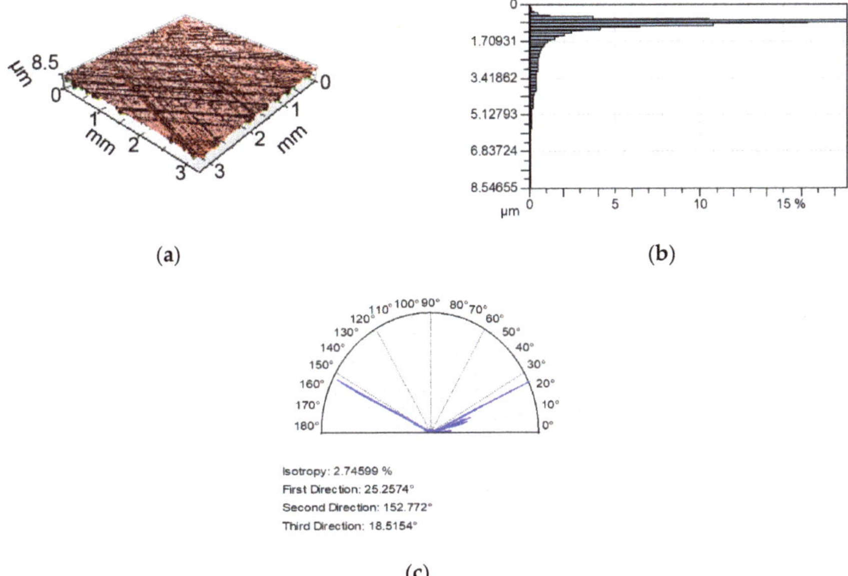

(a)

(b)

(c)

Figure 5. Isometric view (**a**), ordinate distribution (**b**), and directionality plot (**c**) of surface texture after plateau honing.

3. Selection of Surface Texture Parameters

The Sq parameter is a root mean square height of the surface.

$$Sq = \sqrt{\frac{1}{A} \iint_A z^2(x,y)\,dxdy} \tag{4}$$

Where:

A—area,

z—surface height in position x, y,

x, y—lengths in perpendicular directions.

Sq is a statistical parameter of comparatively small sensitivity on the measurement errors [27]. It is frequently used in measurements of optical surfaces. This parameter is related to the standard deviation of asperity heights, which is frequently used in contact mechanics [9–13].

Sa is arithmetical mean surface height.

$$Sa = \frac{1}{A} \iint_A |z(x,y)|\,dxdy \tag{5}$$

The Sa parameter is also non-sensitive to measurement errors. It is frequently used in machining.

Sdq, which is a root mean square gradient, is the hybrid parameter. Surface slope is used in assessing surface ability to the plastic deformation [28–30]. Slope is also related to friction, hydrodynamic lubrication, reflectance [31], and strength of adhesive joints [32]. The Sdq parameter is computed using the following formula:

$$Sdq = \sqrt{\frac{1}{A} \iint_A \left[\left(\frac{\partial z(x,y)}{\partial x}\right)^2 + \left(\frac{\partial z(x,y)}{\partial y}\right)^2 \right] dxdy} \tag{6}$$

The Sq and Sdq parameters are included in ISO 25178-2 (Geometrical Product Specifications (GPS)—Surface texture: Areal—Part 2: Terms, definitions and surface texture parameters) standard [33].

The other parameters: the average summit curvature, Ssc, and areal density of summits, Sds, are related to summits, therefore they are important in contact characteristics [9–13]. The Ssc parameter enables us to know the mean form of the peaks: according to the mean value of the curvature of the surface at these points. The Ssc and Sds parameters are presented in an older standard, EUR15178N [34]. In calculations of the Sds and Ssc parameters, a point is considered as a summit if its ordinate is higher than those of its 8 neighbors. This definition is in accordance with References [5,14,15]. Therefore, these parameters are useful in the analysis of the contact of rough surfaces. In the new ISO 25178 standard, the Spc parameter (arithmetical mean peak curvature) replaces the Ssc parameter and the Spd parameter (density of peaks) replaces the Sds parameter. The Spc and Spd parameters are calculated in the same way as Ssc and Sds parameters but take into account only those significant summits that remain after a discrimination by segmentation [35]. Therefore, the value of the Spd parameter is smaller than that of the Sds parameter. However, because the Spc and Spd parameters do not consider all the peaks existed on a surface, the Ssc and Sds parameters will be analyzed.

The spatial surface properties are characterized by Sal and Str parameters. The Sal, the autocorrelation length, is the horizontal distance, at which the autocorrelation function slowly decays to 0.2 value. A high value of this parameter indicates that surface has mainly high wavelengths. The Str, the texture-aspect ratio, is the ratio of the shortest to the highest correlation lengths. This parameter has values between 0 (anisotropic surface) and 1 (isotropic surface).

The parameters characterizing the shape of the surface topography ordinate distribution are also important in contact mechanics [20,22,23]. Typically, this shape is characterized by the skewness, Ssk, and the kurtosis, Sku [36–38].

$$Ssk = \frac{1}{Sq^3}\left[\frac{1}{A}\iint_A z^3(x,y)dxdy\right] \tag{7}$$

$$Sku = \frac{1}{Sq^4}\left[\frac{1}{A}\iint_A z^4(x,y)dxdy\right] \tag{8}$$

The skewness, Ssk, characterizes the symmetry of the surface texture. The value of the skewness depends on if the material is above (negative skewed) or below (positive skewed) the mean plane. The kurtosis, Sku, describes the sharpness of the surface ordinate distribution. When Sku < 3, the surface has relative few high peaks and deep valleys, and when Sku > 3, the surface has many high peaks and deep valleys [39]. However, due to the large exponent used, the parameters Ssk and Sku are sensitive to presences of spikes as well as narrow and deep valleys. In addition, for two-process surfaces [16,24–26], these parameters are interrelated. Sp is the maximum peak height and Sz is the maximum surface height. The Rp, Rz, Rq, Ra, Rsk, and Rku are profile equivalents of the parameters Sp, Sz, Sq, Sa, Ssk, and Sku, respectively. It was found [40] that the pair of ratios Rp/Rz, Rq/Ra can replace the pair Rsk, Rku in description of the probability distribution of the roughness profile. The recommended parameters can be used for characterization of both one-process and two-process surfaces. These parameters are, in contrast to the pair (Rsk, Rku), statistically independent. The Rq/Ra ratio is much more stable on surface and has smaller sensitivity to the measurement errors than the kurtosis Rku.

The Sv parameter is the maximum valley depth (Sz = Sp + Sv). Figure 6 presents the graphical interpretation of the Sp and Sv parameters. Sv describes the area under, while Sp the area above the material ratio curve. This curve presents the cumulative distribution of surface ordinates. The Sv parameter presents material, while Sp presents void (emptiness). The Rp/Rz or Sp/Sz ratios are called the emptiness coefficient. One can believe that when this ratio is smaller, the wear intensity is also lower [41].

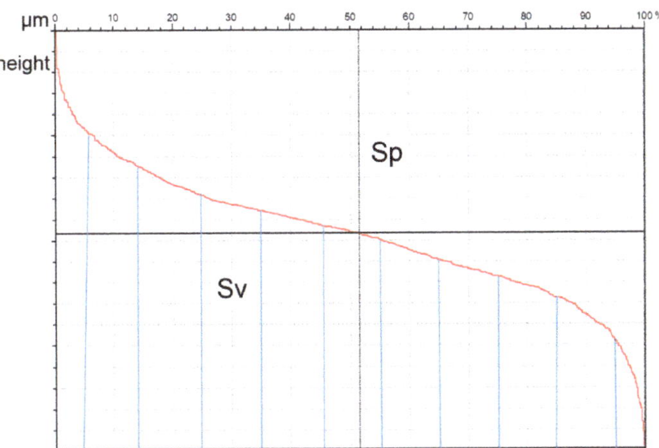

Figure 6. Graphical interpretation of Sp and Sv parameters.

In other to analyze the possibility of using Sp/Sz and Sq/Sa parameters for description of the ordinate distribution of diversified textures, the correlation and regression analysis of forty surfaces was carried out. The linear coefficient of correlation was also used in previous studies [42–45]. Table 1 lists the values of the linear coefficient of correlation among the analyzed parameters.

Table 1. The values of the linear correlation coefficient, r, among surface texture parameters.

Parameter	Ssk	Sku	Sq/Sa	Sp/Sz
Ssk	1			
Sku	−0.77	1		
Sq/Sa	−0.82	0.91	1	
Sp/Sz	0.71	−0.48	−0.45	1

It was assumed that when the absolute value of the coefficient of correlation was higher than 0.7, the parameters were strongly correlated (the coefficient of determination was larger than 0.5). One can see from the analysis of Table 1 that the Ssk parameter is strongly correlated with the Sku parameter ($r = -0.77$), therefore they cannot be used for description of the ordinate distributions of the analyzed textures. In contrast, the suggested parameters Sq/Sa and Sp/Sz are statistically independent ($r = -0.45$). The emptiness coefficient describes similar surface property as the skewness Ssk ($r = 0.71$). The Sq/Sa parameter is correlated with both Ssk ($r = 0.91$) and Sku ($r = -0.82$), however, the relation of Sq/Sa to the skewness, Ssk, is the strongest (Figure 7). On the basis of the presented analysis and due to larger stability and smaller sensitivity to measurement errors of the Sq/Sa ratio compared to the kurtosis, Sku, the pair Sq/Sa and Sp/St is recommended for description of the ordinate distribution of diversified surface textures (one- and two-process, random and periodic, isotropic and anisotropic).

The parameters describing the shape of the ordinate distribution, Ssk, Sku, Sp/Sz, and Sq/Sa, were not correlated with other analyzed parameters. The Sq parameter was proportional to Sdq ($r = 0.75$) and Sds ($r = -0.72$). The Sdq parameter was strongly correlated with the parameters Str ($r = 0.79$), Sds (−0.76), and Ssc ($r = 0.97$).

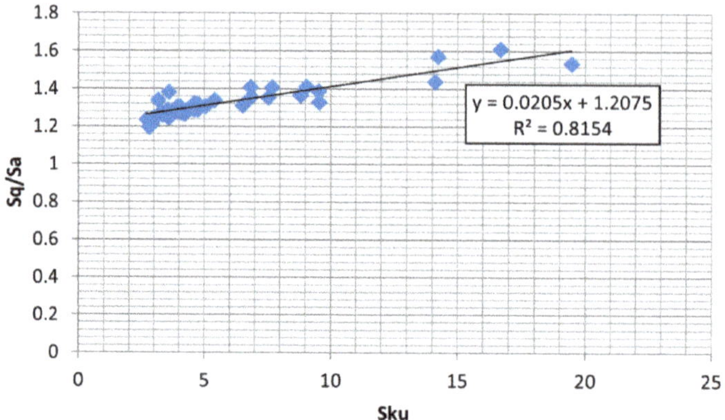

Figure 7. Dependence between parameters Sku and Sq/Sa.

4. Methods of Parameters Prediction

The Sq parameter is related to m_0 spectral moment, which is the profile variance. The Sdq parameter should be related to m_2 moment, which is a square of the profile RMS slope. The Ssc parameter can be related to m_4 moment which is the mean square curvature of the surface profile. For surface of Gaussian ordinate distribution, Equation (2) presents the dependence between m_4 profile spectral moment and the average curvature of summits. Because the spectral moment of the equivalent rough surface is the sum of spectral moments of both surfaces, these parameters of equivalent surfaces were predicted using the following formulae:

$$Sq_{sum} = \sqrt{Sq_1{}^2 + Sq_2{}^2} \tag{9}$$

$$Sdq_{sum} = \sqrt{Sdq_1{}^2 + Sdq_2{}^2} \tag{10}$$

$$Ssc_{sum} = \sqrt{Ssc_1{}^2 + Ssc_2{}^2} \tag{11}$$

It would be difficult to predict the areal summit density, Sds, using Equation (3), because summit density depends on both m_2 and m_4 spectral moments. Furthermore, this formula was proven only for a selected type of textures.

Therefore, the Sds and other analyzed parameters (Sq/Sa, Sp/Sz, Ssk, Sku, Sal, and Str) were predicted as the weighted averages of them and of the Sq parameter of both textures:

$$P_{sum} = \frac{Sq_1 P_1 + Sq_2 P_2}{Sq_1 + Sq_2} \tag{12}$$

where P_1 is a parameter of the first surface, P_2 is a parameter of the second surface, while P_{sum} is predicted parameter of the sum surface. Equation (12) is an original conception of the authors of this paper.

5. Results of Parameters Prediction and Discussion

Table 2 presents the results of this study.

Some examples of surfaces' summations are given below. Surface A after lapping was a smooth anisotropic one-process random structure of a little asymmetrical ordinated distribution. Surface B was an isotropic one-process random texture of Gaussian ordinate distribution characterized by high roughness. Due to different roughness heights, the equivalent rough sum surface was similar to surface B (Figure 8). Table 3 presents parameters of surface A, of surface B, of the sum surface, and predicted parameters of the sum surface. Due to the large difference between amplitudes of two surfaces, the errors of parameters' predictions were rather small. The largest errors were found for the Sal (about 8%) and the Str parameters (about 6%).

Table 2. Relative errors of parameters predictions.

Parameter	Average Error, %	Maximum Error, %
Sq	0.31	0.99
Sdq	0.27	0.97
Ssc	3.1	12.1
Sds	3.5	7.1
Sp/Sz	3.8	12.3
Sq/Sa	1.6	7.1
Ssk	26.2	62.1
Sku	13.1	45.2
Str	16.5	89.9
Sal	11.3	28.2

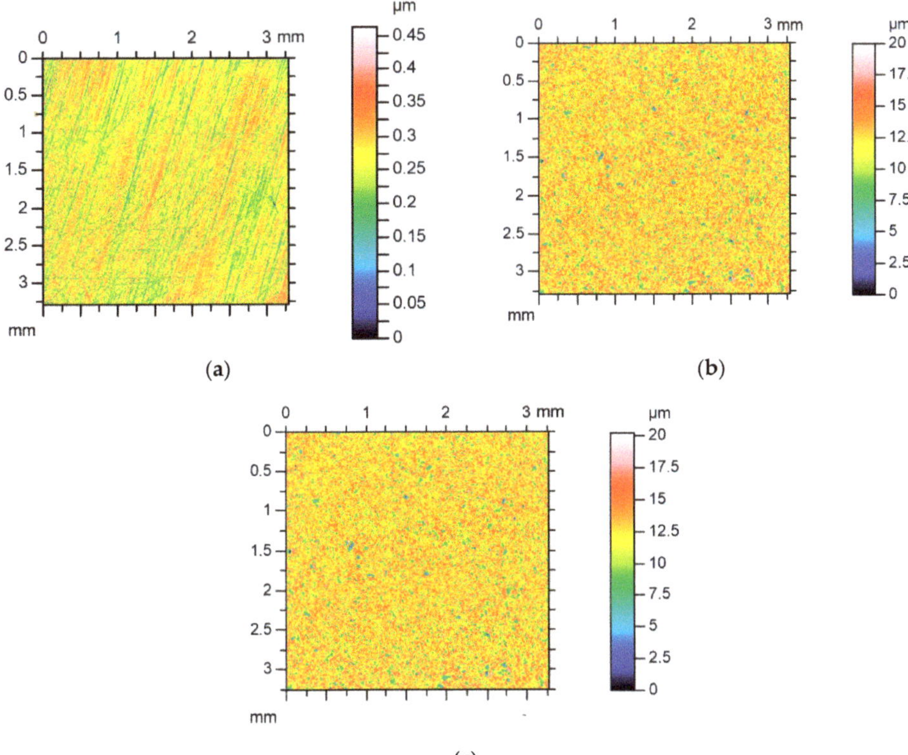

Figure 8. Contour plots of surface A (**a**), of surface B (**b**), and of equivalent sum surface (**c**).

Table 3. Selected parameters of surface A, of surface B, of the sum surface, and of predicted parameters of the sum surface.

Parameter	Surface A	Surface B	Sum Surface	Predicted Sum Surface	Unit
Sq	0.047	1.89	1.89	1.89	μm
Sdq	0.016	0.46	0.46	0.46	-
Ssc	7.28	139.5	139.5	139.7	1/mm
Sds	6094	3593	3612	3654	1/mm^2
Sp/Sz	0.42	0.39	0.39	0.39	-
Sq/Sa	1.31	1.31	1.31	1.31	-
Ssk	−0.3	−0.65	−0.64	−0.64	-
Sku	3.04	4.73	4.68	4.71	-
Str	0.033	0.8	0.83	0.78	-
Sal	0.0078	0.013	0.014	0.013	mm

For the contact between both two-process surfaces C and D of similar characteristics, the predicted skewness, Ssk, was underestimated (Table 4). This was caused by the presence of more deep valleys on the sum surface compared to individual surfaces in the contact (Figure 9). The skewness, Ssk, is more negative for smaller number of deep valleys. The error of skewness prediction was large (40%). This is related to the large error of the kurtosis, Sku, prediction (near 50%). An increase in the number of valleys during creation of the sum surface caused a large increase in the texture aspect ratio, Str (Figure 10), and in the autocorrelation length, Sal. The error of the Str parameter prediction was about 90%, while the error of the Sal parameter expectation was 30%. The errors of other parameters' predictions were smaller, up to 5% (Sds and Sq/Sa).

Table 4. Selected parameters of surface C, of surface D, of the sum surface, and of predicted parameters of the sum surface.

Parameter	Surface C	Surface D	Sum Surface	Predicted Sum Surface	Unit
Sq	0.7	0.79	1.05	1.05	μm
Sdq	0.13	0.16	0.21	0.21	-
Ssc	41.5	57.4	73.2	70.9	1/mm
Sds	5199	6028	5388	5639	1/mm^2
Sp/Sz	0.2	0.31	0.26	0.26	-
Sq/Sa	1.391	1.41	1.34	1.4	-
Ssk	−2.15	−1.9	−1.44	−2.02	-
Sku	9.51	7.68	5.71	8.54	-
Str	0.028	0.019	0.29	0.023	-
Sal	0.016	0.016	0.023	0.016	mm

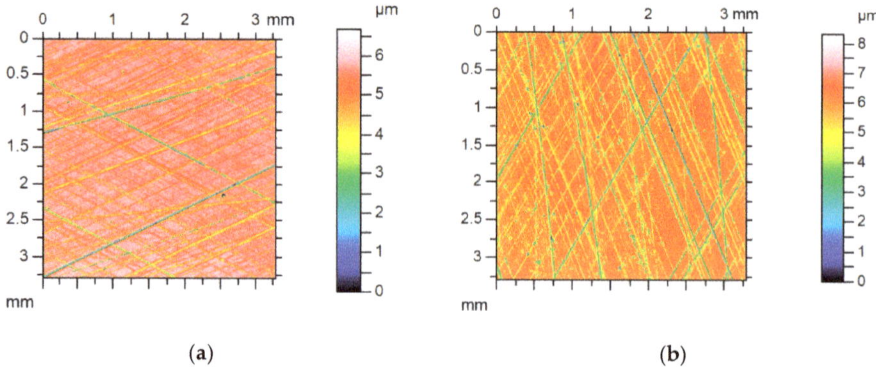

(a) (b)

Figure 9. *Cont.*

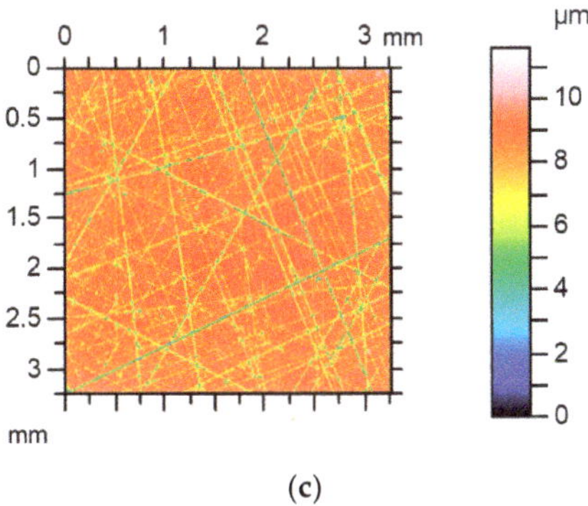

(c)

Figure 9. Contour plots of surface C (**a**), of surface D (**b**), and of equivalent sum surface (**c**).

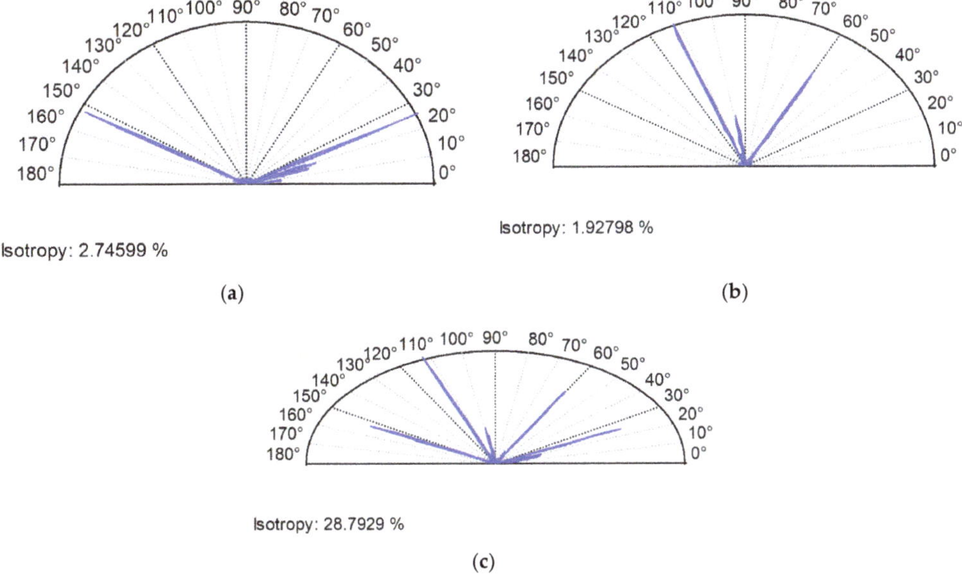

Isotropy: 2.74599 %

(a)

Isotropy: 1.92798 %

(b)

Isotropy: 28.7929 %

(c)

Figure 10. Directionality plots of surface C (**a**), of surface D (**b**), and of equivalent sum surface (**c**).

Comparatively low errors of parameters' predictions were obtained for the contact of anisotropic surface E and isotropic surface F. Both surfaces were characterized by similar surface heights (the values of the Sq parameter were higher than 2). Surface E from cylinder liner was a cross-hatched structure obtained after one-process honing, however, surface B was isotropic texture after vapor blasting. Table 5 presents parameters of these surfaces, of the sum surface, and of predicted parameters of the sum surface.

The errors of the Sq, Sdq, Ssc, Sds, Sp/Sz, Sq/Sa, and Sal parameters' predictions were smaller than 4.5%. Two surfaces were characterized by slightly negative values of skewness. Similar to surfaces analyzed earlier, the skewness of the sum surface was higher than predicted (the relative error was about 30%), but the kurtosis was smaller (the error was about 16%) than predicted. The parameter Str of the equivalent sum surface was higher by about 17% than the predicted parameter. Addition of anisotropic honed surface to isotropic surface after vapor blasting caused a decrease in an isotropy ratio compared to isotropic texture. However, the change was lower than predicted. Figure 11 shows the contour plots and Figure 12 shows the directionality plots of surfaces E and F and of the sum surface.

When isotropic one-process surface G was added to isotropic two-process surface H (Figure 13), the errors of predictions of most parameters (Sq, Sdq, Sds, Str, and Sal) were comparatively low (smaller than 4.5%)—Table 6. However, relative differences among real and predicted parameters describing shapes of the ordinate distributions of analyzed textures were larger (Figure 14). The predicted Ssk parameter was underestimated (the error was near 50%), while the Sku parameter was overestimated. The errors of parameters Sp/Sz and Sq/Sa predictions were smaller and amounted to 11% and 6%, respectively.

Table 5. Selected parameters of surface E, of surface F, of the sum surface, and of predicted parameters of the sum surface.

Parameter	Surface E	Surface F	Sum Surface	Predicted Sum Surface	Unit
Sq	2.06	2.9	3.57	3.56	μm
Sdq	0.23	0.53	0.58	0.58	-
Ssc	62.4	157.9	177.5	169.8	1/mm
Sds	3239	3056	3098	3132	1/mm^2
Sp/Sz	0.56	0.44	0.47	0.49	-
Sq/Sa	1.32	1.33	1.3	1.32	-
Ssk	−0.32	−0.44	−0.3	−0.39	-
Sku	4.39	4.62	3.9	4.52	-
Str	0.024	0.86	0.62	0.51	-
Sal	0.031	0.02	0.024	0.025	mm

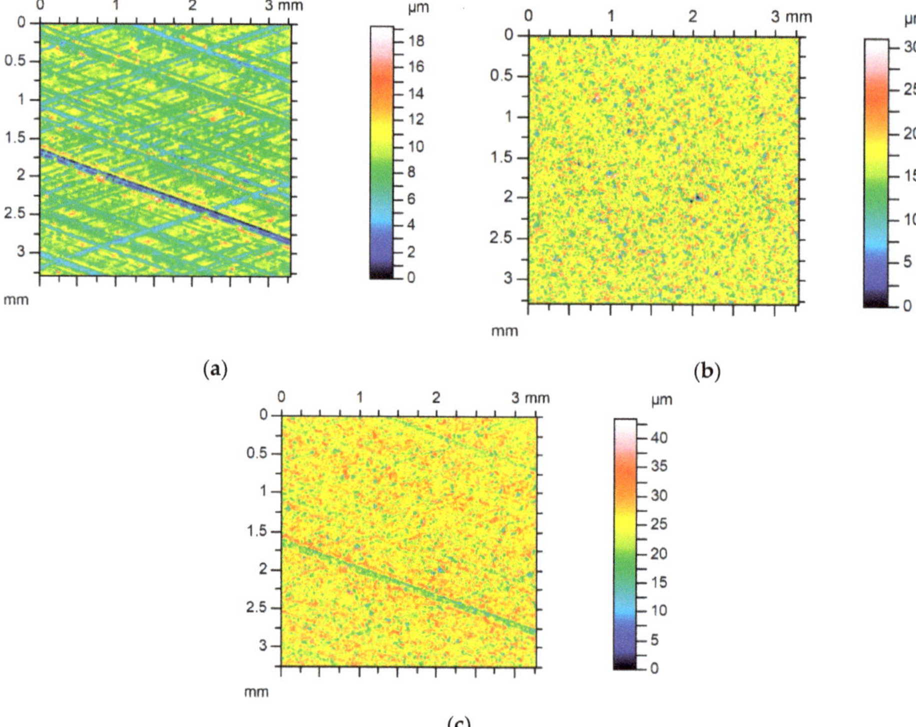

Figure 11. Contour plots of surface E (**a**), of surface F (**b**), and of equivalent sum surface (**c**).

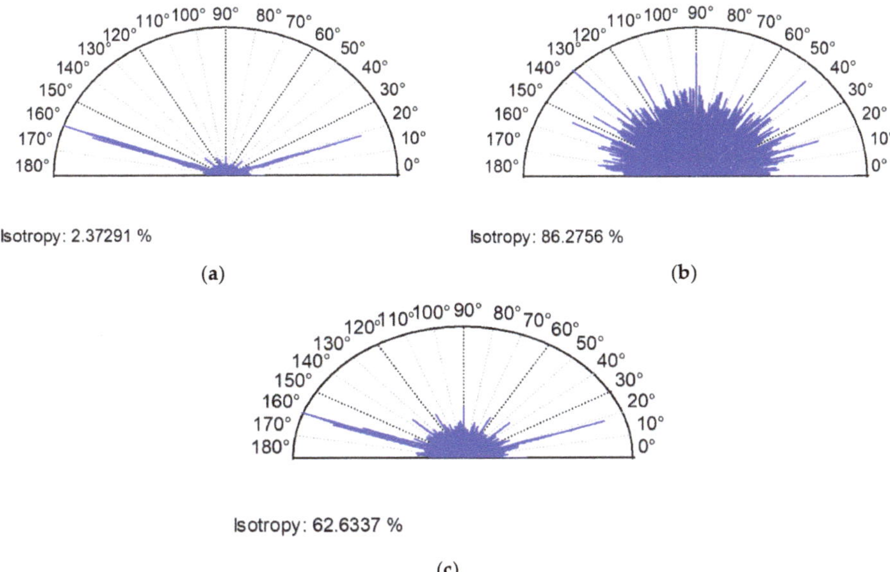

Figure 12. Directionality plots of surface E (**a**), of surface F (**b**), and of equivalent sum surface (**c**).

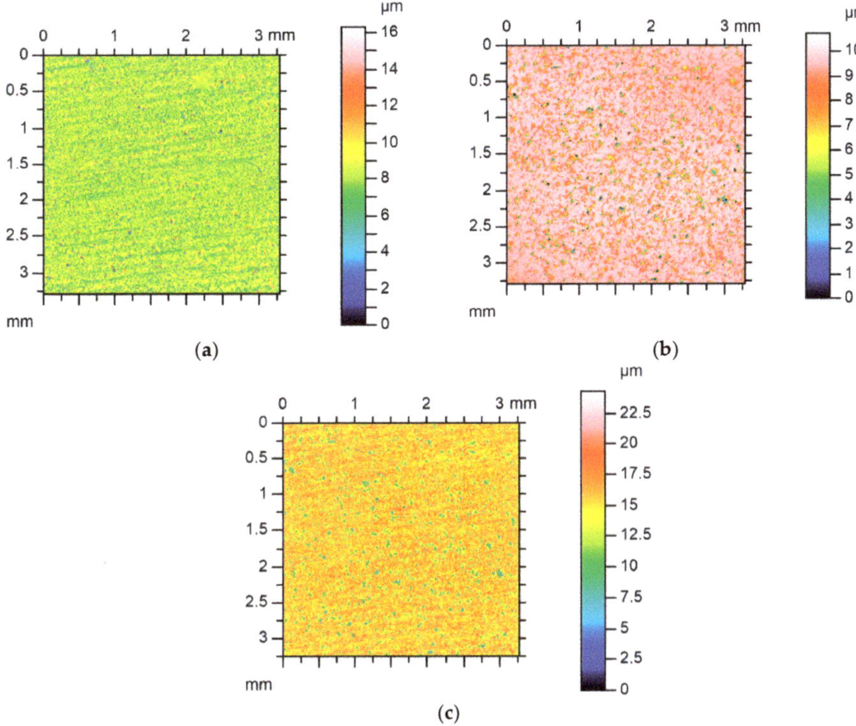

Figure 13. Contour plots of surface G (**a**), of surface H (**b**), and of equivalent sum surface (**c**).

Table 6. Selected parameters of surface G, of surface H, of the sum surface, and of predicted parameters of the sum surface.

Parameter	Surface G	Surface H	Sum Surface	Predicted Sum Surface	Unit
Sq	1.37	1.26	1.86	1.86	μm
Sdq	0.38	0.19	0.42	0.42	-
Ssc	120.3	37.2	131.4	125.9	1/mm
Sds	4137	4136	3978	4136	1/mm^2
Sp/Sz	0.48	0.143	0.36	0.32	-
Sq/Sa	1.39	1.62	1.41	1.5	-
Ssk	−0.52	−3.26	−1.23	−1.84	-
Sku	5.86	16.67	6.82	11.05	-
Str	0.72	0.85	0.79	0.78	-
Sal	0.0091	0.025	0.017	0.017	mm

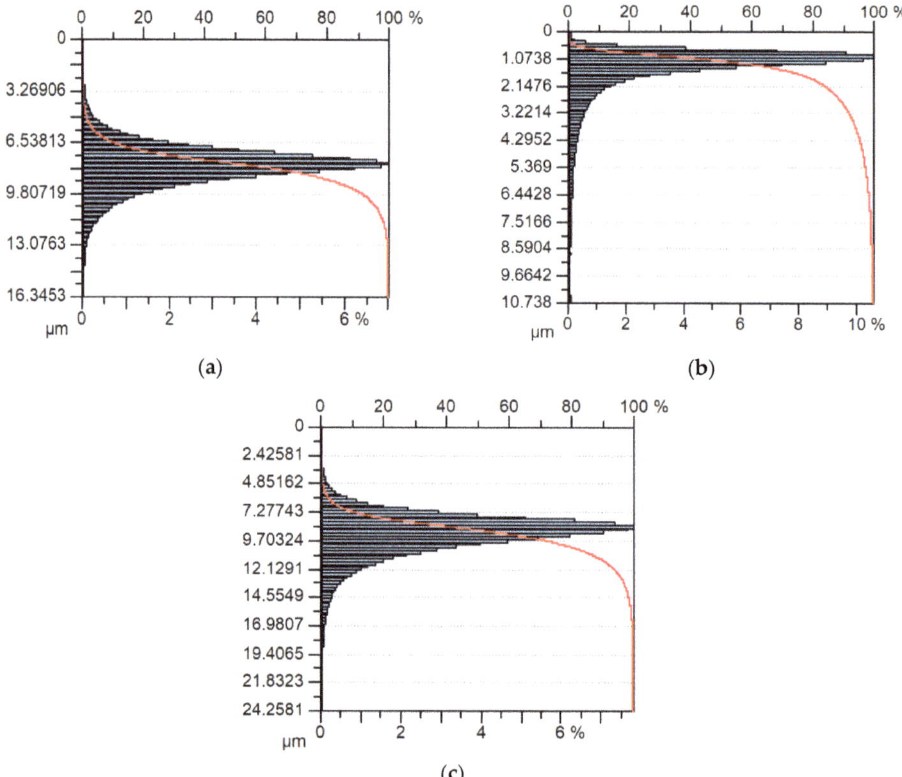

Figure 14. Material ratio curves and ordinate distributions of surface G (**a**), of surface H (**b**), and of equivalent sum surface (**c**).

The square root of m_0 spectral moment is the Pq parameter of the profile or the Sq parameter of isotropic surface topography. When anisotropic one-directional surface texture is analyzed (for example, after grinding or milling), its Sq parameter is equal (or similar) to the mean value of the Pq parameter of the profile measured across the lay (main surface wavelength). In the perpendicular direction (along the lay), the mean value of the Pq parameter is smaller than the Sq parameter of areal surface texture. According to Equation (9), the square of the Sq parameter of the sum surface should be the sum of squares of the Sq parameters of both surfaces in the contact. This assumption was confirmed. For 50 sum surfaces, the maximum error was 0.99%, and the average error was 0.31%.

The analysis of surface slope is more complicated. The square root of m_2 moment is the Pdq parameter, which is the RMS slope of the profile. Of course, this slope can be obtained using various methods (based on 2, 3, or 6 neighboring points). The method based on 2 points gave typically more correct value than the other methods [46].

For profiles of normal ordinate distribution, the rms. slope Pdq is approximately the average slope Pda magnified by 1.25. Areal (3D) slope is much larger than profile (2D) slope. For isotropic surfaces of Gaussian ordinate distribution, the following relation between the average slope of profile Pda and the areal topography slope, Sda, exists, according to Nayak [2]:

$$\text{Sda} = \frac{\pi}{2}\text{Pda} \tag{13}$$

RMS slopes of profile and surface topography are related by similar dependence. For anisotropic surface, spectral moments of second and fourth order should be equal to the square root of the product from moments obtained in two perpendicular directions [5]. However, the Sdq parameter is calculated in a different manner—see Equation (6). When the ratio of larger to smaller average slopes in 2 perpendicular directions was higher than 3.5, the areal slope was similar to the larger profile slope. For two-process surfaces, the ratio of the RMS slope to the average slope is higher than 1.25. The mentioned dependencies were obtained for random surfaces [46].

Therefore, Equation (10) relating the areal (3D) RMS slope of the equivalent sum rough surface to the RMS slopes of the individual surfaces was uncertain. However, this equation was valid in the present research. The maximum error was 0.97%, while the average error was 0.27%. The obtained finding for summation of various types of surfaces is very important.

Equation (2), describing the connection between m_4 spectral moment and the average summit curvature, was obtained for surfaces of Gaussian ordinate distribution. In addition, the m_4 spectral moment is the square of the mean curvature of the whole profile, not only peaks. For surfaces of non-Gaussian ordinate distributions (two-process textures, periodic surfaces), large errors of prediction of average summit curvature of the equivalent sum rough surface using Equation (11) are possible.

The difference between the average summit curvature of the sum surface and predicted by Equation (11) were higher than similar deviations concerning the Sq and Sdq parameters. In most cases, the relative errors of the Ssc parameter predictions were smaller than 8%. In two cases, the relative differences were near 12% and 11%. They corresponded to the contact of comparatively rough one-process isotropic surface (Sq = 1.4 μm, Str = 0.7) after vapor blasting with the also rough isotropic two-process surface after vapor blasting, followed by lapping (Sq = 2.8 μm, Str = 0.89), and also to the contact of one-process anisotropic cylinder surface (Sq = 2.1 μm, Str = 0.024) with two-process surface after vapor blasting and lapping (Sq = 2.5 μm, Str = 0.85). It seems that such errors were caused by the two-process character of one of the surfaces in contact. However, even the errors of 12% are comparatively small compared to errors of surface topography measurement [47]. The average relative error of the Ssc parameter prediction was 3.1%. However, the mean error of the Spc parameter expectation was 33.5%. It was probably caused by a segmentation.

The density of summits is related to spectral moments by Equation (3). However, this equation was developed by Nayak [2] for description of isotropic surfaces of Gaussian ordinate distribution. Its application for other surfaces (anisotropic, two-process, or deterministic) is questionable. For similar values of the Ssc and Sdq parameters, the density of summits is smaller for anisotropic surfaces compared to isotropic textures. The density of summits of two-process random surface is higher compared to the asperity density of one-process textures. However, there are opinions that not all the summits should be taken into consideration in the contact analysis of two-process surfaces [16,48]. Because of the difficulty of using Equation (3), areal summit density of the equivalent sum rough surface was predicted using the weighting average method—Equation (12). All summits were taken into consideration. The maximum error was 7.1%, while the average error was 3.3%. The correct results were obtained, because surfaces of various types were summed.

Prediction of the values of the Str parameter sometimes yield wrong results. The errors depend on the character of the analyzed surface. For example, when anisotropic surfaces were summed, the predicted Str parameter by weighting average method was underestimated. It this case, for similar amplitudes of two surfaces, the Str parameter of the equivalent sum surface was higher than a mean value of the Str parameter of two individual textures—Table 5 and Figure 10. A different situation occurred when isotropic surface was added to anisotropic surface. In this case, the anisotropic character of one surface decided about the Str parameter of the sum surface, and the predicted value of the Str parameter was overestimated (Table 6 and Figure 12). However, after addition of isotropic surfaces, the correct results of surface topography prediction were typically obtained (the errors were smaller than 10%). For all sum surfaces studied, the average error of the Str parameter prediction was 16.5%, while the maximum error was 90%.

A similar situation took place during analysis of the Sal parameter of the equivalent sum surface. When two surfaces of small values of the autocorrelation length were summed, frequently, the Sal parameter of the equivalent surface was higher than the value predicted using the weighted average method—Table 4. Similar to prediction of the Str parameter, when surface of small value of the Sal parameter was added to surface of higher correlation length, the Sal parameter of the sum surface was smaller than the predicted value—Table 5. However, the results were better than those obtained for the Str parameter prediction, the average error of the Sal parameter prediction was 11.2%, while maximum error was 28.2%.

The application of the weighting average method assured good predicted values of the emptiness coefficient Sp/Sz of the equivalent sum rough surfaces. In two cases, errors were higher than 10%— the predicted values were underestimated. This situation occurred for the contact of the one-process surface with the two-process surface. These errors were considerably reduced to values smaller than 8% when the Sq parameter of the two-process surface was replaced by the Spq parameter (plateau root mean square roughness [16,48–51]). The average value of the Sp/Sz ratio prediction was 3.8%. When one-process surfaces were summed, the relative errors were smaller than 6.5%. Good results were obtained, taking into account comparative high sensitivity of the Sp/Sz parameter on measurement errors [47].

Prediction of the Sq/Sa ratio by the weighting average method assured good results. The average error of parameter prediction was 1.6% and the maximum error was 7.1%.

The errors of predicting the values of the skewness, Ssk, of equivalent rough surfaces were comparatively large. Especially, for the contact of two surfaces characterized by negative values of the Ssk parameter, the predicted value of the Ssk parameter was smaller than that of the sum surface—Table 4. A similar situation occurred for the contact of surface with symmetric ordinate distribution and the two-process surface—Table 6 and Figure 14. Large relative errors were also found for the contact of both one-process surfaces. However, they were caused by the value of the Ssk parameter close to 0. The relative error of Ssk parameter prediction was 26.2%, while the maximum error was 62%.

The relative errors of the Sku parameter prediction corresponded to large errors of the Ssk parameter forecasting, which occurred for the contact of both two-process textures (Table 4) and for the contact of the one-process surface and two-process surface (Table 6 and Figure 14). In those cases, the predicted values of the kurtosis, Sku, were overestimated. However, for the contact of two one-process textures, the errors of the Sku parameter prediction were lower than those of the Ssk parameter expectation, because the value of Sku is typically higher than the value of Ssk. The average error of the Sku parameter prediction was 13.1%, while the maximum error was 45.2%.

The question arises, how can the results of parameters' predictions of equivalent sum surface be applied? In some cases, when the statistical models of the contact of rough surfaces are used, important parameters such as the average summit curvature, the areal density of summits, and the standard deviation of asperity heights can be predicted. In other cases, and when the deterministic models are applied, parameters' predictions can be used in initial research. It is important that surface texture, also including texture of the equivalent sum surface, can be computer generated [52,53]. Thanks to this, the cost and time of experimental investigations can be considerably reduced. Based on predicted parameters of the sum surfaces, the plasticity index can be calculated. It describes surface ability to the plastic deformation and hence to wear. The classical version of this index developed by Greenwood and Wiliamson [54] depends on the standard deviation of summit heights and the average summits curvature. According to other works, this index is proportional to surface slope [28–30].

6. Conclusions

1. In this work, we analyzed the relationships among the parameters of two contacted surfaces on parameters of equivalent surface, for which ordinates are sums of ordinates of both surfaces. Surfaces of various types (one- and two-process, isotropic and anisotropic, random or periodic) were studied.

2. Selected parameters: Sq, Sdq, Ssc, Sds, Sp/Sz, and Sq/Sa, of sum surfaces were predicted precisely when the parameters of two individual surfaces are known. The other parameters: Sal, Str, Ssk, and Sku, were anticipated with lower accuracy.

3. The parameters Sq, Sdq, and Ssc were predicted based on the changes of profile spectral moments during surfaces' summation. The results revealed that RMS height Sq and RMS slope Sdq were predicted with very high accuracy. The maximum errors were smaller than one percent, and the average deviations were about 0.3%. In most cases, the relative errors of the Ssc parameter prediction were smaller than 8%, while the average errors were near 3%.

4. The remaining parameters of equivalent sum surface were predicted on the base of parameters of both surfaces in contact, weighted by the values of the Sq parameter of these structures. The maximum errors of the summit density, Sds, and the Sq/Sa ratio predictions were near 7%, while the average errors were near 3% and 1.5%, respectively. In most cases, relative discrepancies between Sp/Sz parameter values of equivalent rough sum surfaces and predictions were smaller than 10% and the average errors were about 4%.

5. During parameters selection, it was found that the pair of parameters, Sp/Sz (the emptiness coefficient) and Sq/Sa, better described the shape of the ordinate distribution of diversified surface textures than the typically applied set, the skewness, Ssk, and the kurtosis, Sku.

Author Contributions: Conceptualization: P.P., R.R. and W.Z.; methodology, investigation, and formal analysis: P.P., R.R. and W.Z.; writing—original draft preparation: P.P., R.R., and W.Z.; writing—review and editing: P.P., R.R. and W.Z. All authors have read and agreed to the published version of the manuscript.

Funding: This research received no external funding.

Conflicts of Interest: The authors declare no conflict of interest.

Nomenclature

Sa (Ra)	arithmetical mean height
Sal	fastest decay autocorrelation length
Sds	areal density of summits
Sdq	root mean square gradient
Sku (Rku)	kurtosis
Sp (Rp)	maximum peak height
Spc	arithmetical mean peak curvature
Spd	density of peaks
Spq	plateau root mean square roughness
Sq (Rq)	root mean square height
Ssc	average summit curvature

Ssk (Rsk)   skewness
Str   texture aspect ratio
Sv (Rv)   maximum valley depth
Sz (Rz)   maximum height

References

1. Whitehouse, D.J. *Handbook of Surface Metrology*; Institute of Physics Publishing: Bristol, UK; Philadelphia, PA, USA, 1994.
2. Nayak, P.R. Random Process Model of Rough Surfaces. *J. Lubr. Technol.* **1971**, *93*, 398–407. [CrossRef]
3. Longuet-Higgins, M.S. Statistical properties of an isotropic random surface. *Philos. Trans. R. Soc. London. Ser. A Math. Phys. Sci.* **1957**, *250*, 157–174. [CrossRef]
4. Longuet-Higgins, M.S. The statistical analysis of a random, moving surface. *Philos. Trans. R. Soc. London. Ser. A Math. Phys. Sci.* **1957**, *249*, 321–387. [CrossRef]
5. Sayles, R.S.; Thomas, T.R. Measurements of the Statistical Microgeometry of Engineering Surfaces. *J. Lubr. Technol.* **1979**, *101*, 409–417. [CrossRef]
6. Greenwood, J.A.; Tripp, J.H. The Contact of Two Nominally Flat Rough Surfaces. *Proc. Inst. Mech. Eng.* **1970**, *185*, 625–633. [CrossRef]
7. O'Callaghan, M.; Cameron, M. Static contact under load between nominally flat surfaces in which deformation is purely elastic. *Wear* **1976**, *36*, 79–97. [CrossRef]
8. Francis, H. Application of spherical indentation mechanics to reversible and irreversible contact between rough surfaces. *Wear* **1977**, *45*, 221–269. [CrossRef]
9. Greenwood, J.A.; Tripp, J.H. The Elastic Contact of Rough Spheres. *J. Appl. Mech.* **1967**, *34*, 153–159. [CrossRef]
10. Chang, W.R.; Etsion, I.; Bogy, D.B. An Elastic-Plastic Model for the Contact of Rough Surfaces. *J. Tribol.* **1987**, *109*, 257–263. [CrossRef]
11. Zhao, Y.; Maietta, D.M.; Chang, L. An Asperity Microcontact Model Incorporating the Transition from Elastic Deformation to Fully Plastic Flow. *J. Tribol.* **1999**, *122*, 86–93. [CrossRef]
12. Kogut, L.; Etsion, I. A Finite Element Based Elastic-Plastic Model for the Contact of Rough Surfaces. *Tribol. Trans.* **2003**, *46*, 383–390. [CrossRef]
13. Jackson, R.L.; Green, I. A Finite Element Study of Elasto-Plastic Hemispherical Contact against a Rigid Flat. *J. Tribol.* **2005**, *127*, 343–354. [CrossRef]
14. Pawlus, P.; Zelasko, W. The importance of sampling interval for rough contact mechanics. *Wear* **2012**, *276*, 121–129. [CrossRef]
15. Pawar, G.; Pawlus, P.; Etsion, I.; Raeymaekers, B. The Effect of Determining Topography Parameters on Analyzing Elastic Contact between Isotropic Rough Surfaces. *J. Tribol.* **2012**, *135*, 011401. [CrossRef]
16. Pawlus, P.; Zelasko, W.; Dzierwa, A. The Effect of Isotropic One-Process and Two-Process Surface Textures on the Contact of Flat Surfaces. *Materials* **2019**, *12*, 4092. [CrossRef]
17. Sahlin, F.; Almqvist, A.; Larsson, R.; Glavatskih, S. Rough surface flow factors in full film lubrication based on a homogenization technique. *Tribol. Int.* **2007**, *40*, 1025–1034. [CrossRef]
18. Sahlin, F.; Larsson, R.; Amqvist, A.; Lugt, P.M.; Marklund, P. A mixed lubrication model incorporating measured surface topography. Part I: Theory of flow factors. *Proc. Inst. Mech. Eng. Part J J. Eng. Tribol.* **2010**, *224*, 335–349. [CrossRef]
19. Stanley, H.M.; Kato, T. An FFT-Based Method for Rough Surface Contact. *J. Tribol.* **1997**, *119*, 481–485. [CrossRef]
20. Yu, N.; Polycarpou, A.A. Combining and Contacting of Two Rough Surfaces with Asymmetric Distribution of Asperity Heights. *J. Tribol.* **2004**, *126*, 225–232. [CrossRef]
21. Bush, A.W.; Gibson, R.D.; Keogh, G.P. Strongly Anisotropic Rough Surfaces. *J. Lubr. Technol.* **1979**, *101*, 15–20. [CrossRef]
22. Yu, N.; Polycarpou, A.A. Contact of Rough Surfaces with Asymmetric Distribution of Asperity Heights. *J. Tribol.* **2001**, *124*, 367–376. [CrossRef]
23. McCool, J.I. Non-Gaussian effects in microcontact. *Int. J. Mach. Tools Manuf.* **1992**, *32*, 115–123. [CrossRef]

24. Jeng, Y.-R. Impact of Plateaued Surfaces on Tribological Performance. *Tribol. Trans.* **1996**, *39*, 354–361. [CrossRef]

25. Dzierwa, A.; Pawlus, P.; Zelasko, W. Comparison of tribological behaviors of one-process and two-process steel surfaces in ball-on-disc tests. *Proc. Inst. Mech. Eng. Part J J. Eng. Tribol.* **2014**, *228*, 1195–1210. [CrossRef]

26. Grabon, W.; Pawlus, P.; Sep, J. Tribological characteristics of one-process and two-process cylinder liner honed surfaces under reciprocating sliding conditions. *Tribol. Int.* **2010**, *43*, 1882–1892. [CrossRef]

27. Pawlus, P. The errors of surface topography measurement using stylus instrument. *Metrol. Meas. Syst.* **2002**, *9*, 273–289.

28. Rosén, B.-G.; Ohlsson, R.; Thomas, T. Wear of cylinder bore microtopography. *Wear* **1996**, *198*, 271–279. [CrossRef]

29. Mikić, B. Thermal contact conductance; theoretical considerations. *Int. J. Heat Mass Transf.* **1974**, *17*, 205–214. [CrossRef]

30. Greenwood, J.A. Contact Pressure Fluctuations. *Proc. Inst. Mech. Eng. Part J J. Eng. Tribol.* **1996**, *210*, 281–284. [CrossRef]

31. Thomas, T. Characterization of surface roughness. *Precis. Eng.* **1981**, *3*, 97–104. [CrossRef]

32. Zielecki, W.; Pawlus, P.; Perłowski, R.; Dzierwa, A. Surface topography effect on strength of lap adhesive joints after mechanical pre-treatment. *Arch. Civ. Mech. Eng.* **2013**, *13*, 175–185. [CrossRef]

33. Leach, R. *Characterisation of Areal Surface Texture*; Springer: Berlin/Heidelberg, Germany, 2013.

34. Stout, K.J.; Blunt, L.; Sullivan, P.J.; Dong, W.P.; Mainsah, E.; Luo, N.; Mathia, T.; Zahouani, H. *The Development of Methods for the Characterisation of Roughness in Three Dimensions*; Publication EUR 15178 EN; Commission of the European Communities: Brussels, Belgium, 1993.

35. Senin, N.; Blunt, L.A.; Leach, R.K.; Pini, S. Morphologic segmentation algorithms for extracting individual surface features from areal surface topography maps. *Surf. Topogr. Metrol. Prop.* **2013**, *1*, 015005. [CrossRef]

36. Krolczyk, G.; Krolczyk, J.; Maruda, R.; Legutko, S.; Tomaszewski, M. Metrological changes in surface morphology of high-strength steels in manufacturing processes. *Measurement* **2016**, *88*, 176–185. [CrossRef]

37. Świrad, S.; Wydrzynski, D.; Nieslony, P.; Królczyk, G. Influence of hydrostatic burnishing strategy on the surface topography of martensitic steel. *Measurement* **2019**, *138*, 590–601. [CrossRef]

38. Mezari, R.; Pereira, R.; Sousa, F.J.P.; Weingaertner, W.L.; Fredel, M. Wear mechanism and morphologic space in ceramic honing process. *Wear* **2016**, *362*, 33–38. [CrossRef]

39. Gadelmawla, E.; Koura, M.; Maksoud, T.; Elewa, I.; Soliman, H. Roughness parameters. *J. Mater. Process. Technol.* **2002**, *123*, 133–145. [CrossRef]

40. Pawlus, P.; Reizer, R.; Wieczorowski, M. Characterization of the shape of height distribution of two-process profile. *Measurement* **2020**, *153*, 107387. [CrossRef]

41. Peters, J.; Vanherck, P.; Sastrodinoto, M. Assessment of surface topology analysis techniques. *CIRP Ann. Manuf. Technol.* **1979**, *28*, 539–554.

42. Nowicki, B. Multiparameter representation of surface roughness. *Wear* **1985**, *102*, 161–176. [CrossRef]

43. Anderberg, C.; Pawlus, P.; Rosén, B.-G.; Thomas, T. Alternative descriptions of roughness for cylinder liner production. *J. Mater. Process. Technol.* **2009**, *209*, 1936–1942. [CrossRef]

44. Klauer, K.; Eifler, M.; Seewig, J.; Kirsch, B.; Aurich, J. Application of function-oriented roughness parameters using confocal microscopy. *Eng. Sci. Technol. Int. J.* **2018**, *21*, 302–313. [CrossRef]

45. Dimkovski, Z.; Cabanettes, F.; Löfgren, H.; Anderberg, C.; Ohlsson, R.; Rosén, B.-G. Optimization of cylinder liner surface finish by slide honing. *Proc. Inst. Mech. Eng. Part B J. Eng. Manuf.* **2012**, *226*, 575–584. [CrossRef]

46. Pawlus, P. An analysis of slope of surface topography. *Metrol. Meas. Syst.* **2005**, *12*, 295–313.

47. Pawlus, P.; Wieczorowski, M.; Mathia, T. *The Errors of Stylus Methods in Surface Topography Measurements*; Zapol: Szczecin, Poland, 2014.

48. Pawlus, P.; Zelasko, W.; Reizer, R.; Wieczorowski, M. Calculation of plasticity index of two-process surfaces. *Proc. Inst. Mech. Eng. Part J J. Eng. Tribol.* **2016**, *231*, 572–582. [CrossRef]

49. Malburg, M.C.; Raja, J.; Whitehouse, D.J. Characterization of Surface Texture Generated by Plateau Honing Process. *CIRP Ann.* **1993**, *42*, 637–639. [CrossRef]

50. Sannareddy, H.; Raja, J.; Chen, K. Characterization of surface texture generated by multi-process manufacture. In Proceedings of the 7th International Conference on Metrology and Properties of Engineering Surfaces, Gothenburg, Sweden, 2–4 April 1997; pp. 111–117.

51. Whitehouse, D.J. Assessment of Surface Finish Profiles Produced by Multi-Process Manufacture. *Proc. Inst. Mech. Eng. Part B Manag. Eng. Manuf.* **1985**, *199*, 263–270. [CrossRef]
52. Pérez-Ràfols, F.; Almqvist, A. Generating randomly rough surfaces with given height probability distribution and power spectrum. *Tribol. Int.* **2019**, *131*, 591–604. [CrossRef]
53. Pawlus, P.; Reizer, R.; Wieczorowski, M. A review of methods of random surface topography modeling. *Tribol. Int.* **2020**, *152*, 106530. [CrossRef]
54. Greenwood, J.A.; Williamson, J.B.P. Contact of nominally flat surfaces. *Proc. R. Soc. London. Ser. A Math. Phys. Sci.* **1966**, *295*, 300–319. [CrossRef]

Publisher's Note: MDPI stays neutral with regard to jurisdictional claims in published maps and institutional affiliations.

materials

Article

Influence of Variable Radius of Cutting Head Trajectory on Quality of Cutting Kerf in the Abrasive Water Jet Process for Soda–Lime Glass

Marzena Sutowska [1], Wojciech Kapłonek [1], Danil Yurievich Pimenov [2], Munish Kumar Gupta [2,3], Mozammel Mia [4,*] and Shubham Sharma [5]

[1] Department of Production Engineering, Faculty of Mechanical Engineering, Koszalin University of Technology, Racławicka 15-17, 75-620 Koszalin, Poland; marzena.sutowska@tu.koszalin.pl (M.S.); wojciech.kaplonek@tu.koszalin.pl (W.K.)

[2] Department of Automated Mechanical Engineering, South Ural State University, Lenin Prosp. 76, 454080 Chelyabinsk, Russia; danil_u@rambler.ru (D.Y.P.); munishguptanit@gmail.com (M.K.G.)

[3] Key Laboratory of High Efficiency and Clean Mechanical Manufacture, Ministry of Education, School of Mechanical Engineering, Shandong University, Jinan 250061, China

[4] Department of Mechanical Engineering, Imperial College London, Exhibition Rd., South Kensington, London SW7 2AZ, UK

[5] Department of Mechanical Engineering, IK Gujral Punjab Technical University, Jalandhar-Kapurthala Road, Kapurthala 144603, Punjab, India; shubham543sharma@gmail.com or shubhamsharmacsirclri@gmail.com

* Correspondence: m.mia19@imperial.ac.uk

Received: 2 September 2020; Accepted: 21 September 2020; Published: 25 September 2020

Abstract: The main innovation of this article is the determination of the impact of curvature of a shape cut out in a brittle material using an abrasive water jet (AWJ) process as an important factor of the machined surfaces. The curvature of a shape, resulting from the size of the radius of the cutting head trajectory, is one of the key requirements necessary for ensuring the required surface quality of materials shaped by the abrasive water jet process, but very few studies have been carried out in this regard. An important goal of the experimental studies carried out here and presented in this work was to determine its influence on the quality of the inner and outer surfaces of the cutting kerf. This goal was accomplished by cutting the shape of a spiral in soda–lime glass. For such a shape, the effect of radius of the trajectory of the cutting head on selected parameters of the surface texture of the inner surface of the cutting kerf (IS) and the outer surface of the cutting kerf (OS) was studied. The obtained results of the experimental studies confirmed that the effect of the curvature of the cut shape is important from the point of view of the efficiency of the glass-based brittle material-cutting process using AWJ. Analyses of the surface textures of the areas located in the upper part of the inner and outer surfaces separated by the use of AWJ machining showed that the OS surfaces are characterized by worse technological quality compared with IS surfaces. Differences in the total height of surface irregularities (given by St amplitude parameter), determined on the basis of the obtained results of the measurements of both surfaces of the cutting kerf, were as follows: $\Delta St_{r\,=\,50} = 0.6$ μm; $\Delta St_{r\,=\,35} = 1$ μm; $\Delta St_{r\,=\,15} = 1.3$ μm. The analysis of values measured in areas located in the more sensitive zone of influence of the AWJ outflow proved that the total height of irregularities (St) of the OS was higher. Differences in the total heights of irregularities for inner and outer surfaces of the cutting kerf were as follows: $\Delta St_{r\,=\,50} = 2.1$ μm; $\Delta St_{r\,=\,35} = 3$ μm; $\Delta St_{r\,=\,15} = 14.1$ μm, respectively. The maximum difference in the total heights of irregularities (St), existing between the surfaces considered in a special case (radius 15 mm), was almost 20%, which should be a sufficient condition for planning cutting operations, so as to ensure the workpiece is shaped mainly by internal surfaces.

Keywords: abrasive water jet machining; cutting kerf; soda–lime glass; radius of the cutting head trajectory; quality

---

1. Introduction

Abrasive water jet (AWJ) machining is a nontraditional advanced hybrid method used for shaping a wide range of modern and conventional materials, which can replace other more traditional machining techniques, an aspect which was presented by Liu et al. [1]. As stated by Hashish [2], AWJ machining consists of shaping materials using a highly concentrated water jet doped with abrasive grains. The most commonly used abrasive is garnet [3]. In addition to numerous advantages, such as those mentioned by Krajcarz [4] (including no thermal distortion, high flexibility, high machining versatility, small machining force, and the absence of a heat-affected zone), this method has some limitations, as reported Wang et al. [5]. While cutting materials using AWJ, two phenomena may be observed which are important in the context of the quality of this process. The first is the deviation of the AWJ in the opposite direction to the movement of cutting head, as described by Hlaváč et al. [6]. This means that, during movement of the cutting head along the workpiece, the outflow of the jet occurs with a delay in relation to its site of entry into the material. This phenomenon is presented graphically in Figure 1a.

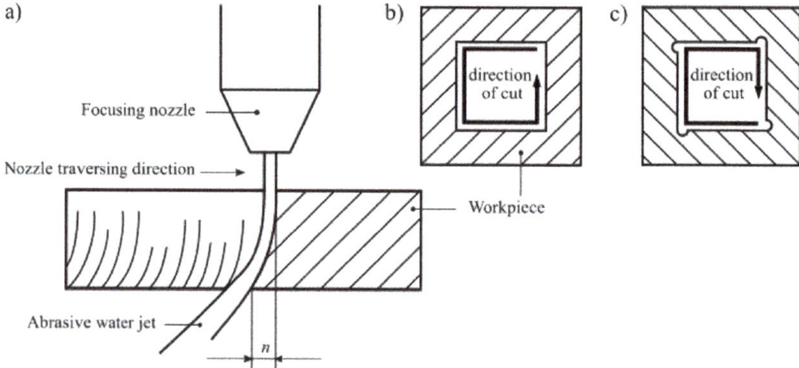

Figure 1. Phenomena occurring during material cutting with high-pressure abrasive water jet (AWJ), important in the context of the quality of this process—deviation of the AWJ: (**a**) graphical interpretation of the phenomenon; (**b**) workpiece top view; (**c**) workpiece bottom view with errors in the shape of corners caused by the jet lag.

The distance between the entry and outflow points of the jet was defined by Hashish [7] as the jet lag. The shape that the jet adopts during the cutting process of materials is expressed on the side surfaces of the workpieces usually in the form of parallel striation whose intensity increases in the area of the lower part of the cutting zone [8]. During rectilinear cuts, the AWJ can move at high speed on the surface of the workpiece, as its deviation does not affect the accuracy of the process. However, in the case of the corners cut, an excessive speed of AWJ movement may cause the creation of shape errors, as reported by Chen et al. [9]. This situation is presented graphically in Figure 1b. For this reason, when cutting objects with complex shapes, the traverse speed must be properly selected to eliminate such kinds of technological errors. It should be emphasized that the phenomena discussed here can be practically eliminated by implementing appropriate algorithms in the controllers of devices intended for cutting materials employing AWJ. The AWJ systems currently used in industry enable carrying out the precise cutting process via the appropriate selection of its parameters, which means the above-discussed machining method may be treated as a good alternative to other methods of cutting materials.

The shape of the AWJ also significantly affects the outline of the cutting kerf, which changes in the area of the cutting zone, as shown by Wang et al. [10]. These differences in the width of the resulting kerf were defined by Hlaváč et al. [11] as a taper. This can be positive or negative, depending on whether the width of the cutting kerf reaches a larger size in the area of the entry or outflow jet from the material. In industrial practice, the width of the cutting kerf is usually smaller in the lower part of cutting zone, taking the shape of the letter V, as presented in Figure 2a.

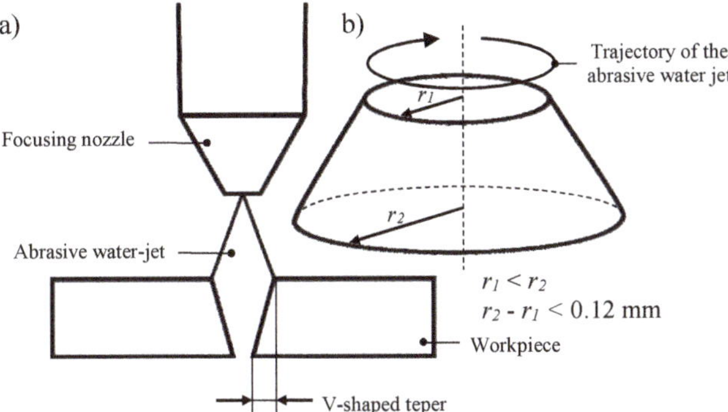

Figure 2. Phenomena occurring during material cutting with a high-pressure AWJ, important in context of the quality of this process: (**a**) formation of a V-shaped taper; (**b**) convergence effect.

The formation of a V-shaped taper during the AWJ cutting process, as shown by Kechagias et al. [12], illusively increases the radius of cutting curve while the cutting head moves along a circular path. This phenomenon is characterized by the differentiation of the diameter of the cut object in the area of AWJ entry and outflow points. This situation is presented graphically in Figure 2b. Higher differences between both diameters denote closer proximity of the value of the traverse speed to the maximum (critical) speed.

The above-described results of the effect of the shape of the jet on the outline of the cutting kerf are presented in Figure 2 on an exaggerated scale. In fact, with properly selected machining parameters, they are almost imperceptible. The results of the experimental studies carried out by Wu et al. [13] clearly indicate that the value of the V-shaped taper of the cutting surface, even in the case of not optimally selected process parameters, does not exceed 0.1 mm. Under normal conditions, this value is lower, which means that the method of machining materials being discussed may be considered as relatively accurate. Given the above, AWJ, which is a universal technological tool, provides itself with a wider range of machining applications, enabling the shaping of many types of materials, such as metals and its alloys [14–19], ceramic, glass [20], composites [21,22], flammable materials, leather, and natural stone-based materials [23], as well as regenerating the cutting ability of grinding tools [24]. Nevertheless, there is still a need for further research related to this technology in order to increase the efficiency of material machining. The obvious method is to generate a water jet with very high pressures reaching up to 700 MPa. An alternative solution to that presented above is eliminating the use of high pressures through the use of a pulsating flow of water jet. These were described in the literature by Hloch et al. [25] and Lehocka et al. [26].

The generation of vibrations, i.e., the vibration spectrum, is one of the accompanying features which indirectly characterizes the abrasive water jet cutting process, its quality, and its expected result. Therefore, it is possible and advisable to apply the analysis of accompanying physical phenomena and the determination of their limit values in the controlling phase of cutting processes through a simple control of at least one quantity which characterizes the process. These were presented in works by Hreha et al. [27].

The quality of surfaces cut with AWJ is assessed similarly to that in the case of other machining methods (for example, Matuszewski et al. [28] discussed the influence of the configuration of the geometric structure of the surface to be treated on the course of the wear process of friction pairs machined using various machining methods, and Bustillo et al. [29] solved the problem of predicting it with the help of artificial intelligence). First of all, indicators of dimensional and shape accuracy (as deviations from the nominal size) are determined. In addition, the shape of surface texture of the cut is also characterized. An important role in the assessment of the technological quality of the cutting process is played by the quality indicators of the shape of the cutting kerf, as well as the quality indicators of the topography and microgeometry of the surface cut. These were described in the literature by Borkowski et al. [30], as well as in the VDI 2906-5:1994 standard developed by the Association of German Engineers [31].

A wide range of two- and three-dimensional (2D and 3D) parameters can be used for a quantitative assessment of cut surface roughness. The most commonly used parameters for assessment of the surface texture in 2D, as presented in works by Löschner et al. [32], Zagórski et al. [8], Hreha et al. [33], and Klichova and Klich [34] and defined in the ISO 4287:1997 standard [35], are the amplitude parameters Ra, Rq, and Rt. The 3D parameters can be divided into the following groups: amplitude, area and volume, spatial, hybrid, and functional; they are included in the ISO 25178-2:2012 standard [36] and the EUR 15178 EN report [37]. The significant influence of cutting parameters (traverse speed, water pressure, abrasive feed rate) was observed in the case of amplitude parameters (especially Sa, Sq, St), as noted by Aich et al. [20] and Borkowski et al. [30]. The shape of the topography of the surfaces cut by AWJ also indicates the presence of waviness, as presented Sutowska [38], whose intensity increases with the distance from the jet's input zone to the material. For a quantitative assessment of this phenomenon, a set of waviness parameters (especially Wt, WSm), defined in the ISO 4287:1997 standard [35], can be used. The general characteristics of above-mentioned roughness and waviness parameters are given in Table 1.

Table 1. Characteristics of roughness and waviness parameters calculated during experimental studies.

Group of Parameters	Symbol	Unit	Description
	Ra	μm	Arithmetical mean deviation of the roughness profile
Roughness (profile)	Rq	μm	Root-mean-square deviation of the roughness profile
	Rt	μm	Total height of the profile on the evaluation length
	Sa	μm	Arithmetic mean deviation of the surface
Amplitude (surface)	Sq	μm	Root-mean-square (RMS) deviation of the surface
	St	μm	Total height of the surface
Spatial (surface)	Sal	mm	Fastest decay autocorrelation length
Areal (surface) [1]	Sds	pks/mm^2	Density of summits
Feature (surface)	SPc	pks/mm^2	Arithmetic mean peak curvature
Waviness (profile)	Wt	μm	Maximum height of waviness profile
	WSm	mm	Mean width of profile elements, within a sampling length

[1] As reported by Blateyron [39], the Sds parameter corresponds to Spd in ISO 25178-2, but the discrimination method is different.

The basic parameters of the process that characterize the cutting of materials by the AWJ are as follows: water jet pressure p; traverse speed v; abrasive feed rate m_a; water jet orifice diameter d_o; focusing tube diameter d_f; standoff distance l [40]. Knowledge of the effect of machining conditions on the quality of the obtained cuts ensures high-quality cutting.

The curvature of the shape cut out, resulting from the determined radius of the trajectory of the cutting head, is one of the essential conditions of the quality of the cut using AWJ. A good example which facilitates the analysis of the impact of this parameter on the quality of the cutting kerf is provided by a spiral. Its shape allows smoothly (continuously) assessing the changing curvature of the outline being cut out by an AWJ. This is advantageous and allows the assessment of the inner and outer surfaces (IS and OS), whose varying shapes may suggest what should be product and what should be waste, due to the level of dimensional deviations and the quality of the surface texture.

2. Methodology of Experimental Studies

2.1. Main Goal

The main goal was to investigate the influence of the radius of the cutting head trajectory on the surface quality of areas adjacent to the upper and lower cutting zone. Such differentiation refers to the shaping of the inner (IS) and outer (OS) surfaces of the cutting kerf. The innovativeness of the described studies is based on the fact that, although the curvature of the shape cut out by an AWJ (resulting from a fixed radius of the trajectory of cutting head) is one of the important conditions for the quality of materials cutting with AWJ, no studies have yet been carried out to determine its impact on the quality of the cuts obtained. In the sections below, details related to conditions in which the experimental studies were carried out, as well as results of the experiments, along with their analyses, are given.

2.2. Characteristics of the Samples

A planned cycle of experimental studies was carried out on soda–lime glass. The selection of this type of material was made deliberately as it is easier to expose the differences in the shape of the IS and the OS. This is due to the greater susceptibility of this brittle material to hydro-jetting erosion, as well as its limitations resulting from machining by the use of conventional machining techniques.

Soda–lime glass is an amorphous body, created as a result of supercooling molten raw minerals and other inorganic substances without the crystallization of ingredients. This relatively inexpensive and widely available glass is a base material for most types of glass (colorless, colored, and patterned). Its chemical composition and selected physical properties are given in Table 2, whereas a general view of the spiral used in the experimental studies is presented in Figure 3. The mass of the spiral was 30.74 g.

Table 2. General characteristics of soda–lime glass.

Chemical Composition							
SiO_2, %	Na_2O, %	CaO, %	MgO, %	Al_2O_3, %	K_2O, %	SO_2, %	Fe_2O_3, %
72.60	13.90	8.40	3.90	1.10	0.60	0.20	0.11

Strength					
Flexural			Compressive		
Annealed, MPa	Heat-strengthened, MPa	Toughened, MPa	Annealed, MPa	Heat-strengthened, MPa	Toughened, MPa
41	83	165	19	39	77

Physical Properties					
Density, kg/m^3 [1]	Mohs hardness	Modulus of elasticity, GPa	Shear modulus, GPa	Poisson's ratio	Coeff. of thermal stress, MPa/°C
2500	5–6	72	30	0.23	0.62
Thermal conductivity, W/m·K	Specific heat, kJ/kg·K	Coeff. of linear expansion, °C	Index of refraction [2]	Softening point, °C	Annealing point, °C
0.937	0.88	8.3×10^{-6}	1.5	715	548

Max. working temperature, °C		Thermal shock Δ, °C	
Not Toughened	Toughened	Not Toughened	Toughened
110	150	50	118

[1] At 18 °C; [2] in visible wavelength range $\lambda = 380–780$ nm.

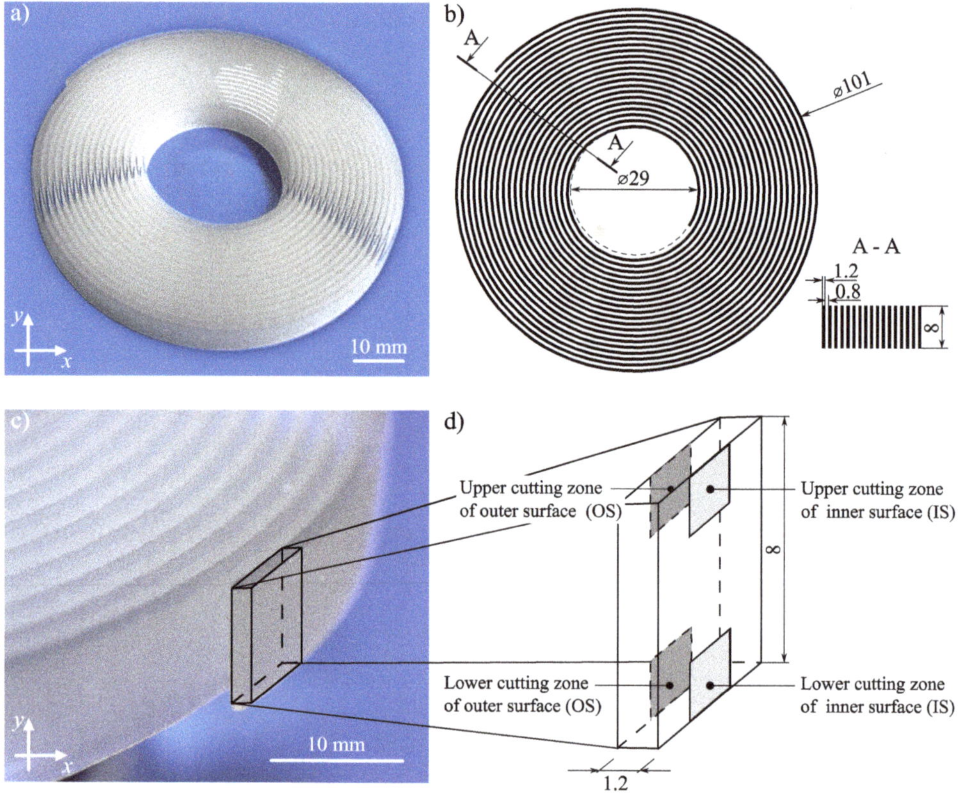

Figure 3. The soda–lime glass spiral used in the experimental studies: (**a**) general view of the spiral; (**b**) basic geometrical dimensions; (**c,d**) a section of the spiral with the upper/lower inner and outer surfaces (IS and OS) marked.

2.3. Conditions and Course of the AWJ Process

The cutting process was carried out using a JetMachining® Center type 55100 cutting machine produced by OMAX (Kent, WA, USA). This precision AWJ system, whose general view is presented in Figure 4, is widely used in many modern applications, as presented by Zhao and Guo [41], Linke et al. [42], and Saurabh et al. [43].

The Jet Machining® Center was equipped with a P4055V plunger pump (Figure 4c), driven by a 30 kW electric motor. Water drawn from the waterworks system was used to power the pump, which produces=d a water jet with the following parameters: p_{max} = 385 MPa; Q_{max} = 4.9 dm3/min. From the pump, water under high pressure was supplied by special tubes to the MAXJET®5 cutting head operating as part of the Tilt-A-Jet® mechanism (Figure 4b). The cutting head body used a set of nozzles (Figure 4d). The center was equipped with an automatic abrasive feeder, used in connection with a small hopper mounted on a movable arm (axis y). This allowed the abrasive particles to be delivered to the cutting head continuously without interrupting the JetMachining® Center. An important element of the AWJ system was a cutting table whose working range was 3200 × 1600 mm. Inside the table frame, a catcher tank was placed equipped with support slats on which the material to be cut was attached. Dedicated computer-aided design (CAD) software, namely, OMAX Layout, was used to prepare drawings of the outline of the cut object and to specify the cutting path, whereas OMAX Make software, installed on a controller equipped with a color monitor and an industrial keyboard, was used to start the cutting process. The controller was used to control the cutting head during the process, the intensity of the abrasive feed, and the water jet pressure. Additionally, the controller screen allows for continuous monitoring of the course of the process.

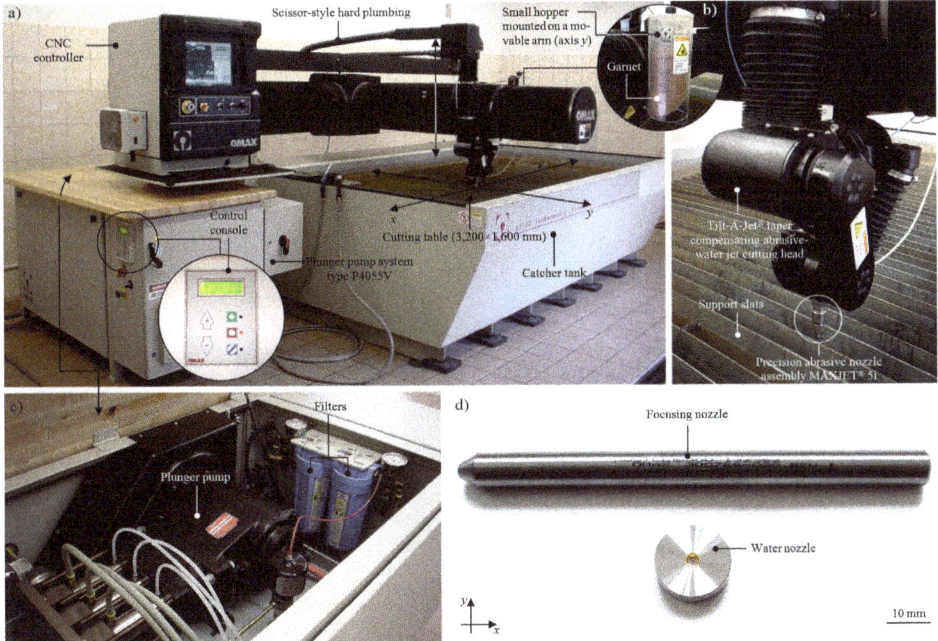

Figure 4. Precision AWJ system used during experimental studies—JetMachining® Center type 55100 (OMAX Corp., Kent, WA, USA): (**a**) general view of the center with main components; (**b**) close-up on the Tilt-A-Jet® cutting head (OMAX Corp., Kent, WA, USA) and abrasive nozzle assembly MAXJET® 5; (**c**) close-up on the plunger pump and filters; (**d**) two type of nozzles used in the cutting head.

After the cutting process started, the geometry of the spiral and the cutting path of the sample was generated using the OMAX Layout software (OMAX Corp., Kent, WA, USA). Then, in the OMAX Make software, the cutting path created in OMAX Layout was opened. In specifying the cutting process conditions, the type and thickness of the processed material (plate glass soda lime silica, $g = 8$ mm) and the cut quality (Quality 1) were chosen. On this basis, the OMAX Make software calculated the value of the traverse speed $v = 427$ mm/min and the abrasive feed rate $m_a = 0.363$ kg/min (Garnet 80 mesh size). A water nozzle with a diameter of $d_o = 0.38$ mm and focusing tube diameter $d_f = 0.76$ mm was mounted in the cutting head body. Next, the material prepared for the cutting process in the form of a soda–lime glass square plate ($150 \times 150 \times 8$ mm) was placed upon the cutting table of the AWJ system. Due to the small size of the spiral, the plates were directly mounted on a special base (a waterjet brick), fixed to the cutting table. The last procedure before the cut process was to determine the working length of the AWJ, namely a distance between the outflow from the focusing nozzle and the material surface of $l = 1.5$ mm. After the cutting process was finished, the samples were thoroughly washed and dried using compressed air.

2.4. Characteristics of Measurement Systems and Course of Measurement Process

After finishing the AWJ cut process, the geometrical shape of both surfaces of the cutting kerf was measured using one of the advanced optical methods involving optical profilometry. In carrying out measurements, a Talysurf CLI 2000 multisensory optical profilometer (Taylor-Hobson, Leicester, Great Britain) was used. This instrument was extensively described by Kapłonek et al. [44], while its selected applications were presented by Yuan et al. [45], Nadolny et al. [46], and Fan [47]. Considering the relatively large variation in the heights of irregularities occurring on shaped surfaces due to the AWJ cutting process, measurement of their surface texture was carried out using a type LK-031 laser sensor (Keyence Corp., Osaka, Japan) [41] installed in the measuring head of the Talysurf CLI 2000.

The microstructure measurements of the surfaces cut by the AWJ were carried out using a Quanta 200 Mark II scanning electron microscope (SEM) (FEI Company, Hillsboro, OR, USA). This high0resolution environmental microscope (ESEM) is intended for the observation of samples at a magnification of 30 × to ~1,000,000 × in high or low vacuum and variable pressure conditions. During measurements, the low-vacuum mode (LowVac) for observing the surface was used. The SEM micrographs were acquired for a surface area of 2.133 × 1.966 mm with a magnification of 140 × at an accelerating voltage of Ua = 15–20 kV. Due to its broad observation and measurement capabilities, this microscope is used in many areas of modern science and technology. Examples of its use were given by Chen et al. [48], Nadolny et al. [49], and Kapłonek and Ungureanu [50].

The general characteristics of the observation/measurement systems used in the experimental studies are presented in Table 3.

Table 3. Characteristics of observation/measurement systems used in experimental studies.

No.	Instrument Type	Model	Producer	Configuration and Features
1.	Multisensory optical profilometer	CLI2000	Taylor-Hobson (Leicester, Great Britain)	Components: laser triangulation sensor LK-031 (Keyence Corp., Osaka, Japan) Features (sensor): scanning frequency: 2000 Hz, measuring range: 10 mm, resolution: 1 μm (vertical), 30 μm (lateral), measuring slope: 40°, speed: 30 mm/s Features (instrument): measuring capacity: 200 × 200 × 200 mm, axis traverse length: 200 mm, axis resolution: 0.5 μm, dimensions: 800 × 800 × 800 mm, measuring speed: 0.5, 1, 5, 10, 15, and 30 mm/s, positioning speed: 30 mm/s
				Software: Talyscan CLI 2000 2.6.1+ TalyMap Silver 4.1.2 (Digital Surf, Besançon, France)
2.	SEM microscope	Quanta 200 Mark II	FEI Company, (Hillsboro, OR, USA)	Components: detectors: SEI (Everhart-Thornley SED, low-vacuum SED (LFD), gaseous SED (GSED)), BEI (solid-state (BSED), gaseous SED (GSED)), specimen stage: eucentric goniometer stage (four-axis motorized) Features: magnification range: 30 × to ~1,000,000 ×, vacuum pressure in the specimen chamber: < 0.0006 Pa (HVM), 10–130 Pa (LVM), accelerating voltage: 0.2–30 kV, resolution (using HVM): 3.0 nm at 30 kV SEI, 4.0 nm at 30 kV BSE, 10 nm at 3 kV SEI, (using LVM): 3.0 nm at 30 kV SEI, 4.0 nm at 30 kV BSE, < 12 nm at 3 kV SEI
				Software: dedicated FEI software

3. Results and Discussion

The analysis of the results of the experimental studies was divided into the following phases:

1. A study of the influence of the curvature of the cut out shape on the IS and OS surface texture shaped using an AWJ, carried out on the basis of the calculated values of roughness and waviness parameters characteristic for this type of machining [24] measured by the Talysurf CLI 2000 multisensory optical profilometer (Taylor-Hobson, Leicester, Great Britain) (Section 3.1).

2. A study of the surface texture of the OS (Section 3.2), as well as the IS (Section 3.3), shaped with an AWJ, using surface microtopographies measured with an optical method using the Talysurf CLI 2000 multisensory optical profilometer (Taylor-Hobson, Leicester, Great Britain) and

SEM-micrographs obtained by a Quanta 200 Mark II SEM microscope (FEI Company, Hillsboro, OR, USA).

3.1. Study of the Influence of the Curvature of the Cut Out Shape on the IS and OS Surface Texture

A graphical interpretation of the results of the experimental studies obtained after data processing by TalyMap Silver 4.1.2 software, depicting the relationship between the curvature of the cut out shape and the mean square deviation of the surface Sq, measured for the IS and OS, is shown in Figure 5a.

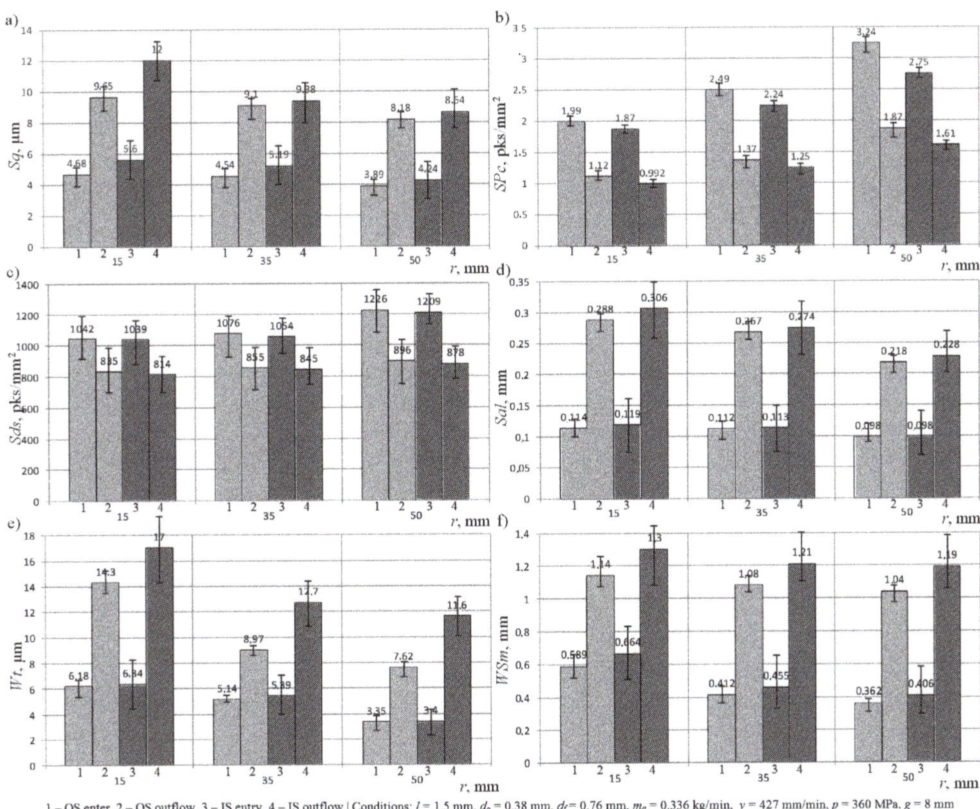

1 – OS enter, 2 – OS outflow, 3 – IS entry, 4 – IS outflow | Conditions: l = 1.5 mm, d_o = 0.38 mm, d_f = 0.76 mm, m_a = 0.336 kg/min, v = 427 mm/min, p = 360 MPa, g = 8 mm

Figure 5. Collection of selected results of experimental studies in graphical form presenting calculated values of selected surface texture parameters using TalyMap Silver 4.1.2 software: (**a**) Sq; (**b**) SPc; (**c**) Sds; (**d**) Sal; (**e**) Wt; (**f**) WSm.

Analyzing the graphs, it may be observed that an increase in the radius of the cutting head trajectory from 15 mm to 50 mm reduced the value of the Sq amplitude (surface) parameter by an average of 16% in the areas located in the zone where the AWJ entered the material, and by an average of 26% in the areas located in the lower part of cutting zone. At the same time, it can be seen that, for each of the considered radii of the cutting head trajectory, the Sq amplitude (surface) parameter took on higher values in the case of measurements of the OS. For example, when the radius r = 15 mm, the mean square deviation of the surface, measured at the lower part of outer surface of the cutting kerf, exceeded by 2.35 μm the value of the Sq amplitude (surface) parameter, determined in the same area on the inner surface.

The results of experimental studies on the influence of the changes in the radius of the trajectory of the cutting head (r) on the arithmetic mean peak curvature (SPc) presented in graphical form (Figure 5b)

indicate the existence of significant interdependencies between the considered parameters. The change in the curvature of the cut out shape from a value of $r = 15$ mm to $r = 50$ mm caused the *SPc* parameter to increase by 1.25 pks/mm^2 (for IS entry). Additionally, analyzing the above graph, it may be seen that the considered surface texture parameter assumed the highest values when measurements were taken for the upper IS (*SPc* = 3.24 pks/mm^2). However, the smallest value of the arithmetic mean peak curvature was observed when measuring the areas located on the lower OS (*SPc* = 0.992 pks/mm^2). In addition, the obtained results indicate that, changing the radius of the cutting head trajectory, the *SPc* feature (surface) parameter adopted higher values in the areas located on the IS.

In Figure 5c, the results obtained for the *Sds* areal (surface) parameter are presented. They indicate that increasing the radius of the cutting head trajectory during the cutting process increased the density of the summits of surface irregularities for the OS and IS. At the same time, when analyzing the results of measurements of the *Sds* areal (surface) parameter, carried out for both surfaces of the cutting kerf, it may be noted that a relatively higher value of density of summits occurred on its inner surface.

The results of the influence of the radius of the curvature on the fastest decay autocorrelation length (*Sal*) are presented in Figure 5d. Analyzing the graph, it may be concluded that the increase in the radius of the trajectory of the cutting head reduced the value of this parameter for both surfaces of the cutting kerf. The presented results of the measurements also indicate that the analyzed spatial (surface) parameter reached the maximum value when measuring the lower part of the OS (*Sal* = 0.306 mm), whereas its minimum value was observed when the AWJ entered the material (IS) (*Sal* = 0.098 mm).

Figure 5e–f present the results of the experimental studies on the influence of the curvature of the shape cut out with the AWJ on the maximum height of waviness profile (*Wt*) and mean width of profile elements, within a sampling length (*WSm*). Analyzing the graph, in the case of the first waviness parameter (Figure 5e), it may be observed that the change in the radius trajectory of the cutting head from $r = 50$ mm to $r = 15$ mm influenced an increase in the value of the *Wt* parameter up to 17 µm (OS outflow), which qualified the obtained surface as an inferior quality class. At the same time, the presented results of measurements for the *Wt* waviness parameter prove that the surface texture of the OS was characterized by poorer technological quality compared with the IS. Analyzing the obtained experimental results for the mean width of profile elements within a sampling length, it may be observed that the reduction of the radius of the curvature of the shape cut out by the AWJ from a value of $r = 50$ mm to $r = 15$ mm caused an increase in the average interval of the waviness profile. In addition, the obtained results indicate that, in this situation, when the changing of the cutting head trajectory radius occurred, the *WSm* waviness parameter assumed higher values in the areas located on the outer surface of the cutting kerf.

Analyzing the obtained results of the experimental studies on the influence of the curvature of the cut out shape on selected parameters of the surface texture of the IS and OS, it may be stated that a change in the radius of the trajectory of the cutting head by over a factor of three from $r = 50$ mm to $r = 15$ mm caused an increase in the value of surface texture parameters such as Sq, Sal, Wt, and *WSm*. The most important of these parameters (*Sq* and *Wt*) undergo unfavorable changes, reaching even more than 40% of the value. The principle analyzed here also functions analogically in the opposite direction. Thus, it becomes obvious that the increase in the value of the curvature of the shape cut out by an AWJ leads to an improvement in the quality of the cutting process. In addition, it should also be noted that there is a positive influence of such a change on the increase in the arithmetic mean peak curvature (*SPc*) and the density of summits (*Sds*).

Clear qualitative differences may be observed between the surfaces of material cut by an AWJ. The values of amplitude (surface) and waviness parameters, measured on the inner and outer surfaces of cutting kerf, indicate that the surface texture of the former was characterized by a much higher quality. The differences occurring reached a level of about 20%.

On the basis of the obtained experimental results, a mathematical model, which allows for predicting values of the *Rq* roughness (profile) parameter using information about the location of the considered area (A_1 and A_2) depending on cutting kerf and radius values of the trajectory of the cutting head r, was developed. This model is given by the following dependence:

$$Rq = (-0.01158 - 0.03645\gamma_1 - 0.03975\gamma_2)r + 4.65903 + 2.04845\gamma_1 + 6.12725\gamma_2. \qquad (1)$$

The estimation error of the developed mathematical model, in relation to the experimental results, was relatively low and did not exceed 5%. The values of coefficients γ_1 and γ_2 are given in Table 4.

Table 4. Method for designation of measured area on kerf cutting.

Area location	γ_1	γ_2
IS enter	1	0
IS outflow	1	1
OS enter	0	0
OS outflow	0	1

3.2. The Shaping Quality of Inner Surface of Cutting Kerf

In practice, the influence of the curvature of the cut out shape on the quality of the inner surface of the cutting kerf corresponds to the formation of cylindrical surfaces by the AWJ (the cutting of cylinders). In order to carry out their qualitative assessment, an analysis of the surface texture of areas adjacent to the upper IS (smooth and regular) and to the lower IS (more ridged and jagged) should be carried out.

The surface texture of the inner surface of the cutting kerf, being the result of the influence of the machining conditions ($r = 50$ mm) when the AWJ enters the material, is presented in Figure 6a. The height analysis of the surface texture allows one to state that the total height of the surface irregularities was $St = 38.6$ µm. Approximately 70% of the analyzed cut out surface occupied areas were located at heights from 15 µm to 25 µm (yellow and green). Irregularities of a height exceeding 25 µm (red) occurred only on about 30% of the inner surface of the cutting kerf. Similar conclusions can be obtained by carrying out an analysis of the amplitude of the averaged waviness profile. As shown in the lower part of Figure 6a, the areas located near the average line of the waviness profile played an important role in its formation.

A graphical form of irregularities of the inner surface of the cutting kerf, located in the upper part of the IS, is shown in Figure 6b. Analyzing its spatial form, it may be seen that the total height of the cut out surface had a value of over 44 µm. The shaping of the surface topography presented in this image also points to the fact that the percentage of areas whose height exceeded 30 µm (red) was about 40% of the total area. In addition, by analyzing the shape of the averaged waviness profile of the IS, a locally increasing waviness may be noted whose maximum increase above the mean line was over 5 µm.

In Figure 6c, the surface texture of the upper IS, created as a result of the influence of the machining conditions ($r = 15$ mm) on the material being cut, is presented. Analyzing its geometrical shape, a large difference in the height of the irregularities ($St = 46.6$ µm) measured between the highest peak and the lowest valley can be noted. In addition, the maximum increase in the average waviness profile above the average line exceeded 8 µm, while the minimum decrease thereof below the average value was about 7 µm.

In Figure 6, the shape of the inner surface of the cutting kerf, occurring in a situation when the AWJ entered the material, is presented. Upon analyzing the obtained changes in the experimental study's results caused by the modification of the curvature of the cut shape from $r = 50$ mm to $r = 15$ mm, visible differences in the height of the surface texture could be observed. These differences testify to the decreasing effectiveness of the influence of the AWJ on the cutting surface.

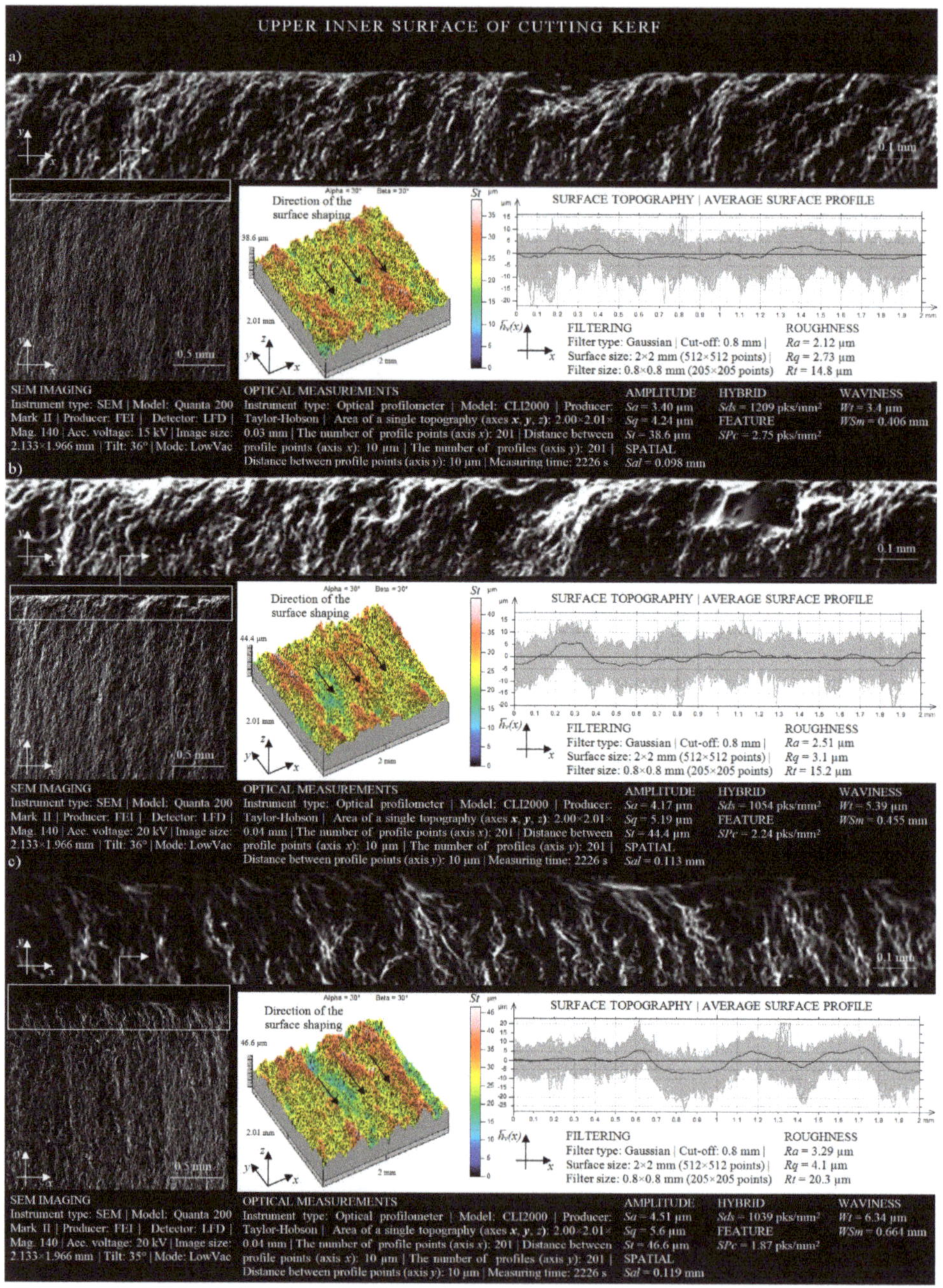

Figure 6. Collection of selected results of experimental studies obtained for the upper IS in a situation when the AWJ entered the sample surface (l = 1.5 mm, d_o = 0.38 mm, d_f = 0.76 mm, m_a = 0.336 kg/min, v = 427 mm/min, p = 360 MPa, g = 8 mm): (**a**) r = 50 mm; (**b**) r = 35 mm; (**c**) r = 15 mm.

A graphical form of the irregularities of the inner surface of the cutting kerf, located in the lower part of the IS, is shown in Figure 7a. Analyzing its irregularities, resulting from the effect of machining conditions for r = 50 mm on the material being cut, it was possible to extract the profiles of slightly curved machining marks (striation). These marks were created as a result of the loss of a part of the energy produced by abrasive particles during the penetration of the AWJ into the material. The difference in the height occurring between the highest peak and lowest valley, in the case of the considered surface irregularities, was 62.3 µm. Geometric shaping of the inner surface of the cutting kerf, located in the lower part of the inner surface, indicates the fact that approximately 50% of its area was characterized by surface irregularities in levels of more than 30 µm. At the same time, significant local deviations in the amplitude of the averaged waviness profile of the analyzed surface can be observed. These deviations oscillated between the limit values of −12 µm and 8 µm.

In Figure 7b, the surface texture of the inner surface of the cutting kerf adjacent to the lower kerf, created as a result of the influence of the machining conditions ($r = 35$ mm) on the material being cut, is presented. Analyzing the surface texture of the cut out surface, it may be concluded that the total height of its irregularities was 65.6 µm. In addition, in about 80% of the area considered, the height of irregularities exceeded 30 µm. Moreover, the maximum increase in the average waviness profile above the average line was approximately 11 µm, while the minimum decrease thereof below the average value was about 9 µm.

Geometric shaping of the lower IS, resulting from the machining conditions ($r = 15$ mm) on the material being cut, is presented in Figure 7c. Analyzing the obtained microstructure of the cut out surface, one may observe an increased intensity in the machining marks appearing on it, which directly influenced the increase of both roughness and waviness of the area considered. In consequence, the height difference between the highest peak and the lowest valley was over 70 µm.

The amplitude of the averaged waviness profile of the inner surface of the cutting kerf covered, within its range (from 11 µm to −9 µm), up to six limited increases, uniformly distributed over the entire analyzed surface. Analyzing the results of the experimental studies on the influence of the trajectory of the cutting head on the roughness and waviness of the areas located in the lower surface of the cutting kerf (Figure 7), a clear influence of the curvature of the shape cut out by the AWJ on the shaping of surface texture of the analyzed surface can be observed.

Changing the radius from $r = 50$ mm to $r = 15$ mm, while maintaining the remaining parameters of the cutting process at a constant level, caused an increase in the height of irregularities (given by the *St* amplitude parameter) of 8.5 µm. A relatively high number of local increases of the amplitude of the averaged waviness profile is also observed. A deterioration in the quality of the IS, resulting from the reduction of the radius of the cutting head trajectory *r*, is also evidenced by the clear differences in its shaping, observed on SEM micrographs of the microstructure of the surface under consideration (Figure 7).

From an analysis of the height of the irregularities of the inner surface of the cutting kerf, which occurred in the zone of AWJ entry and outflow, it may be observed that the areas located in the upper part of the cutting zone were characterized by a higher quality. This fact is also confirmed by surface topographies, the amplitudes of averaged waviness profiles, and the SEM micrographs of areas adjacent to two opposite kerfs of the cut. Differences in the height of irregularities (given by *St* amplitude parameter), determined on the basis of the obtained results of surface texture measurements of areas located in the upper and lower part of the cutting zone, were as follows: 23.7 µm for $r = 50$ mm; 21.2 µm for $r = 35$ mm; 24.2 µm for $r = 15$ mm.

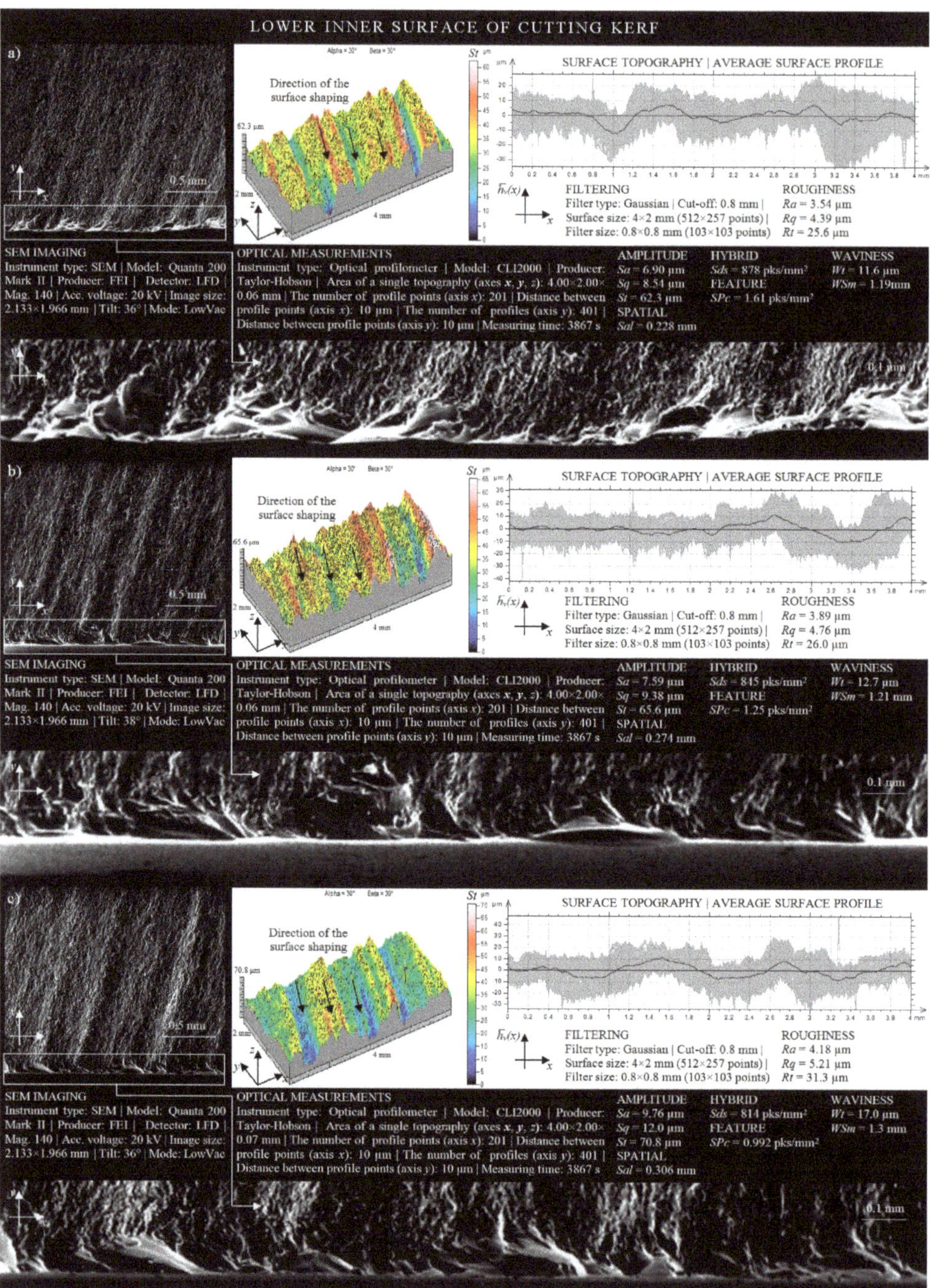

Figure 7. Collection of selected results of experimental studies obtained for the lower IS in a situation where the AWJ entered the sample surface (l = 1.5 mm, d_o = 0.38 mm, d_f = 0.76 mm, m_a = 0.336 kg/min, v = 427 mm/min, p = 360 MPa, g = 8 mm): (**a**) r = 50 mm; (**b**) r = 35 mm; (**c**) r = 15 mm.

3.3. Shaping Quality of Outer Surface of Cutting Kerf

The influence of the radius of the cutting head trajectory on the quality of the OS concerns the shaping of cylindrical surfaces (holes) which may have a different diameter. In their qualitative assessment, as in the case of the cylindrical surfaces presented in Section 3.2, the analysis of the geometrical shaping of areas adjacent to two opposite surfaces of the cutting kerf is particularly important.

The surface texture of the outer surface of the cutting kerf, formed by the AWJ, is presented in Figure 8a. Analyzing its geometrical shaping, resulting from the influence of the cutting process parameters with the material being processed ($r = 50$ mm), it may be observed that the highest peak of irregularities (given by St amplitude parameter) had a value of 39.2 μm. Subsequent SEM micrographs and surface topographies (Figure 8b) also present the geometric shaping of the upper area of the OS.

When analyzing the obtained surface topography, one can notice that the change in the radius of the cutting head trajectory from $r = 50$ mm to $r = 35$ mm caused an increase in the maximum height of the peaks appearing on it up to the value of 45.4 μm. In addition, surface irregularities with a height exceeding 30 μm occurred on 50% of the analyzed area of the OS. Considering the averaged surface waviness profile presented in Figure 8b, it may be stated that its maximum increase above the mean line reached a height of 4 μm, while the minimum decrease was of −4 μm.

The texture of the surface located in the upper area of the outer surface of the cutting kerf, which was the result of the machining conditions ($r = 15$ mm) on the material being cut, is presented in Figure 8c. Analyzing its spatial form, one may observe a large variation in the heights of irregularities occurring between the highest peak and the deepest valley. The maximum height of the surface microstructure (given by St amplitude parameter) was approximately 47.8 μm. In addition, the geometrical shaping of the upper area of the outer surface of the cutting kerf indicates that approximately 80% of its irregularities were characterized by a height whose value was more than 30 μm. However, the height of amplitude of the average waviness profile presented in Figure 8c oscillated between two extreme values: −5 μm and 5 μm.

Analyzing the results of experimental studies on the influence of the curvature of the shape cut out by AWJ on the surface texture of areas located in the upper part of the outer surface of the cutting kerf (Figure 8), it may be concluded that the change in the trajectory of the cutting head from $r = 15$ mm to $r = 50$ mm caused a reduction in surface irregularities in the area considered by 8.6 μm. At the same time, the values of the average waviness profile $\overline{h_v}(x)$ decreased by approximately 2 μm. Differentiations in roughness and waviness of the OS, caused by the increase in the radius of the trajectory of the cutting head during the cutting process, were also revealed by the SEM micrographs of the microstructure presented in Figure 8.

In Figure 9, the geometrical shaping of the lower areas of the OS is presented. The surface texture, formed as a result of the effect of the influence of the curvature of the cut shape $r = 50$ mm on the material being processed is shown in Figure 9a. Analyzing the shaping of the area located in the more sensitive zone of AWJ outflow, it is possible to discern the outlines of machining marks (striation) that arose as a result of the AWJ losing some of its energy during removal of the upper surface of the material being cut. The total height of the peaks obtained in this way in the microstructure of the OS (given by St amplitude parameter) was 64.3 μm. About 20% of the analyzed area had surface irregularities, the height of which exceeded 45 μm. The amplitude of the averaged waviness profile of the outer surface of the cutting kerf $\overline{h_v}(x)$ oscillated between two extreme values of −6 μm and 5 μm in relation to the mean line.

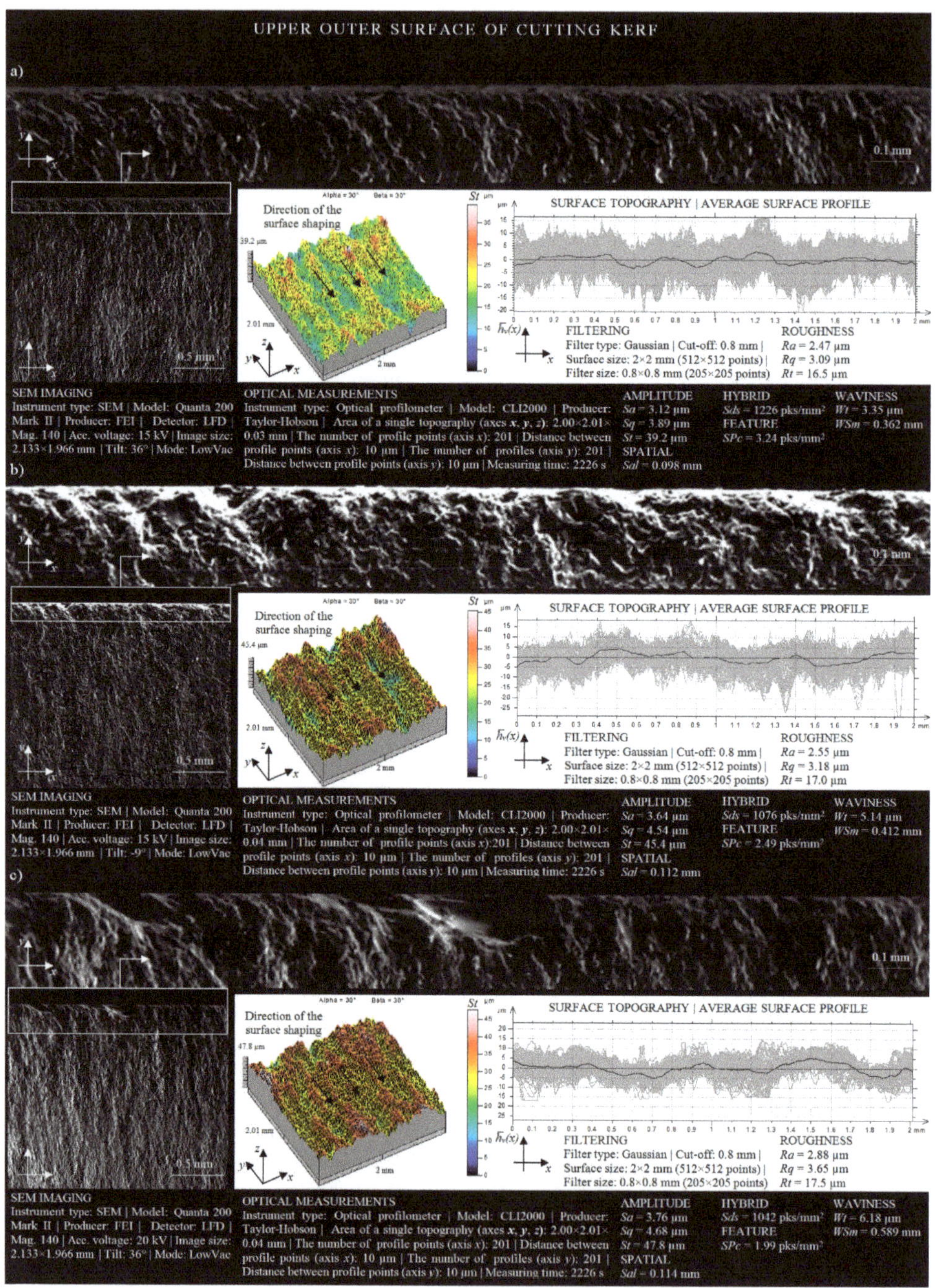

Figure 8. Collection of selected results of experimental studies obtained for the upper OS in a situation where the AWJ entered the sample surface ($l = 1.5$ mm, $d_o = 0.38$ mm, $d_f = 0.76$ mm, $m_a = 0.336$ kg/min, $v = 427$ mm/min, $p = 360$ MPa, $g = 8$ mm): (**a**) $r = 50$ mm; (**b**) $r = 35$ mm; (**c**) $r = 15$ mm.

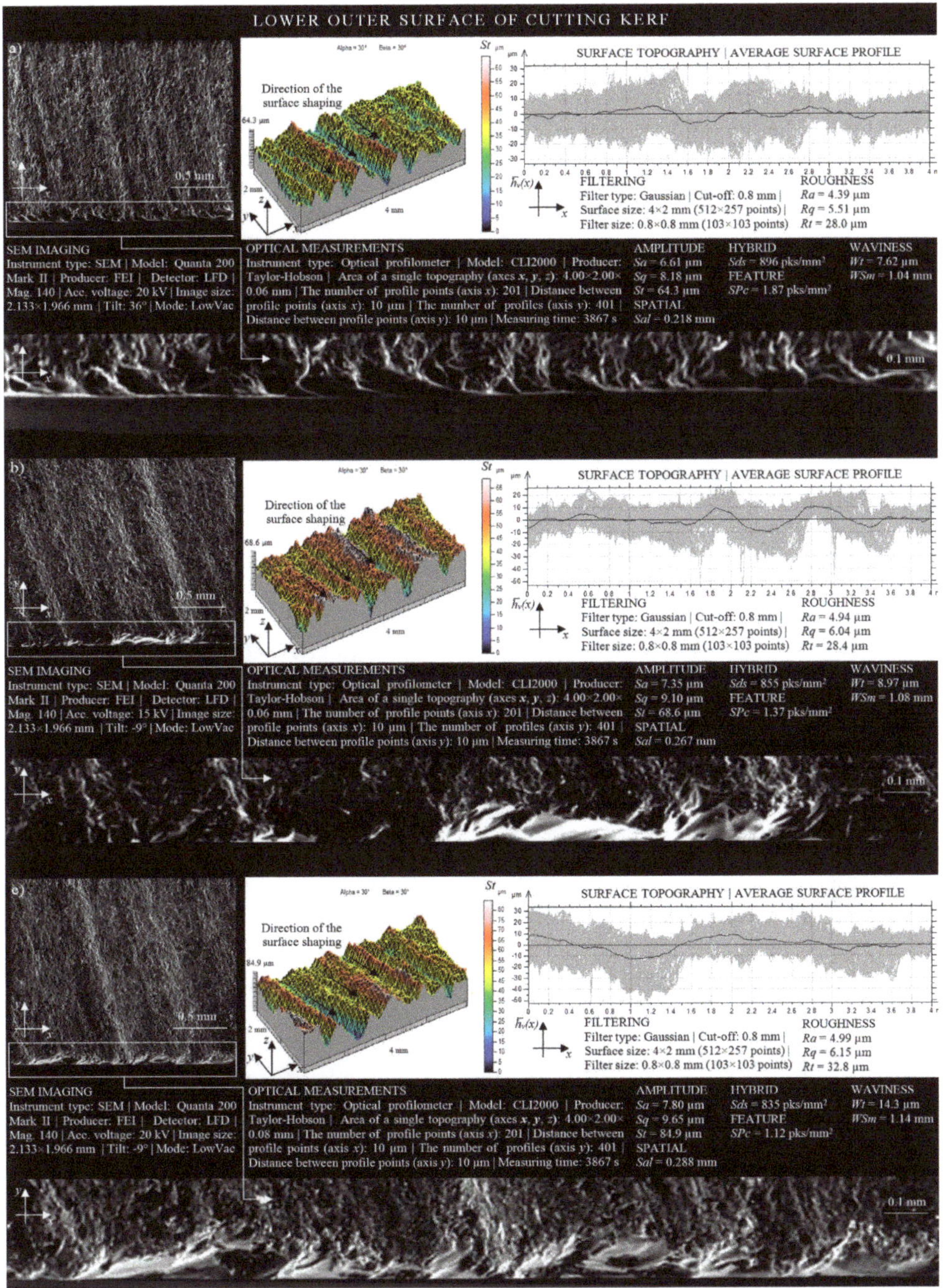

Figure 9. Collection of selected results of experimental studies obtained for the lower OS in a situation where the AWJ entered the sample surface (l = 1.5 mm, d_o = 0.38 mm, d_f = 0.76 mm, m_a = 0.336 kg/min, v = 427 mm/min, p = 360 MPa, g = 8 mm): (**a**) r = 50 mm; (**b**) r = 35 mm; (**c**) r = 15 mm.

The texture of the surface located in the lower area of the outer surface of the cutting kerf is presented in a graphical form in Figure 9b. Analyzing its shaping, it may be observed that the change in the radius of the curvature of the cut out shape from $r = 50$ mm to $r = 35$ mm caused an increase in the highest peak of the irregularities to a height of over 68 µm. However, when analyzing the surface topography of the considered area occurring in the lower part of the outer surface of the cutting kerf, it may be concluded that the percentage of surface irregularities whose height exceeded 45 µm (red) was about 50% of the average waviness profile $\overline{h_v}(x)$, which covered the range from −10 µm to 10 µm.

The surface texture of the area of the OS, adjacent to the lower surface of the cutting kerf, is presented in Figure 9c. Considering its geometrical shaping, resulting from the influence of the process parameter $r = 15$ mm on the material being processed, one may clearly observe the machining marks created as a result of the interaction of the AWJ on the material being processed. In addition, the highest peak of irregularities of the analyzed cut surface occurred at a height (given by St amplitude parameter) of 84.9 µm at the averaged waviness profile contained in ranges from about −12 µm to 10 µm relative to the mean line.

Analyzing the shaping of the areas occurring in the lower part of the OS, as presented in Figure 9, resulting from the curvature of the cut out shape, it may be concluded that the change in the radius of the cutting head trajectory from $r = 15$ mm to $r = 50$ mm reduced surface irregularities of 20.5 µm. It also changed the average waviness profile $\overline{h_v}(x)$, whose amplitude decreased with the increase in the curvature of the shape cut out by the AWJ. Differences in roughness and waviness occurring in the areas located in the lower part of the outer surface of the cutting kerf, caused by the change in the radius of the cutting head trajectory, are clearly observed in the SEM-micrographs. The increase in the curvature of the cut out shape (r) during the AWJ cutting process improved the quality of the outer cut surface by limiting the intensity of the occurrence of striation on it.

Comparing the height of the areas of the external surface of the cutting kerf located in the zone of AWJ entry and outflow from the material, it may be concluded that the areas adjacent to the upper cutting kerfs were characterized by a relatively better technological quality. The sample SEM micrographs, surface topographies, and averaged waviness profiles presented in Figures 8 and 9 provide evidence for the information given above. Differences in the height of the irregularities (given by St amplitude parameter), determined on the basis of the surface texture measurements carried out for areas located in the upper and lower part of the cutting zone, were as follows: 23.7 µm for $r = 50$ mm; 23.2 µm for $r = 35$ mm; 37.1 µm for $r = 15$ mm.

Analyzing the shaping of the surface texture of areas located in the upper part of the inner and outer surface of the material cut out by the AWJ (Figures 6 and 8), it may be observed that the outer surfaces were characterized by much worse technological quality in comparison with inner surfaces. Differences in the height of irregularities (given by St amplitude parameter), determined on the basis of the obtained results of measurements of both surfaces of the cutting kerf, were as follows: $\Delta_{Str = 50} = 0.6$ µm; $\Delta_{Str = 35} = 1$ µm; $\Delta_{Str = 15} = 1.3$ µm.

When comparing the obtained values of the amplitude (surface) parameter St, measured in areas located in the more sensitive outflow zone of the AWJ for both cut out surfaces of the lower IS and lower OS (Figures 7 and 9), it may be clearly observed once again that the height of irregularities of the outer surface were higher. Differences in the heights of irregularities measured for the inner and outer surfaces of the cutting kerf were as follows: $\Delta_{Str = 50} = 2.1$ µm; $\Delta_{Str = 35} = 3$ µm; $\Delta_{Str = 15} = 14.1$ µm.

4. Conclusions

The curvature of a shape cut out by an AWJ, resulting from the size of the radius of the cutting head trajectory, is one of the key requirements necessary for ensuring the required surface quality of materials shaped by AWJ machining. An important goal of the experimental studies carried out and presented in this work was to determine its influence on the quality of the inner and outer surfaces of the cutting kerf. This goal was accomplished by cutting the shape of a spiral. In such a form, the sample was used in experiments during which the influence of the radius of the cutting head trajectory on selected surface texture parameters of the inner and outer surface of the cutting kerf located in the zone of AWJ entry and outflow was analyzed. The obtained results of measurements and analyses allowed one to draw the following detailed conclusions:

1. The obtained results of the experimental studies confirmed that the effect of the curvature of the cut shape is important from the point of view of the efficiency of the glass-based brittle materials cutting process using the AWJ. On the basis of the obtained experimental results, it may be concluded that the feed speed should be limited when $r < 35$ mm.
2. The determined mathematical model in Equation (1), which describes the influence of the cutting head trajectory on the surface quality of the soda–lime glass, describes with approximately 95% accuracy the relationships occurring between the trajectory radius of the cutting head and the amplitude Sq (surface) parameter. This means that the model was adequate for the experimental data and could be successfully used to predict the quality of both surfaces of the cutting kerf. The model ran properly in the range of radius variation $r = 15$–50 mm.
3. The determined values of the surface texture parameters for the inner and outer surfaces of cutting kerf (Figures 5–9) clearly indicate that these surfaces were characterized by worse technological quality than cylindrical surfaces. The maximum difference in the total height of the surface (St) existing between the considered surfaces (for $r = 15$ mm) was almost 20%, which should be a sufficient condition for planning cutting operations, so that the workpiece is shaped mainly by internal surfaces.
4. The results of experimental studies presented in this article do not exhaust all the issues related to the problem of curvilinear cutting of brittle materials (glass) using AWJ, particularly the aspects of their surface quality inspection. The authors see a strong need to continue this interesting and promising subject, especially in the context of AWJ process optimization for industrial applications planning subsequent publications in this area in the near future.

Author Contributions: Supervision, M.S. and W.K; conceptualization, M.S.; research methodology, W.K., D.Y.P., M.K.G., and M.M.; investigation, M.S., W.K., and D.Y.P.; formal analysis, M.S., D.Y.P., and M.M.; writing—original draft preparation, M.S., W.K., S.S., and D.Y.P.; writing—review and editing, M.M., M.K.G., and S.S. All authors have read and agreed to the published version of the manuscript.

Funding: This research received no external funding.

Conflicts of Interest: The authors declare no conflict of interest.

Nomenclature

AWJ	Abrasive water jet
BEI	Backscattered electron imaging
BSED	Solid-state secondary electron detector
CAD	Computer-aided design
ESEM	Environmental scanning electron microscope
GSED	Gaseous secondary electron detector
IS	Inner surface of cutting kerf
LFD	Low-vacuum secondary electron detector
OS	Outer surface of cutting kerf
RMS	Root mean square
SED	Secondary electron detector
SEI	Secondary electron image
SEM	Scanning electron microscope
df	Focusing tube diameter, mm
do	Water jet orifice diameter, mm
g	Material thickness, mm
$\overline{h_v}(x)$	Average waviness profile, µm
l	Standoff distance, mm

ma	Abrasive feed rate, kg/min
n	Jet lag distance, mm
p	Water jet pressure, MPa
pmax	Nominal pressure, MPa
r	Machining radius, mm
v	Traverse speed, mm/min
Qmax	Maximum water flow rate, dm^3/min
Ra	Arithmetical mean deviation of the roughness profile, μm
Rq	Root-mean-square deviation of the roughness profile, μm
Rt	Total height of the profile on the evaluation length, μm
Sa	Arithmetic mean deviation of the surface, μm
Sal	Fastest decay autocorrelation length, mm
Sds	Density of summits of the surface, pks/mm^2
SPc	Arithmetic mean peak curvature, pks/mm^2
Sq	Root-mean-square deviation of the surface, μm
St	Total height of the surface, μm
Ua	Accelerating voltage, kV
WSm	Mean width of profile elements, within a sampling length, mm
Wt	Maximum height of waviness profile, μm

References

1. Liu, X.; Liang, Z.; Wen, G.; Yuan, X. Waterjet machining and research developments: A review. *Int. J. Adv. Manuf. Technol.* **2019**, *102*, 1257–1335. [CrossRef]
2. Hashish, M. Waterjet machining process. In *Handbook of Manufacturing Engineering and Technology*; Nee, A.Y.C., Ed.; Springer: London, UK, 2015.
3. Valíček, J.; Harničárová, M.; Hlavatý, I.; Grznárik, R.; Kušnerová, M.; Hutyrová, Z.; Panda, A. A new approach for the determination of technological parameters for hydroabrasive cutting of materials. *Mater. Werkst.* **2016**, *47*, 462–471. [CrossRef]
4. Krajcarz, D.; Bańkowski, D.; Młynarczyk, P. The effect of traverse speed on kerf width in AWJ cutting of ceramic tiles. *Procedia Eng.* **2017**, *192*, 469–473. [CrossRef]
5. Wang, S.; Zhang, S.; Wu, Y.; Yang, F. Exploring kerf cut by abrasive waterjet. *Int. J. Adv. Manuf. Technol.* **2017**, *93*, 2013–2020. [CrossRef]
6. Hlaváč, L.M.; Hlaváčová, I.M.; Geryk, V.; Plančár, Š. Investigation of the taper of kerfs cut in steels by AWJ. *Int. J. Adv. Manuf. Technol.* **2015**, *77*, 1811–1818. [CrossRef]
7. Hashish, M. Enhanced Abrasive Waterjet Cutting Accuracy with Dynamic Tilt Compensation. In Proceedings of the CRIP 2nd International Conference on High Performance Cutting, Vancouver, BC, Canada, 12–13 June 2006; pp. 12–13.
8. Zagórski, I.; Kłonica, M.; Kulisz, M.; Łoza, K. Effect of the AWJM method on the machined surface layer of AZ91D magnesium alloy and simulation of roughness parameters using neural networks. *Materials* **2018**, *11*, 2111. [CrossRef]
9. Chen, M.; Zhang, S.; Zeng, J.; Chen, B. Correcting shape error located in cut-in/cut-out region in abrasive water jet cutting proces. *Int. J. Adv. Manuf. Technol.* **2018**, *102*, 1165–1178. [CrossRef]
10. Wang, S.; Zhang, S.; Wu, Y.; Yang, F. A key parameter to characterize the kerf profile error generated by abrasive water-jet. *Int. J. Adv. Manuf. Technol.* **2017**, *90*, 1265–1275. [CrossRef]
11. Hlaváč, L.M.; Hlaváčová, I.M.; Geryk, V. Taper of kerfs made in rocks by abrasive water jet (AWJ). *Int. J. Adv. Manuf. Technol.* **2017**, *88*, 443–449. [CrossRef]
12. Kechagias, J.; Petropoulos, G.; Vaxevanidis, N. Application of Taguchi design for quality characterization of abrasive water jet machining of TRIP sheet steels. *Int. J. Adv. Manuf. Technol.* **2012**, *62*, 635–643. [CrossRef]
13. Wu, Y.; Zhang, S.; Wang, S.; Yang, F.; Tao, H. Method of obtaining accurate jet lag information in abrasive water-jet machining process. *Int. J. Adv. Manuf. Technol.* **2015**, *76*, 1827–1835. [CrossRef]
14. Sutowski, P.; Sutowska, M.; Kapłonek, W. The use of high-frequency acoustic emission analysis for in-process assessment of the surface quality of aluminium alloy 5251 in abrasive waterjet machining. *Proc. Inst. Mech. Eng. B.* **2018**, *232*, 2547–2565. [CrossRef]

15. Perec, A. Experimental research into alternative abrasive material for the abrasive water-jet cutting of titanium. *Int. J. Adv. Manuf. Technol.* **2018**, *97*, 1529–1540. [CrossRef]

16. Trivedi, P.; Dhanawade, A.; Kumar, S. An experimental investigation on cutting performance of abrasive water jet machining of austenite steel (AISI 316L). *Adv. Mater. Res.* **2015**, *1*, 263–274. [CrossRef]

17. Phokane, T.; Gupta, K.; Gupta, M.K. Investigations on surface roughness and tribology of miniature brass gears manufactured by abrasive water jet machining. *Proc. Inst. Mech. Eng. C* **2018**, *232*, 4193–4202. [CrossRef]

18. Supriya, S.B.; Srinivas, S. Machinability studies on stainless steel by abrasive water jet-Review. *Mater. Today Proc.* **2018**, *5*, 2871–2876. [CrossRef]

19. Niranjan, C.A.; Srinivas, S.; Ramachandra, M. An experimental study on depth of cut of AZ91 magnesium alloy in abrasive water jet cutting. *Mater. Today Proc.* **2018**, *5*, 2884–2890. [CrossRef]

20. Aich, U.; Banerjee, S.; Bandyopadhyay, A.; Das, P.K. Abrasive water jet cutting of borosilicate glass. *Procedia. Mater. Sci.* **2014**, *6*, 775–785. [CrossRef]

21. Dhanawade, A.; Kumar, S. Experimental study of delamination and kerf geometry of carbon epoxy composite machined by abrasive water jet. *J. Compos. Mater.* **2017**, *51*, 3373–3390. [CrossRef]

22. Sasikumar, K.S.K.; Arulshri, K.P.; Ponappa, K.; Uthayakumar, M. A study on kerf characteristics of hybrid aluminium 7075 metal matrix composites machined using abrasive water jet machining technology. *Proc. Inst. Mech. Eng. B* **2018**, *232*, 690–704. [CrossRef]

23. Abdullah, R.; Mahrous, A.; Barakat, A.; Zhou, Z. Surface quality of marble machined by abrasive water jet. *Cogent Eng.* **2016**, *3*, 1178626. [CrossRef]

24. Nadolny, K.; Plichta, J.; Sutowski, P. Regeneration of grinding wheel active surface using high-pressure hydro-jet. *J. Cent. South Univ. Technol.* **2014**, *21*, 3107–3118. [CrossRef]

25. Hloch, S.; Srivastava, M.; Krolczyk, J.B.; Chattopadhyaya, S.; Lehocká, D.; Simkulet, V.; Krolczyk, G.M. Strengthening effect after disintegration of stainless steel using pulsating water jet. *Teh. Vjesn.* **2018**, *25*, 1075–1079.

26. Lehocka, D.; Klich, J.; Foldyna, J.; Hloch, S.; Krolczyk, J.B.; Caracha, J.; Krolczyk, G.M. Copper alloys disintegration using pulsating water jet. *Measurement* **2016**, *82*, 375–383. [CrossRef]

27. Hreha, P.; Radvanská, A.; Hloch, S.; Peržel, W.; Królczyk, G.M.; Monková, K. Determination of vibration frequency depending on abrasive mass flow rate during abrasive water jet cutting. *Int. J. Adv. Manuf. Technol.* **2015**, *77*, 763–774. [CrossRef]

28. Matuszewski, M.; Mikolajczyk, T.; Pimenov, D.Y.; Styp-Rekowski, M. Influence of structure isotropy of machined surface on the wear process. *Int. J. Adv. Manuf. Technol.* **2017**, *88*, 2477–2483. [CrossRef]

29. Bustillo, A.; Pimenov, D.Y.; Matuszewski, M.; Mikolajczyk, T. Using artificial intelligence models for the prediction of surface wear based on surface isotropy levels. *Robot. Comput. Integr. Manuf.* **2018**, *53*, 215–227. [CrossRef]

30. Borkowski, J.; Sutowska, M.; Borkowski, P. Jakościowy model procesu cięcia wybranych materiałów metalowych wysokociśnieniową strugą wodno-ścierną. *Mechanik* **2014**, *9*, 255–258. (In Polish)

31. VDI 2906-5:1994. *Quality of Cut Faces of (Sheet) Metal Parts after Cutting, Blanking, Trimming or Piercing—Fine Blanking*; Verlag des Vereins Deutscher Ingenieure: Düsseldorf, Germany, 1994.

32. Löschner, P.; Jarosz, K.; Niesłony, P. Investigation of the effect of cutting speed on surface quality in abrasive water jet cutting of 316l stainless steel. *Procedia Eng.* **2016**, *149*, 276–282. [CrossRef]

33. Hreha, P.; Radvanska, A.; Knapcikova, L.; Królczyk, G.M.; Legutko, S.; Królczyk, J.B.; Hloch, S.; Monka, P. Roughness parameters calculation by means of on-line vibration monitoring emerging from AWJ interaction with material. *Metrol. Meas. Syst.* **2015**, *22*, 315–326. [CrossRef]

34. Klichova, D.; Klich, J. Study of the effect of material machinability on quality of surface created by abrasive water jet. *Procedia Eng.* **2016**, *149*, 177–182. [CrossRef]

35. ISO 4287:1997. *Geometrical Product Specifications (GPS)—Surface Texture: Profile Method: Terms, Definitions and Surface Texture Parameters*; International Organization for Standardization: Geneva, Switzerland, 1997.

36. ISO 25178-2:2012. *Geometrical Product Specifications (GPS)—Surface Texture: Areal—Part 2: Terms, Definitions and Surface Texture Parameters*; International Organization for Standardization: Geneva, Switzerland, 2012.

37. Stout, K.J.; Sullivan, P.J.; Dong, W.P.; Mainsah, E.; Luo, N.; Mathia, T.; Zahouani, H. *The Development of Methods for the Characterization of Roughness in Three Dimensions*; Publication No. EUR 15178 EN (Final Report); BCR: Brussels, Belgium, 1993.

38. Sutowska, M. Jakościowy model procesu cięcia AWJ wybranych materiałów konstrukcyjnych. *PAK* **2014**, *60*, 901–903. (In Polish)
39. Blateyron, F. The areal field parameters. In *Characterisation of Areal Surface Texture*; Leach, R., Ed.; Springer: Berlin/Heidelberg, Germany, 2013.
40. Nair, A.; Kumanan, S. Multi-performance optimization of abrasive water jet machining of Inconel 617 using WPCA. *Mater. Manuf. Process.* **2017**, *32*, 693–699. [CrossRef]
41. Zhao, W.; Guo, C. Topography and microstructure of the cutting surface machined with abrasive waterjet. *Int. J. Adv. Manuf. Technol.* **2014**, *73*, 941–947. [CrossRef]
42. Linke, B.; Garretson, I.; Jan, F.; Hafez, M. Integrated design, manufacturing and analysis of airfoil and nozzle shapes in an undergraduate course. *Procedia Manuf.* **2017**, *10*, 1077–1086. [CrossRef]
43. Saurabh, S.; Tiwari, T.; Nag, A.; Dixit, A.R.; Mandal, N.; Das, A.K.; Mandal, A.; Srivastava, A.K. Processing of alumina ceramics by abrasive waterjet—An experimental study. *Mater. Today Proc.* **2018**, *5*, 18061–18069. [CrossRef]
44. Kapłonek, W.; Sutowska, M.; Ungureanu, M.; Çetinkaya, K. Optical profilometer with confocal chromatic sensor forhigh-accuracy 3D measurements of the uncirculated and circulated coins. *J. Mech. Energy Eng.* **2018**, *2*, 181–192. [CrossRef]
45. Yuan, Z.; Zheng, P.; Wen, Q.; He, Y. Chemical kinetics mechanism for chemical mechanical polishing diamond and its related hard-inert materials. *Int. J. Adv. Manuf. Technol.* **2018**, *95*, 1715–1727. [CrossRef]
46. Nadolny, K.; Kapłonek, W.; Królczyk, G.; Ungureanu, N. The effect of active surface morphology of grinding wheel with zone-diversified structure on the form of chips in traverse internal cylindrical grinding of 100Cr6 steel. *Proc. Inst. Mech. Eng. B.* **2018**, *232*, 965–978. [CrossRef]
47. Fan, W.C.; Cao, P.; Long, L. Degradation of joint surface morphology, shear behavior and closure characteristics during cyclic loading. *J. Cent. South Univ. Technol.* **2018**, *25*, 653–661. [CrossRef]
48. Chen, S.; Carlson, M.A.; Zhang, Y.S.; Hu, Y.; Xie, J. Fabrication of injectable and superelastic nanofiber rectangle matrices ("peanuts") and their potential applications in hemostasis. *Biomaterials* **2018**, *179*, 46–59. [CrossRef] [PubMed]
49. Nadolny, K.; Sutowski, P.; Herman, D. Analysis of aluminum oxynitride AlON (Abral®) abrasive grains during the brittle fracture process using stress-wave emission techniques. *Int. J. Adv. Manuf. Technol.* **2015**, *81*, 1961–1976. [CrossRef]
50. Kapłonek, W.; Ungureanu, M. SEM-based imaging and analysis of surface morphology of the Trizact™ advanced structured abrasives. *J. Mech. Energy Eng.* **2018**, *2*, 17–26. [CrossRef]

 *materials*

Article

An Efficient and Adaptable Path Planning Algorithm for Automated Fiber Placement Based on Meshing and Multi Guidelines

Hong Xiao [1,*], Wei Han [1], Wenbin Tang [1,2] and Yugang Duan [1]

[1] State Key Lab for Manufacturing Systems Engineering, Xi'an Jiaotong University, Xi'an 710049, China; jx18hanwei@stu.xjtu.edu.cn (W.H.); tangwb@xpu.edu.cn (W.T.); ygduan@xjtu.edu.cn (Y.D.)
[2] School of Mechanical and Electrical Engineering, Xi'an Polytechnic University, Xi'an 710048, China
* Correspondence: xiaohongjxr@xjtu.edu.cn

Received: 11 August 2020; Accepted: 18 September 2020; Published: 22 September 2020

Abstract: Path planning algorithms for automated fiber placement are used to determine the directions of the fiber paths and the start and end positions on the mold surfaces. The quality of the fiber paths determines largely the efficiency and quality of the automated fiber placement process. The presented work investigated an efficient path planning algorithm based on surface meshing. In addition, an update method of the datum direction vector via a guide-line update strategy was proposed to make the path planning algorithm applicable for complex surfaces. Finally, accuracy analysis was performed on the proposed algorithm and it can be adopted as the reference for the triangulation parameter selection for the path planning algorithm.

Keywords: fiber-reinforced polymers; automated fiber placement; path planning

1. Introduction

Fiber-reinforced polymers (FRPs), especially carbon fiber reinforced polymers (CFRPs), are widely used in the aerospace and automobile industry as well as other fields because of their high specific strength, high specific modulus, excellent corrosion- and fatigue-resistance, and outstanding designability [1]. Automated fiber placement combines special manufacturing equipment with computerized numerical control (CNC) systems, which enables good forming quality, strong adaptability, and high efficiency. As an advanced composite manufacturing technology of large and complex components, automated fiber placement has become one of the most advanced and cutting-edge technologies for composite forming [2]. Path planning algorithms for automated fiber placement are used to specify the direction of a fiber path and determine the start and end positions of the fiber on the mold surface. The quality of the fiber path plays a crucial role for the efficiency and quality of the fiber placement process.

At present, the typical fiber path planning algorithms include the geodesic method and the meshing method based on parametric surfaces. The path planning algorithm of geodesic method based on parametric surfaces includes analytical and numerical methods [3]. The fiber path, which is obtained via the numerical solution of geodesics, is more aligned with practical engineering requirements. Lewis et al. [4] proposed a path planning method based on natural path, which is also widely used for path planning of tape laying. Shirinzadeh et al. [5–7] put forward the method of intersecting lines and surfaces to build the initial path, which improves the adaptability of the path to the surface curvature. However, there are certain disadvantages with the numerical solution of geodesics, e.g., the difficulty of finding the solution and the low efficiency associated with the calculation [8,9]. The meshing method, which uses the polygonal meshes to describe the complex parametric surfaces, can reduce computational complexity. The deviation of the fiber path correlates with the accuracy of the meshing [10,11]. Shinno et al. [12] proposed an iterative geodesic algorithm for a quadrilateral mesh surface to obtain the fiber path. Li et al. [13] proposed a path planning algorithm which used the mesh information contained in the STL file. However, there is a lack of quantitative analysis of the efficiency of the fiber path planning algorithm based on meshed surfaces. Also, no deviation analysis was performed for the generated fiber paths. Meanwhile, for complex components (surfaces), the design of angle reference direction datum is too simple, which leads to the poor applicability of the algorithm, fiber wrinkles, and eventually, affects the quality and mechanical properties of the fabricated FRP components. In this paper, a path planning algorithm for automated fiber placement based on meshing and multi guide-lines was proposed. Both the efficiency and accuracy of the algorithm were analyzed. The outline of the proposed algorithm is shown in Figure 1.

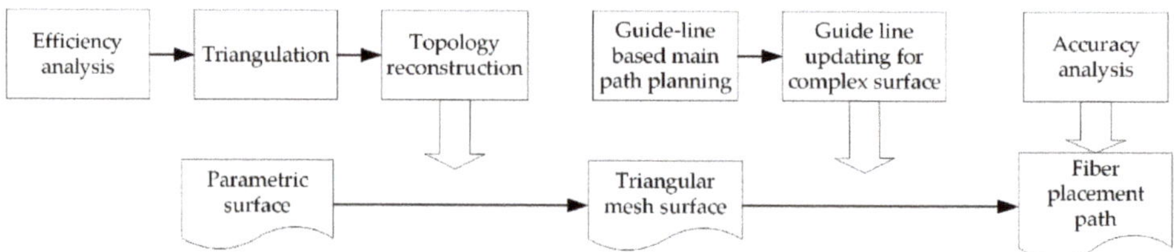

Figure 1. The outline of the proposed algorithm for automated fiber placement.

2. Efficiency Analysis and Topology Reconstruction

2.1. Efficiency Analysis of the Proposed Algorithm

A traditional path planning algorithm based on the geodesic method of parametric surface generates fiber paths by solving the direction of each point on the surface. It mainly uses the two geometric numerical solution operations, i.e., the intersection between a surface and a plane and the parallel offset of a curve. Essentially, the parallel offset of a curve is realized by the intersection of the surface at the sampling points and the corresponding offset direction planes.

Commercial CAD/CAM softwares generally use NURBS (Non-Uniform Rational B-Splines) for modelling, and the NURBS surface is represented by the following parameter equations:

$$\begin{cases} x = s_x(u,v) \\ y = s_y(u,v) \\ z = s_z(u,v) \end{cases} \tag{1}$$

$$s(u,v) = \frac{\sum_{i=0}^{C_u} \sum_{j=0}^{C_v} W_{i,j} P_{i,j} N_{i,p}(u) N_{j,q}(v)}{\sum_{i=0}^{C_u} \sum_{j=0}^{C_v} W_{i,j} N_{i,p}(u) N_{j,q}(v)} \tag{2}$$

where Ω_s is the domain of s, $P_{i,j}$ is the control point, $W_{i,j}$ is the weight, $N_{i,p}(u)$ is the basis function of the pth-order B-spline in u direction, and $N_{j,q}(v)$ is the basis function of the qth-order B-spline in v direction.

Let the plane be represented by an implicit surface equation:

$$h(x,y,z) = 0 \tag{3}$$

where Ω_h is the domain of h.

Combining Equation (1) with Equation (3) yields the intersection of plane h and surface s in domain $\Omega = \Omega_h \cap \Omega_s$. The parameters u and v of the intersection line satisfy the equation for the intersection line [14]:

$$h\big(s_x(u,v), s_y(u,v), s_z(u,v)\big) = l(u,v) = 0 \tag{4}$$

Intersection line l is a plane curve in the plane of parameter fields for u and v. If the surface s is the $p \times q$-th order, the intersection line l is the $p \times q$-th order. Because of the low efficiency of high-precision floating-point calculations, the frequent and high-precision solution of the intersection equation l consumes a lot of resources. This decreases both efficiency and stability of the numerical algorithms for the intersection between surfaces and planes. In the path planning algorithm based on the parametric surfaces, the intersection between surfaces and planes needs to be solved frequently. This decreases the efficiency of the path calculation, and for complex surfaces, the equation order is too high to be solved, which leads to the failure of path planning.

For triangular mesh surfaces, the surface is approximately a plane within the mesh cell [15]. The intersection between the surface and the plane can be converted into an intersection between planes, and the straight line in the plane is the result of the intersection.

The intersection line equation of a triangular patch $A_i(x, y, z) = 0$ and the plane h is:

$$l'(x, y, z) = 0 \tag{5}$$

$$\begin{cases} x = x_0 + mt \\ y = y_0 + nt \\ z = z_0 + pt \end{cases} \tag{6}$$

The intersection line l' is the first order equation about t, and the solution complexity decreases significantly. Since the triangulation algorithm will generate discrete grid planes of different sizes with different model complexity, the algorithm will not increase the order and complexity of equation solution during path generation due to the increase of model complexity. Therefore, the fiber path planning algorithm based on triangular mesh surface can obtain stable numerical solution quickly and efficiently.

2.2. Triangulation Algorithm for Parametric Surface

In the triangulation algorithm used in this paper, the edge and inner surfaces of the cell surface are approximated by straight line segments. Hence, the surface is discretized into area strips, and it is further divided into plane triangles. The triangulation results for the surface are generally given as strips of triangles. A strip of triangles is a list of points, such that any three consecutive points define a triangle. A parametric surface can be approximated by a series of triangle strips.

Due to the approximation process of replacing a curved surface with plane surfaces, the discrete error occurs in the triangulation process for parametric surface. This error is mainly reflected in the distance between the meshed plane surfaces and the original parametric surface. Two triangulation parameters are used to constrain the approximation error: (a) D_{l2s}: The maximum distance from the straight-line segment of the discrete triangular area to the original parametric surface. (b) L_{seg}: The maximum length of the straight-line segment in the discrete triangular area.

2.3. Sub-surface Boundary Splicing and Surface Topology Reconstruction

NURBS curves and surfaces, which are widely used in CAD/CAM software, have exact mathematical expressions, strong expression ability, good quality, and are easy to control. However, for complex surfaces, NURBS surfaces need to be split, spliced and trimmed. A complex surface is usually composed of multiple cellular surfaces. To reconstruct the topology of the whole complex surface, in addition to the mesh reconstruction of the cellular parameter surfaces, boundary splicing between cellular patches should be performed to obtain the global geometric topology.

2.3.1. Topology Reconstruction Algorithm for the Triangular Mesh of a Cellular Patch

As mentioned above, a cellular surface can be split into feature sampling points. To restore the surface information, it is necessary to obtain the segmentation results and establish the data structure for the triangular meshes by the vertex aggregation algorithm for triangulation (Algorithm 1).

Algorithm 1: Vertex aggregation algorithm for triangulation

Input: All vertex sets after the triangulation of a cellular parametric surface.

Output: Face ID List, Edge ID List and Point ID List for this cellular parametric surface.

1: *Loop all strip, fans and triangles:*

2:　　According to the right-hand rule, the vertices in the current discrete cell are stored to form the Point ID List of the current cell parameter surface.

3: *Loop all points in the Point ID List*

4:　　Every three points in the point table form a triangle patch to summarize the Face ID list and store the indexes of the three points of the current triangle patch.

5:　　Point ID List update, add triangle patch ID index.

6: Build the edge ID list, update the two-point indexes of the edge, update the Point ID list to add the edge index, update the Face ID list to add the included edge index.

7: Calculate the normal vector of the current triangular patch according to the right-hand rule, and update it in the Face ID list.

Through the above process, the local Point ID list, Edge ID list and Face ID list are established, which reconstruct the topology information for the current NURBS cellular surface.

2.3.2. Algorithm for Subsurface-Boundary Splicing

Because the mesh discretization results for each cellular parameter surface are independent of each other, and two adjacent cellular surfaces share the same edge, there are duplicate vertices for the adjacent cellular surfaces. It is necessary to remove the duplicate vertices. The subsurface-boundary splicing algorithm includes removing duplicate vertices and updating the global index of the vertices in all the cellular surfaces.

The brute force algorithm, which traverses the whole Point ID List for the duplicate vertices with the same coordinates with a given vertex and hence delete the duplicate ones, is the most straightforward method. Suppose that the surface is made up of k quadrilateral cellular NURBS surfaces with similar area and all the NURBS surfaces are smooth. Following the triangulation, it was assumed that the triangulation results are all given as strips of triangles to obtain a uniform triangular network. If the number of vertices on the boundary is a and b, the number of vertices on the cellular surface is $a \times b$, and the number of all vertices is $N = k \times a \times b$. The time complexity of the brute force algorithm is $O(N^2)$.

However, all the duplicate vertices are located on the boundary line because they are formed by two adjacent cellular surfaces sharing a common edge. For the vertices on the boundary line, they are in a semi-closed state and not surrounded by all triangles. On the other hand, the vertices inside the surfaces are in a fully-closed state, surrounded by several triangles (Figure 2). When the vertex is in the fully-closed state, the number of adjacent patches is equal to the number of edges, while in the semi-closed state, the number of patches is not equal to the number of edges. Therefore, all triangle vertices can be divided into two types: boundary semi-closed vertices and internal fully-closed vertices. Thanks to this feature, the boundary vertex set can be filtered out. Then, duplicate vertices can be removed from the boundary vertex set, and Point ID list, Face ID list and Edge ID list can be updated simultaneously.

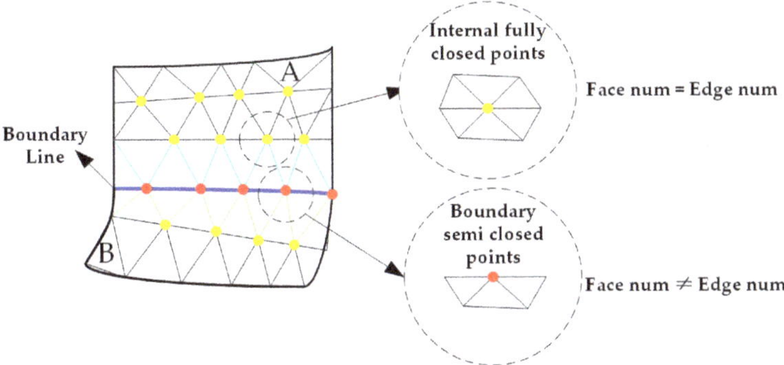

Figure 2. Internal fully-closed points and boundary semi-closed points in cell parametric surfaces.

De-duplication algorithm based on the vertex closure checking was described in Algorithm 2.

Algorithm 2: De-duplication algorithm based on vertex closure checking

---

Input: Point ID list before de-duplication.
Output: Point ID list after de-duplication.
1: Find semi-closed state vertices
2: *Loop all points in all Point ID List:*
3:     *if point->edgenum != point->facenum*
4:         save this point to duplicate Point ID List;
5: *Loop all point1 in duplicate Point ID List:*
6:     *Loop all point2 in duplicate Point ID List:*
7:         *if point1.distanceto (point2) < eps*
8:             delete point2 in Point ID List;
9:             refresh Face ID List & Edge ID List;

The time complexity of the algorithm is $O\ (k \times (a + b)^2)$, which is far less than $O\ (N^2)$ of the brute force algorithm, and the efficiency of the algorithm is greatly improved.

At the end of the above process, both sub-surface boundary splicing and surface topological relation reconstruction are completed. This enables fast search of adjacent triangles and efficient path calculation. The topological information of the discrete mesh surfaces is contained in the Point ID list, Face ID list and Edge ID list. The forms and requirements of the three data structure list are shown in Table 1 [13].

Table 1. The forms and requirements of the three data structure lists.

Class	Contents	Features
Vertex	int vIndex; double p [3]; int EdgeIndexList [cur edge index]; int FaceIndexList [cur face index];	Given the index number of a vertex, one can quickly find the global index of the triangle patch and the global index for the edge, where the current vertex belongs.
Edge	int eIndex; int VertexIndexList [2]; int FaceIndexList [2];	Given the number of a side, one can quickly find the global index of the face, where the current edge belongs and the global indices of the two vertices at the current edge.
Face	int fIndex; doubla n [3];int VertexIndexList [3]; int EdgeIndexList [3];	Given the number of a triangle patch, one can quickly find the global index and the global index of the three edges of the three vertices on the triangle patch.

3. Main Path Planning Algorithm Based on Guidelines on Triangular Mesh Surfaces

The path planning process needs to generate the main motion paths of the head of the automated fiber placement equipment and the corresponding fiber paths which are offset by the main motion paths. The fiber path generation is relatively simple and this paper focuses on the main path planning algorithm. The guidelines are extracted from the CAD model of the mold of the to-be-fabricated FRP component to reflect the skeleton and unique appearance of the component. According to the guide-lines, the direction vector of the main path can be calculated and adjusted adaptively, so that the main path and the corresponding fiber paths can comply with the shape of the component. It is beneficial to improve the mechanical performances of the FRP component while satisfying the constraints of the minimum steering radius of the fiber materials, even distribution of the cutting points, etc.

3.1. Main Motion Path Generation Algorithm

According to the fiber placement process, three geometric parameters (initial starting point P_0, guide line L, parametric surface Π) as well as the ply angle θ should be provided as input to the main motion path generation algorithm, which outputs the main motion path l_i. Basically, the algorithm needs to calculate the direction vectors and subsequently generate the continuous points on the main motion path, as described in Algorithms 3 and 4, respectively.

Algorithm 3: The direction vector calculation algorithm for the main motion path.

Step 1: Find the projection point P_0' for the initial starting point P_0 on the triangle patch A of the triangular mesh surfaces Σ (triangulation of the parametric surface Π) and obtain the normal vector n of triangle patch A. Then, redefine P_0' as the starting point for the current main motion path.

Step 2: Calculate the tangent vector t' for the projection point P_0'' of P_0' on the guide line L, and generate the parallel vector t of t' at point P_0'.

Step 3: Calculate the orthogonal vector k of the normal vector n and the parallel tangent vector t, where $k = n \times t$.

Step 4: Calculate the orthogonal vector m of normal vector n and vector k, where $m = k \times n$. The vector m is the datum direction vector for the current path point, based on the guide-line and corrected by the curvature feature of the surface, where the path point is located (as shown in Figure 3).

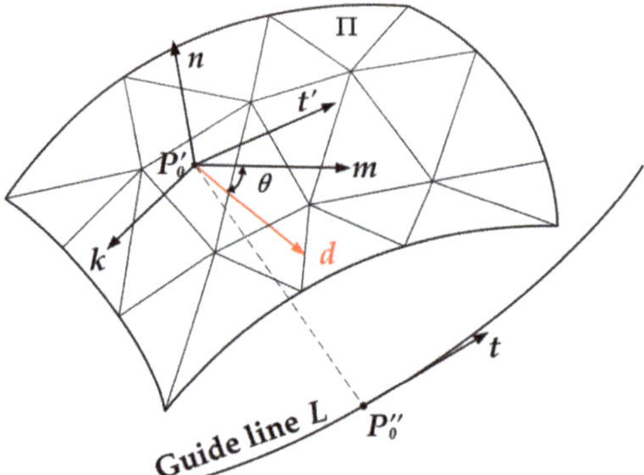

Figure 3. Illustration of the direction vector calculation algorithm.

Step 5: Using normal vector n as the rotation axis, rotate the datum direction vector m for θ degree to obtain the laying direction vector d for the current path point P_0'.

$$d = \begin{bmatrix} \cos\theta & -\sin\theta & 0 \\ \sin\theta & \cos\theta & 0 \\ 0 & 0 & 1 \end{bmatrix} \times m \tag{7}$$

In the current triangle patch A, use point $P_{0'}$ and vector d to construct a straight line and hence obtain the intersection point P_1 of the straight line and the three sides of triangle patch A.

The intersection point P_1 is the next path point, and all the path points can be obtained after continuous iteration for the main motion path generation.

Algorithm 4: The continuous path point generation algorithm.

Step 1: Take the starting point P_i and direction vector d_i in the current triangle patch A_i, construct a ray $P_a(\lambda)$, which is defined as the solution line for the path point. Its parameter equation is:

$$P_a(\lambda) = P_i + \lambda \cdot d \tag{8}$$

Step 2: The vertices V_a and V_b of the triangle patch form a straight line $P_b(\lambda)$, and the parameter equation is:

$$P_b(\lambda) = (1 - \lambda) \cdot V_a + \lambda \cdot V_b \tag{9}$$

Step 3: Using $P_a(\lambda) = P_b(\lambda)$, we can find the unique solution λ:

$$\lambda = \frac{V_a - P_i}{d + \frac{\overrightarrow{V_a - V_b}}{\|\overrightarrow{V_a - V_b}\|}} \tag{10}$$

If $0 \le \lambda \le 1$ is true, this means that the next path point is on the current sideline. If it is not on the current sideline, then take two points to form a straight line for the calculation. Next, take $V_c V_b$ and $V_a V_c$ to form a straight line for the calculation. Repeat steps 1, 2, and 3 to get the next path point.

The solution line for the path point may overlap with the three side lines of the triangle in step 3, and the next path point P_{i+1} cannot be calculated using the above steps. At this time, the endpoint of the edge $V_m V_k (m, k \in \{a, b, c\})$ of the triangle, where the point P_i is located, is used as the next path point P_{i+1}. The endpoint selection rules are described in Equation (11), where d is the laying direction vector.

$$P_{i+1} = \begin{cases} V_k, & d * \overline{V_m V_k} = 1 \\ V_m, & \text{otherwise} \end{cases} \tag{11}$$

Step 4: According to the topological relationship for the triangular mesh surface, obtain the adjacent triangles on the other side of the edge, where P_{i+1} is located, and update the patch index as A_{i+1}.

Step 5: Update the normal vector n_{i+1}. When the next path point P_{i+1} lies on the triangle edge or vertex, use the area weighing method to reduce the deviation.

$$n = \frac{\sum_{k=1}^{m} A_k \cdot n_k}{\left| \sum_{k=1}^{m} A_k \cdot n_k \right|} \tag{12}$$

In Equation (12), m is the total number of adjacent triangles to which P_{i+1} belongs, and A_k is the area of the triangles. The area can be determined using $A_k = \frac{\|AB \times AC\|}{2}$, where points A, B, C are the three vertices of a triangle A_k.

Step 6: Update the parallel tangent vector t_{i+1}, and solve the direction vector d_{i+1} of the next path point P_{i+1} using Algorithm 3. The projection vector $d_{i+1'}$ of d_{i+1} on patch A_{k+1} is the laying-direction vector of point P_{i+1}.

$$d'_{i+1} = d_{i+1} - \frac{d_{i+1} \cdot n}{\|n\|^2} \cdot n \tag{13}$$

Step 7: Repeat Step 1 to Step 6 to update and calculate the next path point until reach the triangular mesh surface boundary and no adjacent patch can be found and the partial path on the one single direction is constructed.

Step 8: Go back to the path initial point P_0, take the inverse of the laying-direction vector d_0 and repeat Step 1 to Step 7 until the whole main motion path is completely constructed.

The continuous path point generation algorithm is shown in Figure 4. After completing the above steps, the main motion path can be derived using the cubic spline interpolation algorithm.

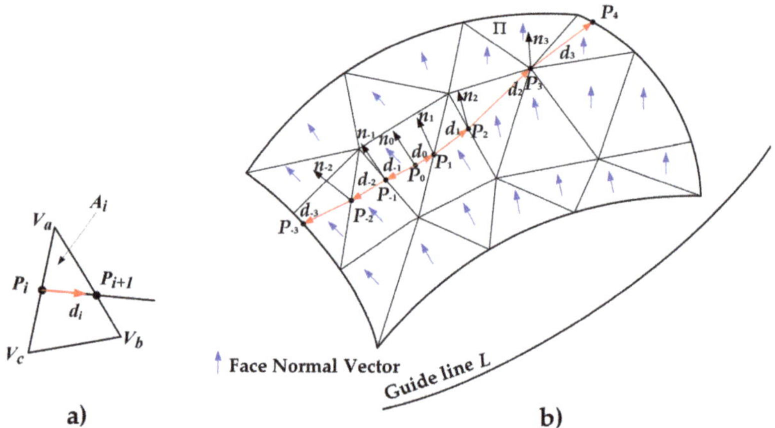

Figure 4. Continuous path point generation algorithm. (**a**) Calculation of next point in single triangle patch. (**b**) Calculation of all points of one path on the triangular mesh surfaces.

3.2. Guide-line Update Algorithm for Complex Surfaces

If the curvature of the surface of certain components changes substantially, the path, which was calculated and planned using a single guide line, cannot meet the requirements for the laying ability in each surface area. This leads to wrinkling of the fiber during the laying process, which degrades the mechanical properties of the final fabricated components. To address this problem, this paper proposed a guide-line algorithm for the update of the tangent vector t and datum direction vector m for the path points according to the surface shape and the distribution of multi guide-lines, as demonstrated in Figure 5 and described in Algorithm 5.

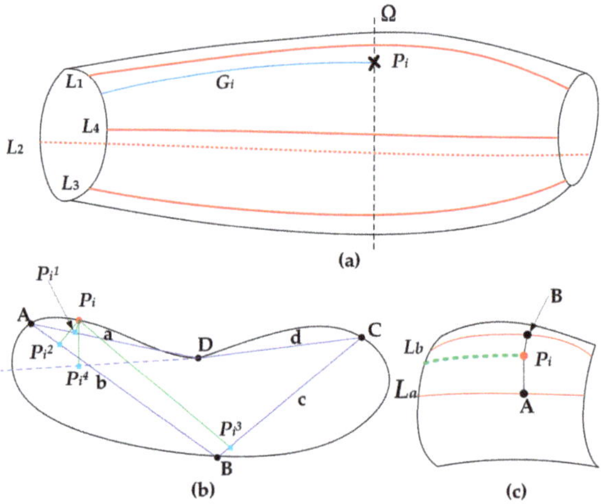

Figure 5. Guide-line update algorithm for complex surfaces: (**a**) Distribution diagram for multi guide-lines. (**b**) Polygonal section formed by the projection point of the point on the guide-line. (**c**) Diagram of tangent vector calculation.

Algorithm 5: Guideline update algorithm for complex surfaces.

Step 1: Project the current path point P_i on each guide-line L_i to obtain projection points A, B, C and D. Connect the projection points to form a polygonal section plane Ω, the four sides of the polygon on the section plane Ω are a, b, c and d.

Step 2: Project the current path point P_i to the edge of the polygon, and find the projection points P_i^1, P_i^2, P_i^3 and P_i^4.

Step 3: If the projection point is on the extension line of the edge, the edge is excluded. As shown in Figure 5b, the projection point P_i^4 is not on the edge d, so the edge d is not considered in the algorithm below.

Step 4: Calculate the distance from P_i to the projection points on the not excluded edge lines, where $dist = \|P_iP_i^x\|$.

Step 5: Determine the minimum distance $dist$ and its edge x. The two guide-lines L_a and L_b, at the end of x, are the guide-lines for the current path point P_i.

Step 6: As shown in Figure 5c and Equation (14), calculate the tangent vectors on L_a and L_b, respectively. The smooth transition for the path direction vector from L_a to L_b is realized using the distance-weighing method:

$$t = \frac{t_1 \cdot dist_1 + t_2 \cdot dist_2}{dist_1 + dist_2} \tag{14}$$

where $dist_1$ and $dist_2$ are the distances of P_i to L_a and L_b.

Update the tangent vector t and calculate the datum direction vector m of P_i in Algorithms 3 and 4. The path, which is generated by Algorithm 5, is more adaptive to the changing curvature of the surface and the scalability is improved. Figure 6 shows the paths generated based on single guideline and multi-guidelines for a panel and a curved surface models. It can be seen that the fiber direction transits smoothly between the guide lines and this makes the placed fibers adapt to the shape of the moulds of the final parts.

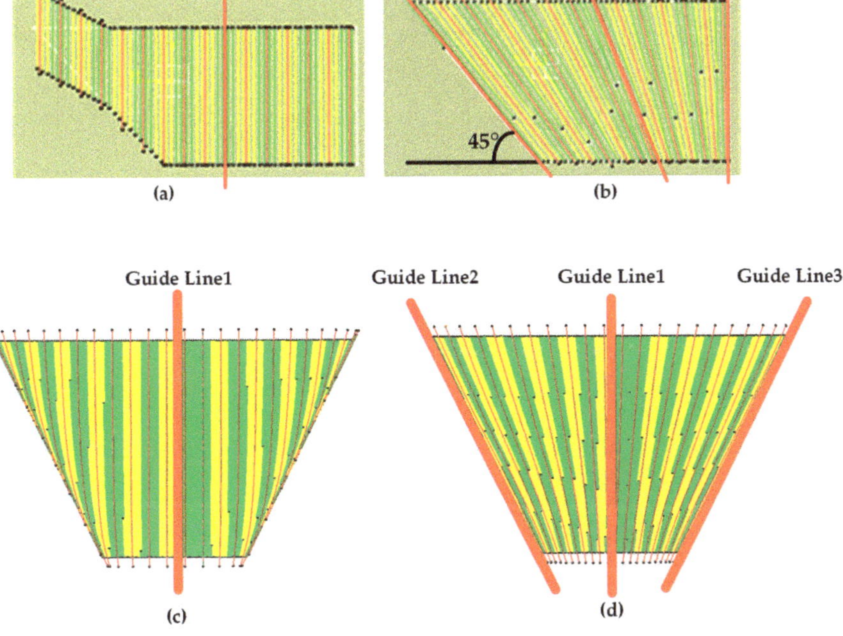

Figure 6. The paths generated based on single guide line and multi-guide lines for a panel (**a,b**) and a curved surface (**c,d**) models.

4. Accuracy Analysis of the Generated Path Based on Surface Meshing

In the triangulation process in Section 2, D_{12s} and L_{seg} are the main parameters that affect both the mesh density and the approximation accuracy. When the two parameters are set to larger values, the mesh density is small, the surface approximation accuracy is low, the subsequent path planning process data volume is small, and the algorithm efficiency is high. If the two parameter values are continuously reduced, on the other hand, the efficiency is lower.

It is generally believed that the path generated using the geodesic method on the parametric surface is used as the standard path when the planning accuracy and error of the algorithm need to be verified. The points on the discrete surface of the mesh are used to replace the path points on the original parametric surface. Furthermore, the normal vector of the path points on the triangular mesh surface can be used to replace the normal vector on the original parametric surface. As a result, a cumulative error can occur during the iteration process of the proposed path planning algorithm, which can cause the angle deviation from the design datum and the distance deviation from the path generated on the original parametric surface.

A complex surface with positive and negative curvature was adopted to conduct accuracy analysis of the proposed path planning algorithm, as shown in Figure 7. The surface consists of 29 independent cellular patches. Each cellular parameter surface represents a NURBS surface, and the order of each cellular parameter surface in both U and V directions is 6 degrees.

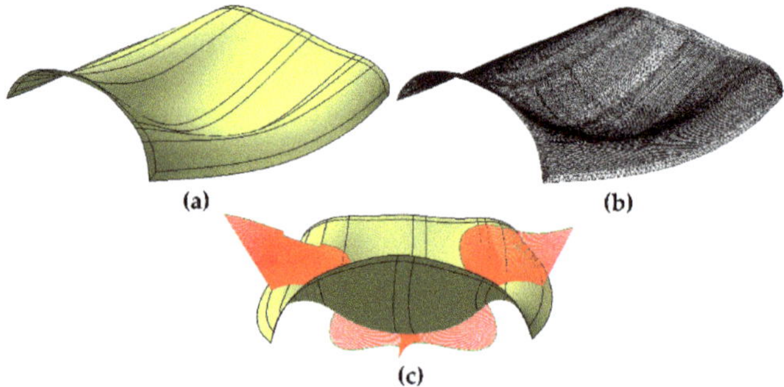

Figure 7. Illustration of the surface for accuracy analysis. (**a**) Original parametric surface. (**b**) Triangular mesh surface with $D_{12s} = 0.2$ and $L_{seg} = 100$. (**c**) Surface curvature variation diagram for the path-deviation-error test area.

4.1. Distance Deviation Analysis

The distance deviation of the path generated on the triangular mesh surface can be decomposed into the normal distance deviation and the geodesic distance deviation. In this paper, a uniform orthogonal test was carried out for the appropriate ranges of D_{12s} and L_{seg}. The path was uniformly sampled (with a distance of 5 mm) to evaluate the distance deviation, as shown in Figure 8.

During the triangulation process, both D_{12s} and L_{seg} did not reach 0, which means that the parameter value (without approximation error) could not be determined. Therefore, in the actual application, the D_{12s} range was 0.2–1.0 mm, and the L_{seg} range was 20–200 mm. The experiment was carried out by the $L_{25}(5^6)$ orthogonal design. The path-generation time, the distance deviation, the normal distance deviation and the geodesic distance deviation were recorded.

The Euclidean distance deviation d from the sample point on the generated path to the standard reference path was calculated and decomposed into the normal distance deviation d_N and the geodesic distance deviation d_T.

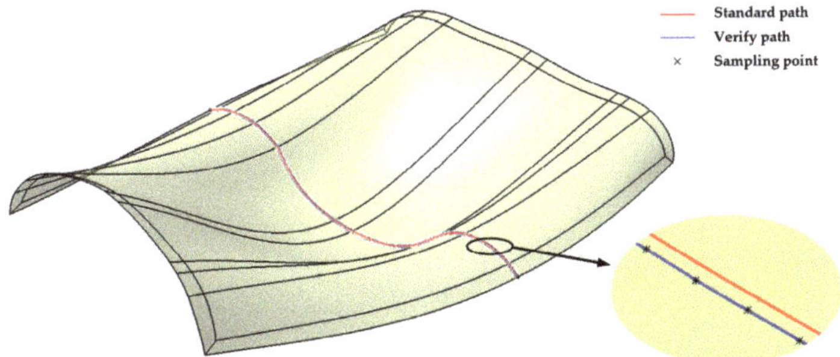

Figure 8. Standard path, verification path generated by the new algorithm, and sample points for $D_{12s} = 0.4$ and $L_{seg} = 100$.

The project vector from the sample point to standard reference path is V, and the normal vector for sample point P_i on the original parametric surface is n. d_N and d_T can be calculated as follows:

$$\begin{cases} d = DistBetween\left(P_i, P'_i\right) \\ d_N = d \times \cos < V, n > \\ d_T = \sqrt{d^2 - d_N^2} \end{cases} \tag{15}$$

According to the $L_{25}\left(5^6\right)$ orthogonal design, 25 sets of experiments were carried out. Four of them, which were $D_{12s} = 0.2$ and $L_{seg} = 65$, $D_{12s} = 0.4$ and $L_{seg} = 110$, $D_{12s} = 0.6$ and $L_{seg} = 155$ and $D_{12s} = 0.8$ and $L_{seg} = 200$, were selected to analyze the distribution of distance deviation along the generated paths.

The initial path point was located near the 350th sample point. According to Figure 9, the closer the sample point was to the initial point, the smaller was the distance deviation. During path generation, the normal vector for the path points on the triangular mesh surface was used to (approximately) replace the normal vector of the path points on the original parameter surface. Therefore, the direction datum of the path point was deviated, and it caused geodesic distance deviation between the generated path and the standard reference path. It also shows that the geodesic distance deviation d_T was the main deviation and it increased as the parameters D_{12s} and L_{seg} increased.

To analyze the relationship between the parameters (i.e., D_{12s} and L_{seg}) and the distance deviation of the generated path and the algorithm efficiency, the mean distance deviation, the normal mean distance deviation, the geodesic mean distance deviation, the maximum distance deviation, the maximum normal distance deviation, the maximum geodesic distance deviation, and the path generation time were calculated. This was done when the generation time started from the beginning of the surface triangulation to the end of the path generation, as shown in Table 2.

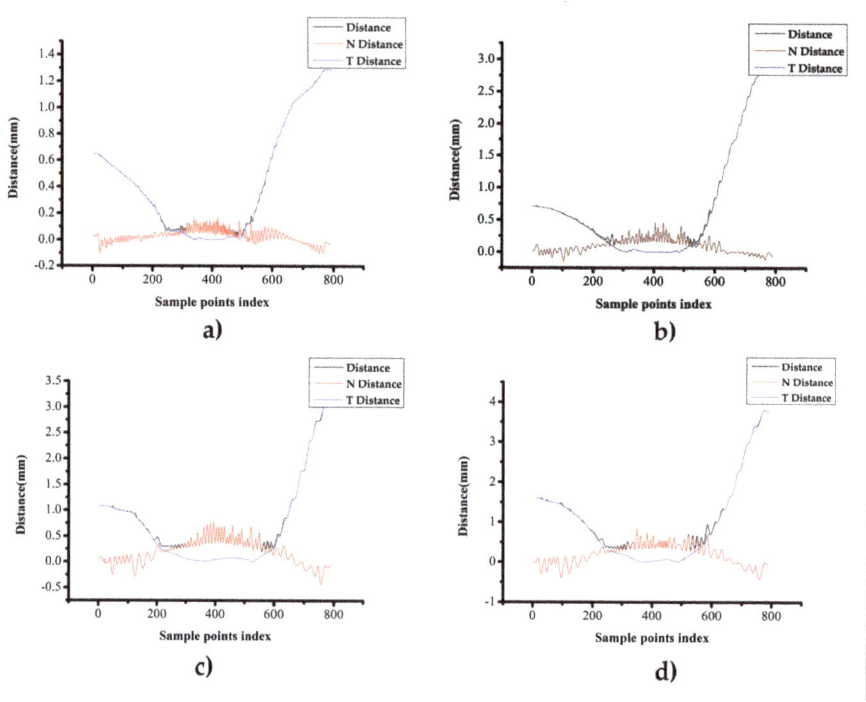

Figure 9. The distance deviation distribution (N Distance is d_N and T Distance is d_T.) for different parameters (**a**) $D_{12s} = 0.2$, $L_{seg} = 65$, (**b**) $D_{12s} = 0.4$, $L_{seg} = 110$, (**c**) $D_{12s} = 0.6$, $L_{seg} = 155$ and (**d**) $D_{12s} = 0.8$, $L_{seg} = 200$.

Table 2. Distance deviation and generation time.

D_{12s} /mm	L_{seg} /mm	Mean Distance Deviation /mm	Normal Mean Distance Deviation /mm	Geodesic Mean Distance Deviation /mm	Maximum Distance Deviation /mm	Maximum Normal Distance Deviation /mm	Maximum Geodesic Distance Deviation /mm	Generation Time /s
0.2	20	0.41072	0.00824	0.40834	0.81222	0.05218	0.81219	5.506
0.2	65	0.45268	0.02694	0.4329	1.29759	0.16468	1.29745	0.549
0.2	110	0.57411	0.02534	0.55406	1.60354	0.16447	1.60342	0.508
0.2	155	0.57298	0.02557	0.55294	1.60354	0.16447	1.60342	0.508
0.2	200	0.57309	0.02547	0.55304	1.60354	0.16447	1.60342	0.505
0.4	20	0.41072	0.00824	0.40834	0.81222	0.05218	0.81219	5.514
0.4	65	0.54446	0.09072	0.47922	2.12237	0.36747	2.12172	0.296
0.4	110	0.78375	0.09412	0.71355	3.00629	0.46252	3.00561	0.214
0.4	155	0.78403	0.09052	0.71476	2.94274	0.46246	2.94204	0.209
0.4	200	0.78403	0.09052	0.71476	2.94274	0.46246	2.94204	0.213
0.6	20	0.41072	0.00824	0.40834	0.81222	0.05218	0.81219	5.495
0.6	65	0.62143	0.1036	0.55255	2.14027	0.55588	2.13962	0.231
0.6	110	0.65292	0.20369	0.4915	1.62566	0.77333	1.62391	0.142
0.6	155	0.86012	0.18893	0.69804	3.09987	0.77349	3.0993	0.13
0.6	200	0.8602	0.18895	0.6985	3.08182	0.77349	3.08124	0.128
0.8	20	0.41072	0.00824	0.40834	0.81222	0.05218	0.81219	5.499
0.8	65	0.62315	0.10188	0.55424	2.19272	0.55588	2.19207	0.213
0.8	110	0.78546	0.20771	0.63916	1.94612	0.84012	1.94459	0.115
0.8	155	1.18547	0.19293	1.03511	3.91933	0.84004	3.91922	0.098
0.8	200	1.15638	0.19065	1.00657	3.8031	0.84004	3.80299	0.103
1	20	0.41072	0.00824	0.40834	0.81222	0.05218	0.81219	5.491
1	65	0.62147	0.10314	0.55262	2.14027	0.55588	2.13962	0.203
1	110	0.83477	0.2216	0.68092	2.01082	1.10312	2.00955	0.105
1	155	1.33796	0.21533	1.16792	3.97448	1.10346	3.97407	0.082
1	200	1.29659	0.20791	1.12927	3.81894	1.10346	3.8185	0.083

The distribution of mean distance deviation are shown in Figure 10, while the maximum distance deviation are shown in Figure 11, and the generation time are shown in Figure 12.

Supplementary Video S1 further demonstrates the high efficiency of the proposed algorithm, which can complete the path planning for one layer of a complex surface in just only tens of seconds.

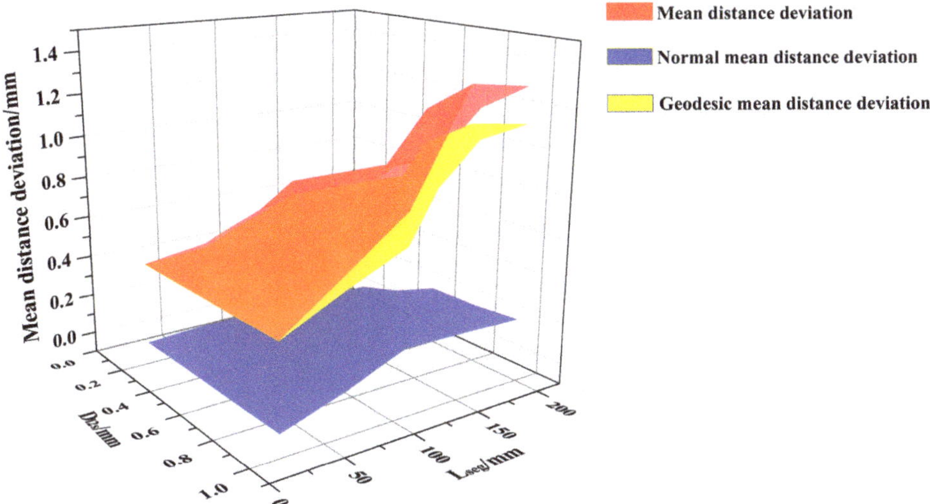

Figure 10. Mean distance deviation and its decomposition.

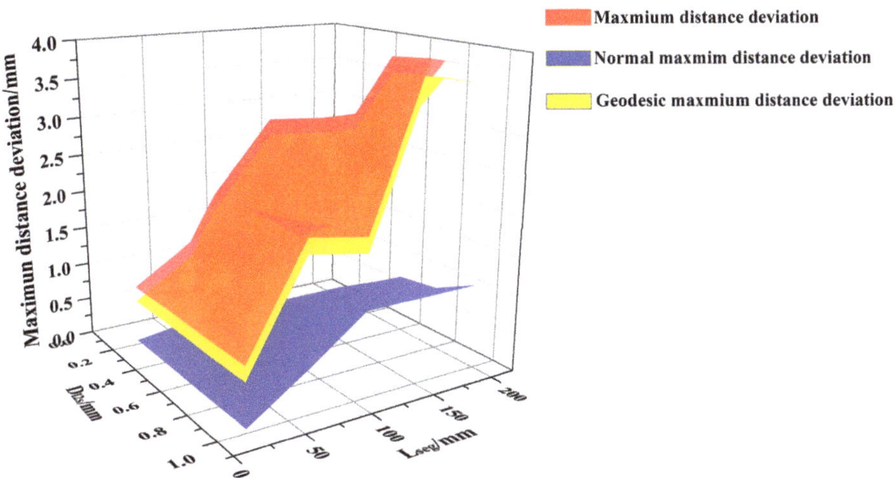

Figure 11. Maximum distance deviation and its decomposition.

According to Figures 10 and 11, for different parameters, the distance deviation mainly consisted of geodesic distance deviation, and the effect of normal distance deviation on the overall distance deviation is relatively small. Meanwhile, it can be observed that D_{l2s} and L_{seg} restricted each other in the triangulation process. When the two parameters cannot be satisfied simultaneously, the algorithm will adopt the parameter that makes the mesh more precise. For example, when L_{seg} is 20 mm, the triangulation algorithm generated the same triangular mesh surfaces for a D_{l2s} of 0.4 mm, 0.6 mm, 0.8 mm and 1.0 mm as for the D_{l2s} of 0.2 mm.

Table 2 shows that, when D_{l2s} exceeds 0.8mm and L_{seg} exceeds 155 mm, the mean distance deviation surpasses 1mm, and the maximum distance deviation exceeds 2 mm. When D_{l2s} is between

0.2 and 0.6mm, and L_{seg} is 65 to 155 mm, the average distance deviation remains within 1mm, while the maximum distance deviation is within 2 mm. The generation time of the algorithm is within 0.5 s.

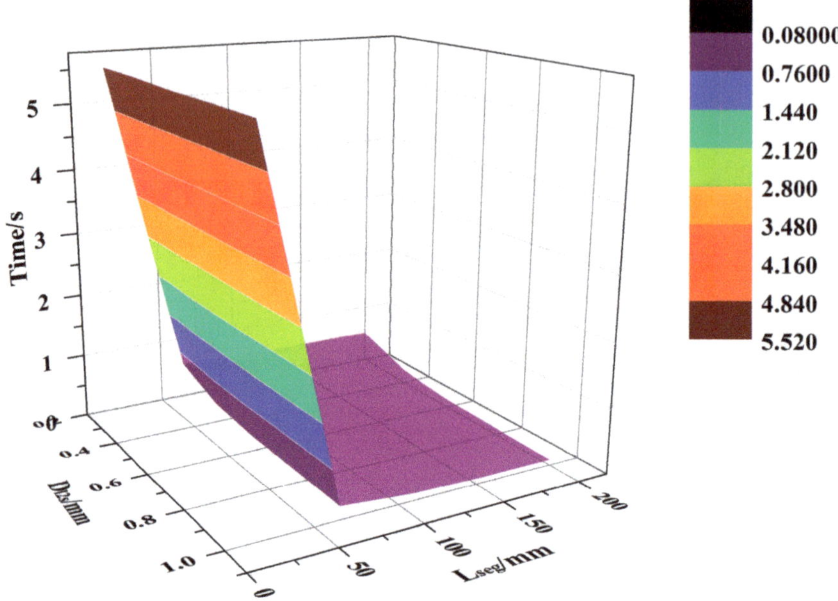

Figure 12. The generation times.

4.2. Angle Deviation Analysis

To ensure the orthogonal arrangement of fibers in different layers, the angle distribution between different layers is strictly defined. This ensures quasi-isotropic mechanical behavior for the components [3]. To analyze the angle deviation between the generated paths and the design datum, an angle deviation analysis was performed in this section.

The generated path was sampled uniformly with a step of 5 mm. For each sample point, the forward direction of the current path and the design direction datum of the path were calculated using both the guide-line and surface normal vector at the point. Subsequently, the angle deviation was obtained. The forward direction vector d_i at each sample point is the tangent vector of the sample point on the path. The datum direction vector d_i' can be calculated using:

$$\begin{cases} k_i' = n_i' \times t_i' \\ d_i' = n_i' \times k_i' \end{cases} \tag{16}$$

where n_i' is the normal vector of the sample point on the surface, and t_i' is the tangent of the projection point for the sample point on the guide-line. The angle deviation δ_i is calculated using:

$$\delta_i = arcos< d_i, d_i' > \tag{17}$$

The four sets of experiments selected in Section 4.1 were also adopted here to analyze the distribution of the angle deviation.

As mentioned above, the initial path point was near the 350th sampling point. According to Figure 13, for different triangulation parameters, the angle deviation does not depend on the distance between the sampling point and the initial path point but depend on the curvature of the original surface. The larger the curvature of the surface, where the sample points are located on the path, the greater is the angle deviation. This confirms that the angle deviation is due to the approximation of the normal vector of the triangular mesh surface during the execution of the algorithm, and there is no cumulative error.

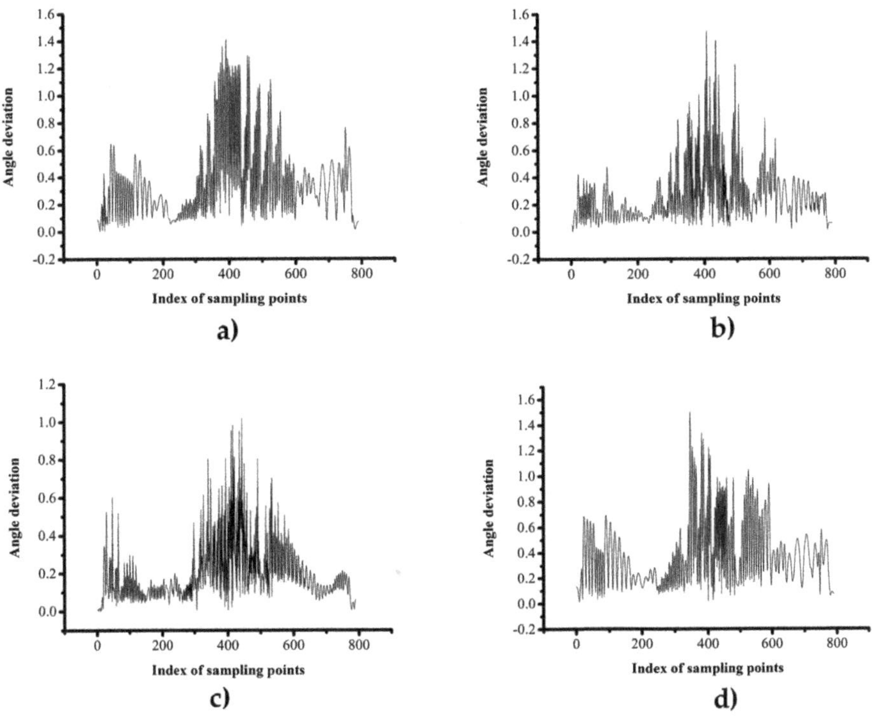

Figure 13. The angle deviation with different parameters: (**a**) D_{12s} = 0.2, L_{seg} = 65; (**b**) D_{12s} = 0.4, L_{seg} = 110; (**c**) D_{12s} = 0.6, L_{seg} = 155 and (**d**) D_{12s} = 0.8, L_{seg} = 200.

To study the effect of D_{12s} and L_{seg} on the angle deviation of the path, the mean angle deviation, the mean square error, and the maximum angle deviation were calculated using the experimentally obtained data—see Table 3.

Table 3. Data used for the calculation of the angle deviation.

D_{12s} /mm	L_{seg} /mm	Mean Angle Deviation /deg	Mean Square Error /deg^2	Maximum Angle Deviation /deg
0.2	20	0.10162	0.09583	0.43632
0.2	65	0.19909	0.1562	1.019
0.2	110	0.21098	0.15039	1.01133
0.2	155	0.21159	0.15056	1.01142
0.2	200	0.21204	0.15059	1.01141
0.4	20	0.10162	0.09583	0.43632
0.4	65	0.22812	0.16134	1.1537
0.4	110	0.27022	0.21115	1.47317
0.4	155	0.27248	0.20807	1.47248
0.4	200	0.27248	0.20807	1.47248
0.6	20	0.10162	0.09583	0.43632
0.6	65	0.2696	0.24432	1.47683
0.6	110	0.33184	0.27956	1.41166

Table 3. *Cont.*

D_{12s} /mm	L_{seg} /mm	Mean Angle Deviation /deg	Mean Square Error /deg^2	Maximum Angle Deviation /deg
0.6	155	0.36801	0.27632	1.4119
0.6	200	0.36668	0.27738	1.4119
0.8	20	0.10162	0.09583	0.43632
0.8	65	0.27083	0.2446	1.47683
0.8	110	0.3184	0.25209	1.50431
0.8	155	0.39048	0.25162	1.50424
0.8	200	0.38589	0.2524	1.50424
1	20	0.10162	0.09583	0.43632
1	65	0.26882	0.24516	1.47683
1	110	0.33021	0.29158	2.0499
1	155	0.45625	0.32954	2.05242
1	200	0.44346	0.32784	2.05242

The distribution of average angle deviation data is shown in Figure 14, and the maximum angle deviation data distribution is shown in Figure 15.

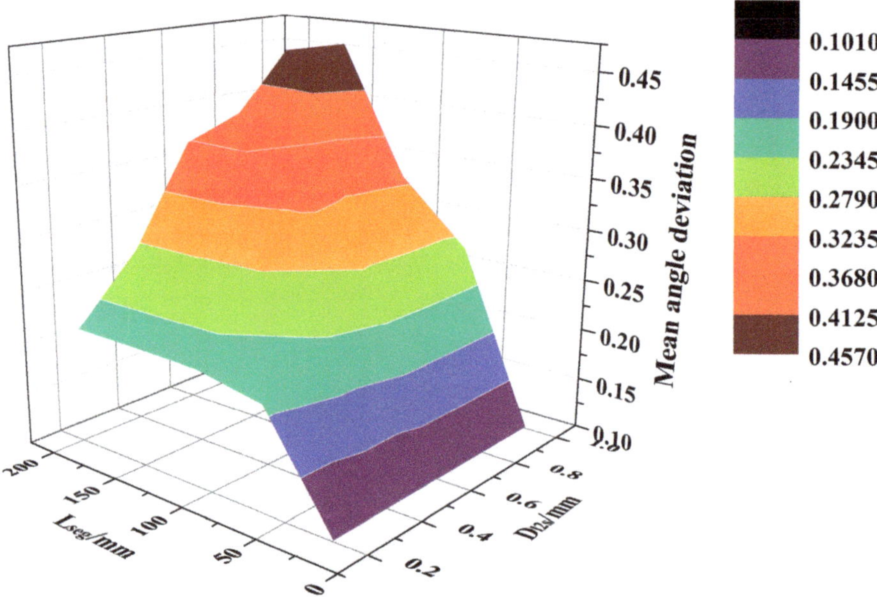

Figure 14. Distribution of the average angle deviation.

According to Figures 14 and 15, as the D_{12s} and L_{seg} increased, both the mesh density and approximation accuracy of the triangular mesh surface decrease. In addition, the normal vector of the path points on the triangular mesh surface deviate significantly from the normal vector on the original parametric surface. Hence, the angle deviation increases. According to Table 3, when D_{12s} exceeds 0.6 mm and L_{seg} exceeds 65 mm, the mean angle deviation surpasses 0.25 deg, and the maximum deviation is more than 1.4 deg.

Based on the above distance deviation analysis, when D_{12s} ranges between 0.2 and 0.6 mm and L_{seg} is 65 to 110 mm, the mean distance deviation remains within 1mm. Furthermore, the maximum distance deviation stays within 2 mm, the mean angle deviation is less than 0.25 deg, and the maximum angle deviation is below 1.4 deg. At the same time, the path generation time stays within 0.5s. In this case, D_{l2s} and L_{seg} can be selected according to the complexity of the surface and the acceptable path error. Within the above range of the parameters, a high-precision triangular mesh surface and fiber path with small error can be obtained, while the generation efficiency of the algorithm is high.

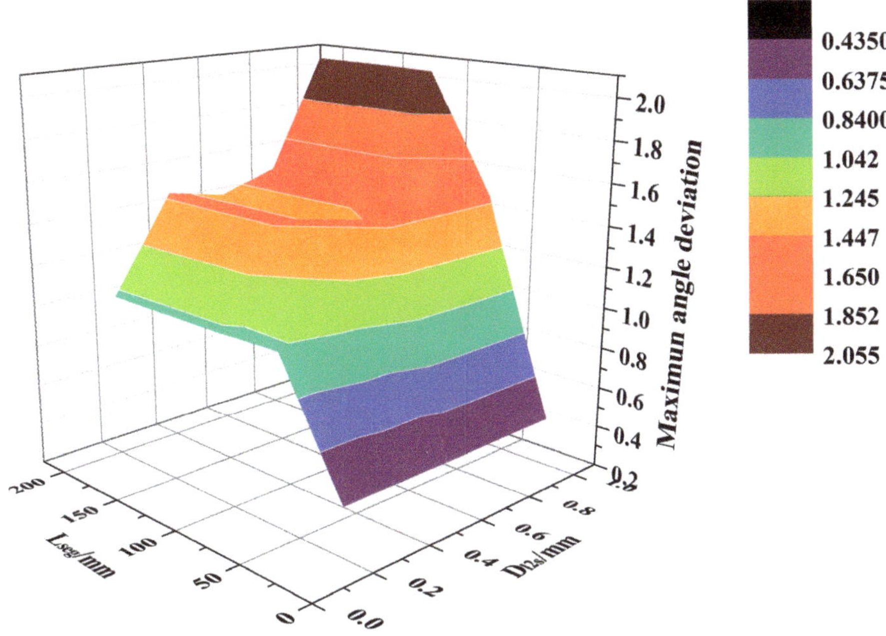

Figure 15. Distribution of the maximum angle deviation.

5. Conclusions

To improve the efficiency of automated fiber path planning process, a new path planning algorithm based on meshing and multi guide-lines were investigated. The original parameter surface of the CAD model of the FRP component was discretized into triangular mesh surface via surface discretization and triangulation. Sub-surface boundary splicing and surface topology reconstruction algorithm was proposed, and both the computational complexity reduction and the efficiency improvement of the algorithm were analyzed. The proposed automated fiber path planning algorithm consists of a main motion path direction vector algorithm and a continuous path point generation algorithm. An updating method for the datum direction vector via the guide-lines update algorithm was also introduced for complex surfaces. It improves the laying ability of the fibers and surface adaptability for the planned path. Accuracy analysis was conducted to investigate the relationship between the triangulation parameters and distance deviation, angle deviation and algorithm efficiency. The analysis indicated that by choosing appropriate triangulation parameters, the fiber path can be generated with high accuracy and efficiency.

More research efforts in the future work should be devoted to conduct experiments to test the mechanical properties of the fabricated FRP components by using the multi-guide-line planned paths.

Supplementary Materials: The following is available online at http://www.mdpi.com/1996-1944/13/18/4209/s1, Supplementary Video S1 demonstrates the high efficiency of the proposed algorithm, which can complete the path planning for one layer of a complex surface in just only tens of seconds.

Author Contributions: Conceptualization, H.X., W.H., W.T. and Y.D.; methodology, W.H. and W.T.; software, H.X., W.H. and W.T.; validation, H.X. and W.H.; investigation, W.H.; writing—original draft preparation, H.X. and

W.H.; writing—review and editing, H.X. and W.H.; supervision, H.X.; project administration, Y.D. All authors have read and agreed to the published version of the manuscript.

Funding: This research was funded by the National Natural Science Foundation of China (Grant No. 51875440), China Postdoctoral Science Foundation (Grant No. 2019M663686) and the Open Fund of the State Key Laboratory for Manufacturing Systems Engineering (Grant No. sklms2020003).

Conflicts of Interest: The authors declare no conflict of interest.

References

1. Kozaczuk, K. Automated fiber placement systems overview. *Trans. Inst. Aviat.* **2016**, *245*, 52–59. [CrossRef]
2. August, Z.; Ostrander, G.; Michasiow, J.; Hauber, D. Recent developments in automated fiber placement of thermoplastic composites. *SAMPE J.* **2014**, *50*, 30–37.
3. Rousseau, G.; Wehbe, R.; Halbritter, J.; Harik, R. Automated Fiber Placement Path Planning: A state-of-the-art review. *Comput.-Aided Des. Appl.* **2018**, *16*, 172–203. [CrossRef]
4. Lewis, H.; Romero, J. Composite Tape Placement Apparatus with Natural Path Generation Means. U.S. Patent 4,696,707, 29 September 1987.
5. Shirinzadeh, B.; Foong, C.W.; Tan, B.H. Robotic fibre placement process planning and control. *Assem. Autom.* **2000**, *20*, 313–320. [CrossRef]
6. Shirinzadeh, B.; Alici, G.; Foong, C.W.; Cassidy, G. Fabrication process of open surfaces by robotic fibre placement. *Robot. Comput.-Integr. Manuf.* **2004**, *20*, 17–28. [CrossRef]
7. Shirinzadeh, B.; Cassidy, G.; Oetomo, D.; Alici, G.; Ang, M.H. Trajectory generation for open-contoured structures in robotic fibre placement. *Robot. Comput.-Integr. Manuf.* **2007**, *23*, 380–394. [CrossRef]
8. Peng, Z.; Ronglei, S.; Xueying, Z.; Lingjin, H. Placement suitability criteria of composite tape for mould surface in automated tape placement. *Chin. J. Aeronaut.* **2015**, *28*, 1574–1581.
9. Zhang, P.; Sun, R.; Huang, T. A geometric method for computation of geodesic on parametric surfaces. *Comput. Aided Geom. Des.* **2015**, *38*, 24–37. [CrossRef]
10. Savio, G.; Meneghello, R.; Concheri, G. Geometric modeling of lattice structures for additive manufacturing. *Rapid Prototyp. J.* **2018**, *24*, 351–360. [CrossRef]
11. Zhang, Q.; Sabin, M.A.; Cirak, F. Subdivision surfaces with isogeometric analysis adapted refinement weights. *Comput.-Aided Des.* **2018**, *102*, 104–114. [CrossRef]
12. Shinno, N.; Shigemat, T. Method for Controlling Tape Affixing Direction of Automatic Tape Affixing Apparatus. U.S. Patent 5,041,179, 20 August 1991.
13. Li, L.; Wang, X.; Xu, D.; Tan, M. A Placement Path Planning Algorithm Based on Meshed Triangles for Carbon Fiber Reinforce Composite Component with Revolved Shape. *Int. J. Control Syst. Appl.* **2014**, *1*, 23–32.
14. Shen, J.; Buse, L.; Alliez, P.; Dodgson, N.A. A line/trimmed NURBS surface intersection algorithm using matrix representations. *Comput. Aided Geom. Des.* **2016**, *48*, 1–16. [CrossRef]
15. Lo, S.H.; Wang, W.X. An algorithm for the intersection of quadrilateral surfaces by tracing of neighbours. *Comput. Methods Appl. Mech. Eng.* **2003**, *192*, 2319–2338. [CrossRef]

materials

Article

Conditions of the Presence of Bimodal Amplitude Distribution of Two-Process Surfaces

Pawel Pawlus [1], Rafal Reizer [2,*] and Michal Wieczorowski [3]

[1] Faculty of Mechanical Engineering and Aeronautics, Rzeszow University of Technology, Powstancow Warszawy 8 Street, 35-959 Rzeszow, Poland; ppawlus@prz.edu.pl
[2] College of Natural Sciences, University of Rzeszow, Pigonia Street 1, 35-310 Rzeszow, Poland
[3] Faculty of Mechanical Engineering and Management, Poznan University of Technology, Piotrowo Street 3, 61-138 Poznan, Poland; michal.wieczorowski@put.poznan.pl
*   Correspondence: rreizer@ur.edu.pl; Tel.: +48-17-8518582

Received: 5 June 2020; Accepted: 9 September 2020; Published: 11 September 2020

Abstract: Two-process surfaces are functionally important. They contain plateau and valley parts. They are created by superimpositions of two one-process textures of Gaussian probability height distributions. It is expected that the resulting two-process surface would have bimodal height probability distribution. However, typically two-process textures have unimodal ordinate distribution. The present authors developed limiting conditions of presence of bimodal ordinate distribution. These conditions depend on the material ratio at the plateau-to-valley transition (the Smq parameter), and on the ratio of heights of the plateau and valley surface parts (Spq/Svq). Generated stratified textures and measured two-process surfaces of cylinder liners were taken into consideration.

Keywords: two-process surface; bimodal distribution; material ratio; parameters

1. Introduction

Surface topography is the fingerprint of a manufacturing process. It affects functional properties of machine elements, such as contact, sealing, friction, and wear [1]. Typically, one-process random surfaces are taken into consideration. Some natural and engineering surfaces have self-affine properties. The fractal analysis should be applied carefully. Not all random surfaces exhibit self-affine properties. During a low wear (within the limits of the machined surface topography), the one-process surface changes and a two-process surface is created. Previously, a surface subjected to low wear was simulated by truncation of the peaks positioned above a given threshold parallel to the mean plane [2–5]. However, this performance seldom occurs. In practice, the obtained transitional surface topography consists of the machined surface and of created during the wear a fine surface with a Gaussian height probability ordinate distribution. This fine surface became smoother as wear progressed [6,7]. Tribological properties of two-process surfaces were found to be better than those of one-process textures [8–10]. Therefore, two-process surfaces were machined. The plateau-honed cylinder surface is the practical example of machined two-process textures [4,7,11,12]. The stratified property of the two-process surface substantially affects the contact behavior [13–18]. The contact stiffness and the normal deformation are governed by the fine part of two-process surface.

The analysis of multi-process textures is more difficult than the study of one-process surfaces. Parameters frequently used to describe one-process surface cannot be applied for two-process topography, such as cylinder liner surface after plateau honing [19]. Many parameters are related to the material ratio curve, which presents the cumulative distribution of surface ordinates. The horizontal axis characterizes the bearing (material) ratio and the vertical axis characterizes the depth. According to ISO 13565-2 standard a surface consists of three parts: core (central), peaks, and valleys. Therefore, three parameters are applied to characterize the heights: core roughness depth which describes the central part of the material ratio curve, reduced peak height and reduced valley depth. The division between zones is performed by sliding 40% of material ratio wide window through the material ratio curve and finding the minimum secant slope [20–23].

In the other approach, based on material probability curve (ISO 13565-3 standard), two parts of two-process surface appear: plateau (peak) and valley [24–27]. Material probability curve is a representation of the material ratio curve; the material ratio is expressed as Gaussian probability in standard deviation values, plotted linearly on the horizontal axis (−3s = 0.13%, −2s = 2.28%, −s = 15.8%, 0 = 50%, s = 84.13%, 2s = 97.72%, 3s = 99.87%). The probability plot of two-process surface presents two straight non-parallel lines. The Spq parameter is the rms. height of the plateau part and the Svq parameter is the rms. height of the valley part. The Spq and Svq parameters are the slopes of the linear regressions through the plateau and valley parts, respectively. This approach contains also the Smq parameter, which is the material ratio at the transition point between plateau and valley parts—Figure 1. Since these parameters are statistical, similar values can be obtained for 2D profile and 3D (areal) texture. Because this method is based on theoretical presumptions, it can be applied in multi-process surface modeling. This material ratio curve is related to tribological properties of functional elements such as a load-carrying capacity and a wear resistance [28].

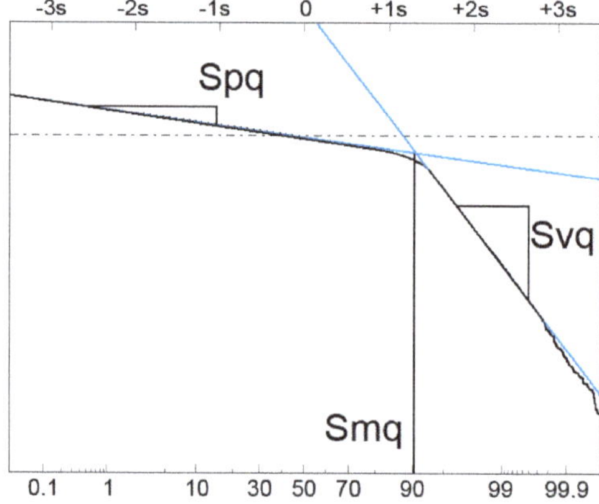

Figure 1. Graphical interpretation of the probability parameters: Spq, Svq and Smq.

During the creation of a two-process random surface, peak (plateau) Gaussian surface is superimposed on the valley Gaussian surface. It is expected that two-process surface would have bimodal ordinate distribution. A bimodal distribution is a probability distribution having two different modes; distinct local maxima exist in the height probability density function. However, the probability distribution of two-process texture can be also unimodal. The conditions of bimodal amplitude distribution of two-process surface should be developed.

2. Theoretical Considerations

Figure 2 presents two different cases of probability plots of material ratio curves of two-process surfaces. In Figure 2a the Smq parameter is smaller than 50% (negative values of the standard deviation s). However, in Figure 2b the Smq parameter is higher than 50% (positive values of s). The plateau depth Pd, which is the distance between the mean planes of two Gaussian surfaces: plateau and valley, is also shown in Figure 2.

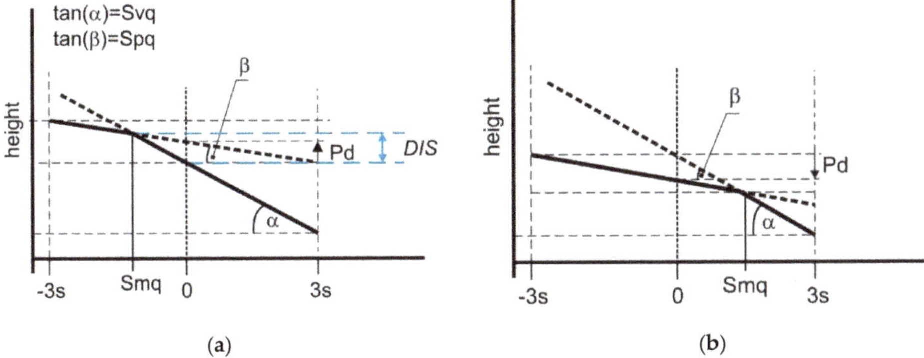

Figure 2. Probability plots of material ratio curves for different two-process surfaces; the Smq parameter is smaller than 50% (**a**) and the Smq parameter is higher than 50% (**b**) with plateau depth Pd and the vertical distance between two modes DIS.

The parameters Spq, Svq, Smq are related to the plateau depth Pd (Figure 2) by the following relation:

$$Pd = Smq \cdot (Spq - Svq), \tag{1}$$

Description of the probability plot of the material ratio curve is related to computer generation of two-process surfaces, which relies on superimposition of two Gaussian textures. The first of them (plateau) is characterized by the standard deviation of height Sq equal to the Spq parameter of a two process surface, while the Sq parameter of the second (valley) texture is equal to the Svq parameter of the two-process surface. The vertical distance between mean planes of these Gaussian structures is the plateau depth Pd. It should be noted that the rms. height Sq does not completely characterize each Gaussian surface, the correlation length CL (the distance, at which the autocorrelation function decays to 0.1 value) must be specified. For an isotropic surface, the correlation lengths in perpendicular directions are the same. From two Gaussian surfaces, smaller ordinates are chosen to generate two-process texture. This method was described in detail in [29]. The authors of papers [30–33] used similar procedures. Of course, generation of two-process surface structures across multiple scales is a further topic that requires attention.

Figure 3 presents an example of a computer generated two-process surface.

During creation of two-process surface, typically two Gaussian surfaces are superimposed; therefore, two-process texture can be called bi-Gaussian surface. However, the plateau surface is random, but the valley surface can be random or deterministic. Therefore, the two-process surface is a more general expression. Sometimes, multi-process textures are created, for example during wear of plateau-honed cylinder surface. However, in this paper two-process textures are analyzed.

Bimodal probability distribution of a two-process surface is only possible when the Smq parameter is smaller than 0 (50%) (Figure 2a). In the other case (Figure 2b), the Smq parameter is higher than 0 (50%), the modal value of the valley part is located above the modal value of the plateau part, so unimodal distribution is obtained; the peak (maximum) corresponds to the material ratio of 50%. Unimodal height probability also takes place when the Smq parameter is equal to 50%—when the means of the two normal distributions are equal, and the combined distribution is unimodal.

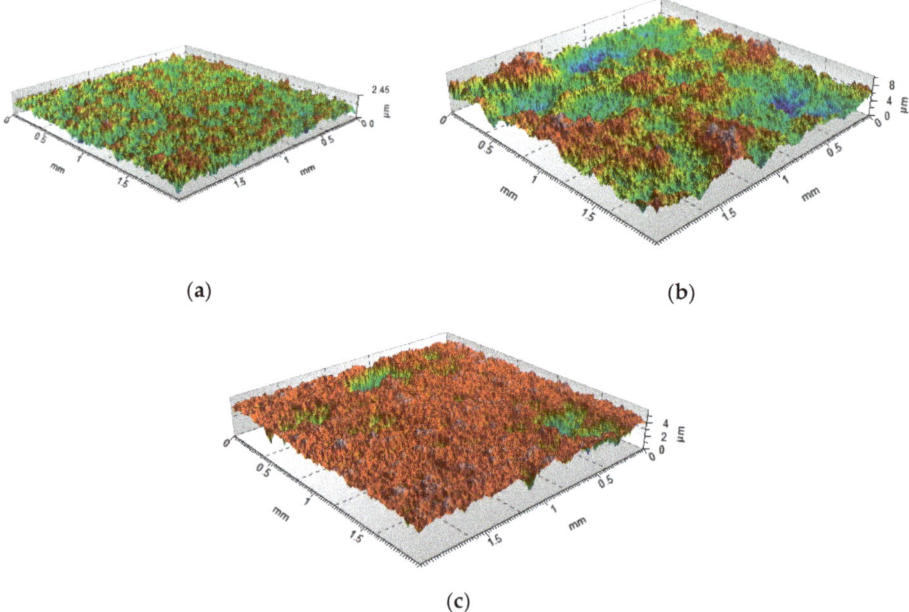

(a)   (b)

(c)

Figure 3. Example of generation of two-process surface topography: the plateau surface: Sq = 0.3 μm, CL = 20 μm (**a**), the valley surface: Sq = 1.7 μm, CL = 150 μm (**b**), two-process surface: Spq = 0.3 μm, Svq = 1.7 μm, Smq = 72.7% (**c**), Sq is the rms. height, CL is the correlation length.

For bimodal probability height distribution (Figure 2a) the lower mode corresponds to the material ratio of 50% (s = 0), however, the upper mode corresponds to the Smq parameter. The vertical distance between two modes should be:

$$DIS = Pd - Spq{\cdot}Smq = -Smq{\cdot}Svq, \tag{2}$$

However, sometimes when DIS is higher than 0, only unimodal distribution can be obtained, particularly when the modes of two Gaussian surfaces are close to each other. The conditions of presence of bimodal distribution will be developed in this work.

The vertical position of the maximum value of the probability height distribution corresponds to the smallest slope of the material ratio curve–inflection point, which is probably tribologically important.

3. The Analysis of Generated Surfaces

Two-process random isotropic topographies were generated. The superimposition method was used. Each Gaussian surface was modeled using procedure developed by Wu [34]. Each surface contained 256 × 256 points. The sampling interval was 1 μm, the correlation lengths in perpendicular directions were 10 μm. The Svq/Spq ratios were in the range: 4–30.

The aforementioned assumptions in Section 2 were confirmed. The upper peak of the bimodal probability distribution corresponded to the material ratio at the transition point Smq. The lower peak corresponded to the material ratio of 50%. Therefore, the vertical distance between two modes DIS were equal to −Smq·Svq. The mean error was 1.8%. This distance was typically higher that the plateau depth Pd. The average relative difference between the vertical distance between two modes DIS and the plateau depth Pd was 19%. This difference was smaller for higher value of the Smq material ratio and also for higher values of the Svq/Spq ratio.

Figures 4–6 present contour plots of computer-generated surfaces, their probability plots of material ratio curves, and material ratio curves with probability height distributions. The larger mode is called the major mode and the other mode is called the minor mode. The green line shows the position of the major mode while the red line shows the location of the minor mode. In addition, the smallest slope of the material ratio curve is marked by a green circle.

The major mode corresponds to the smallest slope of the material ratio curve. Typically, the upper peak (local maximum) is the major mode. However, for small values of the Svq/Spq ratio, and for low material ratio at the transition point Smq, the lower peak can be the major mode (Figure 4c). The ratio of the amplitudes of higher and lower peaks is called the bimodal ratio [35]. When the Smq parameter increased, the bimodal ratio also increased and the vertical distance between two modes DIS decreased.

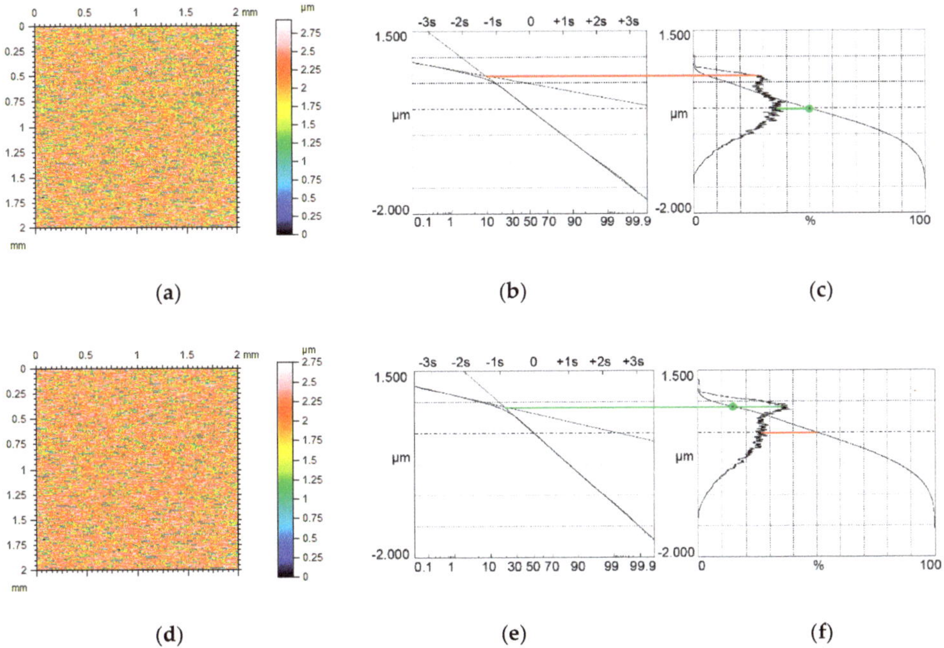

Figure 4. Contour plots (a,d) surface probability plot (b,e), material ratio curve and probability distribution (c,f) of two-process isotropic modeled surfaces of the Smq parameter of 10% (a–c) and 20% (d–f), the other parameters of both surfaces are Spq = 0.12 μm, Svq = 0.5 μm, both surfaces have bimodal ordinate distribution.

When the Smq material ratio was 10% and the Svq/Spq ratio was 4.17, the smallest slope of the material ratio curve was obtained for the material ratio of 50% (green circle in Figure 4c). Due to an increase in the Smq parameter to 20%, the smallest slope of the material ratio curve corresponded to the Smq parameter. Owing to an increase in the Smq material ratio, the skewness Ssk decreased (from −0.35 to −0.57). When the Smq parameter further increased to 25%, only one peak was visible in the ordinate distribution—unimodal distribution took place.

A similar situation occurred when the Svq parameter increased to about 0.7 μm (Figure 5). The amplitudes of both modes are similar for the Smq parameter of 10% (Figure 5c); therefore, the bimodal ratio increased compared to Figure 4c. This ratio is also higher in Figure 5f, compared to Figure 4f, when the Smq parameter increased to 20%. When the Smq parameter increased to 25%, the amplitude probability distribution was still bimodal, but when Smq was 30% unimodal distribution occurred. Similar to lower Svq/Spq ratio, the smallest slope of the material ratio curve corresponded to the material ratio of 50% for the Smq parameter of 10% and to the material ratio of 20% when the Smq parameter increased.

When the Svq/Spq ratio further increased to a value of 8.3 (Figure 6) the upper peak was the major mode even for the Smq parameter of 10%. The bimodal ratio increased for the same Smq parameter compared to textures shown in Figures 4 and 5. Bimodal distribution existed even for the Smq parameter near 30%. One can find in Figure 6 points in material ratio curves, which correspond to changes from the plateau parts to the valley surface portions. In these points, the first derivatives are discontinuous. These points, indicated by green arrows, are located under the major mode. Because the tribological properties of surfaces depend on the material ratio curve, the presence of those points can have functional significance. The described points were also visible for higher Svq/Spq ratios.

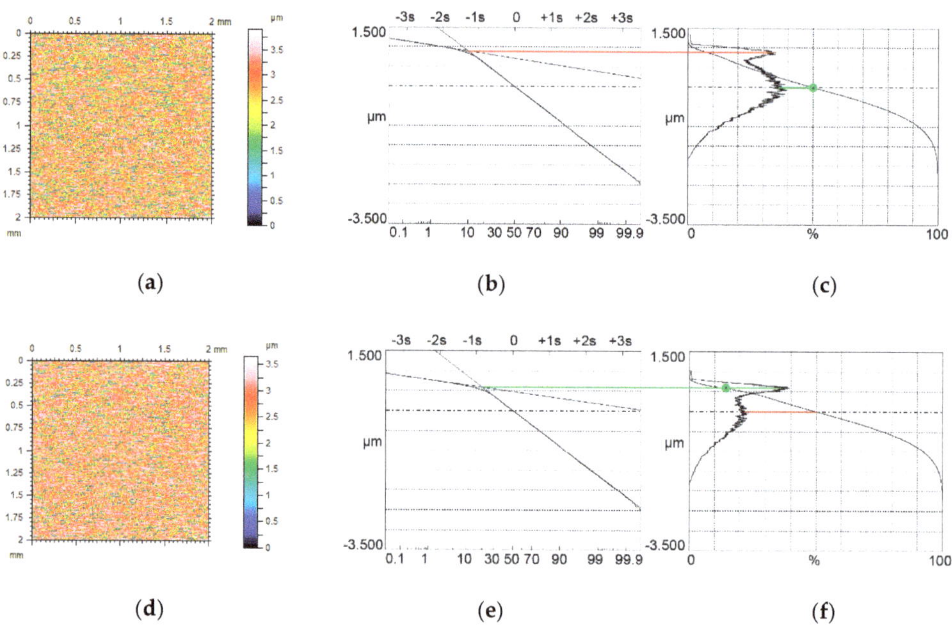

Figure 5. Contour plots (**a,d**) surface probability plot (**b,e**), material ratio curve and probability distribution (**c,f**) of two-process isotropic modeled surfaces of the Smq parameter of 10% (**a–c**) and 20% (**d–f**), the other parameters of both surfaces are Spq = 0.12 μm, Svq = 0.7 μm, both surfaces have bimodal ordinate distributions.

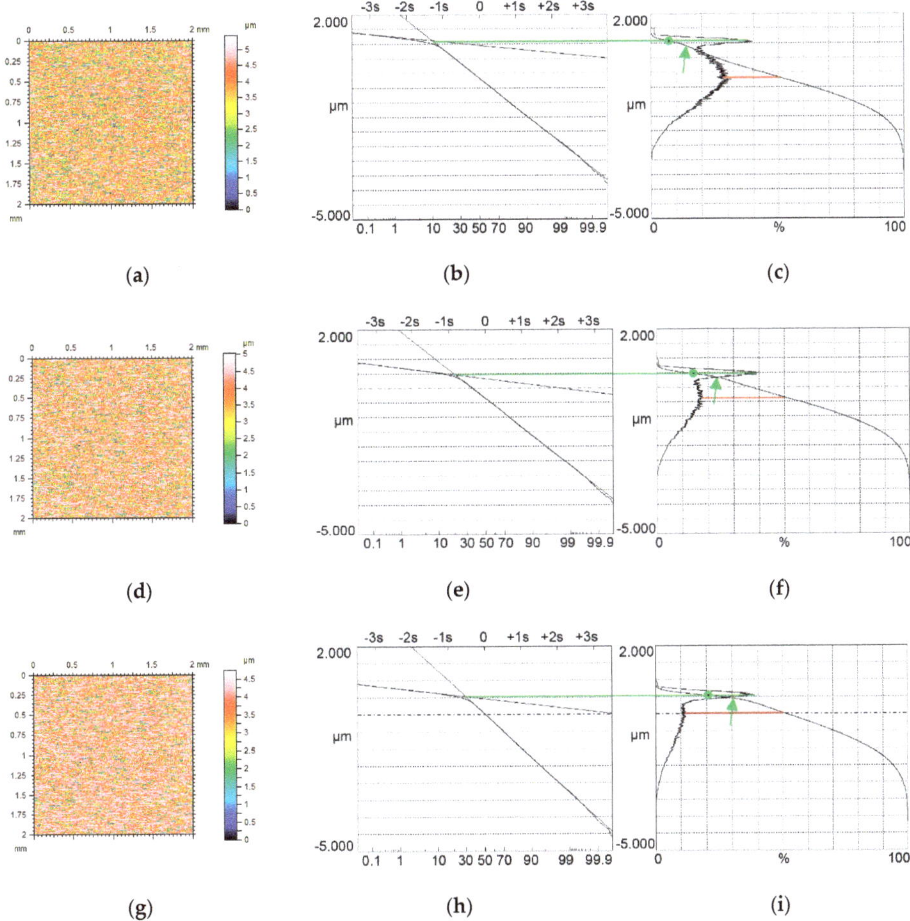

Figure 6. Contour plots (**a,d,g**) surface probability plot (**b,e,h**), material ratio curve and probability distribution (**c,f,i**) of two-process isotropic modeled surfaces of the Smq parameter of 10% (**a–c**), 20% (**d–f**) and 30% (**g–i**), the other parameters of both surfaces are Spq = 0.12 μm, Svq = 1.0 μm, all surfaces have bimodal ordinate distributions.

The question arises: what is the limiting condition for presence of bimodal height probability distribution? Conditions of unimodality or bimodality depend typically on the height standard deviations of two Gaussian height distributions [35–40]. Near the transition point of the Smq material ratio, there is a mixture of two Gaussian distributions. The analysis of many generated two-process surfaces revealed that this mixture also depends on the height standard deviations of two distributions. When the Svq parameter is much higher than the Spq parameter, the following condition of bimodal height probability distribution presence was established:

$$DIS > Spq + 0.5 \cdot Svq,$$ (3)

It was found from the analysis of simulated surfaces that this condition was valid for the Svq/Spq ratio not smaller than 4. The highest analysed Svq/Spq ratio in this work was 30. On the basis of the analysis of many modeled two-process textures it was found that the limiting condition of presence of bimodal ordinate distribution depended on the Svq/Spq ratio. The probability height distribution of a two-process surface is bimodal when the Smq material ratio is lower than the values shown in Figure 7. The shaded area corresponds to bimodal height distribution. One can see that the limiting value of the Smq ratio is higher when the Svq/Spq ratio is higher. This dependence is stronger for smaller Svq/Spq ratio. It was also found that unimodal ordinate distribution occurred for the Smq parameter was higher than the values presented in Figure 7.

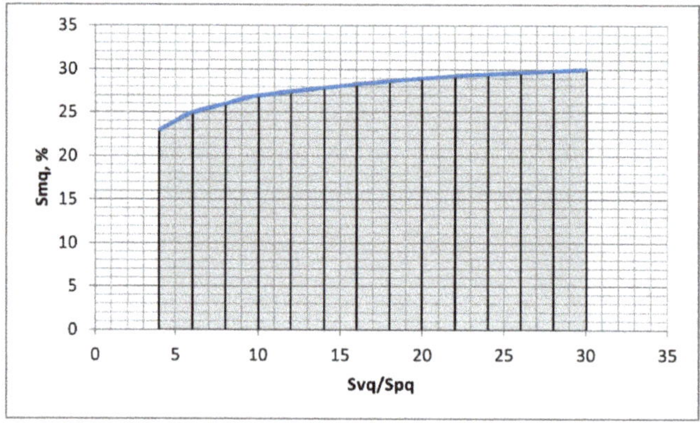

Figure 7. Conditions of bimodal height distributions of two-process surfaces; bimodal distribution appears for the Smq parameter within the shaded area.

When the Smq parameter is higher than that presented in Figure 7, but smaller than 50% then only one mode exists for the Smq material ratio (Figure 8a–c).

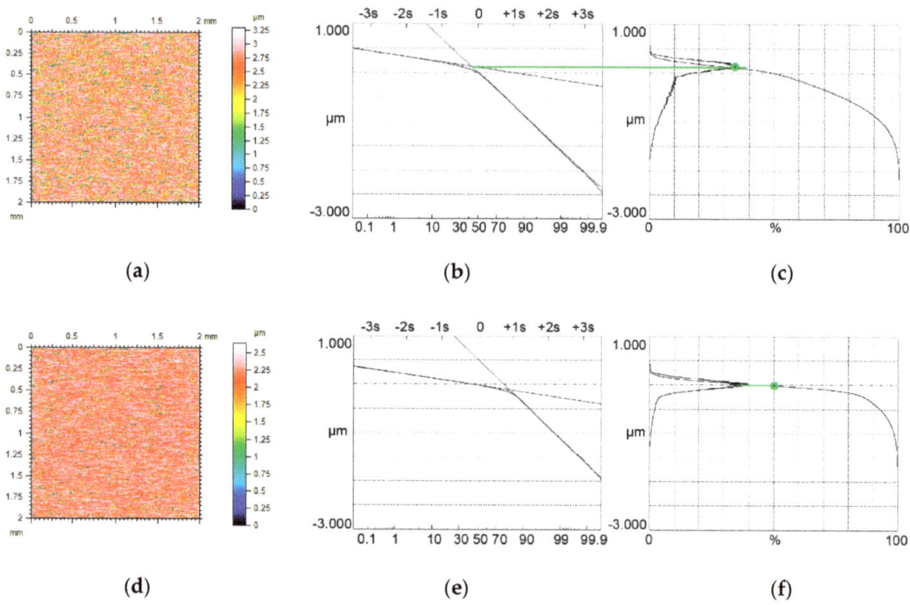

Figure 8. Contour plots (**a,d**) surface probability plot (**b,e**), material ratio curve and probability distribution (**c,f**) of two-process isotropic modeled surfaces of the Smq parameter of 40% (**a–c**) and 80% (**d–f**), the other parameters of both surfaces are Spq = 0.12 µm, Svq = 0.7 µm, both surfaces have unimodal ordinate distributions.

However, when the Smq parameter is similar or higher than 50% the mode corresponds to the material ratio of 50% (Figure 8d–f). The assumptions presented in Section 2 were confirmed for all modeled textures.

The presented analysis can be extended for multi-process surfaces, when the number of processes is higher than 2, especially for three-process structures. The material ratio at the transition point between the second and the third part would be very small for three-modal ordinate distribution. If the number of processes were higher than 3, the analysis would be more difficult.

4. The Analysis of Measured Surfaces

Cylinder surfaces after plateau honing are the typical examples of two-process textures. However, these cylinders are characterized by the Smq parameter higher than 50%, so their probability distributions are unimodal. Cylinder liners with bimodal surface topography were created during tests using an Optimol SRV5 (produced by: Optimol Instruments Prüftechnik GmbH, München, Germany) oscillating wear tester under lubricated conditions at high and low temperatures. This tester allows for precise control of the normal load, temperature and stroke. A chromium-coated piston rings were in contact with cylinder liners. These liners were initially one-process honed by diamond tools. One can find the operating conditions in References [41,42]. These surfaces were measured by a white light interferometer Talysurf CCI Lite (produced by: Taylor Hobson Ltd., Leicester, UK). The measuring area 3.29×3.29 mm^2 contained 1024×1024 data points. Forms were removed by polynomials of the second degree. A digital filtration was not used.

Figures 9 and 10 present contour plots of computer-generated surfaces, their probability plots of material ratio curves and material ratio curves with probability height distributions.

The surface shown in Figure 9a–c was tested in high temperature (80 °C). The Svq/Spq ratio was higher than 5. The upper peak was the major mode, therefore this mode corresponded to the Smq material ratio of 24%. The second and the third surfaces were tested in negative temperature (−20 °C). The Svq/Spq ratio of the surface shown in Figure 9d–f was near 6. The Svq parameter of this surface was comparatively high (2.89 μm). However, due to a low value of the Smq parameter, the lower peak became the major mode and therefore this mode occurred for the material ratio of 50%. The Svq/Spq ratio of the third surface shown in Figure 9g–i was the smallest (near 4). However, due to the material ratio at the transition point equal to 20%, the upper peak was the major mode, the position of which corresponded to the Smq parameter.

When the Smq parameter was higher than 30%, cylinder liner textures had unimodal probability height distribution. Similar to modeled surfaces, when the Smq parameter was smaller than 50%, the mode appeared for the material ratio of Smq (Figure 10a–c). In the other cases, such as that shown in Figure 10d–f, this mode corresponded to the material ratio of 50%.

Generally, the assumptions presented in Section 2 were confirmed for measured surfaces. In addition, Equation (3) was found to properly discriminate between bimodal and unimodal amplitude probability distributions.

In this work we analyzed the vertical position of the major mode of amplitude distribution, which is also the position of the smallest slope of the material ratio curve of two-process surface. This point is of substantial practical significance, since the material ratio curve is strongly related to tribological properties of machine elements, such as a load-carrying capacity and a wear resistance. Perhaps the deformation at the smallest slope of the material ratio curve would be very low, which may be related to high wear resistance. Therefore, the results of this work are tribologically important.

For example, it was found [6,7] that during low wear the Spq parameter decreased and the Smq parameter increased. Therefore, one can predict that during wear of initial one-process random surface, initially the major mode and, perhaps more importantly, the smallest slope of the material ratio curve would be obtained for the material ratio of 50%. Then, during wear, the smallest slope material ratio would be decreased to the Smq parameter of new created two-process surface. Then during the test, the material ratio of the smallest slope would be increased to 50%, and it would be still 50% even for further

Smq parameter increases. This prediction is important from a tribological point of view. Of course, during low wear, the smallest slope of the material ratio curve was reduced as the test progressed.

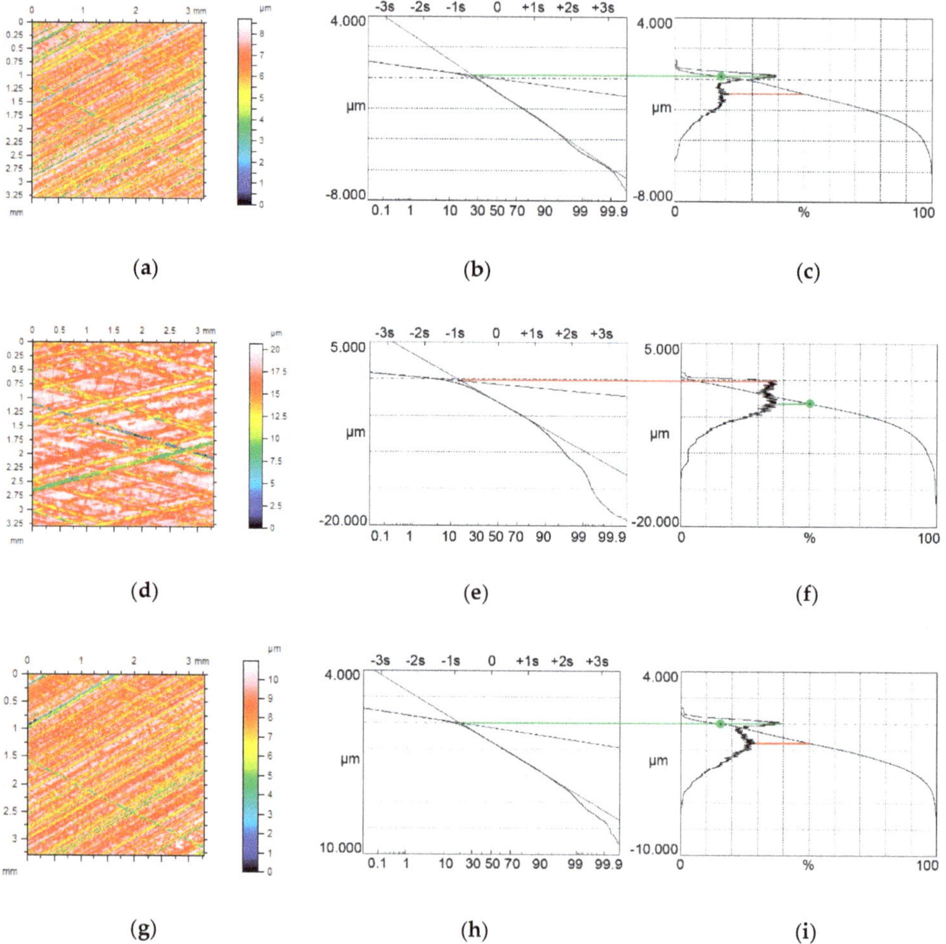

Figure 9. Contour plots (**a,d,g**) surface probability plot (**b,e,h**), material ratio curve and probability distribution (**c,f,i**) of two-process measured surfaces characterized by the following parameters: Spq = 0.31 µm, Svq = 1.61 µm, Smq = 24% (**a–c**), Spq = 0.49 µm, Svq = 2.89 µm, Smq = 15% (**d–f**) and Spq = 0.43 µm, Svq = 1.69 µm, Smq = 20% (**g–i**), three surfaces have bimodal ordinate distributions.

The findings obtained for areal 3D surface texture are also valid for 2D profile. However, due to the higher number of measuring points, the analysis of amplitude probability distribution of areal surface topography is easier compared to that of the 2D profile.

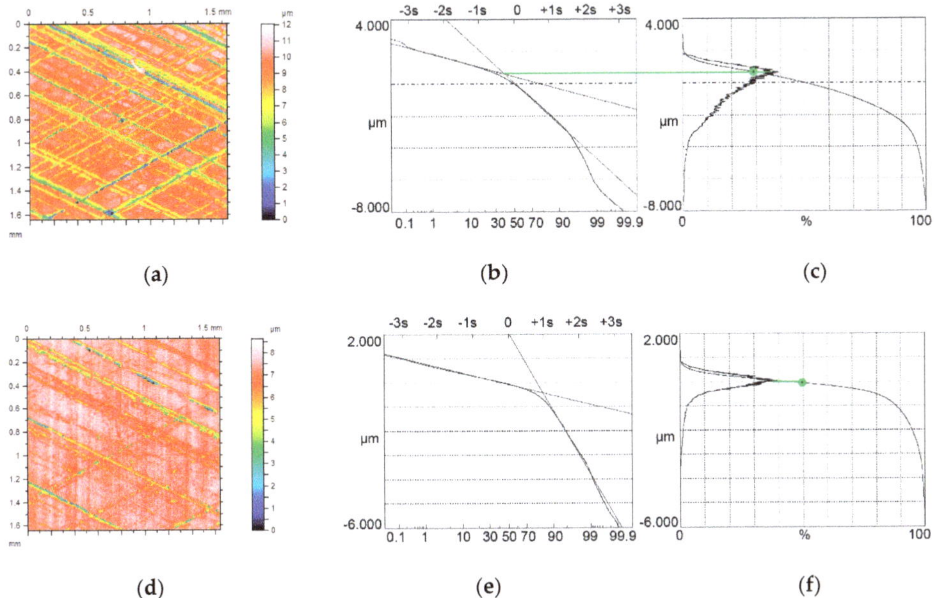

Figure 10. Contour plots (**a,d**) surface probability plot (**b,e**), material ratio curve and probability distribution (**c,f**) of two-process measured surfaces characterized by the following parameters: Spq = 0.61 µm, Svq = 2.04 µm, Smq = 38% (**a–c**) and Spq = 0.43 µm, Svq = 2.51 µm, Smq = 84%, both surfaces have unimodal ordinate distributions.

5. Conclusions

1. Limiting conditions of bimodal height distribution of two-process surface topography were developed. They depend on the ratio of the standard deviations of the valley and plateau parts Svq/Spq and on the material ratio at the transition between plateau and valley portions Smq. Based on these conditions, bimodal and unimodal height probability distributions were correctly discriminated for modeled and measured surfaces.

2. The bimodal ratio increased when the Svq/Spq ratio increased. Typically, the upper peak is the major mode. However, for low values of the Smq parameter and for low Svq/Spq ratio, the lower peak, which corresponds to the material ratio of 50%, can be the major mode.

3. When the Smq parameter is not lower than 50%, unimodal amplitude distribution exists. The mode and the smallest slope of the material ratio curve appear at the material ratio of 50%.

4. For unimodal height distribution and the value of the Spq parameter smaller than 50%, the mode corresponds to the Smq material ratio.

5. The results are functionally important because of the high tribological significance of the material ratio curve. In particular, the position of its smallest slope deserves attention.

Author Contributions: Conceptualization: P.P., R.R., M.W.; methodology, investigation and formal analysis: P.P., R.R., M.W.; software: P.P., R.R., M.W.; writing—original draft preparation: P.P., R.R., M.W.; writing—review and editing: P.P., R.R., M.W. All authors have read and agreed to the published version of the manuscript.

Funding: This research received no external funding.

Conflicts of Interest: The authors declare no conflict of interest.

References

1. Whitehouse, D.J. *Handbook of Surface Metrology*; Institute of Physics Publishing: Bristol, UK; Philadelphia, PA, USA, 1994.
2. Thomas, T.R. Computer simulation of wear. *Wear* **1972**, *22*, 84–90. [CrossRef]
3. Stout, K.J.; Spedding, T.A. The characterization of the combustion engine bore. *Wear* **1982**, *83*, 311–326. [CrossRef]
4. Stout, K.J.; Davis, E.J. Surface topography of cylinder bores—The relationship between manufacture, characterization and function. *Wear* **1984**, *95*, 111–125. [CrossRef]
5. Stout, K.J.; Davis, E.J.; Sullivan, P.J. *Atlas of Machined Surfaces*; Chapman & Hall: London, UK, 1990.
6. Thomas, T.R. *Rough Surfaces*; Imperial College Press: London, UK, 1999.
7. Grabon, W.; Pawlus, P.; Wos, S.; Koszela, W.; Wieczorowski, M. Evolutions of cylinder liner surface texture and tribological performance of piston ring-liner assembly. *Tribol. Int.* **2018**, *127*, 545–556. [CrossRef]
8. Campbell, J.C. Cylinder bore surface roughness in internal combustion engines: Its appreciation and control. *Wear* **1972**, *19*, 163–168. [CrossRef]
9. Jeng, Y.R. Impact of plateaued surfaces on tribological performance. *Tribol. Trans.* **1996**, *39*, 354–361. [CrossRef]
10. Dzierwa, A.; Pawlus, P.; Zelasko, W.; Reizer, R. The study of the tribological properties of one-process and two-process textures after vapour blasting and lapping using pin-on disc tests. *Key Eng. Mater.* **2013**, *527*, 217–222. [CrossRef]
11. Buj Corral, I.; Calvet, J.V.; Salcedo, M.C. Use of roughness probability parameters to quantify the material removed in plateau-honing. *Int. J. Mach. Tools Manuf.* **2010**, *50*, 621–629. [CrossRef]
12. Rosen, B.G.; Anderberg, C.; Ohlsson, R. Parameter correlation study of cylinder liner roughness for production and quality control. *Proc. Inst. Mech. Eng. Part B J. Eng. Manuf.* **2008**, *222*, 1475–1487. [CrossRef]
13. Hu, S.; Huang, W.; Brunetiere, N.; Song, Z.; Liu, X.; Wang, Y. Stratified effect of continuous bi-Gaussian rough surface on lubrication and asperity contact. *Tribol. Int.* **2016**, *104*, 328–341. [CrossRef]
14. Hu, S.; Huang, W.; Shi, X.; Peng, Z.; Liu, X.; Wang, Y. Bi-Gaussian stratified effect of rough surfaces on acoustic emission under a dry sliding friction. *Tribol. Int.* **2018**, *119*, 308–315. [CrossRef]
15. Hu, S.; Huang, W.; Shi, X.; Peng, Z.; Liu, X. Mechanism of bi-Gaussian surface topographies on generating acoustic emissions under a sliding friction. *Tribol. Int.* **2019**, *131*, 64–72. [CrossRef]
16. Pawlus, P.; Zelasko, W.; Reizer, R.; Wieczorowski, M. Calculation of plasticity index of two-process surfaces. *Proc. Inst. Mech. Eng. Part J J. Eng. Tribol.* **2017**, *231*, 572–582. [CrossRef]
17. Pawlus, P.; Zelasko, W.; Dzierwa, A. The effect of isotropic one-process and two-process surface textures on contact of flat surfaces. *Materials* **2019**, *12*, 4092. [CrossRef]
18. Pawlus, P.; Reizer, R.; Wieczorowski, M.; Zelasko, W. The effect of sampling interval on the predictions of an asperity contact model of two-process surfaces. *Bull. Pol. Acad. Sci. Tech. Sci.* **2017**, *65*, 391–398. [CrossRef]
19. Pawlus, P.; Reizer, R.; Wieczorowski, M. Characterization of the shape of height distribution of two-process profile. *Measurement* **2020**, *153*, 107387. [CrossRef]
20. Bohm, H.J. Parameters for evaluating the wearing behaviour of surfaces. *Int. J. Mach. Tools Manuf.* **1992**, *32*, 109–113. [CrossRef]
21. Mummery, L. *Surface Texture Analysis—The Handbook*; Hommelwerke GmbH: Muehlhausen, Germany, 1990.
22. Schneider, U.; Steckroth, A.; Rau, N.; Hubner, G. An approach to the evaluation of surface profiles by separating them into functionally different parts. *Surf. Topogr.* **1988**, *1*, 343–355.
23. Nielsen, H.S. New approaches to surface roughness evaluation of special surfaces. *Precis. Eng.* **1988**, *10*, 209–213. [CrossRef]
24. Malburg, M.C.; Raja, J.; Whitehouse, D.J. Characterization of surface texture generated by plateau–honing process. *CIRP Ann.* **1993**, *42*, 637–640. [CrossRef]
25. Sannareddy, H.; Raja, J.; Chen, K. Characterization of surface texture generated by multi–process manufacture. In Proceedings of the 7th International Conference On Metrology and Properties of Engineering Surfaces, Gothenburg, Sweden, 26–29 June 1998; pp. 111–117.
26. Whitehouse, D.J. Assessment of surface finish profiles produced by multiprocess manufacture. *Proc. Inst. Mech. Eng. Part B J. Eng. Manuf.* **1985**, *199*, 263–270. [CrossRef]

27. Zipin, R.B. Analysis of the Rk surface roughness parameter proposals. *Precis. Eng.* **1990**, *12*, 106–108. [CrossRef]
28. Pawlus, P.; Reizer, R.; Wieczorowski, M.; Krolczyk, G. Material ratio curve as information on the state of surface topography—A review. *Precis. Eng.* **2020**, in press. [CrossRef]
29. Pawlus, P. Simulation of stratified surface topography. *Wear* **2008**, *264*, 457–463. [CrossRef]
30. Peng, W.; Bhushan, B. Modelling of surfaces with a bimodal roughness distribution. *Proc. Inst. Mech. Eng. Part J J. Eng. Tribol.* **2000**, *214*, 459–470. [CrossRef]
31. Hu, S.; Brunetiere, N.; Huang, W.; Liu, X.; Wang, Y. Bi-Gaussian surface identification and reconstruction with revised autocorrelation functions. *Tribol. Int.* **2017**, *110*, 185–194. [CrossRef]
32. Hu, S.; Brunetiere, N.; Huang, W.; Liu, X.; Wang, Y. Continuous separating method for characterizing and reconstructing bi-Gaussian stratified surfaces. *Tribol. Int.* **2016**, *102*, 454–462. [CrossRef]
33. Perez-Rafols, F.; Almqvist, A. Generating randomly rough surfaces with given height probability distribution and power spectrum. *Tribol. Int.* **2019**, *131*, 591–604. [CrossRef]
34. Wu, J.J. Simulation of rough surfaces with FFT. *Tribol. Int.* **2000**, *33*, 47–58. [CrossRef]
35. Zhang, C.; Mapes, B.E.; Soden, B.J. Bimodality in tropical water vapour. *Q. J. R. Meteorol. Soc.* **2003**, *129*, 2847–2866. [CrossRef]
36. Schilling, M.F.; Watkins, A.E.; Watkins, W. Is Human Height Bimodal? *Am. Stat.* **2002**, *56*, 223–229. [CrossRef]
37. Eisenberger, I. Genesis of Bimodal Distributions. *Technometrics* **1964**, *6*, 357–363. [CrossRef]
38. Holzmann, H.; Vollmer, S. A likelihood ratio test for bimodality in two-component mixtures with application to regional income distribution in the EU. *AStA Adv. Stat. Anal.* **2008**, *92*, 57–69. [CrossRef]
39. Sitek, G. The modes of a mixture of two normal distributions. *Silesian J. Pure Appl. Math.* **2016**, *6*, 59–67.
40. Behboodian, J. On the modes of a mixture of two normal distributions. *Technometrics* **1970**, *12*, 131–139. [CrossRef]
41. Grabon, W.; Pawlus, P.; Wos, S.; Koszela, W.; Wieczorowski, M. Effects of honed cylinder liner surface texture on tribological properties of piston ring-liner assembly in short time tests. *Tribol. Int.* **2017**, *113*, 137–148. [CrossRef]
42. Grabon, W.; Pawlus, P.; Wos, S.; Koszela, W.; Wieczorowski, M. Effects of cylinder liner surface topography on friction and wear of liner-ring system at low temperature. *Tribol. Int.* **2018**, *121*, 148–160. [CrossRef]

 *materials*

Article

Analysis of Surface Microgeometry Created by Electric Discharge Machining

Tomasz Bartkowiak *, **Michał Mendak, Krzysztof Mrozek and Michał Wieczorowski**

Institute of Mechanical Technology, Poznan University of Technology, 60-965 Poznań, Poland;
michal.mendak@put.poznan.pl (M.M.); krzysztof.mrozek@put.poznan.pl (K.M.);
michal.wieczorowski@put.poznan.pl (M.W.)
* Correspondence: tomasz.bartkowiak@put.poznan.pl

Received: 22 July 2020; Accepted: 28 August 2020; Published: 30 August 2020

Abstract: The objective of this work is to study the geometric properties of surface topographies of hot-work tool steel created by electric discharge machining (EDM) using motif and multiscale analysis. The richness of these analyses is tested through calculating the strengths of the correlations between discharge energies and resulting surface characterization parameters, focusing on the most representative surface features—craters, and how they change with scale. Surfaces were created by EDM using estimated energies from 150 to 9468 µJ and measured by focus variation microscope. The measured topographies consist of overlapping microcraters, of which the geometry was characterized using three different analysis: conventional with ISO parameters, and motif and multiscale curvature tensor analysis. Motif analysis uses watershed segmentation which allows extraction and geometrically characterization of each crater. Curvature tensor analysis focuses on the characterization of principal curvatures and their function and their evolution with scale. Strong correlations ($R^2 > 0.9$) were observed between craters height, diameter, area and curvature using linear and logarithmic regressions. Conventional areal parameter related to heights dispersion were found to correlate stronger using logarithmic regression. Geometric characterization of process-specific topographic formations is considered to be a natural and intuitive way of analyzing the complexity of studied surfaces. The presented approach allows extraction of information directly relating to the shape and size of topographic features of interest. In the tested conditions, the surface finish is mostly affected and potentially controlled by discharge energy at larger scales which is associated with sizes of fabricated craters.

Keywords: EDM; craters; multiscale analysis; surface topography; microgeometry

---

1. Introduction

The topography of a manufactured surface is a direct product of the physical phenomena occurring during its formation, and contains information critical to comprehend and reconstruct what happened. The investigation of the representative surface features created by the manufacturing process is, therefore, particularly valuable for those processes which are still not fully understood. This involves EDM (electric discharge machining), in which energy is transferred between tool and workpiece via electric discharge.

There exist multiple ways of describing surface topography of engineered surfaces. The most typical approach is the characterization by ISO 25178-2 areal texture parameters [1] calculated for the measured regions after form removal and filtration. Those analyses are aimed at quantifying the properties of an entire analyzed region as a set of texture parameters, which are mostly based on statistical measures such as average, standard deviation or rms (root mean square) [2]. Another approach is to describe process-specific topographic features which are inherent to their fabrication such as ridges and valleys for milling or turning, and craters for EDM. Some feature-based parameters, included in ISO standard, are potentially relevant for characterization of the formations created by electric discharge machining. However, they do not provide the information about the scales of those features. In other words, topographic features of a given size are best discernible when observed at particular scales. This phenomenon is the principle of the third approach, i.e., multiscale methods [3]. The importance of scale lies in the characterizations of physical interactions between formation process and resulted surface topography. These can occur at multiple scales during fabrication.

Geometric quantities of engineered surfaces, such as length [4] or area [5] usually change with the scale of observation. Recently developed multiscale curvature method allows studying the evolution of principle, mean and Gaussian curvatures calculated at a given location on the surface with scale. These measures are indicative of local shape, e.g., can determine concavity or convexity and the amount by which surface bends in particular directions [6,7].

In EDM, fabricated surface can be conceived as a composition of craters and plateaus, which individual geometric properties, i.e., depth, radius, volume and curvature strongly depend on processing parameters. Knowledge about the correlations between formation and surface topographies, i.e., roughness or finish, is vital for process design and control [8,9]. There are essentially two components of the value added to a workpiece by the EDM process: form and surface finish. Those can be controlled by technological parameters (e.g., voltage, current, pulse time and many more), electrode material and shape and dielectric fluid [10,11]. There is probable not a single universal technological parameter which can be used to determine the microgeometry of resulted surface. Craters dimensions strongly depend on the amount of energy that is transferred from the electrode to the workpiece via electric discharges, in a stochastic manner. Klocke et al. found that the depth of recast layer was influenced by resistance and capacity in circuit, both of which impact on the discharge energy and higher energy led to thicker recast layer [12]. Giridharan et al. developed so-called "anode model" in which the energy that reaches the workpiece and forms the crater is proportional to the discharge energy [13]. The presented numerical and experimental results were shown to be in a good agreement in terms of crater morphology. Other studies have indicated proportional relations between discharge and crater volume [14], area [15], diameter [16] or size [17]. Ding et al. showed that in micro wire electrical discharge machining, the spark energy directly influenced both the average diameter and the maximum depth of craters. They found that relationship to follow a logarithmic trend [18]. Some studies indicate linear relations between craters diameter and depth [14,19]. All of those studies show strong functional relations between certain crater dimensions and discharge energy for constant material properties (e.g., physical properties of electrode, workpiece and dielectric fluid). This study concentrates on the energy of electric discharge as a unifying technological parameter that strongly correlates with microgeometry of resulted surface topographies taking into account the above assumption.

In the literature, surface topographies are often characterized using basic ISO parameters: average roughness, Ra (profile) or Sa (areal) [1]. Those parameters were mostly developed for an analysis of conventionally machined surfaces (milled, turned or ground) and they lack the ability to exploit the complexity of non-traditionally manufactured surfaces [2]. They were used in EDM to show the affect of smaller pulse duration on the creation of smaller craters, characterized by lower values of Ra [11]. Masuzawa et al. showed that low open circuit voltage produced small craters and, hence, lesser surface roughness expressed also in average roughness [20]. Guu stated that greater pulse currents and longer pulse durations produced textures of higher Ra. [21]. In micro-EDM, average areal roughness (Sa) seemed to correlate strongly with discharge energy [15]. Bäckemo et al. created a predictive computer supported model of surface topography created as a result of impacts caused by electric discharges between metallic substrate and an electrode [22]. They found that the roughness parameters followed an inverse exponential trend as a function of impact number, and that the strongly concave curvatures reached equilibrium at an earlier impact number for lower depth to radius ratios. The size of each individual impact can depend on the charge that builds up before each spark, which, in turn, is seen as a function of the electric parameters of the process [23]. Klink et al. showed that basic profile roughness parameters are not sufficient to describe the complexity of EDM-created topographies and they proposed the use of average groove width RSm and the average profile gradient $R\Delta q$ for a more sophisticated surface topography description [24].

The traditional height parameters do not consider the spacing nor the sequence of the heights, as well as they do not characterize characteristic surface features. The development of feature parameters addresses the latter. These parameters rely on a technique called *segmentation*, which is based upon the application of a watershed algorithm [25], associated with an algorithm for simplifying graphs that describe the relationships between individual points [26]. Segmentation is useful in identifying significant peaks and pits, and can be used for calculating peak density and peak curvature. In addition, specific parameters were created to quantify the area and mean volume of motifs identified by segmentation, distinguishing between open and closed motifs, depending upon whether or not they are in contact with the edge of the microscopic image. Such distinction is necessary, since open motifs do not provide full information about a particular crater that they describe. This analysis appears to be a prospective candidate to analyze geometric quantities of craters.

Electric discharge machining is a manufacturing process in which material is removed from the workpiece by a series of rapidly recurring current discharges between a tool and workpiece electrodes, separated by a dielectric fluid (liquid or gas) and in response to voltage pulses. The physical phenomenon occurring between the electrodes in EDM when manufacturing surface features on the micrometer scale is not entirely understood [27]. For short pulses and energies, there is not enough time for material to be adequately heated for removal and therefore almost none takes place. In that case, electrostatic force which acts on the surface becomes an essential factor in the removal of metal for short pulses [28]. The acquisitions of the resulting surface topographies and the multiscale analysis of the geometry of created microfeatures by EDM can provide evidence of the material response to the discharge [15]. Relations between curvature and discharge energies (between 18 and 16,500 nJ) in micro-EDM were studied by Bartkowiak and Brown [7]. They also suggested, by having analyzed the principal, mean and Gaussian curvature, that different formation processes governed the creation of surfaces created by higher energies.

This study aims at characterizing the geometric properties of fabricated surface topographies in micrometer scales. This is demonstrated by the use of motif and multiscale analysis to characterize surfaces of hot-work tool steel created by electric discharge machining (EDM), and then to study correlations between the discharge energies and the resulting surface topographies, focusing on microgeometry of craters. In particular, the strengths of the correlations (R^2) between motif and curvature characterizations (i.e., principal, Gaussian or mean curvature) and discharge energies is sought as a function of scale. Motifs are used here to derive geometrical properties of created craters (e.g., area, depth and diameter), whereas curvature allows characterization of their shapes in multiple scales. The proposed approach is feature-based and focuses on the geometric specificity of the topographies created by EDM. As a comparison, additional conventional analysis of surface texture using ISO 25178 standard and its areal characterization parameters are performed. Geometric characterizations of process-specific topographic objects is considered to be a natural and intuitive way of analysis the complexity of EDM surfaces. In contrast to analysis of surface topography through areal texture parameters (as in ISO 25178 2), the presented approach allows extraction of information directly relating to the shape and size of topographic features of interest. The richness of this information is tested via correlations with processing parameter.

The approach here is to calculate the curvature tensors on a surface as functions of position and scale. Statistical characterization parameters of principal, mean and Gaussian are compared with the pulse energies in EDM as calculated from technological parameters. This is accompanied with motif analyses, in which the height or depth, diameter, area and volume of detected motifs are evaluated. According to our best knowledge, this is the first study, in which areal motifs are used to characterize the topographic features created by EDM. Other studies that covered some geometrical aspects of craters, used manual measurements of those quantities from measured datasets. Motif and curvature analysis is supplemented with a comprehensive study using ISO 25178 standard areal parameters, in the analysis of process–surface-texture interactions in this manufacturing process. The differences and similarities between those methods are discussed with a reference to discharge energies and characterization capabilities. Application of both methods reveals relations between microgeometry of surface topographies and formation process, which is impossible to be observed using a conventional approach.

2. Materials and Methods

2.1. Samples Preparation

In this work, roughing and finishing EDM processes of twelve flat surfaces were performed (Figure 1a) with different technological parameters. Machined surfaces were prepared separately with dimensions of 30 mm × 30 mm and height of 3 mm on a 1.2363/X100CrMoV5 (EN 10027-2/ISO4957) on a steel block hardened to 52 HRC, with dimensions of 135 mm × 100 mm × 13 mm. Chemical composition of the steel used in the study is shown in Table 1. Hardness was tested prior to machining, using Rockwell method using C scale and N3A testing machine (EMCO-TEST Prüfmaschinen GmbH, Kuchl, Austria) equipped with 120° diamond spheroconical intender. Minor load was set to 98.07 N and load was 1471.0 N.

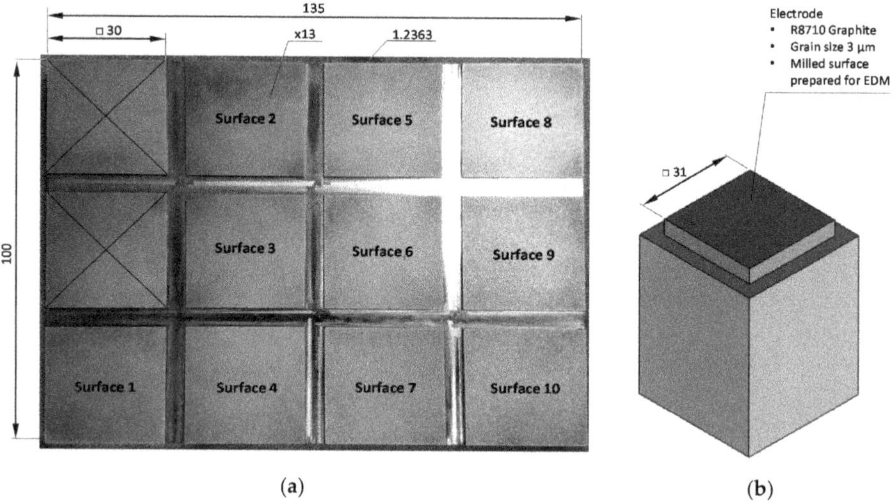

Figure 1. Materials prepared for the study: (**a**) 52 HRC workpiece made of 1.2363 steel as machined; (**b**) graphite electrode with 3 μm grain size.

Table 1. The percentage chemical composition of 1.2363 steel according to EN 10027-2.

C	Si	Mn	P	S	Cr	Mo	V
0.95–1.05	0.10–0.40	0.40–0.80	Max 0.030	Max 0.030	4.80–5.50	0.90–1.20	0.15–0.35

Prior to EDM, the twelve surfaces and base area of the steel block were ground to a height of 13 ± 0.01 mm. The purpose of this process is to obtain the same initial surface texture and height of the workpiece for each of the machined surfaces. Basic areal height parameters according to ISO 25178 2 of as-ground surfaces are shown in Table 2. Waviness and roughness parameters were calculated using robust Gaussian filter with cutoff wavelength of 0.25 mm applied to each of primary surfaces after levelling. Five measurements were done at representative regions using a focus variation microscope (FVM)—Alicona Infinite Focus G5 (Alicona Imaging GmbH, Raaba/Graz, Austria).

Table 2. Basic areal height parameters calculated for pre-electric discharge machined (EDM) (as-ground) surfaces according to ISO 25178 2. Data is presented as mean ± standard deviation for roughness and waviness.

Parameter	Sq (μm)	Sp (μm)	Sv (μm)	Sz (μm)	Sa (μm)
Roughness	0.541 ± 0.281	9.852 ± 4.866	3.752 ± 1.356	13.604 ± 5.448	0.418 ± 0.218
Waviness	0.232 ± 0.116	0.733 ± 0.389	0.577 ± 0.254	1.310 ± 0.635	0.184 ± 0.091

Each surface was then machined with two types of electrodes—roughing and finishing, which had identical geometric parameters (Figure 1b). A total of 24 graphite electrodes with the same geometric and material characteristics were made for the experiment (see physical properties in Table A1 (Appendix A). Two of the dozen created surfaces were made with test parameters and were excluded from the further analysis.

The electric discharge machining was conducted using the high-end EDM die-sinking machine Form X400 (Agie Charmilles, Losone, Switzerland) and was divided into two stages (see Table A2 in Appendix A for machine parameters and dielectric fluid). In the first roughing stage, 1.5 mm layer of material was removed in the z-direction (Figure 2a). This reflects to the standard industrial process in which allowances for this material and geometry are chosen by machine tool control unit automatically to ensure that any residual form, waviness and roughness is fully removed. For as-ground surfaces maximum height (*Sz*) for waviness and roughness is two orders of magnitude lower than layers removed by electric discharge machining. During the rough EDM process the electrode movement was limited to the z-axis as reciprocating motions. There were two types of movement in z-direction—the working travel, which allows the tool electrode to get closer to the machined material and to let the electrical discharges occur, and the idle travel, during which flushing and removal of the eroded material (by-products) from the spark gap take place. The second stage of the processing was finishing, which consisted of two types of movements (Figure 2b,c). The first one was a motion on the *xy* plane in a circle with a diameter of 0.1 mm above the machined surface. After removal of subsequent layers of material, the tool electrode moved forward along z-axis and repeated circular motion. The second type was a motion in a circular trajectory with a diameter of 0.1 mm on planes perpendicular to the *xy* plane. In both EDM processes, a commercially available synthetic hydrocarbon fluid (108 MP-SE by Novotec BV, Reuver, the Netherlands) was used.

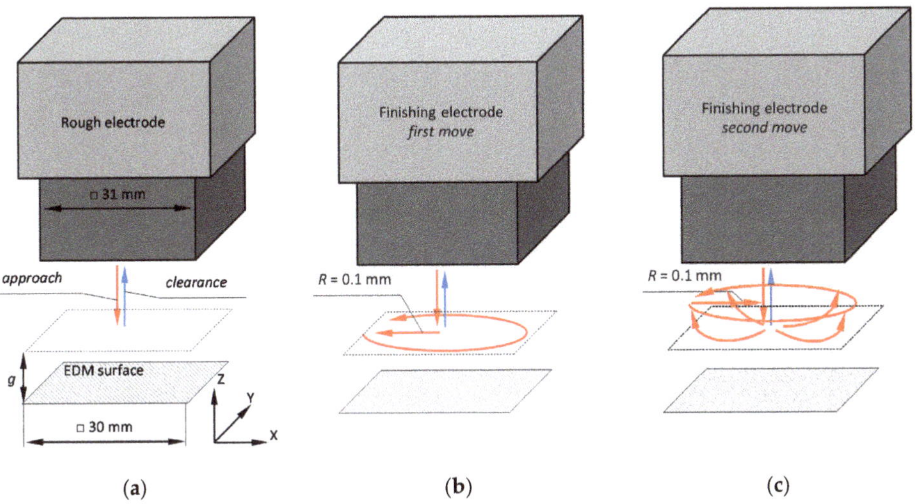

Figure 2. Experiment: (**a**) rough EDM; (**b**) finishing EDM—first move; (**c**) finishing EDM—second move.

In this study, a primary focus is given on the surface topography in relation to the technological parameters of the finishing process. The most important parameters of the process for each of ten analyzed surfaces (S1–S10) are included in Table 3. These include: spark voltage (U), current (I), on- and off-time (T_{on} and T_{off}) as well as face and side gap. Based on the given technological parameters we estimate the discharge energy by applying the formula:

$$E = U{\cdot}I{\cdot}T_{on},$$ (1)

Table 3. Technological parameters of the finishing EDM.

Surface	U (V)	I (A)	T_{on} (µs)	T_{off} (µs)	Face Gap (mm)	Side Gap (mm)	Discharge Energy (µJ)	Theoretical VDI Class
S1	100	3.0	0.5	6.9	0.0126	0.126	150	16
S2	100	3.0	0.9	7.1	0.0150	0.0150	270	17
S3	100	3.0	1.8	7.5	0.0155	0.0155	540	18
S4	100	3.0	3.0	8.0	0.0160	0.0160	900	19
S5	100	3.0	4.8	8.8	0.0164	0.0164	1440	20
S6	100	3.0	7.3	9.9	0.0244	0.0206	2190	21
S7	100	3.0	10.9	11.5	0.0333	0.0253	3270	22
S8	100	3.0	16.1	13.8	0.0432	0.0304	4830	23
S9	100	3.2	22.8	16.6	0.0543	0.0361	7296	24
S10	100	3.6	26.3	19.7	0.0567	0.0376	9468	25

Without considering the actual values of voltage and current over time. This was done for the practical reason as a commercial EDM machine tool either does not measure those quantities over time or simply does not allow to extract this kind of information directly from the control unit without interfering with hardware and software. Still those technological parameters can be adjusted to achieve different surface finish. The approach to calculate the energy straight from technological parameters of pulses as offered by the control system of EDM machine was also done by Ramasawmy and Blunt [29]. Although, no information about the exact values of voltage and current over time is given, it seems not to interfere with achieving strong correlations with areal texture parameters. This might suggest that this effect is either compensated internally by control system, e.g., by averaging at least one technological parameter, or that it is similar for all discharge energies.

In this process, they were selected to achieve different surface finish according to VDI 3400 standard. This standard distinguishes 45 classes (VDI K) depending on average roughness (*Ra*) which can be determined by the following formula:

$$VDI\ K = 20{\cdot}lg(10{\cdot}Ra).$$ (2)

In the presented case, processing parameters were determined automatically by the machine tool control system to obtain surface finish between VDI 16 and VDI 25 for a given electrode (size and material) and workpiece (material). Those classes were chosen to maintain the constant polarity during machining and to achieve different topographies. An operator cannot manipulate any detailed parameters of electric discharge for a selected surface finish. This study reflects the real industrial process and the presented tests were performed in a manner corresponding to the technological processes as indicated by the machine tool manufacturer.

Due to the financial constraints of this study only a single sample was manufactured. However, the stability of the discussed EDM process can be supported by the fact that the machine tool produces inserts for injection molding on a regular basis. Those are regularly tested by the owner of the machine so to check their dimensional and surface quality. The machine is fairly new, and is high-end and well-maintained, which can also testify for the very probable replicability of the results.

2.2. Measurements of Surface Topography

Surfaces (S1 to S10) were measured with a focus variation microscope (FVM), Alicona Infinite Focus G5, using Single Imagefield mode and a 50× objective. Measurement settings are presented in Table 4. In order to handle a potential non-homogeneity of the results, each surface was measured at five independent locations in a cross-like pattern with the central area located in the center of the machined surface. A combination of coaxial light source and an external lighting was used, making only slight adjustments, as lighting conditions varied among the measured surfaces. Surfaces were treated with isopropyl alcohol to remove dielectric fluid and dust.

Table 4. Setup parameters of the focus variation microscope.

Parameter	Unit	Value
Magnification	-	50×
Area Dimensions	μm	323×323
Est. Vertical Resolution	μm	0.016
Est. Lateral Resolution	μm	2.31
Sampling intervals in x- and y-directions	μm	0.176

Each measured surface underwent the following steps, using the exact same processing parameters, therefore producing comparable results:

- Dataset leveling using least squares method;
- Outliers removal;
- Filling in the non-measured points [30];
- Calculation of areal ISO standard parameters, curvature tensor analysis and motif analysis from primary surface;
- Extraction of roughness and waviness surface with gaussian filter;
- Motif analysis and areal parameters extraction from both roughness and waviness surface.

In FVM, height registration is based on contrast estimation from a given region, providing a proper lighting is used [31]. When measuring surfaces with smooth, reflective areas the FVM often miscalculates the height of the surface resulting in "the plateau-like formation", as described by Senin et al. [32]. The same effect was observed in this study and the example is presented on Figure 3a. Therefore, the raw measurement (Figure 3a), had to be subjected to preliminary filtration in order to remove the plateau-like artefacts. Three different approaches were considered: threshold method, robust gaussian filter and morphological filter. The choice of these three methods was dictated by both budget and software limitations. Nevertheless, various filtration methods available in the MountainsMap software were tested, including spatial and standard filters. The three aforementioned methods yielded most promising results, with the least significant, or even negligible, effect on standard roughness parameters. The first method, which was based on a manual choice of lower and upper thresholds for each surface, did remove the plateau-like regions, leaving just the steep slopes, which could be later removed using a dedicated outlier removal technique. Areas that were filtered out were replaced with non-measured points, which were then filled as a smooth shape (calculated from neighboring points). This method was discarded as, for several datasets, a significant portion of other pits from primary surface was also removed, resulting in disqualifying distortions at the later stage of the analysis. The other method utilized the robust gaussian filtration, which is characterized by its mean plane following the general trend of the surface, without being affected by outliers. Its use in extraction of surface features was extensively examined by Lou et al. [33]. In that study it proved to be ineligible, because it either overly smoothed the surface out or failed to exclude the outliers. The last method was a morphological closing filter using sphere with 16 μm diameter. This value was set intentionally as it compromised between artefacts filtration and good surface mapping. The effects of using such filter on surface profile is presented in Figure 3b.

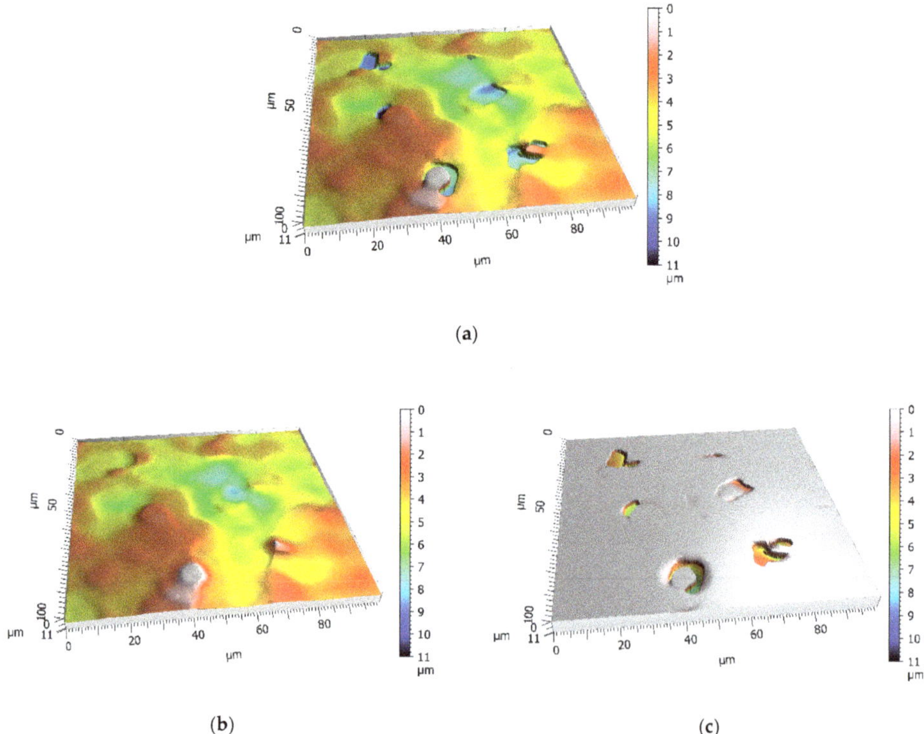

Figure 3. Morphological surface filtration: (**a**) raw surface; (**b**) after filtration using 16μm sphere, closing filter; (**c**) residue surface, all outliers clearly visible.

The downside of morphological closing filter is, that, on several surfaces, it left a few isolated peaks which had to be removed manually. This could be avoided, by adding the opening operation, however the authors' intention was to limit the influence of outliers filtration on the measured surface. No additional form or noise removal was used. Form removal was not applicable due to relatively small measurement area and it was feared it might influence the craters geometry representation. Only levelling using least squares method was applied.

There is little consistency regarding surface or profile filtration among researchers, partly due to lack of information on the filters, cut-off lengths or nesting index used to extract waviness and roughness profile. Some researchers [34,35] provided the Ls or cut-off length (2.5 mm) and the standards name. Some [24] simply stated compliance with the standard.

Areal parameters do not differentiate between roughness, waviness or even primary surface, and therefore a nesting index value must be stated before presenting the results. In this analysis, all three types surfaces were taken into account. As stated by Townsend et al. [36], "filtering is based on the roughness or scale of the largest significant feature". In case of the predicted and calculated roughness values, it indicated that the correct nesting index value should be 0.25 mm. Choosing a smaller value would lead to an excess transfer of roughness information into the waviness surface. The filtering is required, as further stated by Townsend et al. [36], because of significant variations between the results of filtered and unfiltered surface. Given the relatively small measurement area, which limits the value of nesting index, an unfiltered surface was also subjected to areal parameter analysis.

2.3. Analysis of Microgeometry

In this study, each surface is treated as a composition of craters and plateaus resulted from electric discharging with electrode. In order to fully characterize the geometry of those features, three types of analyses were conducted: conventional approach with ISO standard areal parameters, motif analysis and multiscale curvature tensor analysis. The last two directly focus on the characterization of geometrical features or parameters of surface.

2.3.1. Conventional Approach with ISO Parameters

Influence of EDM parameters on surface topography was studied by numerous authors. Most publications concern profile roughness [24,29,34,37,38], and only few considered areal parameters [29,35] (mainly the height class, e.g., Sa or Sq), analogical to the most widely used profile roughness characteristics. Most of them considered at least three parameters, and two examined just one [34,39]. The complexity of EDM surfaces may require more complex analysis to comprehensively study all aspects of their topography. For this analysis, MontainsMap software (version 7, Digital Surf, Besancon, France) is used to calculate areal parameters described in ISO 25178, divided into seven groups:

- Height, which is a class of parameters, that quantify the information on the z-axis of the surface;
- Functional, derived from the Abbott–Firestone curve, which describes the height cumulative distribution on the surface;
- Spatial, which describe topographic characteristics and quantify the lateral information of the surface;
- Hybrid, a class of surface finish parameters, that consider both the amplitude and spacing between heights;
- Functional (volume), which involves volume parameters calculated from the Abbott-Firestone curve;
- Feature, derived from the segmentation of surface into motifs; and
- Functional (stratified surfaces), which includes parameters designed for automotive industry, considering certain aspects of a surface interactions, such as lubrication and grinding.

All parameters from those groups, used in this study are presented in Table 5. In study, a special focus is given to height, feature and functional groups. Height parameters are the most widely used in the academia and the industry, as they are the most intuitive and simple to calculate for given dataset. These parameters, such as Sa (average roughness) correlated strongly with technological parameters in EDM [15]. Feature parameters characterize geometrical properties of motifs which may be associated with craters or dales created by electric discharge. Functional parameters allow to characterize geometric parameter—volume of the void and material with respect to surface core, peaks or valleys, what might be associated with craters [40].

Table 5. ISO areal parameters examined in the study.

Parameter Group	Parameter Symbol
Height Parameters	Sq, Ssk, Sku, Sp, Sv, Sz, Sa
Functional Parameters	Smr, Smc, Sxp
Spatial Parameters	Sal, Str, Std
Hybrid Parameters	Sdq, Sdr
Functional Parameters (Volume)	Vm, Vv, Vmp, Vmc, Vvc, Vvv
Feature Parameters	Spd, Spc, S10z, S5p, S5v, Sda, Sha, Sdv, Shv
Functional Parameters (Stratified surfaces)	Sk, Spk, Svk, Smr1, Smr2

2.3.2. Motif Analysis

Different manufacturing techniques lead to creation of a vast variety of surface textures, each having its own unique artefacts and characteristics. In EDM, one of the characteristic features of the manufactured surface is a crater, created by electric discharge. This kind of features are hard to examine using standard roughness analysis, although several functional correlations were found [29,35].

The term "motifs" can be used to describe either hills or dales. Its extreme points are called peaks and pits, respectively. They are limited by lines called course (hills) or ridge line (dales). Motifs are established using method called segmentation, which utilizes a watershed algorithm [25].

This method is commonly used by researchers, mainly in the analysis of surfaces created by additive manufacturing techniques [32,33,36] and in applications, where grain or fault detection is desired, such as protruding diamond grains in grinding wheels [41].

In this study, watershed segmentation and motifs analysis were used to identify craters and perform a morphological characterization of the detected features, including:

- Height—distance between the lowest saddle point and pit;
- Area—horizontal area limited by the ridge line;
- Volume—volume of the void below the plane of the lowest saddle point;
- Equivalent diameter—diameter of a disk which area is equal to that of a grain;
- Mean diameter—average diameter of a disc constructed at the center of the gravity of a grain.

There was no pre-processing (areal filter) used. Pruning criteria were established as follows:

- Height—<0.75% Sz (maximum height),
- Area—<0.25% of surface area

The above values provided the most consistent and reliable segmentation, without signs of "over-" or "under-segmentation", where visual boundaries of the craters were consistent with calculated ridge lines. Calculation method was set to pit detection using MountainsMap 7 software.

2.3.3. Multiscale Curvature Tensor Analysis

In general, the curvature tensor T is a symmetric 3×3 matrix that can be expressed as a product:

$$T = D \cdot P \cdot D, \tag{3}$$

where: $P = (k_1, k_1, n)$ and

$$D = \begin{bmatrix} \kappa_1 & 0 & 0 \\ 0 & \kappa_2 & 0 \\ 0 & 0 & 0 \end{bmatrix}, \tag{4}$$

The eigenvalues κ_1 and κ_2 represent the principal curvatures, maximal and minimal magnitudes respectively. The sign is used to designate concave surfaces as positive and convex as negative. The eigenvectors k_1, k_2 are the corresponding principal directions of maximum and minimum curvature and n is the surface normal unit vector at the location of the calculated curvature. Mean (H) and Gaussian curvature (K) can be calculated from principle curvatures.

Measurement data is usually a discrete set of point coordinates described in Cartesian coordinate system. In this paper, the curvature tensor is estimated from datasets with regular spacing in x- and y-direction.

The entire concept of multiscale curvature method bases on the principle that the shape of objects, as well as their other geometric characteristics, depend on the scale of observation. For finer scales, surfaces seem to contain more geometrically complex details which curvature is high, whereas for larger scales, they appear flat or only form is visible. Considering the analyzed datasets, the geometry of craters should manifest itself at particular ranges of scales associated with their sizes. This is examined by calculation of curvature tensor at each location for range of scales between 0.352 and 13.716 μm.

The result of multiscale curvature tensor analysis is a curvature tensor calculated for each triangular patch whose size is scale-dependent. As presented in [42], distributions of maximum, minimum, mean and Gaussian curvature can be derived at a particular scale and simple statistical measures can be calculated from that distribution. These include average (signed with "a") and standard deviation (with "q") of curvature distributions: $\kappa 1a$, $\kappa 1q$, $\kappa 2a$, $\kappa 2q$, Ha, Hq, Ka and Kq. They describe central tendencies and variabilities including sign of curvature (positive or negative). In this work, new characterization parameters are proposed which quantify curvature distribution regardless the sign by taking absolute values. They are defined at a particular scale s as below:

- $\kappa_1 a_{abs}$—average absolute maximum curvature

$$\kappa_1 a_{abs} = \frac{1}{n} \sum_{i=1}^{n} |\kappa_{1i}|, \tag{5}$$

- $\kappa_1 q_{abs}$—standard deviation of absolute maximum curvature

$$\kappa_1 q_{abs} = \sqrt{\frac{\sum_{i=1}^{n} \left(|\kappa_{1i}| - \kappa_1 a_{abs}\right)^2}{n}}, \tag{6}$$

- $\kappa_2 a_{abs}$—average absolute minimum curvature

$$\kappa_2 a_{abs} = \frac{1}{n} \sum_{i=1}^{n} |\kappa_{2i}|, \tag{7}$$

- $\kappa_2 q_{abs}$—standard deviation of absolute minimum curvature

$$\kappa_2 q_{abs} = \sqrt{\frac{\sum_{i=1}^{n} \left(|\kappa_{2i}| - \kappa_2 a_{abs}\right)^2}{n}}, \tag{8}$$

- Ha_{abs}—average absolute mean curvature

$$Ha_{abs} = \frac{1}{n} \sum_{i=1}^{n} |H_i| \tag{9}$$

- Hq_{abs}—standard deviation of absolute mean curvature

$$Hq_{abs} = \sqrt{\frac{\sum_{i=1}^{n} \left(|H_i| - Ha_{abs}\right)^2}{n}}, \tag{10}$$

- Ka_{abs}—average absolute Gaussian curvature

$$Ka_{abs} = \frac{1}{n} \sum_{i=1}^{n} |K_i|, \tag{11}$$

- Kq_{abs}—standard deviation of absolute Gaussian curvature

$$Kq_{abs} = \sqrt{\frac{\sum_{i=1}^{n} \left(|K_i| - Ka_{abs}\right)^2}{n}}. \tag{12}$$

where i symbol corresponds to a particular i-patch for which the curvature tensor is calculated at scale s and n is a total number of patches. The scale is associated with the side of the triangular patch for which the curvature is estimated. For the finest scale, it is equal to the original sampling interval in x or y (they are both equal in the presented measurements). With the increasing scale, the dimensions of the triangular patch increase as well. The detailed procedure of how to calculate curvature tensor in multiple scales is shown in [6,7,42].

In this work, parameters (3)–(10) as well as other given in [42], were calculated for a series of scales and regressed linearly and logarithmically with discharge energy. All curvature computations were performed using Wolfram Mathematica software (version 12, Wolfram Research, Oxfordshire, UK).

3. Results

This part is structured in four subsections. The first subsection is dedicated to the measurements of surface textures and their visual impressions of topographic structures as a function of discharge energy. This is followed by the subsequent subsection focusing on ISO standard areal parameters. Not all the parameters are entirely related to the geometric characterization of manufactured features. However, they are most widely used and well understood by the academia and industry, therefore they deserve their place in the characterization of EDMed surfaces in relation to their fabrication parameters. Yet their ability to analyze the microgeometry of created surface features, i.e., craters, is limited. Thus, results of two novel methods are introduced in the next two subsections: motifs and multiscale curvature analysis. They focus on the geometric quantities (area, diameter, height, volume, curvature) and are believed to better describe the nature of interactions between physical phenomena occurring during machining and resulted topographies. Every presented characterization parameter is analyzed with respect to the discharge energy.

3.1. Measurements

Renderings of the measured regions (six exemplary regions from S1, S3, S5, S7, S9 and S10) created using different technological parameters are depicted in Figure 4. The peak to valley roughness clearly increases with the discharge energy.

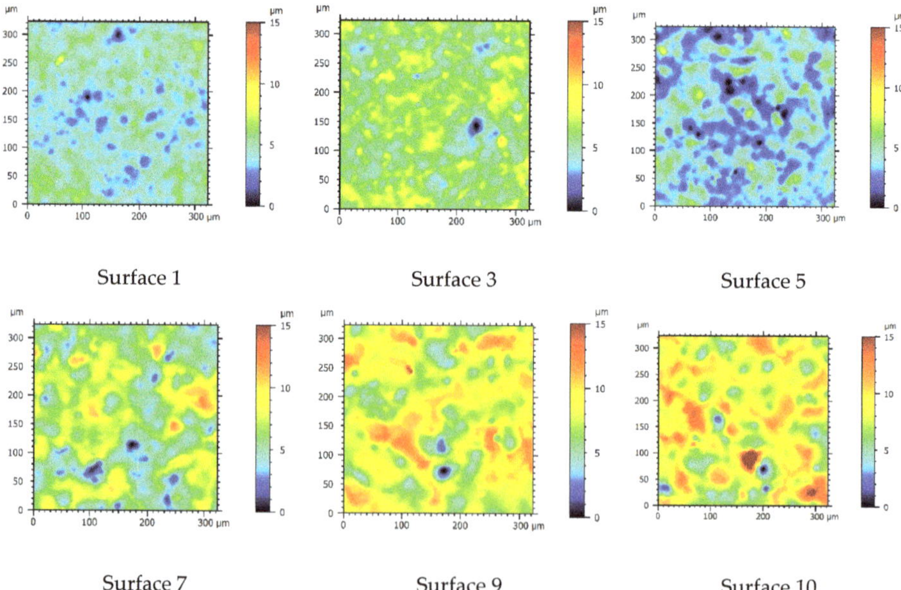

Figure 4. Six out of ten representatives of measured surfaces randomly chosen after dataset preparation step. Renderings of other surfaces are available in Supplementary Materials (Figure S1).

A total of 50 measurements were performed on 10 test surfaces using a focus variation microscope with setup parameters described in Section 2.2. Figure 4 depicts each measured surface after dataset preparation process. For comparison measures all color scales on vertical axes were scaled up to the same range. There are evident discrepancies between the surface topographies, which might be associated with different discharge energies. Craters become larger (in x- and y-directions) and there are more significant differences between minimal and maximal height value (z-direction).

3.2. ISO Parameters

As described in Section 2.3.1, a primary focus is given on three groups of areal parameters. These were presented in were calculated from five measured regions per each surface. Figure 5 presents the scatter plots of selected ISO 25178 areal parameters versus discharge energy value—mean of five measured regions and ±1 standard deviation (SD). Values of coefficient of determination for linear and logarithmic regressions are shown in Table A3 (Appendix B). Presented values are derived from unfiltered surface. The trend lines are also shown to indicate the best fitted functional relations between linear and logarithmic. All ISO 25178 [1] parameters calculated for all ten surfaces are shown in the supplementary spreadsheet to this study.

3.2.1. Height ISO Parameters

Height parameters are in general correlation with discharge energy changes, except for Skewness (Ssk) and Kurtosis (Sku) (Figure 5a). Both of these parameters remained at relatively stable range. Ssk value varied from −0.233 to 0.320 indicating a nearly symmetrical height distribution in relation to the mean plane. Sku varies from 3.005 to 3.867. A higher value (>3) indicates an increased spikiness of the surfaced, which is clearly seen in Figure 5a for S3 (Sku = 3.86).

Basic parameters, such as Sq and Sa changed their values, decreasing from 0.87 or 0.70 μm, respectively, in S1 to 0.799 or 0.63 μm, for S4, and then increasing to 2.23 or 1.75 μm for S10. Sv, apart from a general tendency to increase its value, does not grow with each increment of discharge energy. Both Sz and Sp increase in what appears to be a logarithmic manner. Strong correlations ($R^2 > 0.9$) were observed for all parameters except for Ssk and Sku. Skewness exhibit the moderate correlation when applying linear regression for an unfiltered surface. No correlations can be noted in residual roughness and waviness surfaces. Sa, Sq and Sv show strongest correlation when calculated using linear regression. Sp and Sz behave similarly for logarithmic regression. Surface S2 seems to deviate significantly from the trend for Sv, but this effect appears to be incidental. Similar observations were made for most of the others, but not all (Vmp, Spc and Spd), ISO and motif parameters. From visual impressions of captured topographies, it seems not to be radically different than S1 or S3. No changes in measurement conditions were also noted for measurements of S2. In order to fully investigate the reason for which S2 deviates significantly from the trend, other samples machined with same parameters would have to be manufactured and analyzed.

Calculated areal average roughness corresponds to the desired VDI surface finish in 5 out of 10 surfaces, with others differing by a single class. It should be emphasized that VDI 3400 standard is widely used in the industry but it was originated in early 1970s, where only available roughness measurement instrument was contact profilometry. Different measurement and filtration techniques can significantly influence the values of surface texture parameters [31,43–45]. None of these important issues, for obvious reasons, is addressed in VDI 3400, which may cause direct referencing to surface finish in this standard disputable.

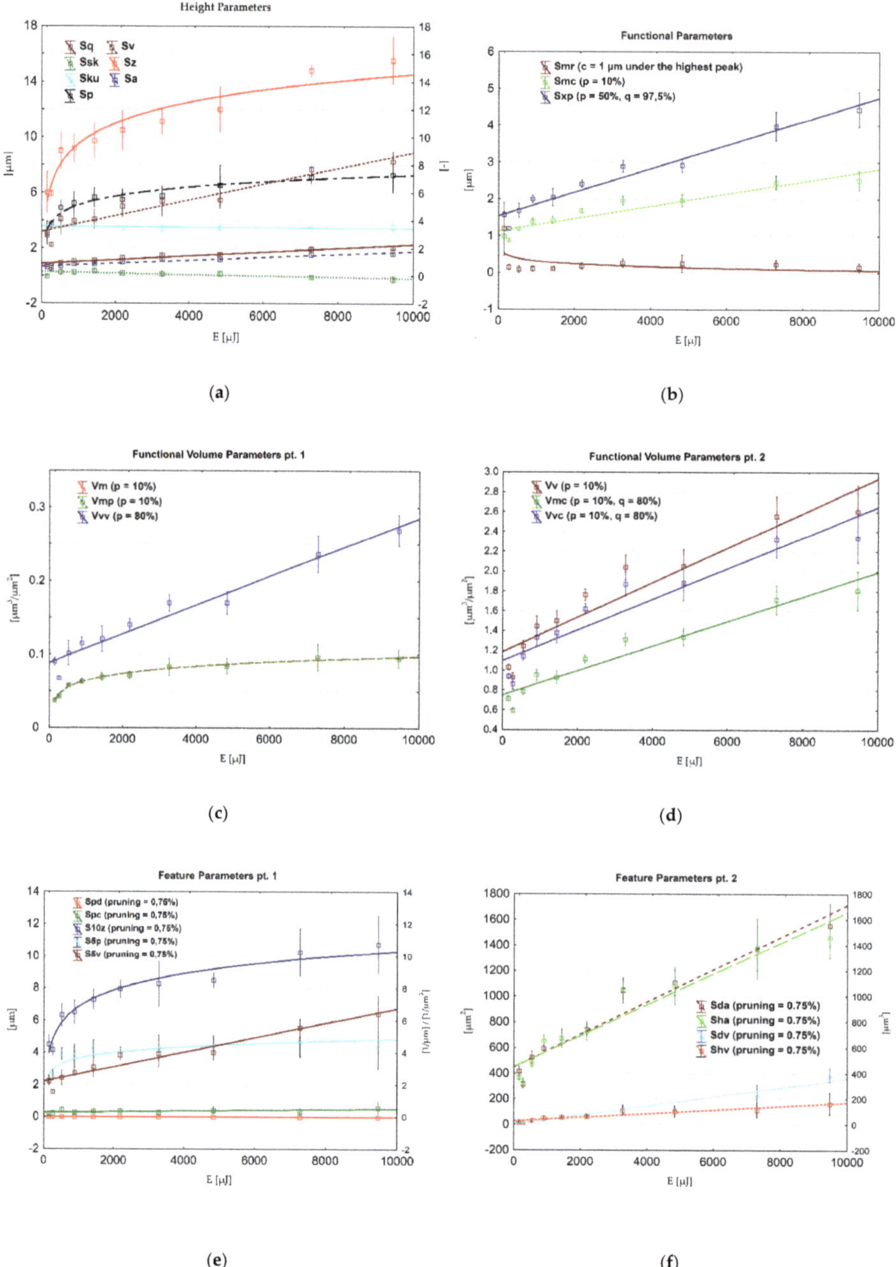

Figure 5. Scatter plots of ISO 25178 areal parameters correlated with discharge energies using linear or logarithmic regression. (**a**) height ; (**b**) functional ; (**c,d**) functional volume; (**e**) and (**f**) feature parameters.

3.2.2. Functional ISO Parameters

Smr depends on the c parameter, while Smc and Sxp depend on p and q. In this study, software default settings for this parameter, i.e., $c = 1$ µm under the highest peak, $p = 10\%$ for Smc, $p = 50\%$ and $c = 97,5\%$ for Sxp were used. For Smr, a steep decline was observed between S1 and S2, followed by flattening of the trend, which may suggest its logarithmic nature. Smc and Sxp also showed a slight decline between S1 and S2. The way these parameters change their values is very similar to that of the functional volume parameters, as seen in Figure 5c,d. Smc and Sxp show strong correlations (R^2 = 0.923 and 0.953, respectively) in both primary (unfiltered) and residual roughness surface.

3.2.3. Spatial ISO Parameters

Selected spatial parameters Sal (autocorrelation length), Str (texture aspect ratio) and Std (texture direction) did not show any significant trends following consecutive increments of discharge energy. This however may be strongly limited by relatively small measurement area. Spatial parameters show poor to medium correlation with discharge energy. The highest values of R^2 were achieved for Str (R^2 = 0.566, linear regression, unfiltered) and Sal (R^2 = 0.815), logarithmic regression, residual roughness).

3.2.4. Hybrid ISO Parameters

Sdq and Sdr show signs of correlation with increased discharge energy. However, for S1 to S3 their values fluctuate and stabilized their inclining trends from S4 to S10. Both parameters correlate strongly with discharge energy in an unfiltered surface, when regressed logarithmically (R^2 = 0.929 and 0.916, respectively). These values slightly decline (R^2 = 0.900 and 0.884), when calculated from residual roughness surface, however both parameters still correlate better for logarithmic regressions.

3.2.5. Functional Volume ISO Parameters

Functional volume parameters depend on the particular value of the p and q parameters. For all, except Vvv, p = 10%, and q = 80% for Vmc and Vvc. The pit void volume (Vvv) was calculated for p = 80%. All functional volume parameters correlated well with discharge energy. A general trend for these parameters is clearly seen on scatter plots in Figure 5c,d. Vm and Vmp also have the highest correlations (R^2 = 0.973 and 0.977, unfiltered and residual roughness, respectively), both regressed logarithmically. All parameters show good correlation (R^2 > 0.9) for both filtered and residual roughness surface.

3.2.6. Feature ISO Parameters

Feature parameters characteristic features which were part of motif analysis (pits, dales, etc.) and therefore pruning settings were set to the same value. The only parameter that did not show any dominating trend is the arithmetic mean peak curvature Spc. Density of peaks increased with the discharged energy in a probable logarithmic manner. S5v and S10z recorded a significant decline in value for S2. However, as the value of S10z increased significantly for the next surface (by almost 50%), the S5v only just regained its value from before the decline. The similar sudden decrease was observed for Sda, Sha and Sdv. The most stable growth was observed for Shv. All parameters, except for Spc, show high values of the coefficient of determination (R^2 > 0.9) for unfiltered surfaces. Some of these values slightly decline, when calculated from residual roughness surface. However, for S5p, Sda and Sha they increased. Shv and S5v declined most significantly after filtration to R^2 = 0.852 and R^2 = 0.888 respectively.

3.2.7. Functional ISO Parameters for Stratified Surfaces

Any significant trend was observed only for three parameters: Sk, Spk, Svk. Core roughness depth Sk showed certain fluctuations of value. There are three clearly visible points of decline (S2, S5 and S8), where the most significant ones are for S2 and S5. The aforementioned parameters exhibit very good correlations in both unfiltered (R^2 > 0.92) and residual roughness surface (R^2 > 0.92)

3.2.8. General Comments on ISO Parameters

Ding et al. modeled wire EDM process using FEM [18]. They noted that craters average diameter and maximum depth as well as height parameter Sa follow logarithmic trend with discharge energy. In this study, as for ISO standard areal parameters, the logarithmic regression shows evidently stronger correlations for areal parameters calculated for unfiltered surfaces (without decomposing to roughness and waviness) which describe peaks distributions (Ssk, Sku, Spd, Smr1), heights (Sp, Sz, Spk) and volumes (Vmp). This comes in the opposition to parameters that describe pits and valleys, for which stronger correlations are found applying linear regression. These differences are not present for both waviness and roughness surfaces filtered using the proposed nesting index, although for some parameters this dependency can be found. Weak relations between Ssk and Sku for differently EDMed surfaces are also noticed by [46–48].

3.3. Motif Analysis

Exemplary surfaces S1, S5 and S10, for which watershed segmentation is performed, with ridge lines of the detected motifs can be seen in Figure 6. The pit (extreme point) of each detected dale is marked with a cross. Visually, these segments can be intuitively associated with craters. The distribution of geometrical properties of the segments may reflect the randomness of electric discharges over the machined surface and sudden abruption of material, which lead to creation of overlapping craters which are not perfectly round. This can be visualized using equivalent diameter as presented in Figure 7 in which distributions of that parameter are plotted for S1, S5 and S10. With increasing discharge energy, motifs are on average larger and more dimensionally dispersed. Threshold of 18 µm can be noticed, above which diameters may be associated with craters. This threshold results from the processing parameters during segmentation in which the minimum area of the motif is set. The presence of the smaller motif can occur as the height parameter is a superior criterium or some fine-scale motifs could not be merged.

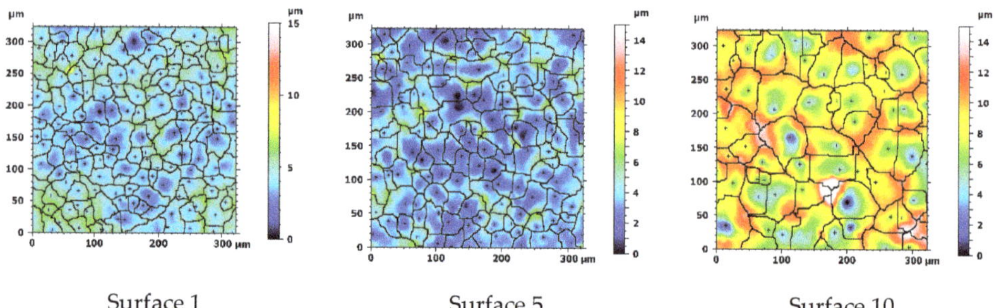

Surface 1   Surface 5   Surface 10

Figure 6. Visualization of effects of the watershed segmentation. Surfaces with watershed boundaries (ridge lines) superimposed on a surface image.

As seen on the scatter plot (Figure 8d), number of motifs decreases as the discharge energy increases. This alone suggests formation of larger, in terms of their volume and diameter, craters on the surface, which conforms to the actual data, presented on other plots on Figure 8b–d. The craters depth or height also increases with the discharge energy. All motifs parameters show a sudden change for S2 what may be considered incidental. All motif parameters, except the number of motifs, exhibit exceptionally strong correlations with discharge energy ($R^2 > 0.96$) using linear regression. Logarithmic regression seems to be more appropriate to model the functional relation between number of motifs and the energy ($R^2 = 0.936$). The strongest correlations can be found for mean motif area, for which coefficient of determination is equal to 0.986. Standard deviations of motif geometrical parameters tend to increase with discharge energy. They also show strong correlations when regressed linearly ($R^2 > 0.95$). This might be interpreted as surfaces machined with higher energies exhibit craters whose dimensions and variability is proportional to this energy. This is also supported by the equivalent diameter distributions presented for two extreme and one middle sets of technological parameters as depicted in Figure 7. More information about the regression analysis can be found in Table A4 (Appendix B).

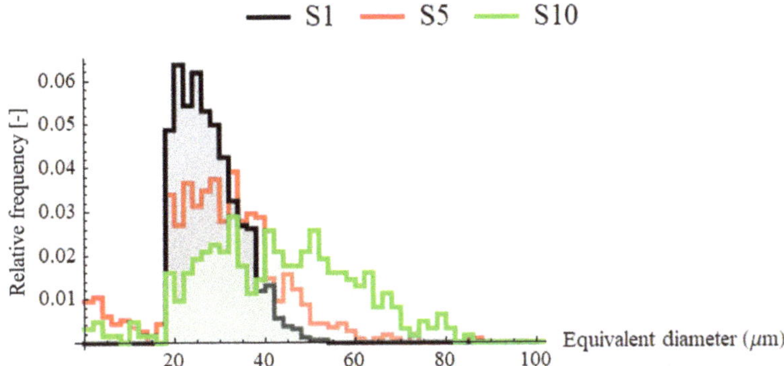

Figure 7. Distributions of equivalent diameters calculated for S1, S5 and S10 surfaces using motif analysis.

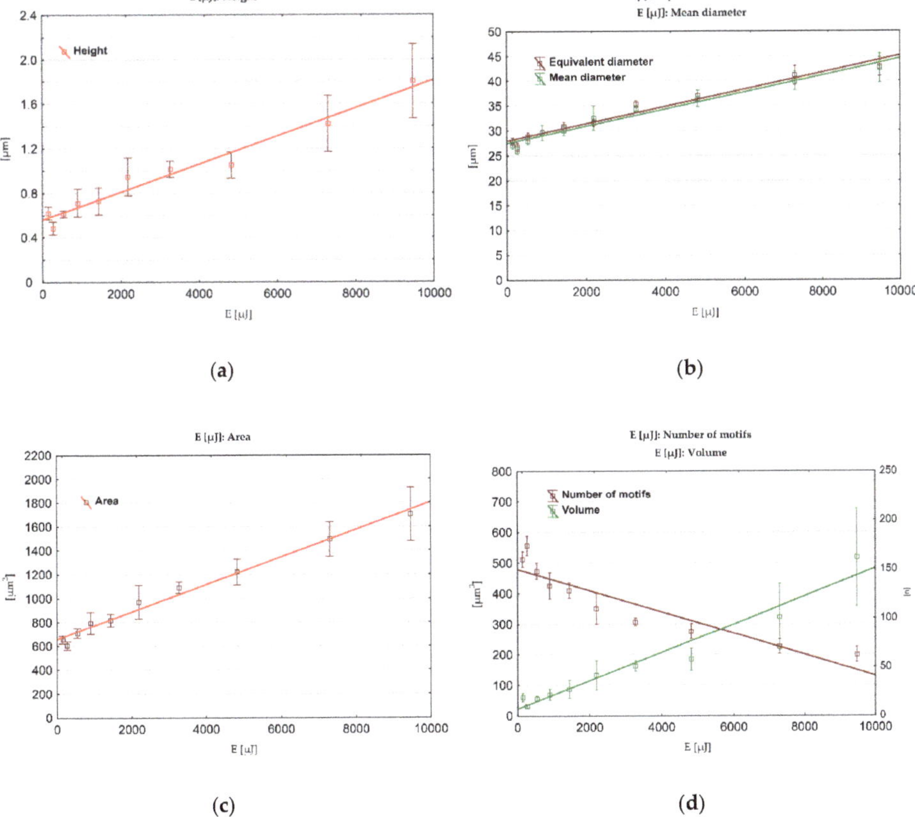

Figure 8. Scatter plots for motif analysis: (**a**) mean height; (**b**) equivalent and mean diameter; (**c**) mean area; (**d**) mean volume and number of motifs correlated with discharge energies using linear regression.

3.4. Multiscale Curvature Analysis

The maximum principal curvature (κ_1), calculated for two different scales from representative regions located on three different surfaces (S1, S5 and S10) are shown in Figure 10. Negative curvatures (values on the figure) represent convex surface features or peaks. Different scales of calculation show different features. Grooves and pores are visible on κ_1 curvature plot as positive curvature regions. The amplitude of curvatures decreases with scale what can be explained by the fact that for larger scale features of larger size (and radius) can be characterized. This effect can be observed for other research examples using this method of characterization [6,7,42].

The calculated values also show the influence of the technological parameters on the surface topography. The maximum curvatures clearly increase with increasing discharge energy for both of the two scales, as shown in Figure 9. However, the differences between curvatures calculated for the smaller scale are less evident in terms of amplitude. The same tendencies were observed for minimum, mean and Gaussian curvatures.

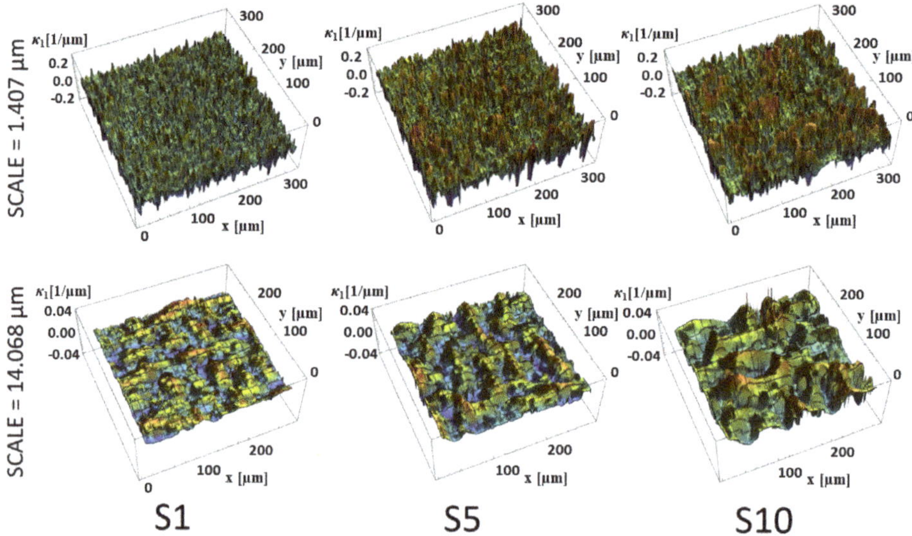

Figure 9. Maximum curvature κ_1 for representative measurements of surfaces: S1, S5 and S10, calculated for two different scales: $s = 1.407$ and $s = 14.068$ μm. Please note that vertical scale (z-axis) for upper row is fivefold greater than for the lower row.

The mean and standard deviation of the maximum curvature κ_1, as a function of scale, for five representative surfaces S1, S3, S5, S8 and S10 are shown in Figure 10. No clear tendency for κ_1a between surfaces can be observed. These values seem to converge with scale to zero but at different rates, what might be associated with the fact that for the larger scale curvature of form is characterized. Since in all analyzed surface, the form is a flat plane, its curvature is null. This parameter can be associated with average shape (convexity or concavity) at certain scale. Considering averages of absolute values maximum, their values decreases with scale for all surfaces. This parameter characterizes mean deviation from zero (flatness), without taking the signs of curvatures into account. It can be used for describing the evolution of curvature magnitude with scale. Fine-scale features are generally characterized with large curvature and the similar observation is made here as a declining trend. Standard deviation of maximum curvature also decreases with scale. This might be explained by the fact that variation of curvatures declines as the scale gets larger. Similar observation can be made for κ_1q$_{abs}$. Similar tendencies were noticed for parameters related minimum, mean and Gaussian curvature.

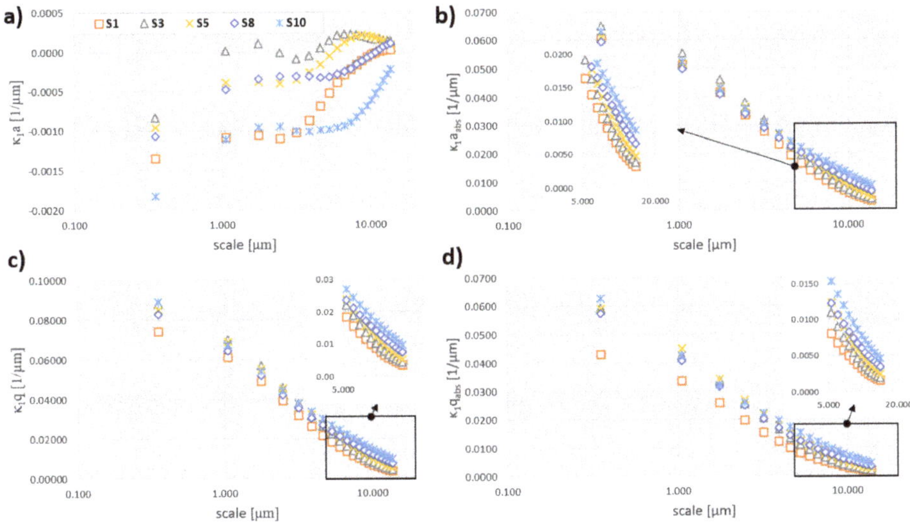

Figure 10. Various statistical parameters related to curvature calculated for surfaces S1, S3, S5, S8 and S10 depicted as a function of scale: (**a**) average of absolute maximum curvature, (**b**) average of absolute maximum curvature, (**c**) standard deviation of maximum curvature, (**d**) standard deviation of absolute maximum curvature.

The discharge energy evidently influences the magnitude of the both principal curvatures and their combinations (H and K). The average parameters ($\kappa_1 a$, $\kappa_2 a$, Ha and Ka) demonstrated to be least influenced by different material processing as no clear tendency was observed for them for all analyzed scales. Their absolute values and standard deviations are more useful in finding functional relations between process parameter and resulted curvature. This is noticed for scales greater than 5 µm, and might suggest that those topographies can be discriminated at larger scales and that they do not differ significantly at the finest scales.

Two different types of regression (linear and logarithmic) were applied to discharge energy versus three groups of characterization parameters: ISO standard areal parameters, motif analysis parameters and curvature tensor statistical parameters (8). The logarithmic functional relation between discharge and topographic parameters was reported by Ding et al. [18] and it is tested here to confirm or deny it. By strong correlations we take R^2 greater than 0.9.

The strengths of the linear and logarithmic regression analyses (R^2) for the curvatures versus the discharge energies are shown as a function of scale in Figure 11. Both average principal curvatures κ_1 and κ_2 correlate do not correlate well ($R^2 < 0.8$) with the discharge energy for the analyzed range of scales (Figure 11a). When considering absolute values of curvature, strong correlations are observed for $\kappa_1 a_{abs}$ and logarithmic regression better than linear reflects that relation as R^2 is greater than 0.9 for $s > 8.089$ µm (total of nine analyzed scales) versus the single largest analyzed scale for linear. Similar tendencies can be noticed for average values of κ_2 for which only $\kappa_2 a_{abs}$ correlates strongly when regressed logarithmically for largest scales (>8 µm) (Figure 11c). Standard deviation measures show strong correlations with both models for larger scales (Figure 11b,d). Logarithmic regression correlates for broader range of scales when compared to linear.

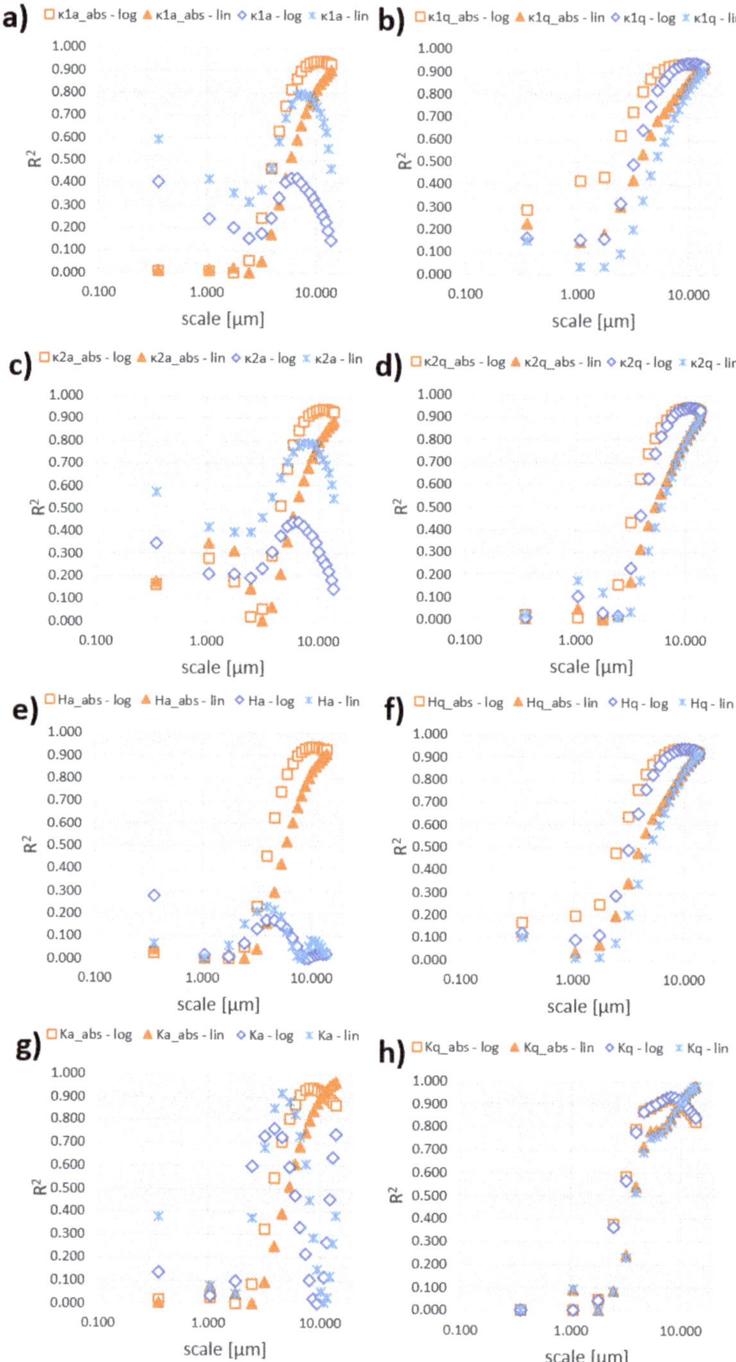

Figure 11. Coefficient of determination for linear and logarithmic regression calculated for statistical parameters calculated for: (**a**) average maximum; (**b**) standard deviation of maximum; (**c**) minimum and (**d**) standard deviation of minimum curvature versus discharge energy; (**e**) average mean curvature; (**f**) standard deviation of mean curvature; (**g**) average Gaussian curvature and (**h**) standard deviation of Gaussian curvature as a function of scale.

The average mean curvature, H, correlates well only at larger scales and when absolute values are considered (Figure 11e). Strong correlations are found also for standard deviation but for larger scales as well (Figure 11f). In comparison with other curvature parameters, the Gaussian curvatures, K, appear to correlate the best as the largest values of R^2 are found using linear regression for Kq (0.977) and Kq_{abs} (0.981), both at the largest scale equal to 13.716 μm (Figure 11g,h). Ka, as only average non-absolute parameter, corelated strongly but at the middle scale (4.572 μm). Gaussian curvature K appeared to correlate the strongest for the widest range of scales. It is the only parameter for which strong correlations were observed when average values are considered. Taking into account the absolute values of curvatures, it significantly improves the strengths of correlations. These parameters might be attributed with the magnitude of curvature, regardless of its sign.

4. Discussion

Two types of topographic feature, i.e., craters and ridges, can be distinguished on images rendered from the measurements with the focus variation microscope on the surfaces created by EDM. The features are consistent with known mechanisms in electric discharge machining. The increase of discharge energy lead to the creation of deeper and larger (in area, radius and volume) craters with a greater magnitude of curvature. The measured surfaces created with different pulse energies can be discriminated clearly using ISO standard parameters, motif analysis and, over wide ranges of scales, using multiscale curvature analysis. Strong correlations can be found between the discharge pulse energies with which the surfaces were created and the texture characterization parameters calculated from the measurements. In six out of seven groups of ISO areal parameters, there are parameters that correlates well with discharge energy. Only spatial parameters could not be used to establish such functional relations, which may result from relatively small measurement area used to calculate autocorrelation function.

Most of modeling approaches, which explain the nature of electric discharge machining, focus on the creation of craters depending on the physical properties of material and controllable technological parameters [10,13,14,18,22]. The ability to characterize geometrical aspects of fabricated features becomes important step in the understanding of the interactions between surface and processing. Although most of ISO parameters correlate strongly with discharge energy, their ability to describe the aspects of craters morphology is generally weak or indirect. Height and functional groups include parameters that quantify, in various ways, height information focusing on z-component only. Spatial parameters involve calculation of auto-correlation function which reference to the craters' geometry is vague. Hybrid parameters, which quantify both amplitude and spacing between heights, try to capture some geometric aspects of the surface (area and slope) but they consider a measured surface as a single entity not as a set of individual topographic features. Only feature parameters, related to dales or valleys, i.e., S5v, Sda and Sdv, have a direct association with crater depth, area and volume and, in addition, they correlate strongly with discharge energy. The similar statement can be formulated for the geometric parameters calculated using motif analysis, which is supported by the fact that they use same segmentation technique. Feature-based characterization is of particular importance in non-traditional manufacturing [47]. It provides additional perspective on the surface and supports better understanding of the phenomena governing manufacturing process and the interactions between controllable technological parameters and resulted surface topography [48]. This study proves that this is also a valuable approach for EDM. Geometrical properties of detected motifs, such as depth or volume, are easy to interpret in relation to their formation process, unlike average, root mean square roughness or skewness.

Multiscale curvature analysis, as presented in this study, is not specifically feature-based, but concentrates on the geometric characterization of shape when considering the surface as a composition of numerous overlapping craters. The key advantage of any multiscale analyses, unlike motifs and conventional studies, is that they can be useful in indicating the scales at which the correlations are the strongest. Knowledge of these scales could help to increase the understanding of the surface processing and function. It was noticed that absolute parameters, although they correlate strongly for the largest scales, do not vary significantly for the finest scales. Considering the three finest scales (<2.462 µm), coefficients of variation calculated for all analyzed surfaces do not exceed 10% for principal, mean and Gaussian curvature (average and SD). It might suggest that the created surfaces do not differ significantly at the finest scales, and also that the EDM process leads to the creation of fine-scale features similarly, regardless of the discharge energy. This might be supported by similar observations done by Hyde et al. [15] using areal multiscale analysis but different instrumentation, discharge energy level, material and outliers-removal method. What is more, curvature parameters start to correlate strongly with the discharge energy ($R^2 > 0.8$) for scales starting from between 8 and 9 µm. This corresponds well to the average equivalent radius of detected motif (mean—1 × standard deviation $\approx$ 10 µm) for the sample created with lowest energy. Starting with those scales the curvature of craters is the most affected by the discharge energy and the fabricated microgeometry is the most adequately characterized as the sizes of features are best discerned at those scales. This follows the concept that scale could be enmeshed with size [3].

Strong correlations are observed for height parameters, which are most sensitive to longest wavelengths. This observation would appear to support similar, previously reported work/findings [15]. Spc (mean peak curvature) does not correlate well as it refers to the curvature at the finest scale as it is calculated for originally sampled data. Sdq and Sdr which might be related to multiscale areal method [5], when calculated also for the finest scale, correlate strongly with discharge energy only when regressed logarithmically. This might support our observation that the characterization of surface topography of EDM texture should be concentrated on the most accurate registration, filtration and analysis of large-scale features if relation between discharge energy is to be established.

Differences in physical processes that took essential role in the fabrication of surface finish might be analyzed through discrepancies in curvatures at similar scales on the surfaces manufactured by different discharge energies. Bartkowiak and Brown [7] found that for micro-EDM, there was a significant change in trend of topographic curvature versus discharge energies for surfaces manufactured with more than 1 µJ. Similar observations were made by Hyde et al. [15]. This study was done at significantly greater energies as well as different materials of both workpiece and electrode, and phenomenon of this kind was not observed. This might suggest that for the considered energies, the physical process that formed surface topography did not change its nature but rather the intensity. The aforementioned studies might have described the shift between micro and standard EDM process. In the former, electrostatic force might be a dominative factor in the creation of surface topography as there is simply not enough time for material to be moltened or vaporized and removed efficiently [28].

There is probably not a single parameter that can functionally describe the relations between technological parameters (discharge energy, current, voltage, gap, polarity) and fabricated surfaces, for all materials (electrode and workpiece), part shapes and machining conditions. Although some theoretical approaches are well known in conventional machining such as modeling Ra or Rz in turning and milling, the randomness and suddenness of electric discharges as well as complexity of physical phenomena make a development of analytical modeling for EDM rather challenging. Some undisclosed relations between VDI roughness and parameters of discharge are incorporated in the control systems of machine tools, as presented in this study. Therefore, establishing credible functional relations for most common materials and conditions are highly anticipated. Geometric characterization which focus on the morphology of craters is shown here to be a prospective candidate as it has a direct association with nature of the electric discharge machining.

Important aspect of surfaces is their designed functional behavior such as lubrication properties, gloss reflectance, corrosion resistance, load bearing capabilities, adhesion or wear. Basic profile or areal texture parameters correlate rarely with their performance parameters, or correlate only if narrow range of band-pass filter is applied [3,7]. Functional volume parameters certainly provide a better way of monitoring the effect of process parameters on the resulted surface texture. This can eventually help in defining proper machining conditions in order to fabricate surfaces according to functional needs [29]. Multiscale methods, including curvature, were found to be successful in establishing functional relations of those kinds for different processes such as friction, adhesion, fatigue, gloss and many more [3]. Motif analysis seem to be also prospective but further research should be conducted to fulfil its full potential. A primary focus should be given to establish functional relations between technological parameters of EDM and the functional behavior of fabricated surface through the adequate characterization of the topography.

From the metrological side, the main challenge for the measurement and dataset preparation steps was a reliable outliers removal process for FVM. As stated in Section 2.2, focus variation microscopy can produce surface- and method-specific outliers, that cannot be removed using standard removal procedures. The method used in this study was successful because of the small dimensions of the plateau-like formations, and therefore cannot be recommended as a general method for outliers removal. Considering all the aforementioned issues, the authors find using this method, in this particular study, justified. In addition, although there was a chance, that this method might affect roughness measurements. However, it introduced only insignificant changes, with its effect being similar to that of the λs -filter (microroughness filter).

Some other measurement techniques might be less susceptible to registering outliers on this kind of surface (locally smooth and reflective), such as CSI (coherent scanning interferometry) [36]. However, they are also burdened with both surface- and method-specific outliers, yet its hardware filtration is advanced, thus reducing the risk of unreliable measurements.

Another challenge of this study was to perform the correct filtration of the surface. A relatively small measurement area left little room for intuitive form and roughness evaluation. Therefore, a nesting index should be suited for the surface characteristic, originating from the manufacturing technology. The authors suggest that in this study, craters and their geometry belong to the roughness spectrum of the surface, thus the nesting index of 250 μm should be used. The resulting roughness surface contains most information regarding craters asperities. It must be noted that a larger measurement area would enable an easier choice of nesting index in future research. However, this poses another challenge for computational capabilities, since datasets derived from focus variation microscopes are quite large.

For the chosen nesting index areal parameters from the roughness surface are significantly influenced by the discharge energy. Waviness surface did not show any strong correlations. This might mean that the crucial information containing EDM-specific topographic features (craters) is in the roughness. Assuming a relatively large nesting index of 250 μm, it led to creation of residue surface of narrower bandwidth when compared to residual roughness. The maximum cut-off wavelength was limited by each measurement size of 323 μm × 323 μm. The primary surface, which contained all wavelengths was also rich in the information about craters geometry, what was evident in strong correlations with the discharge energy.

Conventional analyses using ISO parameters have the advantage that they are included in most commercial software and they can be evaluated with little knowledge of surface metrology principles (aside from noise and form removal). These make them used extensively by the industry and academia. Non-traditional characterization methods, such as motif and multiscale, are more complex and would require more expertise from the users. They will be appreciated once they add value by advancing the understanding of the relations between topographies and phenomena or if they can better exploit the acquired topographic information [2]. This could be facilitated by automatic, more intuitive and easy-to-use computer applications released for industrial and academic purposes. Motif analyses are included in commercial metrological software but they are limited to detection of dales and hills, whereas other geometric shapes can also be important signatures of manufacturing process.

5. Conclusions

The results show experimentally that the microgeometry of surfaces created by EDM is strongly affected by the discharge energy. This was proven by achieving strong correlations for geometric properties of fabricated features (height, area, diameter and volume) and their curvatures. In contrast to analysis of surface topography through ISO areal texture parameters, the presented approach extracted of information directly relating to the shape and size of topographic features of interest. This showed that geometric characterizations of process-specific surface formations were useful in determining functional relations with energy of electric discharges. Although, strong correlations with conventional parameters were also found, they miss the opportunity to study the effects of physical phenomena governing the creation of craters. Thus, geometric characterizations of crater morphology seem to be more natural and intuitive way of analysis of EDM surfaces. Further research in motifs and curvature versus technological parameters could promote better understanding and modeling of topographic response to its formation process.

Some detailed conclusions of this study can also be stated:

- Strong correlations ($R^2 > 0.9$) were found between discharge energies and ISO parameters that were calculated for original surfaces (prior to S- or L-filtration but after morphological filtration) and S-surfaces (roughness). ISO standard parameters did not correlate well when computed for L-surfaces (waviness). This suggests that the creation of topographic features of larger dimensions is affected by the discharge energy. The dimension limit is constrained by cut-off wavelength of 250 μm. The characteristics of fine-scale surface features do not differ significantly. This is also supported by the outcome of multiscale curvature analysis which indicated that curvature correlated strongly also for larger scales.
- In the tested conditions, the surface is mostly affected and potentially controlled by discharge energy at larger scales which is associated with sizes of fabricated craters. For smaller scales, effect of machining with different parameters did not manifest itself.
- Strong correlations ($R^2 > 0.9$) were also observed between motif parameters that characterized height, diameter, area and diameter of the detected motif, which might be associated with craters. This analysis, together with curvature and ISO areal parameters allow comprehensive characterization of surface microgeometry created by EDM.
- ISO areal parameters that describe peaks distributions exhibit higher coefficient of determination than others when regressed logarithmically. Correlations using a logarithmic model were also strong for curvature parameters.
- Registration of surface topography using focus variation microscopy leads to the occurrence of surface- and method-specific outliers which are hard to be removed using Gaussian filtration or thresholding. The application of a morphological filter proved to be successful in the outliers removal what was also evident to achieving strong correlations with discharge energy with parameters of three various types.

Supplementary Materials: The following are available online at http://www.mdpi.com/1996-1944/13/17/3830/s1, Figure S1: renderings of representative regions of all ten analyzed surfaces, spreadsheet with ISO parameters calculated for primary, waviness and roughness surfaces.

Author Contributions: Conceptualization, T.B., M.M. and K.M.; methodology, T.B., M.M., K.M. and M.W.; software, T.B. and M.M.; validation, T.B. and M.M.; formal analysis, T.B. and M.M.; investigation, T.B., M.M. and K.M.; resources, K.M. and M.W.; data curation, T.B. and M.M.; writing—original draft preparation, T.B., M.M., K.M. and M.W.; writing—review and editing, T.B., M.M., K.M. and M.W.; visualization, T.B., M.M. and K.M.; supervision, T.B. and M.W.; project administration, T.B.; funding acquisition, K.M. and M.W. All authors have read and agreed to the published version of the manuscript.

Funding: This research was funded by the Polish Ministry of Science and Higher Education as a part of research subsidy—Project 0614/SBAD/1529.

Acknowledgments: In this section you can acknowledge any support given which is not covered by the author contribution or funding sections. This may include administrative and technical support, or donations in kind (e.g., materials used for experiments).

Conflicts of Interest: The authors declare no conflict of interest. The funders had no role in the design of the study; in the collection, analyses, or interpretation of data; in the writing of the manuscript, or in the decision to publish the results.

Appendix A

Appendix A contains supplementary information related to the sample preparations. This includes properties of graphite electrodes (Table A1) and EDM machine tool and dielectric fluid (Table A2).

Table A1. Graphite electrode parameters.

Properties	Units	Test Standard	Values
Average grain size	μm	ISO 13320	3
Bulk density	g/cm^3	DIN IEC 60413/204	1.88
Open porosity	Vol. %	DIN 66133	10
Medium pore entrance diameter	μm	DIN 66133	0.6
Ambient temperature	cm^2/s	DIN 51935	0.01
Hardness	HR $_{5/100}$	DIN IEC 60413/303	105
Resistivity	μΩm	DIN IEC 60413/402	13
Flexural strength	MPa	DIN IEC 60413/501	85
Compressive strength	MPa	DIN 51910	170
Dynamic modulus of elasticity	MPa	DIN 51915	13.5×10^3
Thermal expansion (20–200 °C)	K^{-1}	DIN 51909	4.7×10^{-6}
Thermal conductivity (20 °C)	Wm^{-1}K^{-1}	DIN 51908	105
Ash content	ppm	DIN 51903	200

Table A2. Machine o dielectric fluid parameters.

Properties	Units	Values
Machine Architecture		C-frame/Fixed table/Drop tank
X, Y, Z travel	mm	$400 \times 300 \times 350$
X, Y axes speed	m/min	6
Z axis speed	m/min	15
Positioning resolution	μm	0.1
Work tank size	Mm	$900 \times 630 \times 350$
Work table size	Mm	600×400
Max. machining current	A	80
Best surface finish Ra	μm	0.08
Dielectric fluid		
Color		Straw yellow
Kinematic viscosity (20 °C/40 °C)	mm^2/s	5.0/3.0
Density	kg/l	0.77

Appendix B

Appendix B presents supplementary information related to the regression analyses for ISO 25178 areal parameters (Table A3) and motif parameters (Table A4). Both of parameter groups were regressed using linear and logarithmic regression. Coefficients of determination (R^2) is given for all analyzed parameters.

Table A3. Coefficients of determination (R^2) between discharge energies and ISO 25178 areal parameters calculated using linear and logarithmic regressions for primary surfaces. Please see the standard for additional settings given in the table.

Parameter	R^2 for Linear Regression	R^2 for Logarithmic Regression
Sq	0.930	0.905
Ssk	0.643	0.291
Sku	0.016	0.001
Sp	0.760	0.961
Sv	0.918	0.849

Table A3. *Cont.*

Parameter	R^2 for Linear Regression	R^2 for Logarithmic Regression
Sz	0.891	0.934
Sa	0.930	0.899
Smr (c = 1 μm under highest peak)	0.024	0.207
Smc (p = 10%)	0.911	0.923
Sxp (p = 50%, q = 97.5%)	0.953	0.846
Sal (s = 0.2)	0.014	0.117
Str (s = 0.2)	0.566	0.531
Std (reference = 0°)	0.113	0.115
Sdq	0.796	0.929
Sdr	0.825	0.916
Vm (p = 10%)	0.790	0.973
Vv (p = 10%)	0.909	0.927
Vmp (p = 10%)	0.790	0.973
Vmc (p = 10%, q = 80%)	0.941	0.881
Vvc (p = 10%, q = 80%)	0.899	0.933
Vvv (p = 80%)	0.952	0.838
Spd (trimming = 0.75%)	0.661	0.947
Spc (trimming = 0.75%)	0.302	0.335
S10z (trimming = 0.75%)	0.853	0.942
S5p (trimming = 0.75%)	0.557	0.901
S5v (trimming = 0.75%)	0.932	0.838
Sda (trimming = 0.75%)	0.932	0.904
Sha (trimming = 0.75%)	0.938	0.890
Sdv (trimming = 0.75%)	0.963	0.716
Shv (trimming = 0.75%)	0.959	0.836
Sk (unfiltered)	0.923	0.887
Spk (unfiltered)	0.772	0.980
Svk unfiltered)	0.953	0.832
Smr1 (unfiltered)	0.292	0.032
Smr2 (unfiltered)	0.631	0.398

Table A4. Coefficients of determination (R^2) between discharge energies and motif parameters calculated using linear and logarithmic regressions.

Motif Parameter	Statistics	R^2 for Linear Regression	R^2 for Logarithmic Regression
Number of motifs	-	0.874	0.936
Height	Mean	0.979	0.766
Height	SD	0.976	0.784
Area	Mean	0.986	0.826
Area	SD	0.975	0.841
Volume	Mean	0.965	0.660
Volume	SD	0.961	0.665
Mean diameter	Mean	0.971	0.861
Mean diameter	SD	0.955	0.877
Minimal diameter	Mean	0.963	0.867
Minimal diameter	SD	0.953	0.846
Maximal diameter	Mean	0.964	0.871
Maximal diameter	SD	0.965	0.891

References

1. ISO 25178-2:2012. *Geometrical Product Specifications (GPS)—Surface Texture: Areal—Part, 2*; International Organization for Standardization: Geneva, Switzerland, 2012.
2. Senin, N.; Thompson, A.; Leach, R. Feature-based characterisation of signature topography in laser powder bed fusion of metals. *Meas. Sci. Technol.* **2017**, *29*, 045009. [CrossRef]
3. Brown, C.A.; Hansen, H.N.; Jiang, X.J.; Blateyron, F.; Berglund, J.; Senin, N.; Bartkowiak, T.; Dixon, B.; Le Goic, G.; Quinsat, Y.; et al. Multiscale analyses and characterizations of surface topographies. *CIRP Ann.-Manuf. Technol.* **2018**, *67*, 839–862. [CrossRef]
4. Underwood, E.E.; Banerji, K. Fractals in Fractography. *Mater. Sci. Eng.* **1986**, *80*, 1–14. [CrossRef]

5. Brown, C.A.; Charles, P.D.; Johnsen, W.A.; Chesters, S. Fractal Analysis of Topographic Data by the Patchwork Method. *Wear* **1993**, *161*, 61–67. [CrossRef]

6. Bartkowiak, T.; Brown, C.A. Multiscale 3D Curvature Analysis of Processed Surface Textures of Aluminum Alloy 6061 T6. *Materials* **2019**, *12*, 257. [CrossRef]

7. Bartkowiak, T.; Brown, C.A. A Characterization of Process-Surface Texture Interactions in Micro-Electrical Discharge Machining Using Multiscale Curvature Tensor Analysis. *J. Manuf. Sci. Eng. Trans. ASME* **2018**, *140*. [CrossRef]

8. Braatz, R.D.; Alkire, R.C.; Seebauer, E.; Rusli, E.; Gunawan, R.; Drews, T.O.; Li, X.; He, Y. Perspectives on the Design and Control of Multiscale Systems. *J. Process Control* **2006**, *16*, 193–204. [CrossRef]

9. Brown, C.A. Axiomatic Design Applied to a Practical Example of the Integrity of Shaft Surfaces for Rotating Lip Seals. In Proceedings of the 1st CIRP Conference on Surface Integrity (CSI), Bremen, Germany, 30 January–1 February 2012; pp. 53–59.

10. Kunieda, M.; Lauwers, B.; Rajurkar, K.P.; Schumacher, B.M. Advancing EDM through Fundamental Insight into the Process. *CIRP Ann.-Manuf. Technol.* **2005**, *54*, 64–87. [CrossRef]

11. Rajurkar, P.; Levy, G.; Malshe, A.; Sundaram, M.M.; McGeough, J.; Hu, X.; Resnick, R.; DeSilva, A. Micro and Nano Machining by Electro-Physical and Chemical Processes. *CIRP Ann.-Manuf. Technol.* **2006**, *55*, 643–666. [CrossRef]

12. Klocke, F.; Lung, D.; Antonoglou, G.; Thomaidis, D. The Effects of Powder Suspended Dielectrics on the Thermal Influenced Zone by Electrodischarge Machining with Small Discharge Energies. *J. Mater. Process. Technol.* **2004**, *149*, 191–197. [CrossRef]

13. Giridharan, A.; Samuel, G.L. Modeling and analysis of crater formation during wire electrical discharge turning (WEDT) process. *Int. J. Adv. Manuf. Technol.* **2015**, *77*, 1229–1247. [CrossRef]

14. Oğun, C.; Tosun, N.; Hasim, P. The Effect of Cutting Parameters on Wire Crater Sizes in Wire EDM. *Int. J. Adv. Manuf. Technol.* **2003**, *21*, 857–865.

15. Hyde, J.M.; Cadet, L.; Montgomery, J.; Brown, C.A. Multi-scale areal topographic analysis of surfaces created by micro-EDM and functional correlations with discharge energy. *Surf. Topogr. Metrol. Prop.* **2014**, *2*. [CrossRef]

16. Gostimirovic, M.; Kovac, P.; Sekulic, M. Influence of discharge energy on machining characteristics in EDM. *J. Mech. Sci. Technol.* **2012**, *26*, 173–179. [CrossRef]

17. Mohanty, C.P.; Mahapatra, S.S.; Singh, M.R. An intelligent approach to optimize the EDM process parameters using utility concept and QPSO algorithm. *Eng. Sci. Technol. Int. J.* **2017**, *20*, 552–562. [CrossRef]

18. Ding, H.; Li, X.; Wang, X.; Guo, L.; Zhao, L. Research on the Topography Model of Micro Wire Electrical Discharge Machining Surface in Proc. In Proceedings of the 2nd International Conference on Advances in Mechanical Engineering and Industrial Informatics (AMEII 2016); Atlantis Press: Paris, France, 2016; pp. 1316–1320.

19. Han, F.; Jiang, J.; Yu, D. Influence of discharge current on machined surfaces by thermoanalysis in finish cut of WEDM. *Int. J. Mach. Tools Manuf.* **2007**, *47*, 1187–1196. [CrossRef]

20. Masuzawa, T.; Yamaguchi, M.; Fujino, M. Surface Finishing of Micropins Produced by WEDG. *CIRP Ann.-Manuf. Technol.* **2005**, *54*, 171–174. [CrossRef]

21. Guu, Y.H. AFM surface imaging of AISI D2 tool steel machined by the EDM process. *Appl. Surf. Sci.* **2005**, *242*, 245–250. [CrossRef]

22. Bäckemo, J.; Heuchel, M.; Reinthaler, M.; Kratz, K.; Lendlein, A. Predictive topography impact model for Electrical Discharge Machining (EDM) of metal surfaces. *MRS Adv.* **2019**, 1–12. [CrossRef]

23. Yu, Z.; Rajurkar, K.P.; Narasimhan, J. Effect of Machining Parameters on Machining Performance of Micro EDM and Surface Integrity. In Proceedings of the Annual ASPE Meeting, Portland, OR, USA, 26–31 October 2003.

24. Klink, A.; Holsten, M.; Hensgen, L. Crater morphology evaluation of contemporary advanced EDM generator technology. *CIRP Ann.-Manuf. Technol.* **2017**, *66*, 197–200. [CrossRef]

25. Barré, F.; Lopez, J. Watershed lines and catchment basins: A new 3D-motif method. *Int. J. Mach. Tools Manuf.* **2000**, *40*, 1171–1184. [CrossRef]

26. Wolf, G.W. Scale independent surface characterisation: Geography meets precision surface metrology. *Precis. Eng.* **2017**, *49*, 456–480. [CrossRef]

27. Ferri, C.; Ivanov, A.; Petrelli, A. Electrical measurements in μ-EDM. *J. Micromechan. Microeng.* **2008**, *18*, 085007. [CrossRef]

28. Singha, A.; Ghosh, A. A thermo-electric model of material removal during electric discharge machining. *Int. J. Mach. Tools Manuf.* **1999**, *39*, 669–682. [CrossRef]

29. Ramasawmy, H.; Blunt, L. Effect of EDM process parameters on 3D surface topography. *J. Mater. Process. Technol.* **2004**, *148*, 155–164. [CrossRef]

30. Pawlus, P.; Reizer, R.; Wieczorowski, M. Problem of non-measured points in surface texture measurements. *Metrol. Meas. Syst.* **2017**, *24*, 525–536. [CrossRef]

31. Pawlus, P.; Reizer, R.; Wieczorowski, M. Characterization of the shape of height distribution of two-process profile. *Measurement* **2019**, *153*, 107387–107395. [CrossRef]

32. Senin, N.; Thompson, A.; Leach, R. Characterisation of the topography of metal additive surface features with different measurement technologies. *Meas. Sci. Technol.* **2017**, *28*. [CrossRef]

33. Lou, S.; Jiang, X.; Sun, W.; Zeng, W.; Pagani, L.; Scott, P.J. Characterisation methods for powder bed fusion processed surface topography. *Precis. Eng.* **2019**, *57*, 1–15. [CrossRef]

34. Kiyak, M.; Çakır, O. Examination of machining parameters on surface roughness in EDM of tool steel. *J. Mater. Process. Technol.* **2007**, *191*, 141–144. [CrossRef]

35. Karmiris-Obratański, P.; Zagórski, K.; Cieślik, J.; Papazoglou, E.L.; Markopoulos, A. Surface Topography of Ti 6Al 4V ELI after High Power EDM. *Procedia Manuf.* **2020**, *47*, 788–794. [CrossRef]

36. Townsend, A.; Senin, N.; Blunt, L.; Leach, R.K.; Taylor, J.S. Surface texture metrology for metal additive manufacturing: A review. *Precis. Eng.* **2016**, *46*, 34–47. [CrossRef]

37. Ramanuj, K.; Soumikh, R.; Parimal, G.; Abhipsa, S.; Divya, D.S.; Rabin, K.D. Analysis of MRR and Surface Roughness in Machining Ti-6Al-4V ELI Titanium Alloy Using EDM Process. *Procedia Manuf.* **2018**, *20*, 358–364. [CrossRef]

38. Vikas; Apurba, K. R.; Kaushik, K. Effect and Optimization of Various Machine Process Parameters on the Surface Roughness in EDM for an EN41 Material Using Grey-Taguchi. *Procedia Mater. Sci.* **2014**, *6*, 383–390. [CrossRef]

39. Assarzadeh, S.; Ghoreishi, M. Prediction of root mean square surface roughness in low discharge energy die-sinking EDM process considering the effects of successive discharges and plasma flushing efficiency. *J. Manuf. Process.* **2017**, *30*, 502–515. [CrossRef]

40. Grochalski, K.; Wieczorowski, M.; H'Roura, J.; LeGoic, G. Optical aspect of errors in measurements of surface asperities with optical profilometry method. *Front. Mech. Eng.* **2020**, *6*. [CrossRef]

41. Ye, R.; Jiang, X.; Blunt, L.; Cui, C.; Yu, Q. The application of 3D-motif analysis to characterize diamond grinding wheel topography. *Measurement* **2016**, *77*, 73–79. [CrossRef]

42. Bartkowiak, T.; Berglund, J.; Brown, C.A. Establishing functional correlations between multiscale areal curvatures and coefficients of friction for machined surfaces. *Surf. Topogr. Metrol. Prop.* **2018**. [CrossRef]

43. Demircioglu, P.; Durakbasa, M.N. Investigations on machined metal surfaces through the stylus type and optical 3D instruments and their mathematical modeling with the help of statistica ltechniques. *Measurement* **2011**, *44*, 611–619. [CrossRef]

44. Vorburger, T.V.; Rhee, H.G.; Renegar, T.B.; Song, J.F.; Zheng, A. Comparison of optical and stylus methods for measurement of surface texture.International. *J. Adv. Manuf. Technol.* **2007**, *33*, 110–118. [CrossRef]

45. Pawlus, P.; Reizer, R.; Wieczorowski, M. Comparison of results of surface texture measurementobtained with stylus methods and optical methods. *Metrol. Meas. Syst.* **2018**, *25*, 589–602.

46. D'Urso, G.; Giardini, C.; Quarto, M. Characterization of surfaces obtained by micro-EDM milling on steel and ceramic components. *Int. J. Adv. Manuf. Technol.* **2018**, *97*, 2077–2085. [CrossRef]

47. Newton, L.; Senin, N.; Chatzivagiannis, E.; Smith, B.; Leach, R. Feature-based characterisation of Ti6Al4V electron beam powder bed fusion surfaces fabricated at different surface orientations. *Addit. Manuf.* **2020**. [CrossRef]

48. Senin, N.; Leach, R. Information-rich surface metrology. *Procedia CIRP* **2018**, *75*, 19–26. [CrossRef]

 *materials*

Article

Multiscale Characterizations of Surface Anisotropies

Tomasz Bartkowiak [1],* , **Johan Berglund [2,3]** and **Christopher A. Brown [4]**

[1]   Institute of Mechanical Technology, Poznan University of Technology, 60-965 Poznań, Poland
[2]   Department of Manufacturing, RISE Research Institutes of Sweden, SE-43153 Mölndal, Sweden; johan.berglund@ri.se
[3]   Department of Industrial and Materials Science, Chalmers University of Technology, SE-41296 Gothenburg, Sweden
[4]   Surface Metrology Lab, Worcester Polytechnic Institute, Worcester, MA 01609, USA; brown@wpi.edu
*   Correspondence: tomasz.bartkowiak@put.poznan.pl; Tel.: +48-61-665-24-52

Received: 22 May 2020; Accepted: 2 July 2020; Published: 7 July 2020

Abstract: Anisotropy can influence surface function and can be an indication of processing. These influences and indications include friction, wetting, and microwear. This article studies two methods for multiscale quantification and visualization of anisotropy. One uses multiscale curvature tensor analysis and shows anisotropy in horizontal coordinates i.e., topocentric. The other uses multiple bandpass filters (also known as sliding bandpass filters) applied prior to calculating anisotropy parameters, texture aspect ratios (Str) and texture directions (Std), showing anisotropy in horizontal directions only. Topographies were studied on two milled steel surfaces, one convex with an evident large scale, cylindrical form anisotropy, the other nominally flat with smaller scale anisotropies; a µEDMed surface, an example of an isotropic surface; and an additively manufactured surface with pillar-like features. Curvature tensors contain the two principal curvatures, i.e., maximum and minimum curvatures, which are orthogonal, and their directions, at each location. Principal directions are plotted for each calculated location on each surface, at each scale considered. Histograms in horizontal coordinates show altitude and azimuth angles of principal curvatures, elucidating dominant texture directions at each scale. Str and Std do not show vertical components, i.e., altitudes, of anisotropy. Changes of anisotropy with scale categorically failed to be detected by traditional characterization methods used conventionally. These multiscale methods show clearly in several representations that anisotropy changes with scale on actual surface measurements with markedly different anisotropies.

Keywords: surface texture; anisotropy; multiscale

1. Introduction

The objective of this paper is to present and study two new multiscale methods for determining anisotropy, i.e., lay or directionality of topographies. One method, based on curvature tensors, is primarily geometric and naturally multiscale. The other method, based on auto correlation and Fourier analyses, usually are applied to large single ranges of scales (ISO25178 [1]) and are implemented as multiscale here by using multiple bandpass filters to cover a multitude of scales. This study compares visual impressions of anisotropy from height maps of four surfaces, selected for their distinctive anisotropic characteristics, with representations of anisotropies by results of these two methods.

This is important because anisotropy at different scales can be an indicator of processing and performance. These indications can be valuable for product and process design and analysis, and for analyzing wear phenomena in engineering, anthropology, and archeology [2]. Currently, we are aware of no multiscale methods for analyzing anisotropies in national or international standards. Many topographically related phenomena are scale specific. Processing can influence different scales differently, and performance can be influenced at different scales differently. Knowing specific scales of interaction for phenomena can improve design and analyses of products and processes and augment the sophistication of many kinds of scientific analyses.

Anisotropy can be an ambiguous term in material science and surface metrology. It is generally related to properties which change with direction [3]. In surface metrology specifically, anisotropy relates to topographic characterization parameters changing with direction of observation, measurement, or calculation. Historically "lay" was used to describe a property of machined surfaces [4]. Lay, apart from its mathematical definition, is intuitively recognized visually. It is perceived as a dominant direction, or directions, in which machining marks seem to the eye to line up with each other. Processes like surface grinding and shaping can produce a surface where a profile parallel to the lay will appear to be nearly smooth. Face turning produces a surface with circular marks, while milling and honing processes produce surfaces with complex patterns. Roughing and finishing operations can create lays in more than one direction. Additively manufactured surfaces contain multidirectional features like hills (with peaks), dales (with pits), ridgelines, courses, and saddle points which directions cannot be described with 2D only [5]. Freeform surfaces present new challenges for topographic characterization, because they should not be required to rely on just one datum [6]. Because curvatures are spatial derivatives of slopes, they do not require a datum. This is an important advantage of curvature for characterization of free forms and internal cavities.

Anisotropies affect functional behaviors of surfaces. Summits which are long and narrow are likely to have different load-bearing properties than symmetrical summits [7] and similarly, leakage between contacting surfaces is likely to be influenced by the direction of the anisotropy [8]. It becomes important, therefore, to find, if possible, some quantitative ways of characterizing anisotropy. There are really two different problems: to find the direction or directions of the anisotropy, and to quantify the degree of anisotropy [9]. Anisotropic features can have different geometrical dimensions or sizes, which means that they can be detectable and/or inspectable at specific scales of observation or calculation, justify application of multiscale methods.

In this paper, "scale" refers to a narrow band of spatial frequencies or wavelengths, and multiscale refers to analyses done systematically over a range of scales [2]. Two multiscale methods are used here. One method uses 3D multiscale curvature tensor analysis. The other uses sliding bandpass filtering [2] prior to the calculation of the texture aspect ratio (Str) and texture direction (Std), measures of anisotropy from ISO 25178 [1]. Sliding bandpass filtering is a multiscale analysis method where bandpass filters are applied to isolate narrow bandwidths in scale at regular, even adjacent or overlapping intervals, to cover a wide variety of scales at which characterization parameters can be computed.

Curvature is a geometric property that naturally varies with scales of observation or calculation, as do length, area, and slope. These can be referred to as multiscale geometric analyses [2]. Curvatures on surfaces can be characterized as second order tensors which vary as functions of scale and position. Curvature tensors can be calculated from areal topographic measurements, i.e., surfaces height maps where heights z vary with position x and y, ($z = z(x,y)$). This kind of areal, multiscale curvature analysis has been recently developed to better understand surface topographies, or textures in mechanical engineering (ASME B46.1 [4]), how they are created and how they perform, based on how they interact with processing, the environment, or other surfaces. Scales of calculation for curvature tensors here are determined by the sizes of the regions over which the height measurements are selected for calculations of curvature to be made.

Topographies are commonly characterized by simplistic height parameters, such as, average roughness, the mean of the absolute values of heights measured from a mean line, Ra (for profiles), or a mean plane, Sa (for surfaces). Typically, roughness characterization analyses are applied to topographic measurements after removing form and waviness with Gaussian filters [1]. The conventional bandwidths for characterizing roughness are large. Sliding bandpass filtering uses different cut-offs to isolate narrow bandwidths. The central spatial frequencies or wavelengths for these narrow bandwidths can be systematically varied over a range of scales. Such multiple, or sliding, bandpass filters can be used to determine scale dependence of conventional texture characterization parameters for finding strong correlations and confident discriminations [2].

Average roughness, like Sa and Ra, is usually most sensitive to the longest wavelengths remaining after filtering form and waviness. This is because amplitudes tend to increase with wavelengths in these scale ranges for many surfaces. Topographic characterizations are valuable for assisting process or product design. When they help to establish correlations, first, between manufacturing processes and the resulting topographies, and, second, between topographies and performance, such as, adhesion or wetting. Conventional height parameters, with conventional filtering, use large bandwidths for waviness and roughness. Because of this coarse treatment of scale, they often fail to find strong correlations with processing parameters or measures of performance. Relations with specific textures, scale-sensitive processing, and performance phenomena, can be specific to certain fine, or narrow, ranges in scale. However, conventional parameters, which show weak correlations with processing and performance when used with traditional filtering, have been shown to correlate strongly when narrow band-pass filters are applied at appropriate scales [10,11]. Conventional height parameters do not provide any insights into the geometric nature of features, their horizontal spacings, or sequences of heights.

Motif analysis was successful in characterizing functional properties of surfaces, especially in friction and contact problems [12–14]. In those studies, parameters were used both to identify and separate anisotropic components by appropriate anisotropic filtering and characterization of surface motifs. Today, motif parameters are less used, although conclusions regarding the relationship between function and specification remain crucial.

Another approach for quantitative, multiscale characterization uses PSD (power spectral density), which is based on Fourier transformations. This approach treats signals as combinations of sinusoidal harmonics with different phases, amplitudes, and frequencies. Michalski used angular diagrams and contour maps of PSD to show that anisotropy on gear teeth flanks discriminates kinds of processing [15]. Jacobs et al. presented three important drawbacks to PSD and proposed strategies to mitigate them, and find aperiodicity, tilt, AFM (atomic force microscope) tip shape, and instrument noise [16]. PSD is commonly used to characterize machined surfaces to determine dominance of feeds, versus tool vibrations, and tool edge wear and lubricants [17], to predict the surface roughness in single point diamond turning [18,19]. On machined and worn surfaces dominant frequencies of PSD indicate anisotropy, and multiscale analysis of morphologies inside wear scars show regularly distributed PSD functions with tendencies to reduce maximum wavelengths with decreasing scale [20].

Multiscale curvature characterization of profiles have been correlated with fatigue life with an R^2 of 0.96 [21], clearly discriminated progressions of edge rounding by mass finishing [22], and found variations in stitching of profiles in measurements of aspheric lenses [23].

Multiscale curvature tensor analyses described complex new types of textures created by additive manufacturing [24], clearly showed fine-scale topographic effects of treating FDM parts with acetone vapor, elucidated differences in topographies of conventionally machined parts [25], and compared microgeometries between milling and grinding and contact interactions to determine relations between coefficient of friction and multiscale curvature [26]. In addition to characterizing principle curvatures, tensor analyses determine their directions, which is essential for characterizing anisotropy.

Anisotropy is important because it can significantly affect interactions between surfaces and phenomena that influence, or are influenced by, topographies. Tribological contacts in sheet forming are dependent on the orientation of microgrooves that influence friction [27]. Characterization of anisotropy is important for understanding topographies of honed piston liner [28]. Anisotropy also influences leakage in ball valves [29].

Fourier transforms and autocorrelation functions (ACF) are commonly used to characterize anisotropy over one scale range. Fourier spectra in polar coordinates show power spectrum in all directions. Directions with the largest amplitudes in these power spectra show anisotropy. Autocorrelations are used for periodic or pseudo-periodic motifs. Anisotropy at one scale can be characterized by surface texture ratios (Str, ISO 25178), which are ratios of lengths of fastest decay of ACF in any direction to lengths of slowest decay of ACF in any direction.

Sliding bandpass filtering can be used for multiscale analyses of traditional parameters. Std, texture direction, is the angle where the angular spectrum is the largest. Str, the texture aspect ratio, characterizes the uniformity of surface textures, determined by the ratio of the autocorrelation decay distances in the directions where the auto-correlation function decays to 0.2, by default, the fastest and the slowest. It is one for isotropic surfaces and zero for highly anisotropic surfaces. To calculate these parameters as a function of scale, measurement data are first filtered multiple times with narrow bands, stepping through a range of wavelengths, then calculating Str and Std for all the wavelengths, thereby creating multiscale characterization of anisotropy. Neither of these methods intrinsically provides multiscale characterizations or describes vertical components of anisotropy.

Thomas et al. applied structure functions, or topothesies, to analyze anisotropy [9]. Topothesy was invariant with orientation for isotropic surfaces. For a strongly anisotropic surface, topothesy was shown to be the same in every direction, except parallel to the lay, where it changes dramatically.

Area-scale and length-scale analyses, two multiscale geometric analyses, stem from fractal geometry. Length-scale has been used to analyze anisotropy in anthropology, to discriminate different kinds of dental microwear of fossil hominis to indicate diet [30,31], and in archeology for multiscale discrimination of types of use wear on stone tools with high confidence [32]. Area-scale has been used for finding strong correlations with loading during use at a particular scale [33].

The current paper includes an introduction of a multiscale curvature tensor analysis method, focusing on its potential to indicate anisotropy. Additionally included is a description of a multiscale bandpass filter. Examples of four measured surfaces with textures that provide different kinds of anisotropy are used to compare characterization techniques. These examples are brought, first, as a test for detecting evident anisotropy, and second, to characterize anisotropy and with respect to different scales of observation. The latter cannot be analyzed with conventional methods that characterize at only one scale or large ranges of scales. Multiscale analyses and characterizations have been shown to be important for understanding interaction with topographies and for establishing strong correlations and confident discriminations [2]. Curvature tensors can be helpful in identifying anisotropy by analyzing principal directions. This is not yet in ISO or ASME standards.

2. Materials and Methods

2.1. Surfaces and Measurements

Four surfaces are studied. Renderings of these surfaces are shown in Figure 2. These are selected to exemplify certain distinct types of anisotropy at different scales. These surfaces are intended to elicit a range of quantitative and qualitative characterization results, that differ scales and that facilitate comparisons and contrasts. These surfaces and their measurements are described below.

MilledC—a convex cylindrical form created by ball-nose end-milling of tool steel. This process creates strongly anisotropic topographies at scales of its cylindrical form and of the stepover between passes in milling. It was measured with a coherence scanning interferometry (CSI), white light interferometer, Wyko RSTPlus (Veeco Instruments, Plainview, NY, USA) with 10× lens. The measured region was 580×430 µm, and x- and y-sampling intervals were 790 and 910 nm respectively.

MilledF—a flat form also created by ball-nose end-milling of tool steel. This process was also expected to create topographies with anisotropies that are strong near the scales of the stepover, although it lacks the larger scale cylindrical form of MilledC. It was also measured with a CSI white light interferometer, Wyko RSTPlus, although with 5×, a magnification lens, giving a measurement region of 1.2 mm (x) and 0.9 mm (y) and sampling intervals of 1.7 µm (x) and 1.9 µm (y), which were, also in contrast to MilledC, resampled to 2 µm in both x and y before any analysis was done.

µEDMed—was created on 316 L stainless steel by µEDM (micro electo-discharge machining) with a discharge energy of 18 nJ. This process creates nominally isotropic topographies at scales larger than the discharge craters. These surfaces were machined by SmalTec (Lisle, IL, USA, www.smaltec.com) using a hydrocarbon-based oil for its dielectric fluid. The surface was measured with a scanning laser confocal microscope equipped with a 405 nm wavelength laser and a 100× objective lens with a numerical aperture of 0.95. The measured regions consist of 1024 × 1024 height samples over 125 × 125 µm, for a sampling interval of 125 nm. The measurements were processed using form removal and modal outlier filtering [34].

L-PBFed—was created using laser powder bed fusion (L-PBF) with 316 L stainless steel powder in a Solutions SLM 125 HL machine. This process creates topographies with subtle anisotropies that are challenging to detect. The measurement was made using CSI (coherence scanning interferometry) with white light on a Sensofar S neox (Sensofar, Barcelona, Spain) instrument with a 50× objective, perpendicular to a wall that had been formed vertically. The measured region was 516 × 516 µm and the sampling interval was 260 nm in both x and y directions.

Original sampling intervals used in multiscale analysis, related to all four surfaces are shown in Table 1, with scales presented in curvature-related figures.

Table 1. Original sampling intervals used in the multiscale analysis.

Surface	MilledC	MilledF	µEDMed	L-PBFed
Original sampling interval [µm]	0.790	2.000	0.125	0.260
5× original sampling interval [µm]	3.950	10.000	0.625	1.300
20× original sampling interval [µm]	15.800	40.000	2.500	5.200
25× original sampling interval [µm]	19.750	50.000	3.125	6.500
40× original sampling interval [µm]	31.600	80.000	5.000	10.400

2.2. Multiscale Curvature Tensor Calculations

Anisotropy is analyzed by using curvature tensors to determine the most prominent directions for ridges and valleys. Curvature tensors are calculated from the measured topographies by a 3D normal-based method, advancing Theisel's work [35]. Topographic measurements are tiled virtually with triangular patches from which normals are calculated. Each triangle is a right-angled isosceles in its (x,y) projection. Scales of calculation are the lengths of the triangle's legs, or catheti [10]. Progressive down-sampling is used to increase scales of calculation regularly for multiscale analysis.

For each group of three tiles, curvature tensors are calculated. The principal directions, $\mathbf{k_1}$ and $\mathbf{k_2}$ are indicative of anisotropy. A principal direction vector can be decomposed into direction cosines, and then to three directional angles: α, β, and γ, between $\mathbf{k_1}$ and the global coordinate system in which the measured topographies are described: $\mathbf{e}_x$, $\mathbf{e}_y$, and $\mathbf{e}_z$ (Figure 1). Distributions of α, β, and γ, over all triangular patches, can be plotted for each scale in spherical coordinates. For anisotropic surfaces, an evident peak or peaks in the distribution is expected to appear, indicating the dominant directions. Whereas, for more isotropic surfaces, more uniform distributions are expected. These results can also be visualized in a horizontal coordinate system, also known as topocentric, bfigurey expressing the orientation of $\mathbf{k_1}$ in angular coordinates: altitude, or elevation, and azimuth [36]. The reason why a spherical coordinate system is not used is because the polar angle, in that case, is measured from a fixed zenith direction (z-axis), whereas in the horizontal coordinate system, elevation is measured from a reference plane, associated with a horizon, which is more intuitive. When principal directions of maximal curvature are coplanar with datum, the elevation is zero, whereas the polar angle would be 90 degrees.

The script that allows calculation of multiscale curvature tensor, including principal directions at each scale and location was created using Mathematica 12 (Wolfram Research, Oxfordshire, UK) computational software. This script is available upon request.

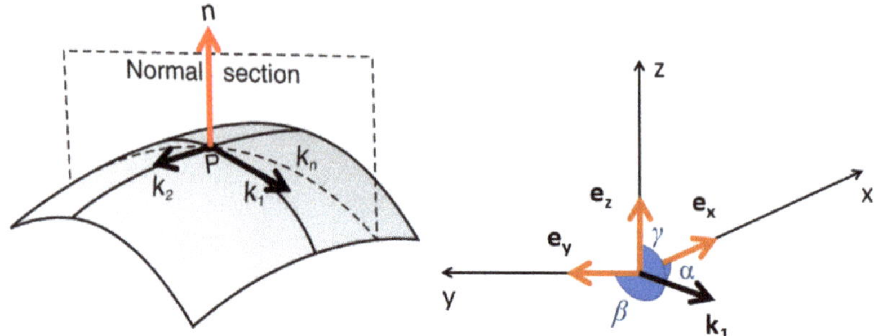

Figure 1. Relations between curvature principal directions $\mathbf{k}_1$ and $\mathbf{k}_2$ and normal vector and visualization of geometrical direction angles for principal direction $\mathbf{k}_1$ in spherical coordinates.

2.3. Bandpass Filtering for Multiscale Analyses

Manipulations of datasets, bandpass filtering, and calculations of characterization parameters were performed with MountainsMap® 7.4 software (DigitalSurf, Besançon, France). Non-measured heights in the measurements were filled-in with using smart shape interpolation.

Step 1: Defining spatial frequency bands (scales) for bandpass filters and filtering.

Bandpass filters divide topographic data into different scales of observation, i.e., spatial frequency or wavelength bands, reciprocals of each other, for calculating ISO 25178 topographic characterization parameters.

Multiscale bandpass filtering, in this study, uses a sequence of robust Gaussian, combining low-pass and high-pass filters, so that only narrow bands of scales between are left. Mean values of upper and lower nesting indices of bands, centers of cutoff wavelengths, are used to represent scales of observation, or calculation. The bands overlap each other. Nesting indexes are selected with ratios of 1 to 2. There is a 50% overlap between the bands as shown in Table 2. The smallest nesting index possible, using a robust Gaussian filter, is three times the sampling interval (the larger one, if the sampling is different in x and y). The longest nesting index possible is the length of the shortest edge of the measurement region. Berglund et al. [37] used a similar method in their approach B, although it was plotted differently and they also used Sq for characterizations, which is not used here. Table 2 shows bandpass filter values with the lowest and highest wavelengths used for nesting indices represented by dashes. Note that low-pass and high-pass refer to spatial frequencies, the inverse of wavelengths.

The nature of scales in bandpass filtering differ slightly from those in multiscale curvature analysis. The latter scales are given as multiples of sampling intervals, achieved by down sampling, acting like step functions. The former uses robust Gaussian filters for bandpass filtering, which are not step functions. Wavelengths slightly beyond low and high cut-off wavelengths are included in a bandpass, although with diminishing amplitudes around the cut-offs.

Step 2: Calculating conventional topographic characterization parameters.

Texture aspect ratios (Str), texture direction (Std) from ISO 25178-2 are calculated separately for each bandpass scale.

Step 3: Creating polar plots from conventional topographic characterization parameters.

Polar plots represent anisotropy symmetrically. Two points are plotted at each scale, with the angle based on Std and the magnitude based Str Distances from the centers are the complement of Str $(1 - \text{Str})$. The angles are Std and its straight angle (St + 180). This provides symmetry, which makes these plots easier to interpret. Using the complement of the texture aspect ratio $(1 - \text{Str})$ portrays stronger anisotropies as larger, i.e., as the magnitudes of the anisotropy with a maximum of one.

Table 2. Wavelengths of the nesting indices, center, low-pass, and high-pass for multiscale (also known as sliding) bandpass filtering, all in [μm].

No.	MilledC			μEDMed			L-PBFed			MilledF		
	Center	Low	High	Center	Low	High	Center	Low	High	Center	Low	High
1	3.0	-	4.0	0.422	-	0.563	1.1	-	1.5	6	-	8
2	4.5	3.0	6.0	0.563	0.375	0.750	1.5	1.0	2.0	9	6	12
3	6.0	4.0	8.0	0.844	0.563	1.125	2.3	1.5	3.0	12	8	16
4	9.0	6.0	12.0	1.125	0.750	1.500	3.0	2.0	4.0	18	12	24
5	12.0	8.0	16.0	1.688	1.125	2.250	4.5	3.0	6.0	24	16	32
6	18.0	12.0	24.0	2.250	1.500	3.000	6.0	4.0	8.0	36	24	48
7	24.0	16.0	32.0	3.375	2.250	4.500	9.0	6.0	12.0	48	32	64
8	36.0	24.0	48.0	4.500	3.000	6.000	12.0	8.0	16.0	72	48	96
9	48.0	32.0	64.0	6.750	4.500	9.000	18.0	12.0	24.0	96	64	128
10	72.0	48.0	96.0	9.000	6.000	12.000	24.0	16.0	32.0	144	96	192
11	96.0	64.0	128.0	13.500	9.000	18.000	36.0	24.0	48.0	192	128	256
12	144.0	96.0	192.0	18.000	12.000	24.000	48.0	32.0	64.0	270	192	348
13	192.0	128.0	256.0	27.000	18.000	36.000	72.0	48.0	96.0	384	256	512
14	270.0	192.0	348.0	36.000	24.000	48.000	96.0	64.0	128.0	522	348	696
15	384.0	256.0	-	48.000	32.000	-	144.0	96.0	192.0	768	512	-
16	N/A	N/A	N/A	N/A	N/A	N/A	192.0	128.0	256.0	N/A	N/A	N/A
17	N/A	N/A	N/A	N/A	N/A	N/A	288.0	192.0	384.0	N/A	N/A	N/A
18	N/A	N/A	N/A	N/A	N/A	N/A	384.0	256.0	512.0	N/A	N/A	N/A
19	N/A	N/A	N/A	N/A	N/A	N/A	576.0	384.0	-	N/A	N/A	N/A

3. Results

3.1. Visual Impressions of Anisotropy

Visual impressions of anisotropy of all four surfaces can be made from the colored height maps, renderings of their topographic measurements in Figure 2. These impressions are to be compared with results of the two kinds of multiscale analyses described above, curvature tensors and bandpass texture aspect ratios (Str) and texture directions (Std), which are shown in Figures 3–8.

Both milled surfaces are clearly anisotropic (Figure 2a,d). Black arrows indicate apparent directionality of characteristic features. MilledC (Figure 2a) exhibits fine scale ridges and valleys that tend to align with the x-axis. Its cylindrical form was not removed to show the anisotropy at larger scales. MilledF (Figure 2d) changes with scale. There are troughs and ridges of different scales. At larger scales, they are oriented in x and most clearly discernible by subtle variations in color. At smaller scales, they are oriented, as stripes, along y and, at even smaller scales, clearly discernible by the distinct repeated cusp shapes of a circular tool nose directed parallel to the x-axis.

The EDMed surface appears isotropic (Figure 2b). There are small sized craters created by low-energy electric discharges that appear to be randomly distributed on the surface. Note that the peak-to-valley roughness, as indicated by the vertical scale, is less than the others.

The L-PBF-ed surface (Figure 2c) has subtle ridges and valleys parallel to the y axis, aligned with the laser scanning direction used to fuse the powders. It also shows columnar features with steep slopes from partly fused powder particles, indicating a vertical component to the anisotropy. Details on steep slopes cannot be measured with conventional instruments, because the illumination and observation direction are nearly parallel to the sides of the columns.

Renderings of bandpass filtered surfaces that were used to calculate Std and Str for all analyzed bands and surfaces are shown in Supplementary Materials.

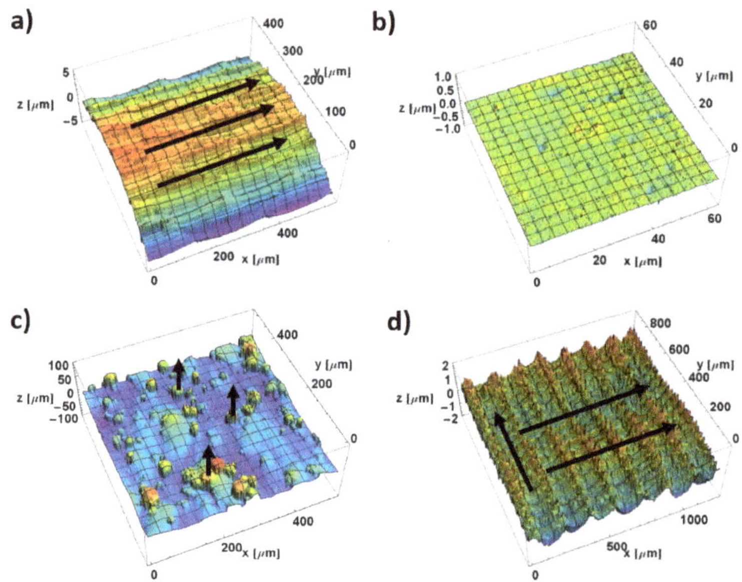

Figure 2. Renderings of measured topographies of (**a**) MilledC, (**b**) µEDMed, (**c**) L-PBFed, and (**d**) MilledF. Please note that black arrows indicate visual impression of apparent anisotropy.

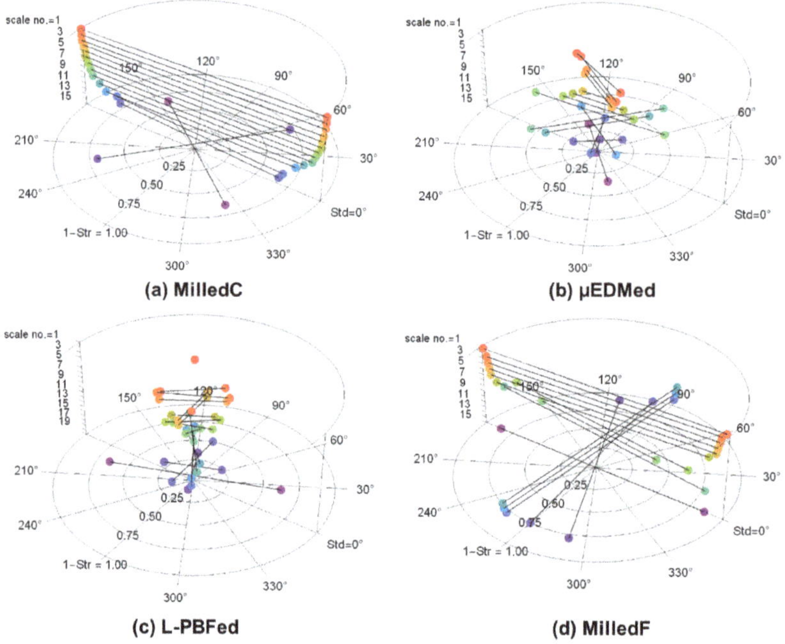

Figure 3. Polar plots showing dominant directions of anisotropy with texture direction (Std) in degrees, magnitude with the complement of the texture aspect ratios (1 − Str), and scales vertically with scale number corresponding to the band numbers from Table 2, with 1 the smallest scale as the highest, calculated for: (**a**) MilledC, (**b**) µEDMed, (**c**) L-PBFed, and (**d**) MilledF.

3.2. Multiscale Characterizations of Surface Anisotropies by Bandpass Filtering

Figure 3 presents polar plots showing anisotropy. A bandpass filter was used for the multiscale decomposition, as described above. The plots show dominant directions, Std, and their magnitudes, as complements of texture aspect ratios (1 − Str), in polar coordinates, for each scale in z. The x axis in Figure 2 corresponds to 0° in these polar plots. For strongly anisotropic surfaces, the complement of Str is close to one, the full radius of the plots. Central wavelengths of the bandpass filters (Table 2) indicating scales, are shown as band numbers, and increase downwards from smallest to largest. These polar plots facilitate analyses of scale-dependent anisotropies.

The polar plot for MilledC (Figure 3a) shows strong anisotropy at 0° for all scales except the two largest. Anisotropy (1 − Str) decreases with increasing scale, suggesting that there are larger, more directionally varied features. Fine scale features are strongly anisotropic. The two largest scales show dominant directions changing from 0° to 50° and to 143°, respectively. Those last bands are the widest, therefore, they include the widest spectrum of feature sizes, including waviness and cylindrical form. Waviness, that can be seen, might be detected by the Std. In addition, in band 15, there is a slope along the x-axis. This due to a misalignment, the surface is not perfectly perpendicular to the z-axis of the microscope. This could affect the change in Std from 50° to 143°. The direction along x-axis is also significant in this band, but not dominating according to Std and Str.

The polar plot for μEDMed (Figure 3b) shows relatively small magnitudes and directions that change with scale. The least anisotropy is noted for scales of 1.688 and 36 μm and the strongest for scales of 4.5 and 6.75 μm. The scales for which the texture is the most anisotropic cover the diameters of the most evident discharge craters of similar directionality. Those features stand off the isotropic background which corresponds to increasing magnitudes.

The polar plot for L-PBFed (Figure 3c) also shows relatively small anisotropy and directions that change with scale. Apart from the largest scale, the largest magnitude of anisotropy varies between 0.06 to 0.36. The largest scale of analysis, band 19 with a low-pass cut-off of 384.0 μm, can be associated with the direction of laser scanning motion during the sintering process. The direction of solidified material is dominant and other differently oriented features do not play key roles for that scale of calculation. Characterization of anisotropies of pillar features is not possible using this method.

The polar plot for MilledF (Figure 3d) shows strong anisotropy around 0° at the finest scales. This changes dramatically for larger scales, which are 90° shifts for scales from 144 to 522 μm. For the largest scales, there is a return to 0°. For all scales, the complements of the texture aspect ratios are greater than 0.84, indicating strong anisotropy.

3.3. Multiscale Characterizations of Surface Anisotropie by the Direction of Maximum Curvature

Figures 4 and 5 show k_1, directions of maximum curvatures, with arrows on height maps surfaces at two scales, five and twenty times their original sampling intervals. Other scales are included in Supplementary Materials. Arrows are plotted for a limited number of regions to improve perceptions. Regions indicating ridges and grooves, that are clearly anisotropic, are visible for MilledC in all scales. This is evident in the similar orientation of k_1 for every location. In contrast, for μEDMed, orientations of principal direction of maximum principal curvature vary with region as expected for isotropic surface. For MilledF finer scale orientations of k_1 are mostly aligned in x, corresponding to anisotropic features created by cutting tool rotation and interactions between the tool edge and workpiece material. For larger scales, arrows are indicative of the bi-directional pattern created by the feed. Visualization of k_1 for L-PBFed topography indicate different types of directional features. At the finest scale, wrinkle-like features created by solidifying melt pools can be seen, as well as vertical grooves on the pillar like structures, and valleys between them.

Distributions of angles for directions of maximum curvatures, at each calculation scale, are plotted as 2D histograms in Figure 6 at their original sampling intervals and at forty times their original sampling intervals. More uniform distributions indicate more isotropy, whereas distributions with distinctive single or multiple nonuniformities in their distributions indicate stronger anisotropies in one or more directions.

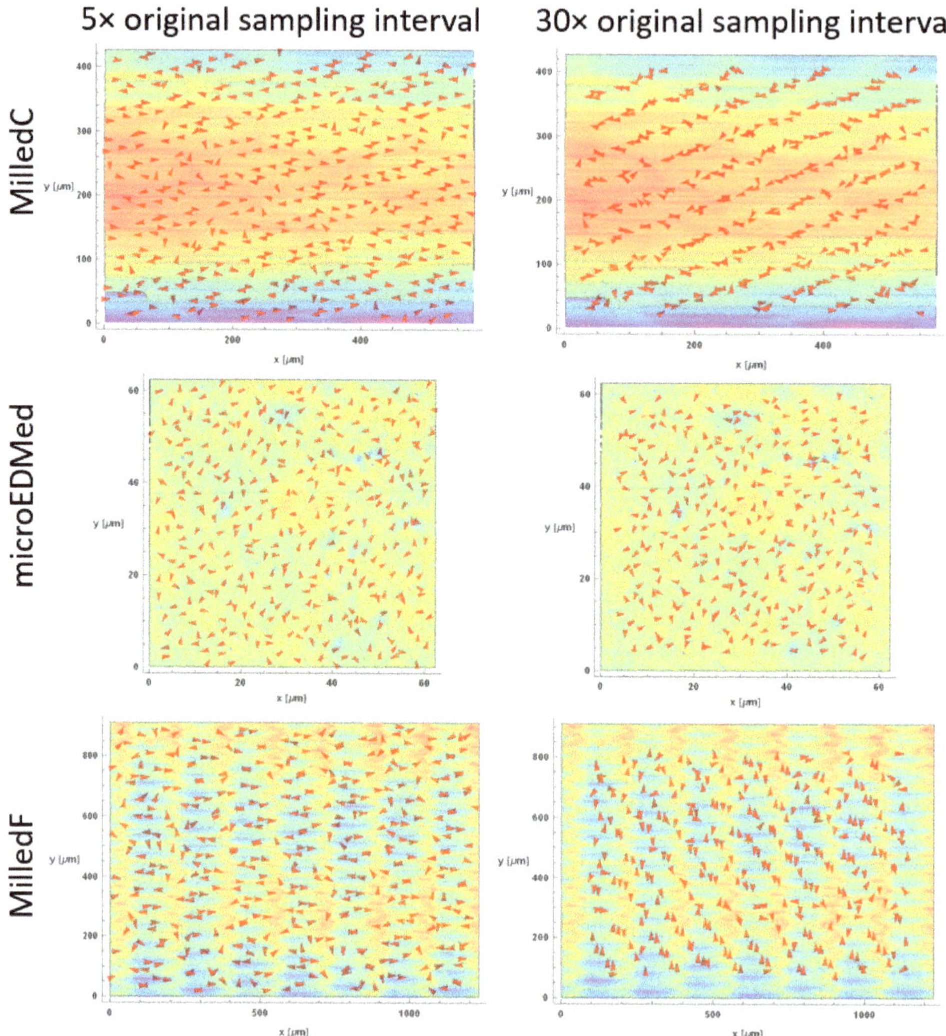

Figure 4. Directions of maximum principal curvatures calculated for MilledC, µEDMed, and MilledF at two scales: 5× and 30× the original sampling interval, plotted together with color-coded height maps. Please note that red arrows indicate a direction of maximum curvature at a given location.

Multiscale curvature approaches facilitate better characterization of surface anisotropy at multiple scales. This can be done by analyzing 2D histograms of direction angles. Anisotropy is evident if those distributions are unimodal, i.e., a single distinctive peak is present. Stronger anisotropies correspond to smaller standard deviation of those distributions. This effect is visible for MilledC (Figure 6a). For all scales analyzed, distributions of direction angles α, β, γ possess corresponding unique modes at 90°, 0°, or 180° and 90°. With increasing scales, distributions become more concentrated around these values. This could indicate that fine scale features are caused by irregularities in chip formation. This includes built-up edge on the tool that is continuously generated and removed during machining and heterogeneity of material microstructures. All those effects create randomly oriented microfeatures. For larger scales, the cylindrical form and waviness, which are directionality consistent, become more dominant.

For isotropic surfaces, the distributions of direction angles are, in contrast to anisotropic textures, multimodal and more uniform. This effect is visible for μEDMed surface for distributions of α and β for all scales (see example data in Figure 6b). For isotropic surface, distributions of α and β are similar. The distribution of γ can indicate how the texture orientation differs from xy-plane. Ideal vertical features should be characterized by $\gamma = 0$ degree or $\gamma = 180$ degree. In case of EDMed topography, distribution of γ are highly concentrated at 90 degrees, indicating features mostly oriented in the xy plane. The opposite effect can be seen for the L-PBFed surface, where γ distributions are more dispersed (Figure 6c). Distributions of α and β indicate isotropy. Some local anisotropies of β equal to 90 degrees, corresponding to laser beam path directions during sintering.

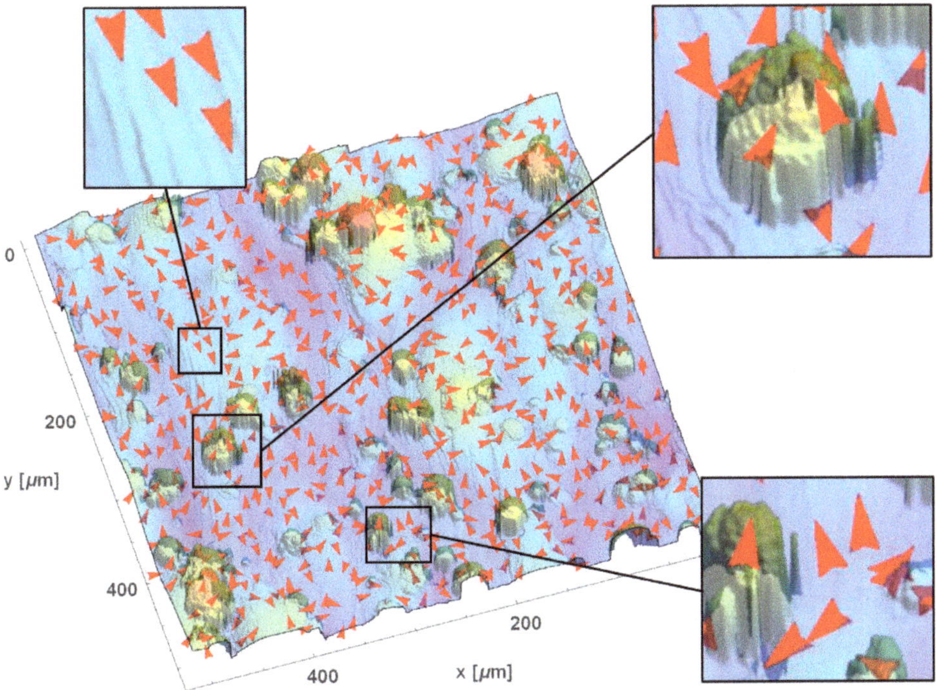

Figure 5. Directions of maximum principal curvature calculated for L-PBFed at the original sampling interval. Please note that red arrows indicate a direction of maximum curvature at a given location.

MilledF (Figure 6d) surface shows an anisotropic character that is represented by unimodal distribution of α, β, and γ, like MilledC. This is observed between scales between 1× and 9× as well 35× and 40× original sampling interval. Clear bimodal distributions of α, β are noted between scales 10× and 25× original sampling interval. The shift in dominant direction by 90 degrees can be seen between scales 26× and 34×, the original sampling interval.

Anisotropy can also be visualized in three-dimensional distributions in the horizontal coordinate system (HCS), i.e., topocentric coordinates. In this method, the entire hemisphere is divided into bins of angular resolutions (5 by 5 degrees Figure 7). Each k_1, is expressed in the HCS and associated certain bin. More k_1 vectors in a direction indicates more anisotropy. Some are shown in Figure 7, other scales are in the Supplementary Materials. An azimuth angle of 90 or 270 degrees corresponds to the alignment with *x*-axis, and 0 or 180 degrees to *y*-axis.

Distribution plots in HCS for MilledC confirm its anisotropic characteristics. The dominant orientation of principal direction k_1 is aligned with 90 and 270 degree azimuths for all scales, while elevation is mostly below 5 degrees. Some fine scale features appear to be inclined at slightly steeper angles, although this is only visible for scales less than 7.846 µm (10× original sampling interval). These features are located around the cylindrical form, which was intentionally not removed to test this effect. Regardless of those fine scale features, the nature of MilledC anisotropy is generally two-dimensional.

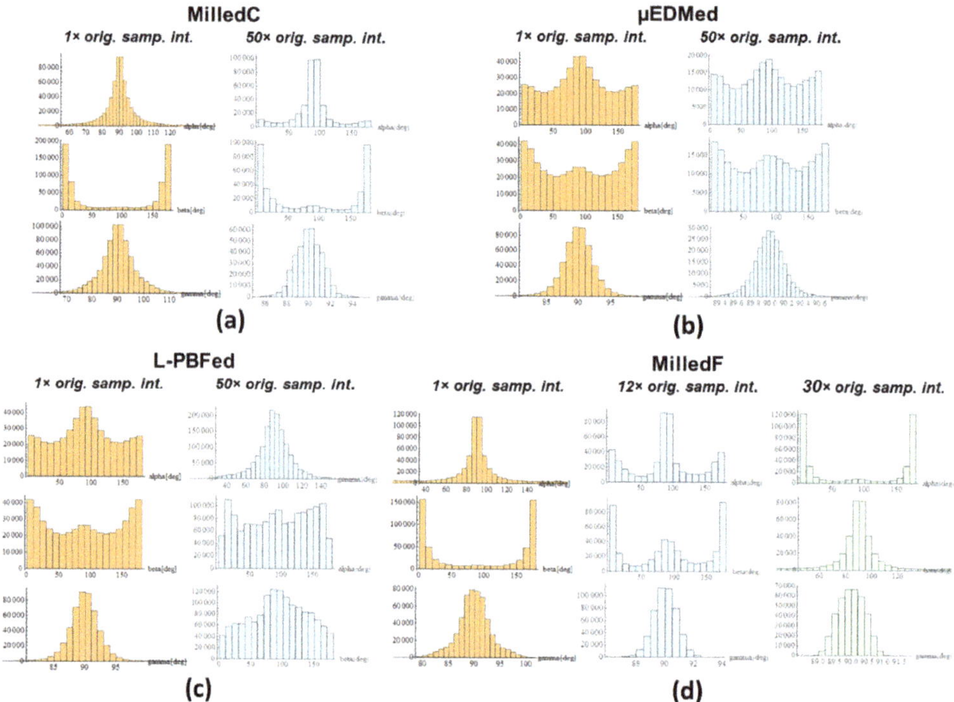

Figure 6. 2D histograms created from direction cosines of maximum curvature calculated for (**a**) MilledC, (**b**) µEDMed, (**c**) L-PBFed, and (**d**) MilledF at the indicated scales.

The distribution of k_1 for µEDMed examples is evidently more dispersed than MilledC. The dominant azimuths of 0 and 180 degrees, as well as 90 and 270 degrees, relate to the scanning direction of the electrodes during electric discharge machining. Elevation is always less than 5 degrees and is related to general flatness of the topography with some slightly inclined slopes of the discharge craters. This confirms the isotropic nature of µEDMed topographies.

Orientation of k_1 in HCS for MilledF show how its anisotropy depends on scale. Between 2 and 18 µm, principal directions of maximum curvatures are aligned with 90 and 270 degree azimuths. Between 20 and 50 µm a second dominant direction is present for azimuth between 0 and 180 degrees. At 60 µm, the first dominant direction is not visible, whereas at 80 µm, the second is not evident. For larger scales, both dominant directions are present. The elevation for all scales analyzed is always close to 0 degrees. This is due to the fact that the form (or the general shape) was removed prior to the analysis.

Other characteristics can be seen for the L-PBFed example. In this case, dominant azimuths of $\mathbf{k_1}$ are 0 and 180 degrees, as well as 90 and 270 degrees, although dispersion is visibly greater than for MilledC and MilledF. The elevation is found to be up to 90 degrees for the finest scales, when distribution is presented in a logarithmic scale for magnitude (Figure 8), and it is related to the curvature of pillar like features. Clearly the anisotropy of this surface topography is 3D and scale-dependent.

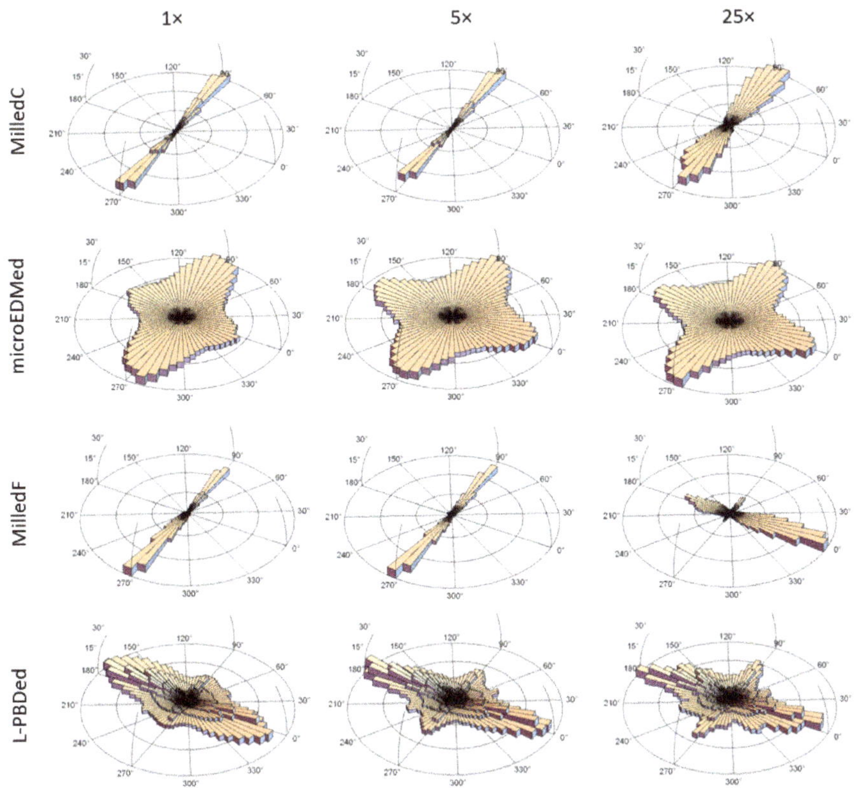

Figure 7. Linear histograms of maximum principal curvature directions in horizontal coordinate system (HCS), i.e., topocentric coordinates at three different scales.

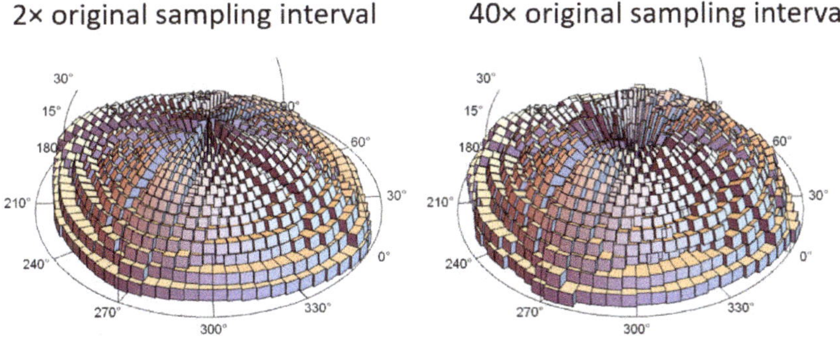

Figure 8. Logarithmic histograms of maximum principal curvature directions in horizontal coordinate system (HCS), i.e., topocentric coordinates at two different scales for the L-PBFed surface.

3.4. Conventional Approach Based on Fourier Transform in Polar Coordinates

Rosette plots created from MilledC and μEDMed topographies can indicate anisotropy and isotropy, which is consistent with indications from bandpass filtering and curvature tensor methods (Figure 9a,b). The limitation of the conventional approach, based on Fourier transform in polar coordinates, becomes more evident for the other two surfaces. For L-PBFed, laser path direction is distinct. No identification and characterization of 3D features is available (Figure 9c). Variation of directionality with scale for MilledF is not shown in Figure 9d, as it appears to be a strongly anisotropic surface, with some weak deviations in perpendicular directions. These results are consistent with expectations based on visual examination of their topographic maps, and intuitive estimations of dominant directions.

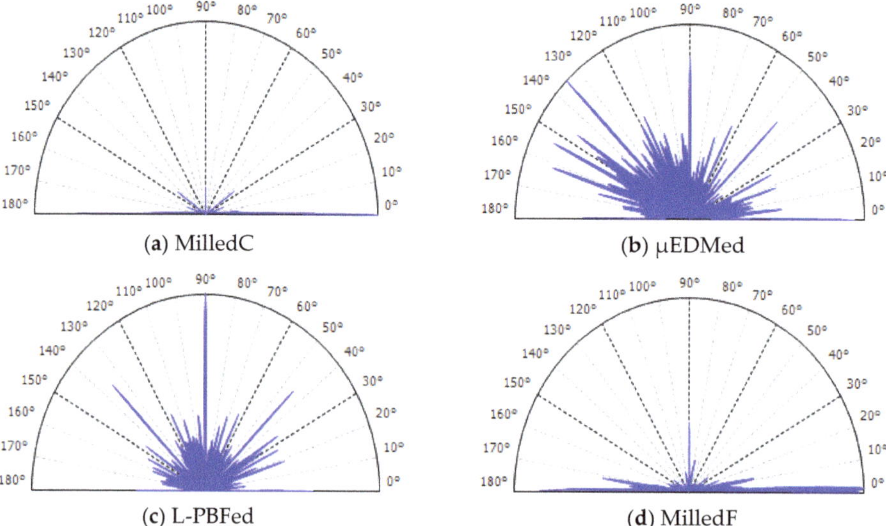

Figure 9. Rosette plots created with a conventional, non-multiscale method using Fourier spectra in polar coordinates, for (**a**) MilledC, (**b**) μEDMed, (**c**) SLMed, and (**d**) MilledF.

4. Discussion

Two new, different methods using multiscale analyses and characterization for quantification and visualization of anisotropy are studied here and tested for their ability to elucidate anisotropies. The analysis algorithms for calculating multiscale characterization parameters for determining anisotropy are described. Results of applications to four measured surfaces, manufactured to have distinctly different anisotropies, are examined critically.

Detecting anisotropies by inspection is important, although not a substitute for algorithms, which can automate detection, remove bias, subjectivity, and tedium of multiple inspections, and might, by its consistency, provide insights that would not be detectable by inspection. Inspection should be used to verify results of new algorithms before they are used to investigate surfaces whose significant structures are revealed only by multiscale curvature analyses.

Validation of these methods for elucidating anisotropies as functions of scale is an issue with a philosophical component. Numerical validation of new characterizations of this kind is problematic because they extend current possibilities beyond current experience. There are no other tests that can produce essentially similar results. Over time, the value of new characterizations might be established, and this could be another kind of validation. For that to happen, papers must be published to disseminate this new knowledge. In an initial work, as here, new methods can be compared with each other. Here, several kinds of representations are used to facilitate these comparisons. Consequently, it is learned that they are consistent with each other. They fulfill reasonable expectations for discriminating measurements from topographies with anisotropies that are known to be different.

These two methods are compared in several different kinds of representations, which are intended to reveal anisotropic properties. These elucidate differences between each surface, with impressions from qualitative visual inspection, and with conventional rosette plots of Fourier analyses. One new multiscale method uses multiple bandpass filters (also known as sliding bandpass filtering) which are applied to measured topographic data prior to calculating conventional anisotropy characterization parameters, texture aspect ratios (Str), and texture directions (Std). These detect anisotropy in two dimensions, parallel to a datum plane. The other uses multiscale curvature tensor analysis to characterize topographies with principal curvatures and their orientation in three dimensions, at each location and multiple scales. The main limitations of sliding bandpass filtering is that it only allows characterization of a single dominant direction at each scale. However, it detects changes in anisotropy with scale, which is not possible with conventional non-multiscale approach using autocorrelation functions, for example. That approach can only be used for 2D anisotropy analysis conducted at a nominal scale, i.e., original sampling interval.

The anisotropies detected by these two multiscale methods correspond well to those that are evident by visual inspection of height maps, and in a more limited sense, to conventional rosette plots of Fourier spectra. Visual inspections can detect changes in anisotropy with scale, whereas conventional rosette plots of Fourier spectra are limited because they cannot.

Determining specific scales of anisotropy is important because, according to Brown et al. [2], some topographically dependent phenomena have certain scales, or narrow scale ranges, over which they interact with topographies. These scales can be advantageous or disadvantageous for certain kinds of performance. Anisotropies can be created at different scales. It is important to be able to recognize these scales for product and process design. They can also be important in physical anthropology and forensics [30].

Multiscale curvature tensor analysis provides the most knowledge about anisotropy of any of the methods studied here. Results of this analysis characterize anisotropy so that it can be represented in horizontal coordinate systems i.e., topocentric coordinates, which show both horizontal and vertical components, providing a true three-dimensional representation of anisotropy.

Multiscale curvature tensor analyses provide characterization and visualization of directionality of subregions. This method, unlike conventional and sliding bandpass filtering, can also be used to characterize features which are directed perpendicularly from nominal or datum plane. This is particularly important for characterizing freeform surfaces and surface manufactured additively by laser fusion. The later can have especially intricate topographies at fine scales due to partly fused particles.

The term "scale" is used in this study differently in the two multiscale analyses. Therefore, a direct quantitative comparison might not be possible. From a qualitative perspective, they both indicate anisotropy changes with scales similarly. In sliding bandpass filtering, scale refers to central wavelengths and widths of bands [11,37]. In geometric multiscale methods, scale is associated with geometrical parameters like length, area, filled volumes or areas, and curvature [1]. In this multiscale curvature analysis, scale is linked with the size of a triangular patch [10] into which the original mesh is divided. The curvature tensor analysis uses what is essentially down sampling. The scale is more specific and tied to multiples of sampling intervals, i.e., pixel sizes. From a qualitative perspective, they both similarly indicate anisotropy changes with scales. The two different manifestations of scale exacerbate precise quantitative comparison of results from these two different methods. However, these two methods similarly indicate the changes in anisotropy with scale in all four analyzed examples.

A limitation of multiscale analysis is its computation complexity, so it can be time-consuming. Maleki et al. showed that curvature tensor calculation required the most time when compared to other existing methods [38]. On the other hand, that study concluded that it performed the best, together with the Bigerelle–Nowicki method, in terms of quality for the analyzed test scenarios.

Additively manufactured surfaces present new challenges for characterizations. The surfaces of metal PBF (powder bed fusion) components are typically highly irregular, with steep sided and re-entrant features [5]. Relevant surface features exist at a wide range of scale, therefore a scale-based characterization becomes of great importance. Being datum-independent, curvature seems to be a prospective candidate for the analysis of AM surfaces in terms of anisotropy. Further research will focus on characterizing complex 3D, freeform structures, and measurements by microCT.

Multiscale curvature tensor analyses are a valuable tool for elucidating changes in anisotropy caused by processing, and for indicating performance, such as sealing, lubrication, and friction. These also require appropriate statistical analyses, which could describe the complexity of the anisotropy expressed as histograms of maximum principal curvature directions in horizontal coordinate system. Potential candidates for visualization should also portray bivariate characters of distributions, as both azimuth and elevation angles are considered together. Potential candidates include measures of distribution, i.e., modes, means and medians, dispersion and associated, such as bivariate mean deviation, total variation, or generalized variance. Higher moments like bivariate skewness and kurtosis can also be relevant [39]. Tracking changes with scale, should be additional tools to enumerate effects of surface processing on anisotropy. This is especially important in physical anthropology, paleontology, and archaeology, where directions of topographic features found in artifacts indicate their function [2,30,31]. Sophisticated indications of anisotropy by multiscale curvature tensors can improve manufacturing processes diagnostics, e.g., in the detection and characterization of tool wear [40], and help to understand its impact on performance of resulting topographies [41].

The main advantage of conventional analyses is that they are incorporated in commercial software, and therefore used extensively by the industry and academia. New characterization methods, like multiscale, will be welcomed, if they add value by advancing the understanding of the relations between topographies and phenomena. This could be facilitated by more intuitive, automated, and easy-to-use software developed and distributed for industrial use. Some efforts have already been made by including multiscale profile and areal analyses and bandpass filtration (profiles only), as easy-to-use features in commercial software. Multiscale curvature analyses, and bandpass filtering for areal datasets prior to calculating traditional parameters, should follow the same path.

5. Conclusions

Two new multiscale methods for quantification and visualization of anisotropy are described, examined critically, and compared logically each other, with impressions from qualitative visual inspection, and with conventional rosette plots of Fourier analyses. One new multiscale method uses multiple bandpass filters (also known as sliding bandpass filtering) which are applied to measured topographic data prior to calculating conventional anisotropy characterization parameters, texture aspect ratios (Str), and texture directions (Std). These detect anisotropy in two dimensions, parallel to a datum plane. The other uses multiscale curvature tensor analysis to characterize topographies with principal curvatures and their orientation in three dimensions, at each location and multiple scales.

The anisotropies detected by these two multiscale methods correspond well to those that are evident by visual inspection of height maps, and in a more limited sense, to conventional rosette plots of Fourier spectra. Visual inspections can detect changes in anisotropy with scale, whereas conventional rosette plots of Fourier spectra are limited because they cannot.

Both these new, multiscale methods can show clearly that anisotropy can change with scale on actual surfaces with markedly different anisotropies.

Changes of anisotropy with scale categorically cannot be detected by traditional characterization methods used conventionally, e.g., Fourier spectra.

Multiscale curvature tensor analysis shows anisotropy in horizontal coordinate systems (HCS), i.e., topocentric, with both horizontal and vertical components, which are a true, three-dimensional representations of anisotropy.

With the bandpass approach, polar plots elucidate anisotropy at specific scales. Multiple plots, at different scales, can be combined and used to show the multiscale nature of different sorts of anisotropies. These polar plots show orientations of anisotropy of texture directions (Std) in degrees, magnitudes in the radial direction are derived from complements of texture aspect ratios (1 − Str), and scales are shown vertically. However, only a single dominant direction can be indicated at each scale.

Directions of principal curvatures superimposed on height maps also elucidate anisotropies at specific scales. Different scales show the multiscale nature of different sorts of anisotropies.

Histograms, showing frequency distributions, created from direction cosines of maximum principal curvatures are another way of elucidating anisotropies as a function of scale. These can either be plotted conventionally in cartesian coordinates, or with hemispherical-type histograms of maximum principal curvatures directions using horizontal coordinate systems (HCS), i.e., topocentric coordinates.

Supplementary Materials: The following are available online at http://www.mdpi.com/1996-1944/13/13/3028/s1, as a supplement to this study, authors include the bandpass filtered surfaces created by robust Gaussian filtration as described in the paper. Additional histograms of maximum principal curvature directions in horizontal coordinate system (HCS), together with plots representing directions of maximum principal curvatures, calculated for all four analyzed surfaces, are also included in the supplement.

Author Contributions: Conceptualization, T.B., J.B. and C.A.B.; methodology, T.B. and J.B.; software, T.B. and J.B.; validation, T.B., J.B.; formal analysis, T.B., J.B. and C.A.B.; investigation, T.B., J.B. and C.A.B.; resources, T.B. and J.B.; data curation, T.B. and J.B.; writing—original draft preparation, T.B., J.B. and C.A.B.; writing—review and editing, T.B., J.B. and C.A.B.; visualization, T.B.; supervision, T.B.; project administration, T.B.; funding acquisition, T.B. and J.B. All authors have read and agreed to the published version of the manuscript.

Funding: This research was funded by the Polish Ministry of Science and Higher, as a part of research subsidy 0614/SBAD/1529.

Acknowledgments: The authors would like to acknowledge previous contributions of Matthew Gleason, a student at WPI, who developed an algorithm for multiscale curvature analysis that led to several papers and discovering a strong correlation with curvature, and of Torbjorn S. Bergstron, who, as a grad student at WPI, developed the first multiscale anisotropy analyses, that are now widely used in physical anthropology.

Conflicts of Interest: The authors declare no conflict of interest.

References

1. International Organization for Standardization. *ISO 25178-2. Geometrical Product Specifications (GPS)—Surface Texture: Areal—Part 2, Terms, Definitions and Surface Texture Parameters*; International Organization for Standardization: Geneva, Switzerland, 2010.
2. Brown, C.A.; Hansen, H.N.; Jiang, X.J.; Blateyron, F.; Berglund, J.; Senin, N.; Bartkowiak, T.; Dixon, B.; Le Goic, G.; Quinsat, Y.; et al. Multiscale analyses and characterizations of surface topographies. *CIRP Ann.* **2018**, *67*, 839–862. [CrossRef]
3. Kirchmayr, H. Magnetic Anisotropy. In *Encyclopedia of Materials: Science and Technology*; Elsevier: Amsterdam, The Netherlands, 2001; pp. 4754–4757.
4. American National Standards Institute; American Society of Mechanical Engineers. *Surface Texture (Surface Roughness, Waviness, and Lay)*; ASME B46.1; American Society of Mechanical Engineers: New York, NY, USA, 2003.
5. Townsend, A.; Senin, N.; Blunt, L.; Leach, R.K.; Taylor, J.S. Surface texture metrology for metal additive manufacturing: A review. *Precis. Eng.* **2016**, *46*, 34–47. [CrossRef]
6. Lou, S.; Pagani, L.; Zeng, W.; Jiang, X.; Scott, P.J. Watershed segmentation of topographical features on freeform surfaces and its application to additively manufactured surfaces. *Precis. Eng.* **2020**, *63*, 177–186. [CrossRef]
7. Bush, A.W.; Gibson, R.D.; Keogh, G.P. Strongly anisotropic rough surfaces. *Trans. ASME J. Lubr. Technol.* **1979**, *101*, 15–20. [CrossRef]

8. Wallach, J.; Hawley, J.K.; Moore, H.B.; Rathbun, F.V.; Gitzen-danner, L.G. *Calculation of Leakage between Metallic Sealing Surfaces*; American Society of Mechanical Engineers Series 68-Lub-15; ASME: New York, NY, USA, 1968.

9. Thomas, T.R.; Rosén, B.-G.; Amini, N. Fractal Characterisation of the Anisotropy of Rough Surfaces. *Wear* **1999**, *232*, 41–50. [CrossRef]

10. Bartkowiak, T. Characterization of 3D surface texture directionality using multiscale curvature tensor analysis. In Proceedings of the ASME 2017 International Mechanical Engineering Congress and Exposition IMECE17, Tampa, FL, USA, 3–9 November 2017.

11. Berglund, J.; Agunwamba, C.; Powers, B.; Brown, C.A.; Rosén, B.-G. On Discovering Relevant Scales in Surface Roughness Measurement—An Evaluation of a Band-Pass Method. *Scanning* **2010**, *32*, 244–249. [CrossRef]

12. Boulanger, J. The "Motifs" method: An interesting complement to ISO parameters for some functional problems. *Int. J. Mach. Tools Manuf.* **1992**, *32*, 203–209. [CrossRef]

13. Lou, S.; Jiang, X.; Scott, P.J. Correlating motif analysis and morphological filters for surface texture analysis. *Measurement* **2013**, *46*, 993–1001. [CrossRef]

14. Zahouani, H. Spectral and 3D motifs identification of anisotropic topographical components. Analysis and filtering of anisotropic patterns by morphological rose approach. *Int. J. Mach. Tools Manuf.* **1998**, *38*, 615–623. [CrossRef]

15. Michalski, J. Surface topography of the cylindrical gear tooth flanks after machining. *Int. J. Adv. Manuf. Technol.* **2009**, *43*, 513–528. [CrossRef]

16. Jacobs, T.D.B.; Junge, T.; Pastewka, L. Quantitative characterization of surface topography using spectral analysis. *Surf. Topogr. Metrol. Prop.* **2017**, *5*, 013001. [CrossRef]

17. Krolczyk, G.M.; Maruda, R.W.; Nieslony, P.; Wieczorowski, M. Surface morphology analysis of duplex stainless steel (DSS) in clean production using the power spectral density. *Measurement* **2016**, *94*, 464–470. [CrossRef]

18. Mishra, V.; Khan, G.S.; Chattopadhyay, K.D.; Nand, K.; Sarepaka, R.G.V. Effects of tool overhang on selection of machining parameters and surface finish during diamond turning. *Measurement* **2014**, *55*, 353–361. [CrossRef]

19. Zhang, Q.; Zhang, S. Effects of Feed per Tooth and Radial Depth of Cut on Amplitude Parameters and Power Spectral Density of a Machined Surface. *Materials* **2020**, *13*, 1323. [CrossRef] [PubMed]

20. Kubiak, K.J.; Bigerelle, M.; Mathia, T.G.; Dubois, A.; Dubar, L. Dynamic evolution of interface roughness during friction and wear processes. *Scanning* **2014**, *36*, 30–38. [CrossRef]

21. Vulliez, M.; Gleason, M.A.; Souto-Lebel, A.; Quinsat, Y.; Lartigue, C.; Kordell, S.P.; Lemoine, A.C.; Brown, C.A. Multi-scale curvature analysis and correlations with the fatigue limit on steel surfaces after milling. *Procedia CIRP* **2014**, *13*, 308–313. [CrossRef]

22. Gleason, M.A.; Kordell, S.; Lemoine, A.; Brown, C.A. Profile Curvatures by Heron's Formula as a Function of Scale and Position on an Edge Rounded by Mass Finishing. In Proceedings of the 14th International Conference on Metrology and Properties of Engineering Surfaces, Taipei, Taiwan, 17–21 June 2013; paper TS4-01, Volume 22.

23. Gleason, M.A.; Morgan, C.; Lemoine, A.; Brown, C.A. Multi-scale calculation of curvatures from an aspheric lens profile using Heron's formula. In Proceedings of the ASPE/ASPEN Summer Topical Meeting, Manufacture and Metrology of Freeform and Off-Axis Aspheric Surfaces, Kohala, HI, USA, 26–27 June 2014; pp. 98–102.

24. Bartkowiak, T.; Lehner, J.T.; Hyde, J.; Wang, Z.; Pedersen, D.B.; Hansen, H.N.; Brown, C.A. Multi-scale areal curvature analysis of fused deposition surfaces. In Proceedings of the ASPE Spring Topical Meeting on Achieving Precision Tolerances in Additive Manufacturing, Raleigh, NC, USA, 26–29 April 2015; pp. 77–82.

25. Bartkowiak, T.; Brown, C.A. Multiscale 3D Curvature Analysis of Processed Surface Textures of Aluminum Alloy 6061 T6. *Materials* **2019**, *12*, 257. [CrossRef]

26. Bartkowiak, T.; Staniek, R. Application of multi-scale areal curvature analysis to contact problem, Insights and Innovations in Structural Engineering, Mechanics and Computation. In Proceedings of the 6th International Conference on Structural Engineering, Mechanics and Computation, SEMC, Cape Town, South Africa, 5–7 September 2016; pp. 1823–1829.

27. Wiklund, D.; Liljebgren, M.; Berglund, J.; Bay, N.; Kjellsson, K.; Rosén, B.-G. Friction in sheet metal forming: A Comparison between milled and manually polished surfaces. In Proceedings of the 4th International Conference on Tribology in Manufacturing Processes, ICTMP, Nice, France, 13–15 June 2010; Volume 2, pp. 613–622.

28. Barre, F.; Lopez, J. On a 3D extension of the MOTIF method (ISO 12085). *Int. J. Mach. Tools Manuf.* **2001**, *41*, 1873–1880. [CrossRef]

29. Lehner, J.T.; Brown, C.A. Wear of abrasive particles in slurry during lapping. In Proceedings of the ASME 2014 International Manufacturing Science and Engineering Conference, MSEC 2014 Collocated with the JSME 2014 International Conference on Materials and Processing and the 42nd North American Manufacturing Research Conference, Detroit, MI, USA, 9–13 June 2014; Volume 45813, p. V002T02A017.

30. Scott, R.S.; Ungar, P.S.; Bergstrom, T.S.; Brown, C.A.; Grine, F.E.; Teaford, M.F.; Walker, A. Dental microwear texture analysis shows within-species diet variability in fossil hominins. *Nature* **2005**, *436*, 693–695. [CrossRef]

31. Arman, S.D.; Ungar, P.S.; Brown, C.A.; DeSantis, L.R.G.; Schmidt, C.; Prideaux, G.J. Minimizing inter-microscope variability in dental microwear texture analysis. *Surf. Topogr. Metrol. Prop.* **2016**, *4*, 024007. [CrossRef]

32. Stemp, W.J.; Childs, B.E.; Vionnet, S.; Brown, C.A. Quantification and discrimination of lithic use-wear: Surface profile measurements and length-scale fractal analysis. *Archaeometry* **2009**, *51*, 366–382. [CrossRef]

33. Stemp, W.J.; Morozov, M.; Key, A.J.M. Quantifying lithic microwear with load variation on experimental basalt flakes using LSCM and area-scale fractal complexity (Asfc). *Surf. Topogr. Metrol. Prop.* **2015**, *3*, 1–20. [CrossRef]

34. Le Goïc, G.; Brown, C.A.; Favreliere, H.; Samper, S.; Formosa, F. Outlier filtering: A new method for improving the quality of surface measurements. *Meas. Sci. Technol.* **2013**, *24*, 015001. [CrossRef]

35. Theisel, H.; Rossi, C.; Zayer, R.; Seidel, H.-P. Normal based estimation of the curvature tensor for triangular meshes. In Proceedings of the 12th Pacific Conference on Computer Graphics and Applications, Seoul, Korea, 6–8 October 2004; pp. 288–297.

36. Clarke, A.E.; Roy, D. *Astronomy Principles and Practice*, 4th ed.; Institute of Physics Pub.: Bristol, PA, USA, 2003; p. 59, ISBN 9780750309172.

37. Berglund, J.; Wiklund, D.; Rosén, B.-G. A method for visualization of surface texture anisotropy in different scales of observation. *Scanning* **2011**, *33*, 325–331. [CrossRef] [PubMed]

38. Maleki, I.; Wolski, M.; Woloszynski, T.; Podsiadlo, P.; Stachowiak, G. A comparison of multiscale surface curvature characterization methods for tribological surfaces. *Tribol. Online* **2019**, *14*, 8–17. [CrossRef]

39. Cain, M.K.; Zhang, Z.; Yuan, K. Univariate and multivariate skewness and kurtosis for measuring nonnormality: Prevalence, influence and estimation. *Behav. Res.* **2017**, *49*, 1716–1735. [CrossRef]

40. Tay, F.E.; Sikdar, S.K.; Mannan, M.A. Topography of the flank wear surface. *J. Mater. Process. Technol.* **2002**, *120*, 243–248. [CrossRef]

41. Liang, X.; Liu, Z. Tool wear behaviors and corresponding machined surface topography during high-speed machining of Ti-6Al-4V with fine grain tools. *Tribol. Int.* **2018**, *121*, 321–332. [CrossRef]

Article

Quantification of the Morphological Signature of Roping Based on Multiscale Analysis and Autocorrelation Function Description

Julie Marteau [1,*], **Raphaël Deltombe [2]** and **Maxence Bigerelle [2]**

[1] Laboratoire Roberval FRE-CNRS, Centre de Recherches de Royallieu, Sorbonne Université, Université de Technologie de Compiègne, CS 60319, 60203 Compiègne Cedex, France
[2] Laboratoire d'Automatique, de Mécanique et d'Informatique industrielle et Humaine (LAMIH) UMR-CNRS 8201, Université Polytechnique Hauts de France, Le Mont Houy, 59313 Valenciennes, France; raphael.deltombe@uphf.fr (R.D.) maxence.bigerelle@uphf.fr (M.B.)
* Correspondence: julie.marteau@utc.fr; Tel.: +33 3442-34603

Received: 26 May 2020; Accepted: 4 July 2020; Published: 7 July 2020

Abstract: Roping or ridging is a visual defect affecting the surface of ferritic stainless steels, assessed using visual inspection of the surfaces. The aim of this study was to quantify the morphological signature of roping to link roughness results with five levels of roping identified with visual inspection. First, the multiscale analysis of roughness showed that the texture aspect ratio S_{tr} computed with a low-pass filter of 32 µm gave a clear separation between the acceptable levels of roping and the non-acceptable levels (rejected sheets). To obtain a gradation description of roping instead of a binary description, a methodology based on the use of the autocorrelation function was created. It consisted of several steps: a low-pass filtering of the autocorrelation function at 150 µm, the segmentation of the autocorrelation into four stabilized portions, and finally, the computation of isotropy and the root-mean-square roughness S_q on the obtained quarters of function. The use of the isotropy combined with the root-mean-square roughness S_q led to a clear separation of the five levels of roping: the acceptable levels of roping corresponded to strong isotropy (values larger than 10%) coupled with low root-mean-square roughness S_q. Both methodologies can be used to quantitatively describe surface morphology of roping in order to improve our understanding of the roping phenomenon.

Keywords: roping; ridging; topography; autocorrelation function; roughness

1. Introduction

Roping or ridging is a visual defect appearing on the surface of defect-free material sheets after drawing or stretching operations. The terms 'roping' and 'ridging' refer to the surface appearance of the material that shows rope-like features parallel to the prior rolling direction and distributed along the transverse direction. This phenomenon was observed in ferritic stainless steels [1,2] as well as aluminum alloys [3,4]. Both materials are often used for exterior applications whose surface appearance is important (e.g., automotive body applications). There is thus a clear need for an objective method for the quantification of roping level. In the literature, roping quantification can be used:

- To assess the differences of predictions made by different models. Wu et al. [5] used a finite element method incorporating measured Electron Back Scattered Diffraction (EBSD) data to simulate the development of roping. They analyzed the changes in the surface profiles to compare different predictions.

- To measure the influence of grain size and shape on roping level. Patra et al. [6] examined the microstructure changes at different steps of the industrial process of 409 L grade ferritic stainless steel and identified a direct correlation between roping and the severity of coarse-grain banding.

- To assess the influence of iron contents on roping phenomenon, Jin and Lloyd [7] investigated the impact of Fe contents on roping. In their study, the examined the evolution of roughness through the use of the arithmetical mean height R_a and total height of the profile R_t but they did not link the roping level (qualitative estimation of roping) with the roughness results.

Different comparison strategies were used to try to assess roping magnitude. As an example, Shi et al. [8] developed a three-dimensional crystal plasticity model based on finite elements to simulate sheet surface roughening after different tensile strain levels. In particular, they assessed the role of the banding of Cube and Goss texture components on roping in AA6111 sheets by examining the roughness profiles given by their model. Engler et al. [9] also used a qualitative description of the roughness profiles obtained with their visco-plastic self-consistent model to discuss the predictive ability of their model. In other studies, the total roping or ridging height is preferred to quantitatively compare roping magnitude. Ma et al. [10] used the ridging height (among other results) to assess the effects of rolling routes on roping magnitude. Shin et al. [11] also used the ridging height to quantify differences of roping between two stainless steel sheets. More recently, Lee et al. [12] used the ridging height to examine the relationship between grain size and ridging for ferritic stainless steel (as-cast and cold-rolled). Other researchers compared roping levels by using standard parameters such as the average surface roughness R_a ([13–16]), the root-mean-square amplitude R_q ([17]), the maximum profile peak height R_p ([14]) or the peak-to-valley roughness ([13]). Lefebvre et al. [17] also computed the Fourier transform of the average two-dimensional roughness profile to identify characteristic wavelengths for roping. Choi et al. [18] preferred to introduce a modified roughness parameter defined as the difference between average heights of the upper N% of peaks and the lower N% of valleys to quantify the degree of surface roughness. They concluded that this parameter was more relevant for the description of roping than the use of R_q. Guillotin et al. [19] computed a roping grade based on the results of the areal power spectral density. They found good agreement with the roping level obtained with visual assessment. However, these computations were made on 'stoned' surfaces. The stoning technique artificially increases the contrast between valleys and peaks by first ink-blackening the surface and then manually grinding it with an abrasive paper.

Thus, many strategies were used to describe the surface topography induced by roping. However, as underlined by Stoudt and Hubbard [20], methods used to interpret roughness data (chosen parameter, use of profiles, etc.) may be sources of error of interpretation.

The aim of this paper is to quantify the morphological signature of roping to understand the link between the surface morphology and the roping levels determined with visual inspection of the surfaces. To do so, a multiscale analysis based on an expert system assessing the best scale and roughness parameter [21] was first used to link a standard roughness parameter at a given scale with the roping levels. Then, a new methodology based on a quantitative description of the autocorrelation function was proposed.

2. Materials and Methods

2.1. Material and Roughness Measurements

Eleven sheets of cold-rolled AISI 445 ferritic stainless steel (20.20%Cr, Aperam, Isbergues, France) were used for this study. Five roping levels were determined by the manufacturer's visual assessment. This visual assessment is based on the recommendations of the quality department established with customer satisfaction. Among these five levels, the first two levels (hereafter called Level 1 and 2) were considered as acceptable while the three other levels (Level 3, 4 and 5) were considered as non-acceptable. The number of cold-rolled sheets per roping level is given in Table 1.

Table 1. Number of cold-rolled sheets per roping level.

Level 1	Level 2	Level 3	Level 4	Level 5
2	2	3	3	1

The topography of the specimens was measured using a white-light interferometer (Zygo NewView™ 7300, Zygo Corp, Middlefield, CT, USA). Roughness measurements were performed before and after 15% tensile tests in the rolling direction. The value of 15% was chosen to match previous works on roping [13,22,23]. Tensile tests were performed at room temperature at a strain rate of 10^{-3} s^{-1} with large tensile test samples (250 mm gauge length by 50 mm gauge width), made from 1.4 mm thick sheets.

Depending on the conducted analysis, different measurement conditions were chosen:

- for the multiscale analysis, 100 measurements of 1188 µm × 891 µm with a step of 1.09 µm were performed on each specimen with a 20× objective (I 200646, Zygo Corp, Middlefield, CT, USA). An example of measurement is shown in Figure 1.
- for the autocorrelation function description, two very large measurements of 84,385 µm × 17,691 µm were performed on each specimen with a 5× objective (CF Plan 427028, Nikon, Tokyo, Japan) (and 0.5× zoom).

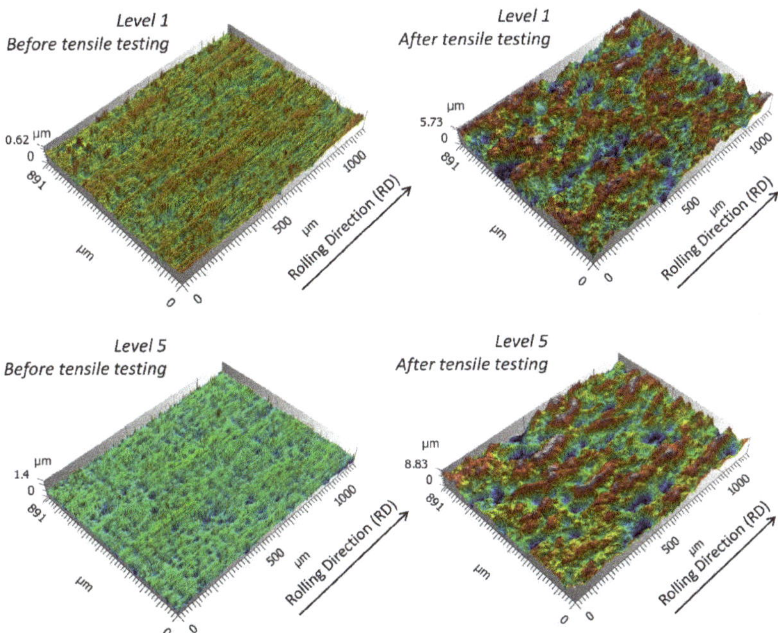

Figure 1. Examples of topography measurements of 1188 µm × 891 µm for roping classified as Level 1 and Level 5, before and after tensile testing.

It should be underlined that the very large measurements (84,385 µm × 17,691 µm) were first used in a preliminary study to assess the capability of detecting roping. Based on these first observations, it was decided to use measurements of lower dimensions (1188 µm × 891 µm) that covered more randomly the surface, with higher accuracy but with a similar total measurement area.

2.2. Multiscale Analysis Methodology

The multiscale analysis was performed using three types of robust Gaussian filters [24]: a low-pass, a high-pass and a band-pass filter (on the 1188 μm × 891 μm measurements). The following eighteen cut-off lengths were used: 8, 9, 11, 14, 17, 20, 25, 31, 38, 48, 59, 74, 99, 132, 170, 238, 396 and 594 μm. This choice was based on a geometric progression. For the band-pass filter, the indicated cut-off length corresponds to the first cut-off of the filter. The cut-off bandwidth is obtained by subtracting the latter value by the next larger cut-off length of the list. As an example, 'Band-pass filter, 17 μm' means that the first cut-off is equal to 17 μm and that the bandwidth is equal to $(20 - 17) = 3$ μm. Following this decomposition of the topography, fifty roughness parameters [25,26] were assessed. These parameters are: height parameters (arithmetical mean height S_a, root-mean-square roughness S_q, kurtosis S_{ku}, etc.), functional parameters (areal material ratio S_{mr}, etc.), spatial parameters (autocorrelation length S_{al}, texture aspect ratio S_{tr}, texture direction S_{td}, etc.), hybrid parameters (root-mean-square gradient S_{dq}, etc.), functional volume parameters, feature parameters, etc.

3. Results and Discussion

3.1. Multiscale Analysis

In this section, the results will be based on the measurements having an area equal to 1188 μm × 891 μm. As previously introduced, visual inspection of the AISI 445 ferritic stainless steel sheets enabled the manufacturer to classify the sheets into five roping levels, after the tensile tests: Level 1 and 2 were acceptable whereas Level 3, 4 and 5 were not acceptable. These different classifications (by levels or by acceptability) led to test two kinds of correlation:

(i)   a correlation between a tested roughness parameter and the five levels of roping, hereafter called 'gradation description',

(ii) a correlation between a roughness parameter values and the acceptable or non-acceptable status of the specimens, hereafter called 'binary description'.

These correlations were made using different types of relationships combining linear and logarithmic parts and the best relationship was chosen as the one giving the highest coefficient of determination.

As the arithmetical mean height is often used to quantify roping level, this roughness parameter was computed for all the specimens. Figure 2a shows the S_a values computed for the five identified levels of roping. These S_a values were computed at full scale, i.e., no filtering was performed on the measured surfaces. The S_a value found for Level 1 is significantly lower than the S_a values computed for the other levels: 0.95 μm for Level 1 while the other levels have values comprised between 1.12 μm and 1.32 μm. However, these differences do not correspond to the manufacturer's categories: Level 1 and Level 2 are considered acceptable, but they have very different S_a values (corresponding to the extrema of the curves). There is no correlation between the S_a parameter and the roping levels defined by the manufacturer. This result is in agreement with the literature: Baczynski et al. [13], who investigated roping in aluminum automotive alloy, found no correlation between height roughness parameters and the visual levels of roping. Similarly, Guillotin et al. [19] found that the height magnitude of the topography was not the most important surface feature for characterizing roping level in aluminum sheets.

Then, the S_a parameter was computed using the multiscale decomposition of the surfaces i.e., it was calculated for all the filtered surfaces listed in Section 2.1. Figure 2b shows the best correlation obtained for a binary description. It was obtained using a low-pass filter and a cut-off length of 200 μm. Again, there is no clear correlation between the roping levels and the S_a values, even when computed at the most relevant scale.

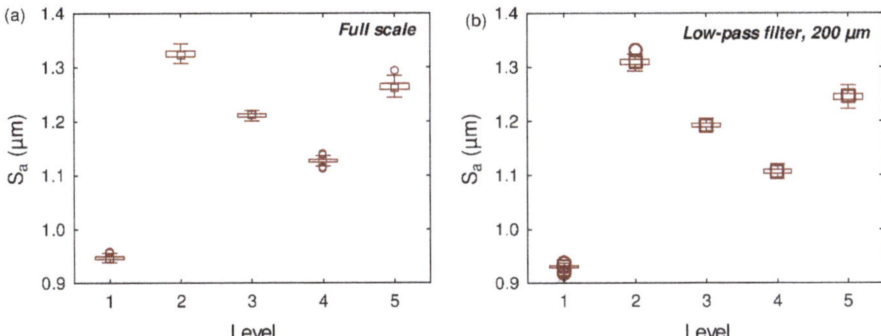

Figure 2. (**a**) Full scale arithmetical mean height S_a values as a function of the visual roping levels, after tensile testing, (**b**) Arithmetical mean height S_a values obtained with a low-pass filter and a cut-off of 200 μm as a function of the visual roping levels, after tensile testing.

The multiscale analysis was then performed using a total of fifty roughness parameters to determine which combination of parameter and scale led to the best level gradation description and to the best binary description of roping. As shown in Figure 3, the best level gradation description was obtained with the bearing index S_{bi} using a band-pass filter with a cut-off of 20 μm and a bandwidth of 5 μm. The bearing index S_{bi} is a functional index defined as the ratio between the root-mean-square parameter S_q and the height at 5% of the bearing surface. Figure 3 shows that there is a gradual increase of the S_{bi} values with the level of roping. However, the median values for all the levels are comprised between 0.45 and 0.48 while the minimum and maximum values are comprised between 0.44 and 0.47, respectively. Thus, the S_{bi} values are globally low with a similar order of magnitude (difference of 7% between the extrema).

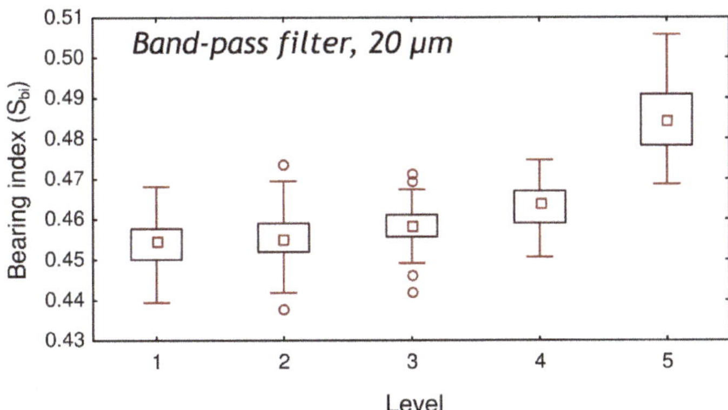

Figure 3. Bearing index S_{bi} values obtained with a band-pass filter with a cut-off of 20 μm and a bandwidth of 5 μm as a function of the visual roping levels, after tensile testing.

Figure 4 shows the best correlation achieved for the binary description of roping: the best combination is obtained with the texture aspect ratio S_{tr} calculated with a low-pass filter and a cut-off length of 32 μm. As observed in Figure 4, there is a clear separation of data: a stronger anisotropy of the autocorrelation function is obtained for the rejected specimens (Level 3 to 5) than for the accepted specimens (Level 1 to 2). According to the previous results, the roping level was not relevantly described by the roughness amplitude but seems to be well described by the oriented and periodical organization of the ropes. This result is in agreement with Guillotin et al. [19] who found that the morphological distribution of surface features was more important than height magnitude to link topography with roping levels in aluminum alloys.

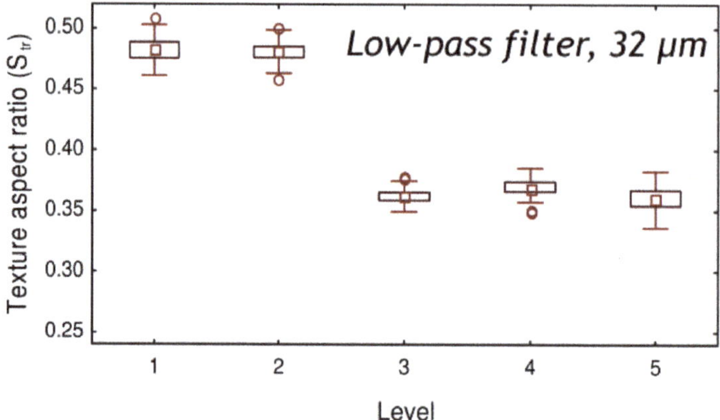

Figure 4. Texture aspect ratio S_{tr} values obtained with a low-pass filter of 32 μm as a function of the visual roping levels after tensile testing.

3.2. Description based on the Autocorrelation Function

Figure 4 showed us that computations based on the autocorrelation function (i.e., the parameter S_{tr}) are promising to establish a relationship between the visual level of roping after tensile testing and the morphology signature of roping. However, to get robust results from the use of the autocorrelation function, large measurement areas are required. This is why the 84,384 μm × 17,691 μm measurements will be used in the following sections. High-pass filtering was performed on these measurements at 25,000 μm to remove waviness caused by tensile testing.

3.2.1. Regularity Parameter

According to the previous results, a certain regularity or order seems to be characteristic of the morphology of the roping phenomenon. Fourier analysis tends to be inadequate for the description of the regularity or order of a surface. This is why Guillemot et al. [27] created a non-standardized 'regularity' parameter. This parameter is based on a normalized autocorrelation function expressed in polar coordinates (R, θ):

$$S_{reg}(\theta, \lambda) = 100 \frac{\sum_{k=1}^{k_{max}(\theta, \lambda)} \left| \int_{k.L(\theta, \lambda)}^{(k+1).L(\theta, \lambda)} ACF(R, \theta) dR \right|}{k_{max} \int_0^{L(\theta, \lambda)} ACF(R, \theta) dR}$$

where λ is the inverse lag length, L is the autocorrelation length and k_{max} the maximum value of index k in any θ direction.

This parameter is equal to 0% for uncorrelated random surfaces whereas it will be equal to 100% for perfect periodic surfaces having no noise. To give relevant results, the autocorrelation function needs to be computed at the appropriate threshold. To find this threshold, the texture aspect ratio S_{tr} was computed for all the threshold values, as displayed in Figure 5. The maximum anisotropy was found when using a threshold equal to 0.5. This value was thus used for the computation of the regularity parameter S_{reg}.

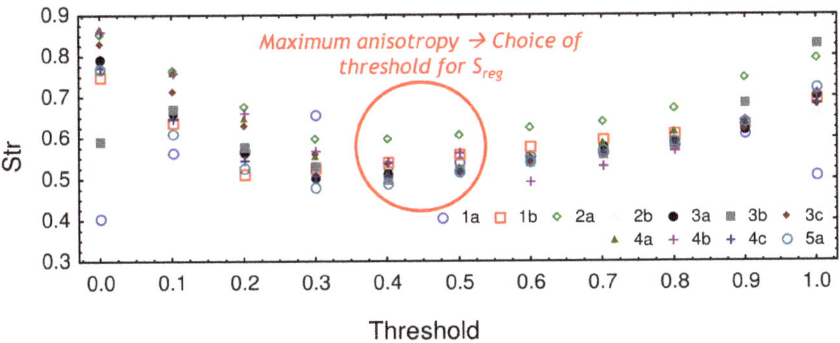

Figure 5. Texture aspect ratio S_{tr} results as a function of the threshold value.

Figure 6 shows the results of the computation of the regularity parameter S_{reg} in polar coordinates. It can be observed that for Level 1 and 2 (acceptable levels of roping), the S_{reg} distribution tends to have a round shape. On the opposite, Level 3, 4 and 5 (rejected sheets) tends to develop a nose along the X-direction. It should be noted that the regularity parameter S_{reg} is mathematically independent of amplitude. It thus confirms our first results: parameters describing the surface order are relevant for the quantification of the roping and more specifically to link topography with roping levels.

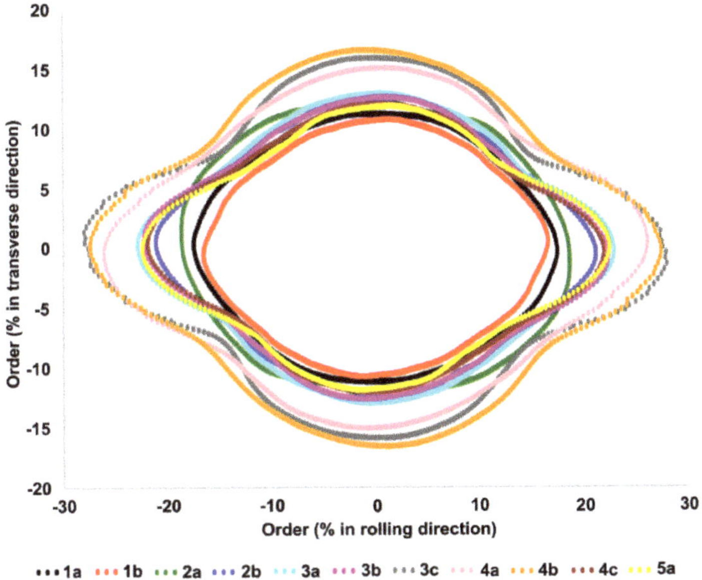

Figure 6. Representation of the regularity parameter S_{reg} results for all the measured sheets in polar coordinates.

The regularity parameter S_{reg} gave very relevant results. However, as it is not a standard roughness parameter, it may limit its use for roping description. This is why we developed another methodology based on standard functions and parameters. The next section is dedicated to the presentation of this methodology.

3.2.2. Quantitative Description of Roping based on the Autocorrelation Function

First, the relevant scale for the computation of the autocorrelation function should be determined. To do so, the autocorrelation length S_{al} is plotted as a function of threshold, as represented in Figure 7. On the latter, it can be seen that for a threshold equal to 0.5, the autocorrelation length S_{al} is equal to 300 µm. As a consequence, the roping phenomenon in this study should be investigated at a scale of 300 µm. However, to avoid any cut-off artefacts, a low-pass filtering at 150 µm will be used hereafter.

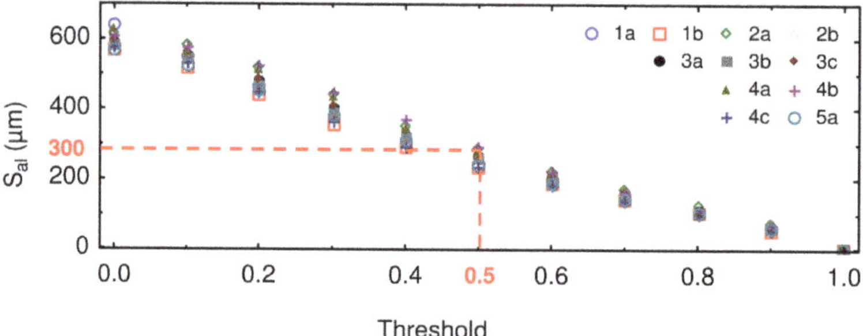

Figure 7. Autocorrelation length S_{al} as a function of the threshold.

The autocorrelation function was then computed on the surface filtered at the appropriate scale. This computed autocorrelation function was then divided into four 'stabilized' quarters (i.e., excluding the central peak), as depicted in Figure 8. The anisotropy of each quarter was then enhanced using a 3×3 gradient filter.

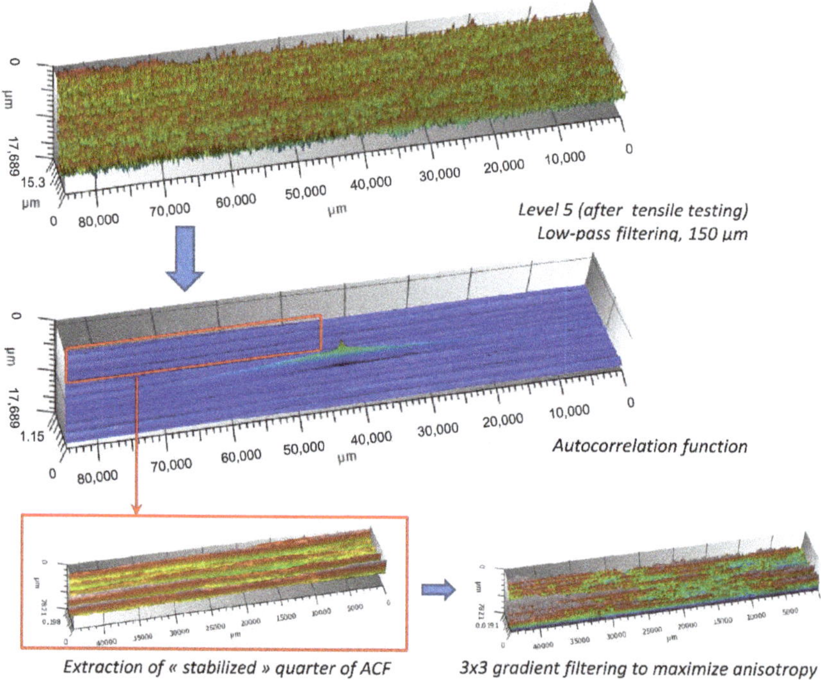

Figure 8. Methodology of extraction of quarters of autocorrelation functions.

Examples of the topography of the quarters of the autocorrelation functions are shown in Figure 9, for all the roping levels. A quick look at the difference of topographies between the levels show that this methodology is promising; clear differences appear between the accepted and rejected sheets. To obtain a quantitative description of these results, height roughness parameters as well as isotropy were computed on the extracted quarters of the autocorrelation function.

The parameters enabling the best binary description of roping were then searched using the same method as the one described in Section 3.1. It was found that the isotropy and the root-mean-square roughness S_q gave the best binary description. Figure 10 shows the mean value obtained with the four corners for each 84,385 µm × 17,691 µm measurement. A clear separation of roping levels can be observed in Figure 10 between the acceptable sheets and the rejected sheets. Large isotropy values (larger than 10%) will guarantee the manufacturer a lack of roping effect on the produced sheets. The combined use of the isotropy and the S_q parameter led to a gradation of the roping response: acceptable levels have low amplitudes as well as large isotropy values while unacceptable levels have larger amplitudes with low isotropy values.

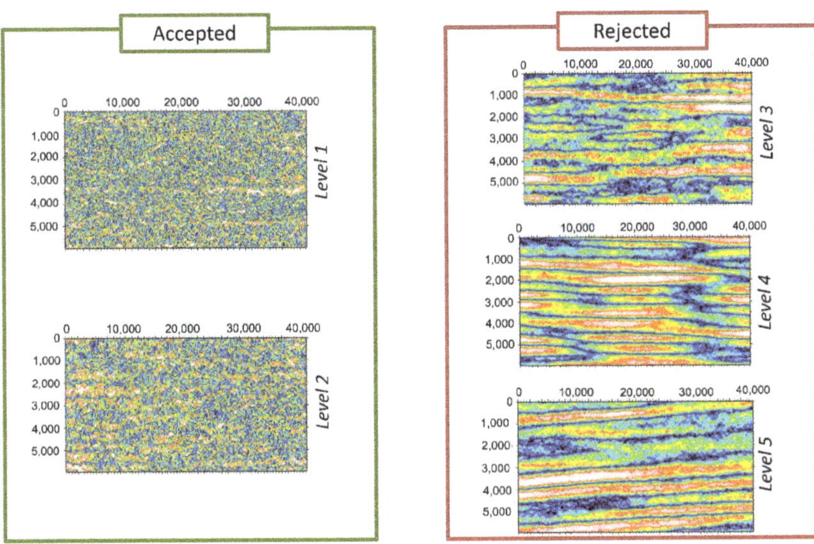

Figure 9. Topography of the autocorrelation function quarters for the five levels of roping (the X and Y axis units are micrometers).

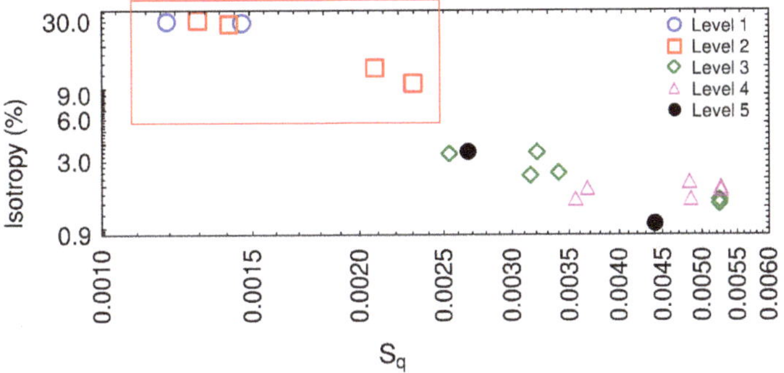

Figure 10. Isotropy as a function of the root-mean-square roughness S_q computed for all the quarters of autocorrelation function.

4. Conclusions

In this work, two main methods were tested to quantify the morphological signature of roping and to link roughness results with the five levels of roping identified with visual inspection.

The first method was based on the use of multiscale analysis to determine the best parameter and scale for the description of roping levels. It was found that the texture aspect ratio S_{tr} computed with a low-pass filter and a cut-off length of 32 μm gave the best binary description: clear separation was obtained between the acceptable levels of roping and the non-acceptable levels. The identified scale may be comparable to the grain sizes (or the sizes of the clusters of grains sharing similar orientations) but further work is required to check this hypothesis. This first method gave interesting results as it enabled the relevant scale of the signal to be identified. Furthermore, it underlined the relevance of the autocorrelation function for the description of the roping phenomenon, through the identification of the S_{tr} parameter.

The second method was based on the use of the autocorrelation function for the quantification of roping. First, the regularity parameter S_{reg} was computed and gave a good detection of roping. However, as this parameter is not standard, its use may be limited. This is a methodology based the description of the autocorrelation function was proposed. First, the relevant scale of the analysis was determined to be 150 μm for this study. After a low-pass filtering, the autocorrelation function was computed and then segmented into four stabilized portions. Different heights parameters and isotropy parameters were computed on these quarters to determine the best quantitative descriptors of roping. It was found that the isotropy combined with the root-mean-square roughness S_q gave a good description of the roping levels. Large isotropy values (larger than 10%) will guarantee the manufacturer a lack of roping effect on produced sheets. Both methodologies can be used to quantitatively describe surface morphology of roping in order to improve our understanding of the roping phenomenon.

Author Contributions: Conceptualization, M.B. and J.M.; methodology, M.B., R.D. and J.M.; software: M.B.; validation, M.B., J.M. and R.D.; formal analysis, M.B.; writing—original draft preparation, J.M.; writing—review and editing, J.M. and M.B.; administration, M.B.; funding acquisition, M.B. All authors have read and agreed to the published version of the manuscript.

Funding: This research received no external funding.

Conflicts of Interest: The authors declare no conflict of interest.

References

1. Sinclair, C.W. Embedded Grain Rotation and Roping of Stainless Steel. *Metall. Mater. Trans. A* **2007**, *38*, 2435–2441. [CrossRef]
2. Ma, X.; Zhao, J.; Du, W.; Zhang, X.; Jiang, L.; Jiang, Z. Quantification of texture-induced ridging in ferritic stainless steels 430 and 430LR during tensile deformation. *J. Mater. Res. Technol.* **2019**, *8*, 2041–2051. [CrossRef]
3. Kusters, S.; Seefeldt, M.; Van Houtte, P. A Fourier image analysis technique to quantify the banding behavior of surface texture components in AA6xxx aluminum sheet. *Mater. Sci. Eng. A* **2010**, *527*, 6239–6243. [CrossRef]
4. Qin, L.; Seefeldt, M.; Van Houtte, P. Analysis of roping of aluminum sheet materials based on the meso-scale moving window approach. *Acta Mater.* **2015**, *84*, 215–228. [CrossRef]
5. Wu, P.D.; Lloyd, D.J.; Bosland, A.; Jin, H.; MacEwen, S.R. Analysis of roping in AA6111 automotive sheet. *Acta Mater.* **2003**, *51*, 1945–1957. [CrossRef]
6. Patra, S.; Ghosh, A.; Sood, J.; Singhal, L.K.; Podder, A.S.; Chakrabarti, D. Effect of coarse grain band on the ridging severity of 409L ferritic stainless steel. *Mater. Des.* **2016**, *106*, 336–348. [CrossRef]
7. Jin, H.; Lloyd, D.J. Roping in 6111 aluminum alloys with various iron contents. *Mater. Sci. Eng. A* **2005**, *403*, 112–119. [CrossRef]
8. Shi, Y.; Jin, H.; Wu, P.D.; Lloyd, D.J. Analysis of roping in an AA6111 T4P automotive sheet in 3D deformation states. *Acta Mater.* **2017**, *124*, 598–607. [CrossRef]

9. Engler, O.; Schäfer, C.; Brinkman, H.J. Crystal-plasticity simulation of the correlation of microtexture and roping in AA 6xxx Al–Mg–Si sheet alloys for automotive applications. *Acta Mater.* **2012**, *60*, 5217–5232. [CrossRef]

10. Shin, H.-J.; An, J.-K.; Park, S.H.; Lee, D.N. The effect of texture on ridging of ferritic stainless steel. *Acta Mater.* **2003**, *51*, 4693–4706. [CrossRef]

11. Ma, X.; Zhao, J.; Du, W.; Zhang, X.; Jiang, Z. Effects of rolling processes on ridging generation of ferritic stainless steel. *Mater. Charact.* **2018**, *137*, 201–211. [CrossRef]

12. Lee, M.H.; Kim, R.; Park, J.H. Effect of nitrogen on grain growth and formability of Ti-stabilized ferritic stainless steels. *Sci. Rep.* **2019**, *9*, 1–11. [CrossRef] [PubMed]

13. Baczynski, G.J.; Guzzo, R.; Ball, M.D.; Lloyd, D.J. Development of roping in an aluminum automotive alloy AA6111. *Acta Mater.* **2000**, *48*, 3361–3376. [CrossRef]

14. Zhang, C.; Liu, Z.; Wang, G. Effects of hot rolled shear bands on formability and surface ridging of an ultra purified 21%Cr ferritic stainless steel. *J. Mater. Process. Technol.* **2011**, *211*, 1051–1059. [CrossRef]

15. Cai, Y.; Wang, X.; Yuan, S. Analysis of surface roughening behavior of 6063 aluminum alloy by tensile testing of a trapezoidal uniaxial specimen. *Mater. Sci. Eng. A* **2016**, *672*, 184–193. [CrossRef]

16. Cai, Y.; Wang, X.; Yuan, S. Surface Roughening Behavior of 6063 Aluminum Alloy during Bulging by Spun Tubes. *Materials (Basel).* **2017**, *10*, 299. [CrossRef]

17. Lefebvre, G.; Sinclair, C.W.; Lebensohn, R.A.; Mithieux, J.-D. Accounting for local interactions in the prediction of roping of ferritic stainless steel sheets. *Model. Simul. Mater. Sci. Eng.* **2012**, *20*, 024008. [CrossRef]

18. Choi, Y.S.; Piehler, H.R.; Rollett, A.D. Introduction and application of modified surface roughness parameters based on the topographical distributions of peaks and valleys. *Mater. Charact.* **2007**, *58*, 901–908. [CrossRef]

19. Guillotin, A.; Guiglionda, G.; Maurice, C.; Driver, J.H. Quantification of roping intensity on aluminium sheets by Areal Power Spectral Density analysis. *Mater. Charact.* **2010**, *61*, 1119–1125. [CrossRef]

20. Stoudt, M.R.; Hubbard, J.B. Analysis of deformation-induced surface morphologies in steel sheet. *Acta Mater.* **2005**, *53*, 4293–4304. [CrossRef]

21. Bigerelle, M.; Najjar, D.; Mathia, T.; Iost, A.; Coorevits, T.; Anselme, K. An expert system to characterise the surfaces morphological properties according to their tribological functionalities: The relevance of a pair of roughness parameters. *Tribol. Int.* **2013**, *59*, 190–202. [CrossRef]

22. Takechi, H.; Kato, H.; Sunami, T.; Nakayama, T. The Mechanism of Ridging Formation in 17%-Chromium Stainless Steel Sheets. *Trans. Jap. Inst. Met.* **1967**, *8*, 233–239. [CrossRef]

23. Wright, R.M. Anisotropic Plastic Flow in Ferritic stainless steels and the "roping" phenomenon. *Met. Trans.* **1972**, *3*, 83–91. [CrossRef]

24. ISO-ISO 16610-49:2015–*Geometrical Product Specification (GPS) – Filtration–Part 49: Morphological Profile Filters: Scale Space Techniques.* ISO: Geneva, Switzerland. Available online: https://www.iso.org/standard/61381 (accessed on 10 March 2020).

25. Stout, K.J.; Matthia, T.; Sullivan, P.J.; Dong, W.P.; Mainsah, E.; Luo, N.; Zahouani, H. *The Developments of Methods for the Characterisation of Roughness in Three Dimensions*; Office for Official Publications of the European Communities: Luxembourg, 1993.

26. ISO-ISO 25178-1:2016–*Geometrical Product Specifications (GPS)–Surface Texture: Areal–Part 1: Indication of Surface Texture.* ISO: Geneva, Switzerland. Available online: https://www.iso.org/standard/46065.html (accessed on 7 July 2020).

27. Guillemot, G.; Bigerelle, M.; Khawaja, Z. 3D parameter to quantify the anisotropy measurement of periodic structures on rough surfaces. *Scanning* **2014**, *36*, 127–133. [CrossRef]

Article

Investigations on Surface Roughness and Tool Wear Characteristics in Micro-Turning of Ti-6Al-4V Alloy

**Kubilay Aslantas [1], Mohd Danish [2], Ahmet Hasçelik [3], Mozammel Mia [4],
Munish Gupta [5], Turnad Ginta [6],* and Hassan Ijaz [2]**

[1]   Department of Mechanical Engineering, Faculty of Technology, Afyon Kocatepe University,
      03200 Afyon, Turkey; aslantas@aku.edu.tr
[2]   Department of Mechanical and Materials Engineering, University of Jeddah, Jeddah 21589, Saudi Arabia;
      mdanish@uj.edu.sa (M.D.); hassan605@yahoo.com (H.I.)
[3]   İscehisar Vocational High School, Afyon Kocatepe University, 03200 Afyon, Turkey; ahascelik@aku.edu.tr
[4]   Department of Mechanical Engineering, Imperial College London, Exhibition Road, South Kensington,
      London SW7 2AZ, UK; m.mia19@imperial.ac.uk
[5]   Key Laboratory of High Efficiency and Clean Mechanical Manufacture, Ministry of Education,
      School of Mechanical Engineering, Shandong University, Jinan 250061, China; munishguptanit@gmail.com
[6]   Mechanical Engineering Department, Universiti Teknologi PETRONAS, Perak 32610, Malaysia
*   Correspondence: turnad.ginta@utp.edu.my

Received: 19 April 2020; Accepted: 6 May 2020; Published: 6 July 2020

Abstract: Micro-turning is a micro-mechanical cutting method used to produce small diameter cylindrical parts. Since the diameter of the part is usually small, it may be a little difficult to improve the surface quality by a second operation, such as grinding. Therefore, it is important to obtain the good surface finish in micro turning process using the ideal cutting parameters. Here, the multi-objective optimization of micro-turning process parameters such as cutting speed, feed rate and depth of cut were performed by response surface method (RSM). Two important machining indices, such as surface roughness and material removal rate, were simultaneously optimized in the micro-turning of a Ti6Al4V alloy. Further, the scanning electron microscope (SEM) analysis was done on the cutting tools. The overall results depict that the feed rate is the prominent factor that significantly affects the responses in micro-turning operation. Moreover, the SEM results confirmed that abrasion and crater wear mechanism were observed during the micro-turning of a Ti6Al4V alloy.

Keywords: micro turning; surface roughness; material removal rate; RSM; Ti6Al4V alloy; tool wear

---

1. Introduction

Micro-mechanical machining, a form of manufacturing used to produce parts with micro dimensions, is noted for higher material removal rates [1,2]. The application of such manufacturing is delicate and requires the use of micro-turning and micro-milling processes. Notable industries are biomedical, defense, aerospace and electronics industries [3]. Though there is a large degree of similarity between traditional turning and micro-turning, turning at the micro scale needs to be accurate as well as precise, causing the necessity for ultra-precision machining [4]. The micro-turning method is especially used in the manufacturing of micro-screws used in orthodontic implants. In this type of small-scale case, the enhancement of surface finish by a secondary method is hard to implement, and thereby process parameters should be chosen so that the desired surface quality is achieved—if possible, the surface finish should be compatible with the grinding process.

Surface roughness is considered as the prominent parameter that significantly affects the mechanical properties as well as the fatigue strength of the machined part [5]. In addition, the surface roughness values in conventional turning highly depend upon the feed rate. Therefore, the required feed rate for the desired surface quality, the cutting speed and depth of cut are taken into account [6]. On the micro scale, the cutting tool nose radius leaves marks on the workpiece surface (Figure 1a). Depending on the feed value and nose radius, the maximum surface roughness value can be estimated. In conventional turning (Figure 1b), the low feed rate can be indispensable for good surface quality. However, this rule does not always work in micro-turning. In the case of micro-turning (Figure 1c), the feed rate is small enough to be compared to the cutting-edge radius. It has been seen that if the value of feed is smaller than the tool edge radius, an increase in surface roughness can be observed [7–9]. The depth of cut has also shown a similar situation. In micro-turning, the depth of cut can approach the workpiece grain size (Figure 1b). This results in both increased thrust force and the deterioration of surface quality.

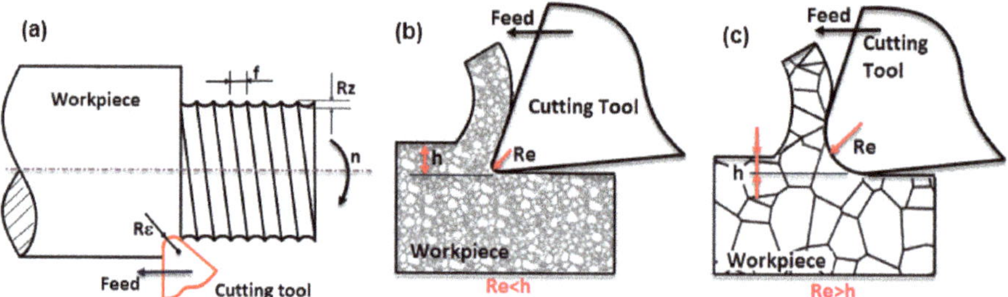

Figure 1. (**a**) Illustration of turning operation showing nose radius (Re), feed rate (*f*), maximum surface roughness (Rz); (**b**) conventional and (**c**) micro-cutting process h: undeformed chip thickness, Re: Edge Radius.

In conventional turning, a good amount of research effort has been made regarding surface roughness characteristics [10–12]. In conventional turning, usually the depth of cut and the feed value are greater than the tool edge and nose radius. Therefore, the difference between the obtained surface roughness value and the theoretical surface roughness value increases. However, in micro-turning, especially in low feed values, the theoretical surface roughness and experimental results do not match. Liu and Melkote [7], while micro-turning the aluminum alloy, established a prediction model for the surface roughness. The effect of plastic side flow was considered in this subjected model. Their model combines cutting parameters and plastic side flow effects with more accurate estimation. Rahman et al. [13] used two different micro-turning techniques for micro-pin production. The responses measured were cutting forces and surface roughness values during the micro-turning operation. The results depict that the average value of surface roughness was 0.1 µm. Alauddin et al. [14] used the second-order polynomials method to establish the surface roughness prediction model. Wang et al. [15] investigated the effect of machining parameters and tool diameter on surface quality in a micro-milling operation. Kuram and Özcelik [16] have developed a model for estimating the surface roughness in the micro-milling process by using a multi-objective optimization technique. The effect of spindle speed, cutting edge radius of the tool, and the roughness of workpiece were investigated. Vipindas et al. [17] examined the surface roughness and top burr formation in the micro-milling of titanium alloy.

In the studies conducted by Aslantas et al. [18] and Ucun et al. [19], the surface roughness and burr width was studied for different process parameters. The result shows that the depth of cut was prominent for surface roughness values, whereas feed per tooth was dominant for burr width. Thepsonthi and Özel [20] optimized the process parameters for surface roughness and burr formation in the micro-milling of a Ti6Al4V alloy. Experiments have been performed and models were obtained by utilizing the particle swarm optimization technique. Kumar [21] also studied the effect of process parameters in the micro-turning process. Cutting speed, depth of cut and feed value were taken as variables, and a C360 copper alloy was used as a workpiece. The analysis shows that the depth of cut is the dominant factor that directly affects the surface roughness and material removal rate values. The influence of cutting parameters on the surface roughness and Material Removal Rate (MRR) is evident from the literature. Optimization of these cutting parameters can greatly help in choosing the optimum cutting parameter for getting the required objective, i.e., minimum surface roughness and high material removal rate.

Response surface method (RSM) is widely used for developing empirical relations between single and multiple responses [22–24]. The most critical factor that affects the output responses can also be determined with this method. Additionally, the multi-response optimization of micro-turning process parameters was performed [25]. Therefore, in this work, the optimization of cutting parameters affecting the surface roughness and MRR was performed in the micro-turning process. Ti6Al4V alloy was used as the workpiece material; average surface roughness of area (Sa) and maximum surface roughness of area (Sz) values were obtained. Single- and multi-objective optimization with the objective of obtaining minimum surface roughness and maximum MRR were carried out by utilizing the response surface methodology. The most dominant factor which affects the surface roughness was also identified. In the end, the SEM was performed on used tools to understand the wear behavior values.

2. Materials and Methods

2.1. Workpiece and Cutting Tool Material

The Ti6Al4V alloy is preferred as the workpiece. This titanium alloy, known as Grade 5, is especially used as an implant material in the biomedical sector. It also has a wide usage area as a screw in dental implant applications. The alloy used in the study was annealed after the manufacturing process and no aging was done. The chemical compositions of the Ti6Al4V alloy are shown in Table 1 and the mechanical properties are given in Table 2. The machining operation was conducted using the cutting tool received from the Kennametal 2-μm-coated tool (ISO name TDHB07T12S0). It has a rake angle of 0°, approach angle of 90° and clearance angle of 15° in the machining condition, nose radius of 40 μm, and edge radius of 7.25 μm (See Figure 2).

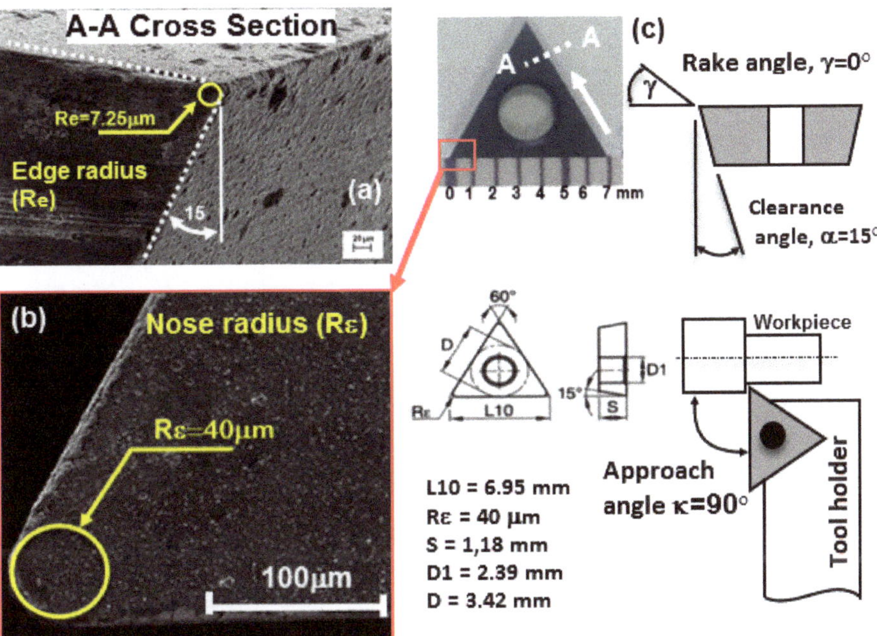

Figure 2. (**a**) Edge radius (**b**) nose radius (**c**) other dimensions of the cutting tool used in micro turning process.

Table 1. Ti6Al4V alloy chemical composition (% by weight).

Element	Al	V	Fe	C	O	N	H	Ti
Wt %	6.40	4.16	0.16	0.028	0.154	0.017	0.001	Balance

Table 2. Mechanical properties of Ti6Al4V Alloy.

Properties	Value
Tensile Strength (MPa)	900–1000
Yield Strength (MPa)	830–910
Elongation (%)	10–18
Elastic Modulus (GPa)	114
Hardness (Brinell)	330–340

2.2. Experimental Setup

Figure 3 shows the experimental setup—a specialized setup for high-precision cutting with high speed. The highest speed achievable is 60,000 rpm, and the highest travel distance of guideway is 150 mm, maintaining a repeatability of 0.4 µm. The cutting tool was placed on the mini dynamometer that is fixed to the x-axis. The feed was applied along the z-axis and the depth of cut was applied along the y-axis. The approach angle of the cutting tool was 90° and a USB microscope was used to more clearly observe the cutting zone in the experiments. The system was maintained as vibration free by using an optical table. A constant cutting distance (75 mm) was used in experiments to observe the effects of cutting parameters and to eliminate the effects of tool wear.

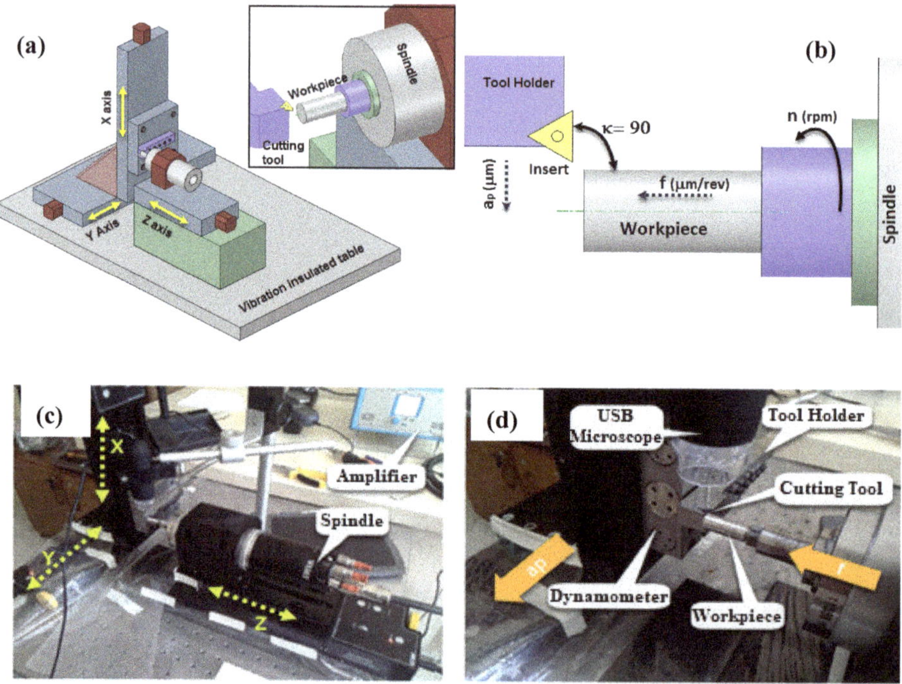

Figure 3. (**a**) Schematic representation of micro-turning setup; (**b**) position of cutting tool and workpiece relative to each other; (**c**) axis definitions of experimental setup; (**d**) close view of cutting setup.

2.3. Surface Roughness Measurement

The representative index for the surface quality after the machining was granted as the surface roughness parameters. These parameters were measured using Nanovea optical profilometer, which works with white light technology. In Figure 4a, the surface roughness tester is shown. The table where the sample is placed is movable in the x and y axis directions and the focusing distance of the optical lens is adjusted with the z axis. Scanning was performed on a 1 × 0.1 mm area (Figure 4b). The surface roughness was measured at four different points with 90° angles on each workpiece. In this study, Sa and Sz values were measured as the surface roughness and taken into consideration. A three-dimensional surface topography of a surface is shown in Figure 4c.

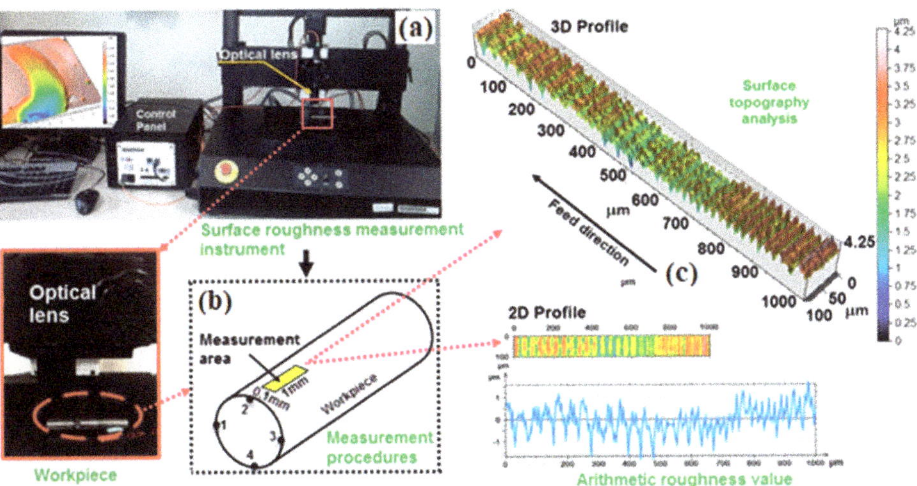

Figure 4. Measurement procedure of surface roughness in current work, (**a**) surface roughness tester; (**b**) area under consideration for the roughness measurement; (**c**) 3D surface topography of the area.

In metal-cutting operations, the material removal rate is expressed as the chip volume removed in one minute and MRR can be calculated by using Equation (1) [26]

$$Q = V_c \times f \times a_p \tag{1}$$

where Q is the MRR in mm^3/min, V_c is the cutting speed, f is feed rate and a_p is depth of cut. MRR is an indication of how slow or fast the machining speed works. It is an important performance parameter for micro-machining. In micro-machining (especially micro-milling), a high MRR value results in high surface roughness, rapid tool wear, and burr formation. It is important to determine the maximum MRR value without compromising surface quality and for longer tool life.

2.4. Design of Experiment

The analysis consisted of determining the influence of each control factor on the surface roughness parameters found after the micro-turning operation. For that purpose, the control factors were defined first, as can be seen in Table 3. As we can see, the feed rate, the cutting speed and the depth of cut have three levels of values. Afterward, these values are oriented among themselves to create the overall design for the experimentation. This was carried out using a face centered composite design; details can be found in Reference [20]. The process parameters with their different levels are given in Table 3. Respective to each experiment, the measurement was conducted. The collected data were then used for the further analysis, using complete a manual of response surface methodology (RSM)—mathematical modeling, analysis and optimization.

Table 3. Process variables used in cutting tests.

Levels	Cutting Speed (V_c) (m/min)	Feed Rate (f) (μm/rev)	Depth of Cut (a_p) (μm)
1	100	25	5
2	250	10	15
3	400	40	25

2.5. Response Surface Methodology

RSM is a complete package for the mathematical modeling, statistical analysis and optimization of the single or multiple responses within the framework of multiple inputs [27,28]. Here, the input parameters, i.e., speed, feed and depth of cut are analyzed to derive the relationship of inputs with the surface roughness parameters (Sa and Sz). Although linear, as well as second-order, polynomial relations can be formed, based on the literature knowledge, it is found that for machining responses, the second-order polynomial relationship works effectively. A general second-order relation is shown in Equation (2)

$$Y = \beta_0 + \sum_{i=1}^{p} \beta_i X_i + \sum_{i=1}^{p} \beta_{ii} X_i^2 + \sum_{i=1}^{p-1} \sum_{j=1}^{p} \beta_{ij} X_i X_j \tag{2}$$

Here, β_0 is the constant. The coefficient β_i is the coefficient term for the linear terms, β_{ii} are coefficients for the square term of the variables, and the β_{ij}, are the coefficients for the interacting terms.

3. Results and Discussion

According to the objectives defined in the introduction section, the experimental results and analysis are given in this section, under a number of sub-sections. Initially, the data found from experiments were collected. Then, the subsequent analysis is reported. For this, full quadratic models were constructed, and then analysis of variance, which shows the influence of each factor on the responses, and, finally, the responses were optimized. Note that significant terms were identified respective to a statistical significance of 0.05. This has been done for the model as well as for the single terms (V_c, f and a_p), square terms (V_c^2, f^2 and a_p^2) and interaction terms ($V_c f$, $V_c a_p$ and $f a_p$). The F-value was marked for the relative influence determination. Table 4 lists the experimental results respective to the 20 experiments, which were oriented as per the description of the design of the experiment. Besides the surface roughness parameters for the micro-turning operation, the material removal rate was considered (calculated).

Table 4. Experimental results on Sa, Sz and material removal rate (MRR) for different cutting parameters.

Sr. NO	Inputs			Outputs		
	Cutting Speed (V_c) (m/min)	Feed Rate (f) (μm/rev)	Depth of Cut (a_p) (μm)	Average Roughness (Sa) (μm)	Maximum Roughness Height (Sz) (μm)	Material Removal Rate (mm³/min)
1	100.00	25.00	15.00	0.72	5.94	37.50
2	400.00	10.00	25.00	0.39	3.35	100.00
3	250.00	10.00	15.00	0.42	3.83	37.50
4	250.00	25.00	15.00	0.70	6.98	93.75
5	100.00	10.00	5.00	0.52	3.48	05.00
6	100.00	40.00	25.00	0.62	6.87	100.00
7	250.00	40.00	15.00	0.91	7.48	150.00
8	250.00	25.00	25.00	0.48	4.23	156.25
9	250.00	25.00	15.00	0.70	6.98	93.75
10	250.00	25.00	15.00	0.69	6.93	93.75
11	400.00	40.00	5.00	0.99	7.12	80.00
12	400.00	40.00	25.00	0.64	5.02	400.00
13	250.00	25.00	5.00	0.62	5.06	31.25
14	250.00	25.00	15.00	0.70	6.95	93.75
15	400.00	10.00	5.00	0.48	4.07	20.00

Table 4. *Cont.*

Sr. NO	Inputs			Outputs		
	Cutting Speed (V_c) (m/min)	Feed Rate (f) (μm/rev)	Depth of Cut (a_p) (μm)	Average Roughness (Sa) (μm)	Maximum Roughness Height (Sz) (μm)	Material Removal Rate (mm³/min)
16	100.00	40.00	5.00	1.02	7.85	20.00
17	250.00	25.00	15.00	0.70	6.85	93.75
18	100.00	10.00	25.00	0.33	3.63	25.00
19	250.00	25.00	15.00	0.69	6.77	93.75
20	400.00	25.00	15.00	0.79	5.59	150.00

3.1. Model of Average Roughness (Sa)

As the first step, the analysis of variance is conducted for the average surface roughness (Sa) and shown in Table 5. It should be noted that, respective to each source, the sum of square term, the degree of freedom term, the mean square, F-value and *p*-value are listed. It can be seen that the model is acceptable, as the F-value is quite high and the *p*-value is less than 0.05. Therefore, the model is statistically significant. Likewise, the feed rate and the depth of cut have been found to be statistically significant. The square term for speed and depth of cut, and the interaction terms for feed-depth of cut were also significant. It can be said that the other terms were statistically insignificant. Based on the F-values, it is possible to claim that the feed rate has the highest value, therefore it is most dominant factor, followed by the influence of depth of cut, then it's square term.

Table 5. Table of ANOVA for average surface roughness (Sa).

Source	Sum of Squares	Degree of Freedom	Mean Square	F-Value	Dominance of Factor	*p*-Value
Model	0.657	9	0.072	56.69	99.55%	<0.0001
V_c	6.084×10^{-4}	1	6.084×10^{-4}	0.48	0.09%	0.5040
f	0.42	1	0.42	330.95	63.64%	<0.0001
a_p	0.14	1	0.14	108.67	21.21%	<0.0001
V_c^2	0.015	1	0.015	12.16	2.27%	0.0059
f^2	4.423×10^{-4}	1	4.423×10^{-4}	0.35	0.07%	0.5676
a_p^2	0.047	1	0.047	37.09	7.12%	0.0001
$V_c f$	7.812×10^{-5}	1	7.812×10^{-5}	0.062	0.01%	0.8088
$V_c a_p$	3.240×10^{-3}	1	3.240×10^{-3}	2.56	0.49%	0.1407
$f a_p$	0.026	1	0.026	20.26	3.94%	0.0011
Residual	0.013	10	1.266×10^{-3}	-	1.97%	-
Total	0.66	19	-	-	100%	-

The arithmetic model developed for average roughness (Sa) is given by Equation (3).

$$Sa = \begin{aligned} &-1.78 \times 10^{-3}\, Vc + 0.0225f + 0.0335ap + 3.325 \times 10^{-6}Vc^2 - 5.64 \times 10^{-5}f^2 \\ &-1.31 \times 10^{-3}ap^2 - 1.39 \times 10^{-6}Vcf + 1.34 \times 10^{-5}Vcap - 3.77 \times 10^{-4}fap + 0.29 \end{aligned} \tag{3}$$

This model, however, includes significant as well as non-significant terms. This means that more refinement is required for this model to improve the model efficiency. This can be done by the removal of non-significant terms by using backward elimination; however, those terms which are required for hierarchy are exempted. After doing this, the new model for ANOVA for average surface roughness is shown in Table 6. This new model then can be analyzed for comparing the R^2.

It is appreciable that the F-value of the new model was increased to 84.73 from 56.69. The *p*-value was found to be under 0.05, which is an indication for a significant model. Interestingly, it is shown that the V_c^2, being a non-significant term, is still in the model. This term is kept to maintain the hierarchy. After such refinement, the R^2 values are compared before and after the backward elimination, listed in Table 7.

Table 6. Table of ANOVA for average roughness (Sa) after refinement.

Source	Sum of Squares	Degree of Freedom	Mean Square	F-Value	Dominance of Factor	*p*-Value
Model	0.64	6	0.11	84.73	96.97%	<0.0001
V_c	6.084×10^{-4}	1	6.084×10^{-4}	0.48	0.09%	0.4999
f	0.42	1	0.42	331.71	63.64%	<0.0001
a_p	0.14	1	0.14	108.92	21.21%	<0.0001
V_c^2	0.016	1	0.016	12.44	2.42%	0.0037
a^2	0.059	1	0.059	46.47	8.94%	<0.0001
$f\,a_p$	0.026	1	0.026	20.31	3.94%	0.0006
Residual	0.016	13	1.263×10^{-3}	-	2.42%	-
Total	0.66	19	-	-	100%	-

Table 7. R^2 parameter for average roughness (Sa) model before and after the backward elimination.

Parameter	Before	After
R^2 (overall)	0.98	0.98
Adjusted R^2	0.96	0.96
Predicted R^2	0.85	0.92
Adeq Precision	27.44	31.37

It is to be noted that the overall R^2 and adjusted R^2 remained the same and they are very close to unity; however, the predicted R^2 value increased from 0.85 to 0.92. Moreover, the adequate precision value increased. This indicates that the refinement of the model improved the efficiency. As such, the final model for Sa, which was used for further analysis and optimization, is shown by Equation (4).

$$Sa = -1.505 \times 10^{-3} Vc + 0.0193 f + 0.0383 ap + 3.114 \times 10^{-6} Vc^2 - 1.354 \times 10^{-3} ap^2 - 3.775 \times 10^{-4} fap + 0.257 \tag{4}$$

3.2. Model for Maximum Roughness Height (Sz)

The maximum roughness height (Sz) has also been analyzed statistically to develop the model. For that purpose, the ANOVA was performed and all important values are given in Table 8.

Table 8. ANOVA table for the maximum roughness height (Sz).

Source	Sum of Squares	Degree of Freedom	Mean Square	F-Value	Dominance of Factor	*p*-Value
Model	40.00	9	4.44	11.29	91.03%	0.0004
V_c	0.69	1	0.69	1.74	1.57%	0.2161
F	25.54	1	25.54	64.87	58.12%	<0.0001
a_p	2.01	1	2.01	5.10	4.57%	0.0475
V_c^2	0.018	1	0.018	0.045	0.04%	0.8368
f^2	0.099	1	0.099	0.25	0.23%	0.6264
a_p^2	3.96	1	3.96	10.06	9.01%	0.0100
$V_c f$	1.04	1	1.04	2.65	2.37%	0.1345
$V_c a_p$	0.50	1	0.50	1.26	1.14%	0.2883
$f\,a_p$	0.79	1	0.79	2.00	1.80%	0.1876
Residual	3.94	10	0.39	-	8.97%	-
Total	43.94	19	-	-	100%	-

The model *p*-value was under 0.05. For roughness parameters, based on *p*-value criteria, most of the terms are statistically non-significant, except three terms (f, a_p and a_p^2). As such, it is imperative to refine the model by using backward elimination. That has been done here, and the new model is listed in Table 9.

Table 9. ANOVA table for the maximum roughness height (Sz) after the backward elimination process.

Source	Sum of Squares	Degree of Freedom	Mean Square	F-Value	Dominance of Factor	p-Value
Model	36.82	3	12.27	27.57	83.80%	<0.0001
f	25.54	1	25.54	57.37	58.13%	<0.0001
a_p	2.01	1	2.01	4.51	4.57%	0.0497
a^2	9.28	1	9.28	20.84	21.12%	0.0003
Residual	7.12	16	0.45	-	16.20%	-
Total	43.94	19	-	-	100%	-

It is admissible that the new mode p-value is less than 0.05 and less than the previous p-value, hence it is significant and obviously improved. In this refined model, the F-value shows that the feed rate is the most dominant followed by the square term of depth of cut. Nevertheless, it is necessary to compare the R^2 parameters in Table 10. Table 10 shows that the overall R^2 value decreased from 0.91 to 0.84. As such, the model has lack of fitness. It can also be noticed that the adjusted R^2 value of the model was closer for the backward elimination compared to the primary model. This shows that the model efficiency was increased. Equation (5), the mathematical model of the maximum height roughness, was achieved and used for further computation and optimization.

$$\begin{aligned} Sz = \ & 6.53 \times 10^{-3} Vc + 0.2202 f + 0.40895 ap - 3.555 \times 10^{-6} Vc^2 - 8.444 \times 10^{-4} f^2 \\ & -0.012 ap^2 - 1.605 \times 10^{-4} Vcf - 1.658 \times 10^{-4} Vcap \\ & -2.092 \times 10^{-3} fap - 0.93 \end{aligned} \tag{5}$$

Table 10. R^2 parameter for maximum roughness height (Sz) model before and after the backward elimination.

Parameter	Before	After
R^2	0.91	0.84
Adjusted R^2	0.83	0.81
Predicted R^2	0.61	0.73
Adequate Precision	10.73	16.78

3.3. Adequacy Tests

The constructed models were put into trial for the adequacy test. It is noted that the residual plot of the data should not follow any type of trend. The residual plots for both Sa and Sz are shown in Figure 5. As it is visible the datapoints for both plots follow a straight-line path and are free from any trend or sequence—an indication that the model is adequate for further analysis, i.e., prediction model and optimization. However, further investigation is required to be certain about the complete adequacy of the models. Thereafter, the models were tested for abnormality—if the datapoints shifted on either side or were distributed fairly on the both sides with respect to the reference line. For that purpose, the outlier's plots were constructed and shown in Figure 6. As a rule, if the data fall outside the ±3.5 permissible range, then they are considered outliers, i.e., abnormal data. Interestingly, all the datapoints for the present study were found within the data range of permissibility. Therefore, the models can be claimed as adequate.

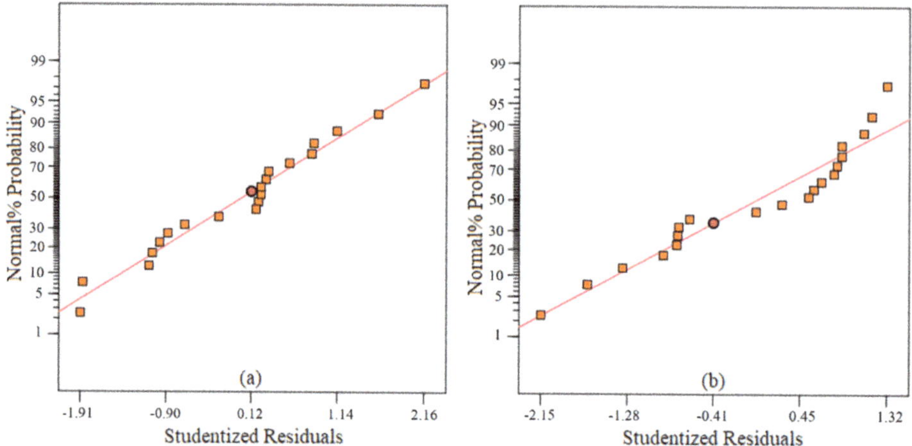

Figure 5. Residual plots for the model developed for (**a**) average surface roughness (Sa); (**b**) maximum roughness height (Sz).

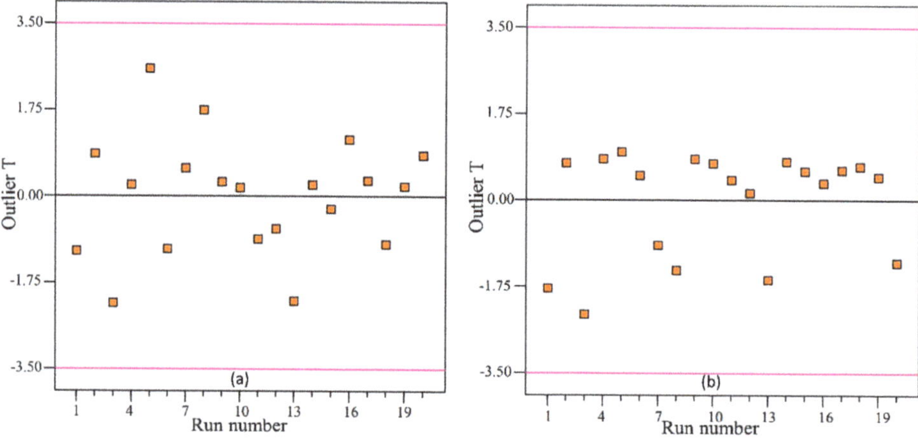

Figure 6. Outlier plot for (**a**) average roughness (Sa); (**b**) maximum roughness height (Sz).

3.4. Experimental Verification Test

The prediction models were developed and tested for verification. This has been done in five random experimental sets of data. For each set, the experimental as well as the predicted data are plotted side by side, as can be seen in Figure 7 (average roughness parameter) and Figure 8 (maximum height roughness parameter). Interestingly, the agreement between the predicted value and the experimental value is quite reasonable, and therefore the models can be accepted. However, the model of average roughness parameter showed better accuracy in the prediction—from 1% to 5.94% error, while that of the maximum height surface roughness parameter ranges from 3.07% to 6.8%.

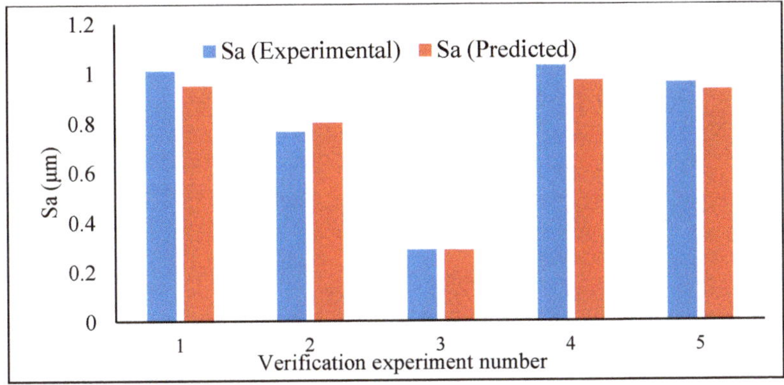

Figure 7. Average roughness parameter (*Sa*) – verification test.

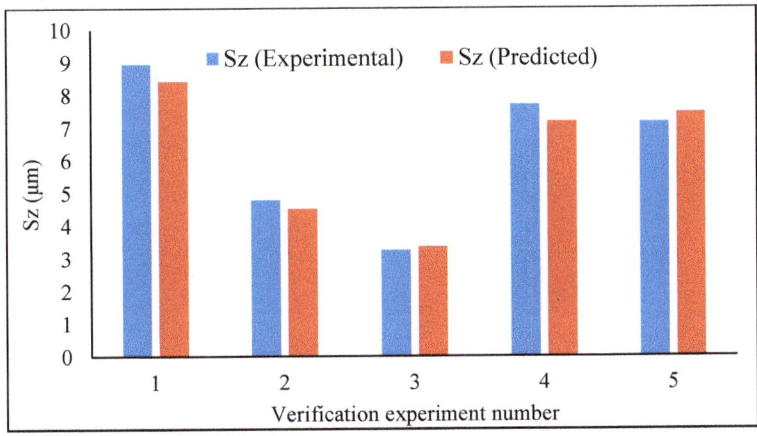

Figure 8. Maximum roughness height (*Sz*) – verification test.

3.5. 3D Response Surface, One Factor Plots and Analysis by SEM

At this stage, 3D response surface plots and one-factor plots were plotted (as shown in Figures 9 and 10) and the effects of cutting parameters on the response were analyzed. The information extracted from the plots was also verified with the experimental results, such as the 3D surface profiles of the tested specimens and scanning electron micrographs (SEM) of the tool and chips. The 3D response surface plot showing the effect of input parameter on the response (average roughness parameter and maximum height roughness parameter) has been shown in Figures 9 and 10. As is evident from Figures 9 and 10, both Sa and Sz increases with every increment in the Feed rate. This trend can be verified by the 3D profiles of the machined surface obtained for varying feed rate, which are shown in Figure 11. It is quite clear from Figure 11 that not only Sa but also Sz become higher for higher values of feed rates. For more details, SEM images for the tool, together with the chips, were also obtained for the varying feed rates, as shown in Figure 12.

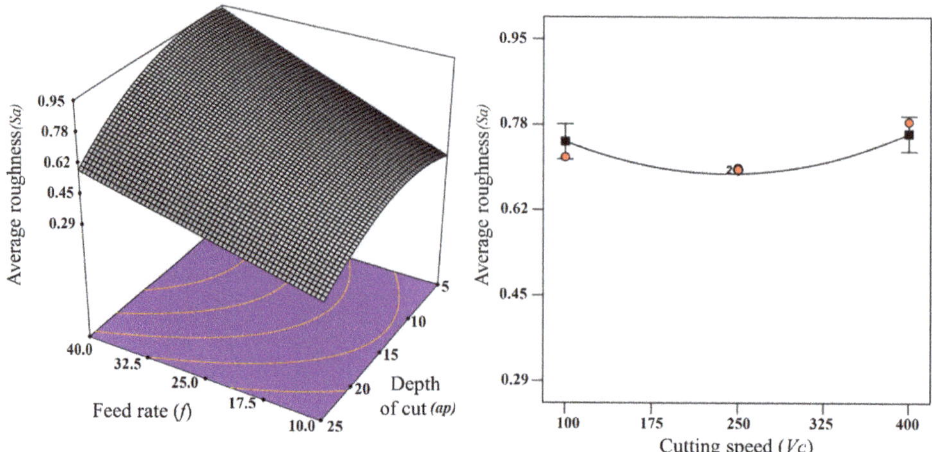

Figure 9. Average surface roughness with 3D plot and one factor plot.

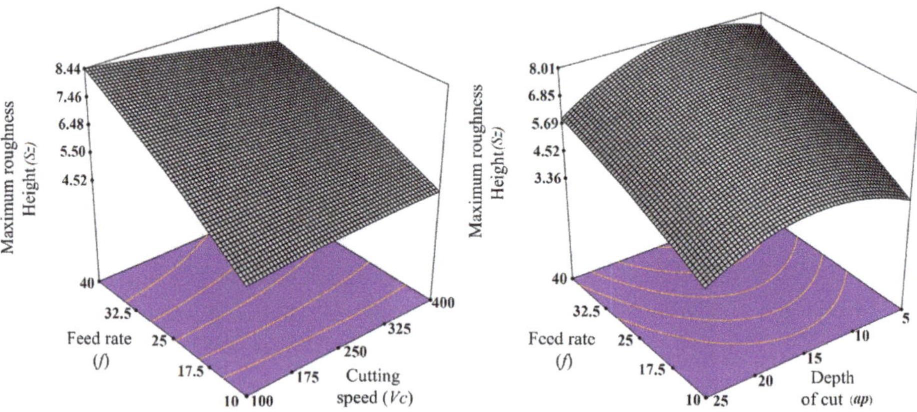

Figure 10. Maximum roughness height with 3D plot and one factor plot.

At a lower feed (f = 5 μm/rev), the adhered work material can be seen on the tool tip, but the tool wear is not significant, thereby giving lower values of surface roughness parameters (Sa and Sz). At f = 15 μm/rev, the increase in crater wear on the tool can be observed and that could be the reason for the increase in the surface roughness values. However, for f = 25 μm/rev, a significant increase in crater wear and also in the adhered work material was observed, as shown in Figure 12. The built-up edges (BUE) on the tool were also observed at this feed rate. All these factors have contributed to the poor surface finish values (higher Sa and Sz values) at this point. The chips' morphology was more or less the same for every variation in the feed rate, as shown in Figure 12. The serrated chips were observed for the cases, however, the serration was clearer at a lower feed rate value. It can be concluded from the SEM images that the lower feed rate values will be best for micro-machining in the present scenario, which is in full agreement with the empirical model and 3D response plots obtained by the RSM method. The effect of cutting speed (V_c) on the Sa factors was found to be mixed, as shown in Figure 9. It was observed that the value of Sa tends to decrease with increases in the cutting speed, but it rises again with further increases in V_c. The effect of cutting speed on the Sz value is not significant, as shown in Figure 10. The depth of cut (a_p) has a significant effect on the surface roughness values (Sa and Sz), which is evident from the surface roughness plots shown in Figures 9 and 10. For both cases, it was observed that surface roughness values first tend to increase when the depth of cut is increased from 5 to 15 μm. However, further rises in the depth of cut tend to produce lower values of surface roughness parameters. This trend was also supported by the SEM images taken for tool and chips for different depths of cut (5 to 15 μm), which are shown in Figure 13. The crater wear on the tool was not significant, except for the depth of cut = 15 μm, which can be observed in Figure 13. As can be seen from the SEM photographs of Figures 12 and 13, tool wear is minimal and BUE and chip plastering occurs mainly at the tool tip. The nose radius is almost unchanged. Therefore, the change in surface roughness is affected by BUE and chip plaster, not by tool wear.

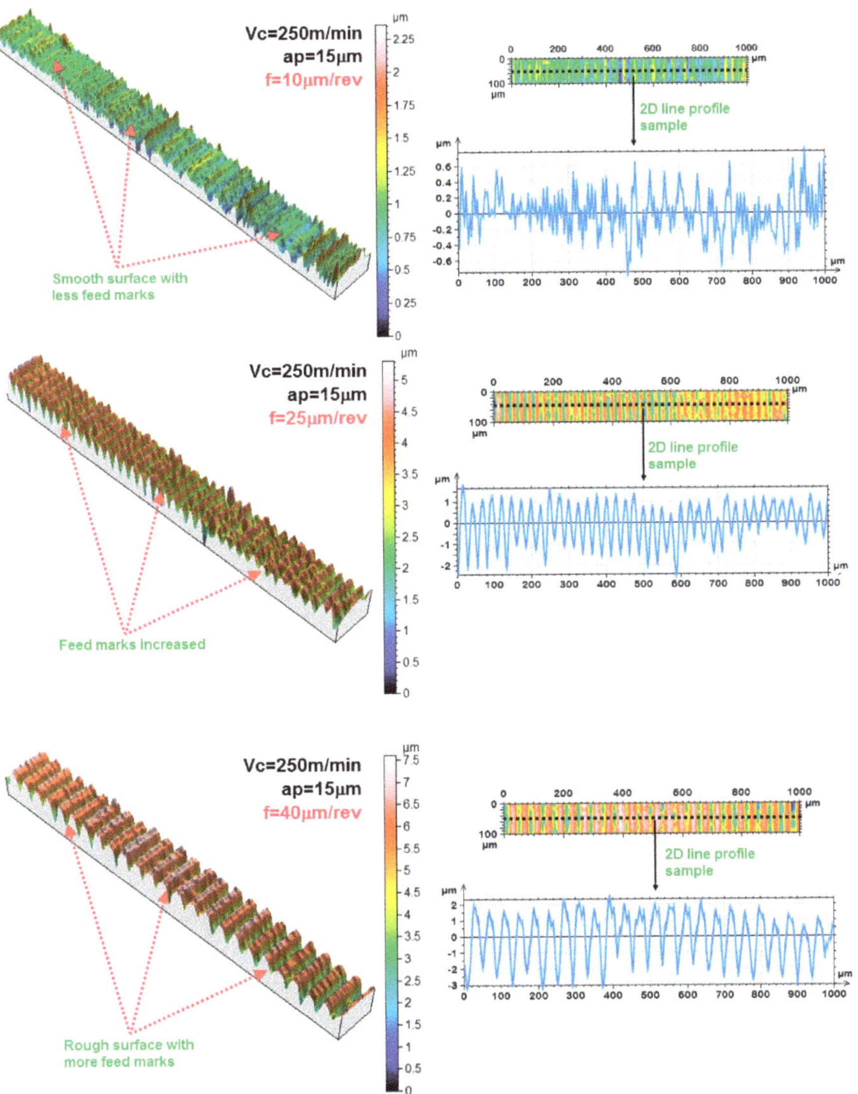

Figure 11. Influence of feed rate on surface 2D and 3D surface profiles.

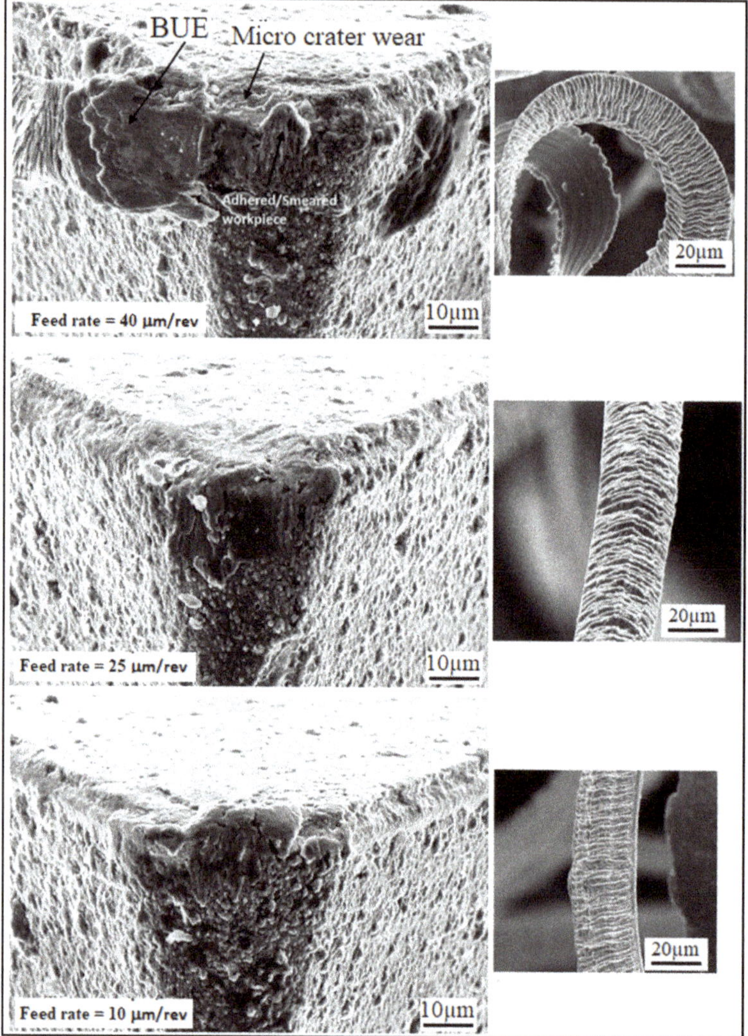

Figure 12. SEM images showing the effect of feed rate on the tool and chip morphology at $V_c = 250$ m/min and $a_p = 15$ µm.

The adhered work material was highest when the depth of cut was 15 µm. These factors may be the reason for the higher surface roughness observed at this point. A significant amount of work material adhered on the tool was also noticed when the depth of cut was only 5 µm. This may be due to the ploughing effect that can occur at a very low depth of cut, which is not desirable for a machining operation as it increases the possibility of tool wear, thereby producing unacceptable surface finish [29]. Therefore, it can be concluded here that, for good surface quality results, the depth of cut should be practiced in the higher range, as this is supportive of both the productivity as well as the finish surface roughness.

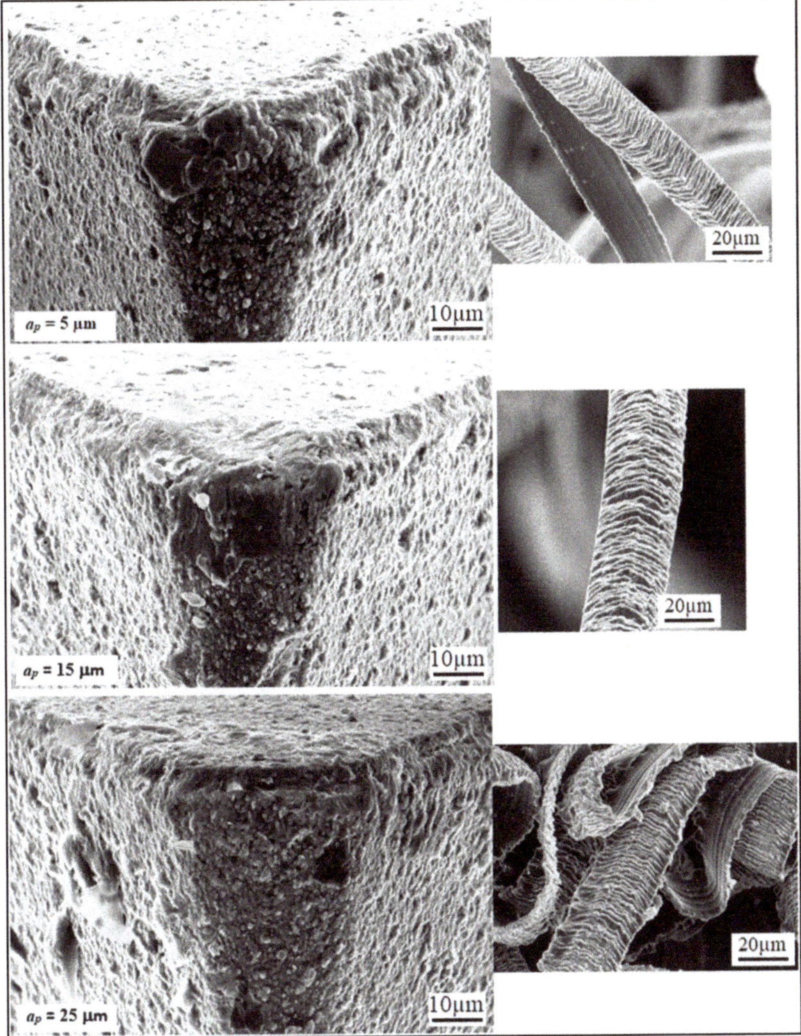

Figure 13. SEM images showing the effect of depth of cut on the tool and chip morphology at $V_c = 250$ m/min and feed rate = 25 µm/rev.

3.6. Optimization of Surface Roughness Parameters and Material Removal Rate

Two surface roughness parameters and the material removal rate have been considered for the system optimization—a multi-objective optimization. For that purpose, the composite desirability approach has been granted. Its details can be found in Reference [30–32]. The common desirability function is presented in Equation (6).

$$D = (d_1 \times d_2 \times d_3 \ldots \ldots \times d_n)^{\frac{1}{n}} = \left(\prod_{i=1}^{n} d_i \right)^{\frac{1}{n}}$$

(6)

Here, the desirability has been represented by the d_i, and the responses are represented by n. Depending on the condition, each response should either have a low value or high value. As the highest is the better value, the desirability is defined as Equation (7).

$$\begin{cases} d_i = 0 \text{ if response} < \textit{low value} \\ 0 \leq d_i \leq 1 \text{ if response in between low and high value} \\ d_i = 1 \text{ if response} > \textit{high value} \end{cases} \qquad (7)$$

However, if the target is to minimize the response, the desirability function becomes Equation (8).

$$\begin{cases} d_i = 1 \text{ if response} < \textit{low value} \\ 0 \geq d_i \geq 1 \text{ if response in between low and high value} \\ d_i = 0 \text{ if response} > \textit{high value} \end{cases} \qquad (8)$$

Four optimization cases (two single-objective and two multi-objective) were considered in the present analysis.

i.   For minimum Sa;
ii.   For minimum Sz;
iii.   For minimization of both Sa and Sz simultaneously;
iv.   For minimum of surface roughness (Sa and Sz) and maximum MRR at the same time.

The input variables, and the responses listed with the goal of optimization, their lower limits and upper limits and the respective importance, are shown in Table 11.

Table 11. Inputs, outputs, ranges and the importance for the optimization.

Name	Goal	Lower Limit	Upper Limit	Importance
Cutting speed (*Vc*)	is in range	100	400	-
Feed rate (*f*)	is in range	10	40	-
Depth of cut (*ap*)	is in range	5	25	-
Average roughness (*Sa*)	Minimize	0.325	1.02	5
Maximum roughness height (*Sz*)	Minimize	3.35	7.85	5
Material removal rate (*MMR*)	Maximize	5	400	5

The target for the multi-objective optimization is to finalize a solution that is supportive of the best possible outcomes from all three responses. The main objective for the multi-objective optimization here is to gain the optimum solution for which minimum surface roughness and maximum MRR can be obtained simultaneously. The solutions for the multi-objective optimization, i.e., minimum surface roughness (Sa and Sz) and maximum MRR case, are summarized in Table 12.

Table 12. Multi-objective optimization solution.

Sr. No.	*Vc*	*f*	a_p	*Sa*	*Sz*	*MMR*	Desirability
1	400.00	23.71	25.00	0.50	4.16	239.03	0.714 Selected
2	400.00	23.88	25.00	0.50	4.17	240.45	0.714
3	400.00	23.18	25.00	0.49	4.13	234.525	0.713
4	400.00	22.24	24.97	0.49	4.08	226.261	0.712
5	400.00	22.56	24.94	0.49	4.11	228.645	0.710
6	400.00	33.41	25.00	0.59	4.70	321.503	0.700
7	400.00	10.31	25.00	0.37	3.16	125.099	0.659
8	100.00	10.00	5.01	0.47	3.21	27.4834	0.356

From Table 12, it is suggested that the best possible solution obtained here has a desirability of 0.714. The respective solution is a cutting speed of 400 m/min, feed rate of 23.71 µm/rev and depth of cut of 25 microns. The optimum average surface roughness parameter is 0.50 µm, the optimum maximum height roughness parameter is 4.16 µm and optimum material removal rate is 239.03 mm^3/min.

The contour plot and ramp function plot respective to the optimum solution is shown in Figures 14 and 15, respectively. Moreover, the solutions for all the cases are listed in Table 13.

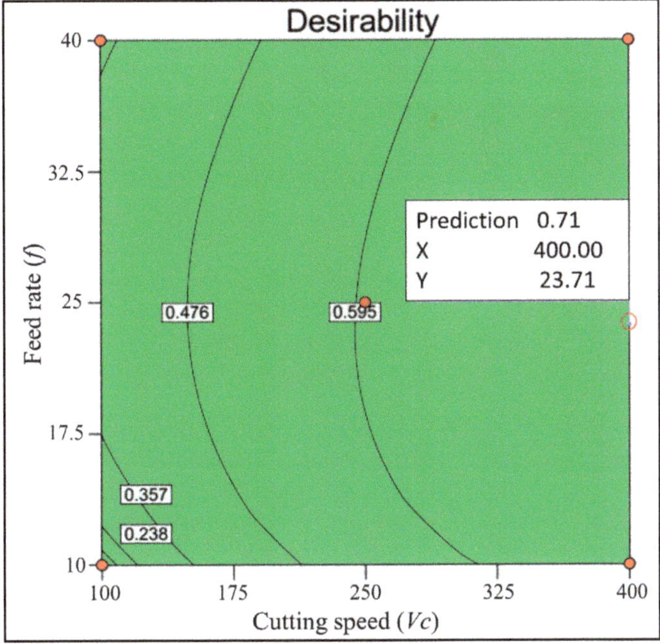

Figure 14. Contour graph for the multi-objective optimization.

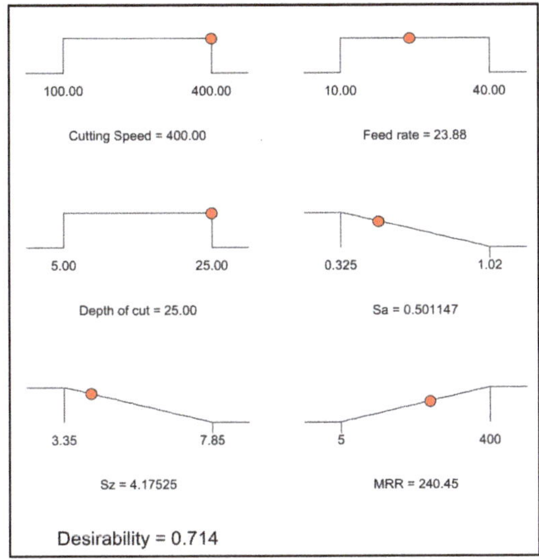

Figure 15. Multi-objective optimization solution as ramp function.

The combinations varied depending on the selectivity of the response. For instance, when the two roughness parameters were considered, leaving the material removal rate out of consideration, the desirability was 1.0 when the cutting speed was 340.49 m/min, feed rate was 10.24 µm/rev and the depth of cut was 24.87 µm.

Table 13. List of solutions for the combinations of target objectives.

Optimization Cases	V_c (m/min)	f (μm/rev)	a_p (μm)	Sa (μm)	Sz (μm)	MMR (mm^3/min)	Desirability
Minimum Sa	156.14	10.44	24.92	0.32	-	-	1.000
Minimum Sz	339.67	10.55	24.87	-	3.34	-	1.000
Minimization of both Sa and Sz	340.49	10.24	24.87	0.32	3.30	-	1.000
Minimization of Sa and Sz and maximization of MMR	400.00	23.71	25.00	0.50	4.16	239.03	0.714

4. Conclusions

1. Empirical relations between cutting parameters and surface roughness (Sa and Sz) of the TiAl4V alloy was successfully developed using RSM for the micro-turning process;

2. The efficiency of both models was checked according to the different R^2 terms. The developed models showed good accuracy in terms of correlation coefficient, close to unity. The residual plots and the outliers plot showed the adequacy of the models. Last but not least, the verification test showed superior accuracy, an error value of less than 7% for both the average roughness parameter and maximum height roughness parameter;

3. With the increase in feed rate, both the Sa and Sz of the TiAl4V alloy were found to be increased, while a mixed trend was observed for other cutting parameters. Overall, the most dominant factor which affects the Sa and Sz of the micro-turned TiAl4V was found to be the feed rate;

4. The tool wear results show that the crater wear is the dominant wear for micro-turned Ti-6Al-4V alloys. Moreover, the higher serrations in the chips were observed at high feed rate values, which is also the reason for the poor surface roughness values;

5. All optimization results are as follows:

   a. Minimum Sa optimization: V_c = 156.14 m/min, f =10.44 μm /rev and a_p = 24.92 μm;

   b. Minimum Sz optimization: V_c = 339.67 m/min, f =10.55 μm /rev and a_p = 24.87 μm;

   c. Minimum Sa and Sz optimization: V_c = 340.49 m/min, f = 10.24 μm /rev and a_p = 24.87 μm;

   d. For minimum of surface roughness (Sa and Sz) and maximum MRR at the same time: V_c = 400 m/min, f = 23.71 μm/rev and a_p = 25 μm;

6. The optimized values for Sa, Sz and MRR obtained by the multi-objective optimization approach were 0.50 μm, 4.16 μm and 239.03 mm^3/min, respectively.

Author Contributions: Conceptualized, K.A., H.I.; Supervision, K.A., H.I., M.M.; Writing—original draft, K.A., M.D., M.M, H.I.; Data curation, M.D., A.H., T.G.; Investigation, M.M., K.A., M.D., M.G.; Formal data analysis, M.M., M.D., M.G., H.I.; Research methodology, K.A., T.G., A.H.; writing—review and editing, M.M., K.A., M.G., H.I., M.D. All authors have read and agreed to the published version of the manuscript.

Funding: This research was funded by Yayasan Universiti Teknologi Petronas (YUTP) Research Fund with the cost centre of 015LCO-052.

Acknowledgments: Authors would like to acknowledge the "Department of Mechanical Engineering, Faculty of Technology, Afyon Kocatepe University" for providing the machining facilities and University Teknologi Petronas for characterization and funding.

Conflicts of Interest: The authors declare no conflict of interest.

References

1. Wu, D.; Wang, B.; Fang, F. Effects of tool wear on surface micro-topography in ultra-precision turning. *Int. J. Adv. Manuf. Technol.* **2019**, *102*, 4397–4407. [CrossRef]

2. Wojciechowski, S.; Matuszak, M.; Powałka, B.; Madajewski, M.; Maruda, R.W.; Królczyk, G.M. Prediction of cutting forces during micro end milling considering chip thickness accumulation. *Int. J. Mach. Tools Manuf.* **2019**, *147*, 103466. [CrossRef]

3.   Yousuff, C.M.; Danish, M.; Ho, E.T.W.; Basha, K.; Hussain, I.; Hamid, N.H.B. Study on the optimum cutting parameters of an aluminum mold for effective bonding strength of a PDMS microfluidic device. *Micromachines* **2017**, *8*, 258. [CrossRef]

4.   Boswell, B.; Islam, M.N.; Davies, I.J. A review of micro-mechanical cutting. *Int. J. Adv. Manuf. Technol.* **2018**, *94*, 789–806. [CrossRef]

5.   Danish, M.; Ginta, T.L.; Abdul Rani, A.M.; Carou, D.; Davim, J.P.; Rubaiee, S. Investigation of surface integrity induced on AZ31C magnesium alloy turned under cryogenic and dry conditions. *Procedia Manuf.* **2019**, *41*, 476–483. [CrossRef]

6.   Piotrowska, I.; Brandt, C.; Karimi, H.R.; Maass, P. Mathematical model of micro turning process. *Int. J. Adv. Manuf. Technol.* **2009**, *45*, 33–40. [CrossRef]

7.   Liu, K.; Melkote, S.N. Effect of plastic side flow on surface roughness in micro-turning process. *Int. J. Mach. Tools Manuf.* **2006**, *46*, 1778–1785. [CrossRef]

8.   Zhang, T.; Liu, Z.; Shi, Z.; Xu, C. Size effect on surface roughness in micro turning. *Int. J. Precis. Eng. Manuf.* **2013**, *14*, 345–349. [CrossRef]

9.   Zhao, M.; He, N.; Li, L.; Liu, Z.Q.; Wu, W.F.; Zheng, K.B. Analyses of Size Effect on Surface Roughness in Micro Turning Process. In Materials Science Forum. *Trans Tech. Publ.* **2012**, *723*, 389–393.

10.   Danish, M.; Yasir, M.; Mia, M.; Nazir, K.; Ahmed, T.; Rani, A.M.A. High speed machining of magnesium and its alloys. In *High Speed Mach*; Elsevier: Amsterdam, The Netherland, 2020; pp. 263–282.

11.   Mia, M.; Morshed, M.S.; Kharshiduzzaman, M.; Razi, M.H.; Mostafa, M.R.; Rahman, S.M.S.; Ahmad, I.; Hafiz, M.T.; Kamal, A.M. Prediction and optimization of surface roughness in minimum quantity coolant lubrication applied turning of high hardness steel. *Measurement* **2018**, *118*, 43–51. [CrossRef]

12.   Mia, M.; Dhar, N.R. Prediction of surface roughness in hard turning under high pressure coolant using Artificial Neural Network. *Measurement* **2016**, *92*, 464–474. [CrossRef]

13.   Rahman, M.A.; Rahman, M.; Kumar, A.S.; Lim, H.S.; Asad, A. Development of micropin fabrication process using tool based micromachining. *Int. J. Adv. Manuf. Technol.* **2006**, *27*, 939–944. [CrossRef]

14.   Alauddin, M.; El Baradie, M.A.; Hashmi, M.S.J. Optimization of surface finish in end milling Inconel 718. *J. Mater. Process. Technol.* **1996**, *56*, 54–65. [CrossRef]

15.   Wang, X.; Lu, X.; Jia, Z.; Jia, X.; Li, G.; Wu, W. Research on the prediction model of micro-milling surface roughness. *Int. J. Nanomanuf.* **2013**, *9*, 457–467. [CrossRef]

16.   Kuram, E.; Ozcelik, B. Multi-objective optimization using Taguchi based grey relational analysis for micro-milling of Al 7075 material with ball nose end mill. *Measurement* **2013**, *46*, 1849–1864. [CrossRef]

17.   Vipindas, K.; Kuriachen, B.; Mathew, J. Investigations into the effect of process parameters on surface roughness and burr formation during micro end milling of TI-6AL-4V. *Int. J. Adv. Manuf. Technol.* **2019**, *100*, 1207–1222. [CrossRef]

18.   Aslantas, K.; Ekici, E.; Çiçek, A. Optimization of process parameters for micro milling of Ti-6Al-4V alloy using Taguchi-based gray relational analysis. *Measurement* **2018**, *128*, 419–427. [CrossRef]

19.   Ucun, İ.; Aslantaş, K.; Gökçe, B.; Bedir, F. Effect of tool coating materials on surface roughness in micromachining of Inconel 718 super alloy. *Proc. Inst. Mech. Eng. Part B J. Eng. Manuf.* **2014**, *228*, 1550–1562. [CrossRef]

20.   Thepsonthi, T.; Özel, T. Multi-objective process optimization for micro-end milling of Ti-6Al-4V titanium alloy. *Int. J. Adv. Manuf. Technol.* **2012**, *63*, 903–914. [CrossRef]

21.   Kumar, S.P.L. Measurement and uncertainty analysis of surface roughness and material removal rate in micro turning operation and process parameters optimization. *Measurement* **2019**, *140*, 538–547. [CrossRef]

22.   Danish, M.; Yahya, S.; Saha, B.B. Modelling and optimization of thermophysical properties of aqueous titania nanofluid using response surface methodology. *J. Therm. Anal. Calorim.* **2019**, *139*, 3051–3063. [CrossRef]

23.   Keshtegar, B.; Mert, C.; Kisi, O. Comparison of four heuristic regression techniques in solar radiation modeling: Kriging method vs RSM, MARS and M5 model tree. *Renew. Sustain. Energy Rev.* **2018**, *81*, 330–341. [CrossRef]

24.   Danish, M.; Ginta, T.L.; Habib, K.; Carou, D.; Rani, A.M.A.; Saha, B.B. Thermal analysis during turning of AZ31 magnesium alloy under dry and cryogenic conditions. *Int. J. Adv. Manuf. Technol.* **2017**, *91*, 2855–2868. [CrossRef]

25.   Chabbi, A.; Yallese, M.A.; Meddour, I.; Nouioua, M.; Mabrouki, T.; Girardin, F. Predictive modeling and multi-response optimization of technological parameters in turning of Polyoxymethylene polymer (POM C) using RSM and desirability function. *Measurement* **2017**, *95*, 99–115. [CrossRef]

26. Groover, M.P. *Fundamentals of Modern Manufacturing: Materials Processes, and Systems*; John Wiley & Sons: Hoboken, NJ, USA, 2007; ISBN 8126512660.

27. Singh, G.; Gupta, M.K.; Mia, M.; Sharma, V.S. Modeling and optimization of tool wear in MQL-assisted milling of Inconel 718 superalloy using evolutionary techniques. *Int. J. Adv. Manuf. Technol.* **2018**, *97*, 481–494. [CrossRef]

28. Mia, M.; Singh, G.; Gupta, M.K.; Sharma, V.S. Influence of Ranque-Hilsch vortex tube and nitrogen gas assisted MQL in precision turning of Al 6061-T6. *Precis. Eng.* **2018**, *53*, 289–299. [CrossRef]

29. Popov, A.; Dugin, A. Effect of uncut chip thickness on the ploughing force in orthogonal cutting. *Int. J. Adv. Manuf. Technol.* **2015**, *76*, 1937–1945. [CrossRef]

30. Mia, M. Multi-response optimization of end milling parameters under through-tool cryogenic cooling condition. *Measurement* **2017**, *111*, 134–145. [CrossRef]

31. Gupta, M.K.; Sood, P.K.; Sharma, V.S. Optimization of machining parameters and cutting fluids during nano-fluid based minimum quantity lubrication turning of titanium alloy by using evolutionary techniques. *J. Clean. Prod.* **2016**, *135*, 1276–1288. [CrossRef]

32. Gupta, M.K.; Mia, M.; Singh, G.R.; Pimenov, D.Y.; Sarikaya, M.; Sharma, V.S. Hybrid cooling-lubrication strategies to improve surface topography and tool wear in sustainable turning of Al 7075-T6 alloy. *Int. J. Adv. Manuf. Technol.* **2019**, *101*, 55–69. [CrossRef]

Article

Thermal Sources of Errors in Surface Texture Imaging

Karol Grochalski [1,*], Michał Wieczorowski [1] , Paweł Pawlus [2] and Jihad H'Roura [3]

[1] Division of Metrology and Measurement Systems, Institute of Mechanical Technology, Poznan University of Technology, Piotrowo 3, PL-60965 Poznan, Poland; michal.wieczorowski@put.poznan.pl

[2] Faculty of Mechanical Engineering and Aeronautics, Rzeszow University of Technology, al. Powstańców Warszawy 12, PL-35959 Rzeszow, Poland; ppawlus@prz.edu.pl

[3] IRF-SIC Laboratory, Department of Mathematics, Ibn Zohr University, Nouveau Complexe Universitaire, 80000 Agadir, Morocco; jihad.hroura@gmail.com

* Correspondence: karol.grochalski@put.poznan.pl; Tel.: +48-61-665-32-23

Received: 6 April 2020; Accepted: 15 May 2020; Published: 19 May 2020

Abstract: This paper presents the influence of thermal phenomena on areal measurements of surface topography using contact profilometers. The research concerned measurements under controlled and variable environmental conditions. The influence of internal heat sources from profilometer drives and their electronic components was analyzed. For this purpose, a thermal chamber was designed and built. Its task was to maintain and control environmental conditions and, at the same time, separate the profilometer from external disturbances. Heat sources and temperature values for elements and systems were determined. It further enabled for the calculation of the displacements in axes as a function of temperature. The largest displacement in the probe due to internal heat sources for the considered cases occurred in the X-axis direction. Its value reached 16.2 μm. However, the displacement in the probe in the Z-axis direction had the greatest impact on the measured surface topography. These displacements for a thermally unstable profilometer reached 7.9 μm in Z, causing results even 90% greater than in the case of a device without such problems. The time after which a proper topography measurement can be started was also determined basing on obtained data. This time for tested profilometers was between 6 and 12 h. It was found that performing thermal stabilization of the profilometer significantly reduced surface irregularity errors. The stabilization time should be determined individually for a specific type of device.

Keywords: contact profilometry; surface topography; thermal disturbance; thermal expansion; thermal chamber

---

1. Introduction

The improvement in the quality of manufacturing processes of machinery and equipment parts is directly related to the development (advancement) of methods to control manufactured products. This is associated with the simultaneous development of quality control departments and the use of more and more advanced measuring devices. One of the most important factors determining the quality of a product is the surface structure of the produced element. It is a consequence of production methods and machining [1].

An important issue in the assessment of the structure condition is broadly understood surface irregularities. They are one of the basic features determining the quality of manufactured parts that also affect their performance, durability, and exchangeability in mechanical engineering [2]. The geometric structure of the surface (GSS), known also as surface texture, is the set of all hills and dales on the surface.

A surface irregularity consists of waviness, roughness, and shape deviation. These components can be assigned to three main groups of deviation scale:

- large scale—shape deviation,
- average scale—waviness,
- small scale—roughness.

There are many methods and devices for surface topography analysis [3], including tactile and optical ones [4]. The stylus method is the oldest yet still the most often used in industry. In this method, the surface texture is often analyzed on the basis of cross-sections—surface asperity profiles. However, three-dimensional (3D) surface analysis provides more important information. There are also sophisticated methods of filtration, enabling the separation of particular components, e.g., the component of a shape deviation or waviness, which, as a result, enables the roughness analysis. They often reach even more and more multiscale analysis for different purposes [5,6].

In contact methods, a measurement is carried out using a diamond measuring tip that moves over the measured surface. These methods are well known, but they are also sensitive to some interfering factors, such as vibrations, thermal phenomena, and geometrical errors of the measuring tip, including the rounding radius and local changes in the tip. Errors in roughness measurements resulting from the tip geometry have been the subject of a number of studies, e.g., performed by Elewa and Koura [7], Bodschwinna [8], Trumpold and Heldt [9], Smith and Chetwynd [10], and Anbari et al. [11]. The shape of the measuring tip changes the actual geometry of asperity. An additional source of error in spatial measurements is the linear displacement of the measuring tip in one axis of the profilometer, causing path synchronization error [12,13], as well as too high a velocity of the tip, causing the detachment of the tip from the measured element [14], for which various options were investigated [15].

Modern-quality control requires the use of devices that correctly return information about the measurement of the selected quantity while maintaining the required accuracy parameters. When the measurement is performed using typical measuring devices, solutions are sought to eliminate the influence of disturbing factors after the test. This is one of the issues related to information-rich metrology [16]. The data set covering many aspects of the measurement, e.g., environmental factors, device status, and signal processing algorithms, allows the most advantageous analysis of the obtained values and their possible correct interpretation.

2. Problem Statement

There are generally a lot of sources of errors in surface topography measurements using optical and tactile techniques [17,18]. In optical techniques, they are mainly connected with non-measured points [19] or sampling intervals [20], in stylus—with the geometry of a probe tip or vibrations and noise. Their impact is described in the references, e.g., work by Haitjema and Morel [21]. In the case of profilometric measurements, this is often associated with the way the environment affects the measuring device. Thermal disturbances are often not considered as an important source of error, which, particularly in the case of 3D measurements, is not true. They may have a form of changes in amplitude and/or frequency. Research on this type of disturbance was carried out, among others, by Miller et al. [22] and Krawiec et al. [23].

When measuring a single profile, provided the time of temperature change is long compared to the measurement time, low-frequency temperature changes do not influence the measurement results. In the case of long-term spatial measurements with a large number of measuring probe traverses, the influence of temperature changes on the measurement system can be significant, particularly when the temperature change occurs directly during the measurement procedure. According to the thermal expansion law, both the specimen and the measuring device undergo geometrical deformations in three dimensions.

The most frequent source of thermal errors is a commonly used, two-step air conditioning (cooling) system. Such a system turns on when the upper temperature limit is reached and turns off when the lower threshold is achieved. Such an operation of the air conditioning system is usually periodic. It also causes the direct influence of cold air (gas) on a device, which may, in turn, result in different temperature gradients. Zhou et al. investigated these phenomena in their work [24].

Furthermore, there are also internal thermal disturbances coming from electronic components, drives, and friction of the moving elements. Disturbances of these types are asymptotic and decrease with time during stabilization of the profilometer. The temperature stabilization process, which may take several hours, should be performed before the beginning of the measurement. Regardless of the nature of the disturbance, thermal phenomena cause errors in the image of the surface and values of parameters, calculated from the measured surface.

Currently, no thermal diagnostics and disturbance compensation are used in practical solutions for contact profilometers; they are carried out only for coordinate measuring machines (CMMs).

A way of compensation of certain errors related to the device geometry when a coordinate measuring technique is involved was proposed by Zha et al. [25]. Errors related to both displacement and deformation of the device components are most often associated with thermal expansion of the device. Thermal errors may also cause residual stress, particularly when construction elements are designed as multilayer ones [26]. Usually, measurements of deformation of this type are performed simultaneously with infrared thermal diagnostics. The methodology of thermal diagnostics of CMMs is presented in the works of Abdulshahed et al. [27], as well as Chenyang et al. [28]. Hao et al. [29] and Muniz et al. [30] also worked on similar issues related to thermal errors. On the contrary, Schwenke et al. tried to compensate these errors [31]. A similar attitude can be found in a publication by Ma et al. [32]. Sladek et al., on the other hand, were modeling the uncertainty changes caused by temperature for CMMs [33]. There are different additional devices used for calculating errors. The use of laser interferometry in deformation measurements of the CMM structure has been discussed by groups directed by Balsamo [34] and Stejskal [35]. Furthermore, the use of thermometry and laser interferometry is also applicable to other types of numerically controlled machines and devices, which makes their application much more versatile. The research concerning this topic has been described by Enming et al. [36], Lo et al. [37], and Miao et al. [38].

The recognition of issues related to temperature changes during measurements causes the need for compensation and thermal conditions control. The first attempts were made by Baird quite a long time ago [39]. Kruth et al. [40] presented the compensation of Thermal Errors on CMMs, dividing them into static and transient ones. Similar works for more generally considered devices were made by Sartori and Zhang [41]. Ge and Ding elaborated a method for thermal error control for precision parts of machine tools [42]. This method was based on the principle of thermal deformation balance. Tang, Xu, and Wang also conducted analyses in order to predict thermal deformations and the impact of environmental conditions on the measuring device using a neural network [43], while Milov et al. created an algorithm and software to identify errors in measuring equipment during the formation of permanent joints [44].

After analyzing the literature, it was decided to carry out the research on the influence of thermal disturbances on the surface asperities using contact profilometers, which will eventually enable compensation of this effect. Additionally, the authors intended to determine the time, after which thermal and geometrical stabilization will take place. The topic is an important issue as it is generally believed that there is no thermal influence on performance of a profilometer, which is commonly used for tactile (and not only) surface topography measurements.

3. Materials, Methods, and Results

The research on the influence of the internal heat sources of a profilometer on the expansion of its structure and elements was based on two types of measurements. First, a static measurement was made in which the measuring head of the profilometer remained stationary. Second, a dynamic measurement was performed, in which surface topography points were collected (measuring head was moving).

3.1. Static Measurement

Test methodology of the static measurement procedure (i.e., measurement without movement of a measuring head) was based on temperature verification of the profilometer components using two independent methods. The temperature measurement started with the moment the device was turned on (with no movement of the measuring probe in direction of X-axis). The power supply influenced the heating of electronic components, drives, and other heat-generating parts placed inside the profilometer structure. Simultaneously to the temperature measurement, the displacement of the measuring head was investigated using a laser interferometer. Temperature values were correlated with displacements, which allowed for the determination of the resultant thermal expansion coefficient of a particular element.

The first method of measuring temperature involved the use of semiconductor temperature sensors DS18B20+ (Maxim Integrated Products, Sunnyvale, CA, USA) localized as close as possible to the drives of individual axes of the profilometer. Sensors allowed to monitor changes in temperature as a function of time during the heating of the profilometer from the moment the device was turned on. A sensor of this type has a measuring range from -55 °C to $+125$ °C and a maximum permissible measurement error of ±0.5 °C in the measuring range from -10 °C to $+85$ °C.

The second method of temperature measurement involved the use of a diagnostic thermal imaging camera FLIR T620 (FLIR Systems, Inc., Wilsonville, Oregon, USA) which enabled detection of heat sources (profilometer components) in the entire system, presented in the form of a colored thermal map. The thermal imaging camera provided possibilities of monitoring the temperature distribution over time on all the profilometer surfaces and measuring the temperature in a certain point on a surface.

The use of the thermal imaging camera enabled the collection of more information about the thermal condition of the profilometer and very precise determination of heat sources.

The spatial (geometric) resolution of the T620 camera in the sense of the instantaneous field of view (IFOV) was 0.62 mrad and the thermal resolution in the sense of noise-equivalent temperature difference (NETD) was less than 0.05 °C. The maximum permissible measurement error of the camera was equal to ±2 °C or $\pm2\%$ of the temperature readings (the greater of these two values is taken as the respective value).

During testing of the profilometer, a laser interferometer LASERTEX LSP30-3D (Lasertex Sp. z o.o., Wrocław, Poland) with a measurement resolution of 0.1 nm was used. Its task was to measure the displacement of the measuring head, caused by changes in the geometrical dimensions of the device structure due to temperature variations from internal heat sources. The reference mirror (stationary) of the interferometer was located on a granite slab. The mobile mirror was attached to the fixing holder of the measuring head of the profilometer. The measurement head displacement was performed for each axis of the profilometer while maintaining the same experimental conditions. A similar method of determining geometrical deviations for coordinate measuring was reported by Hemming et al. [45], Echerfaoui et al. [46], and Schwenke et al. [47]. A scheme of the measuring setup is shown in Figure 1.

Research on the influence of internal heat sources was carried out on three contact profilometers, differing from each other in design and type of drives used for particular axes: T8000 (A), TOPO L50 (B), and S8P/PRK (C). The same measuring equipment settings and the same measurement methodology were used for each of these devices. During testing of the profilometers, the greatest attention was paid to the measurement of the elongation of the Z column. This was due to the direct possibility of a thermal deformation of the column affecting the correctness of the spatial imaging of surface topography.

The distribution of thermal fields on the surface of each device after thermal stabilization, as well as the location and type of internal heat sources, is presented in Figures 2 and 3, respectively. The thermal stabilization state of the device is based on the criterion for which the change in temperature values in the measuring points (column of the profilometer, fixtures of drives, and electronic components) during 3 h does not change more than 0.5 °C. This criterion was adapted from other research conducted for thermal stability for machine elements performed using the same temperature measuring devices [48].

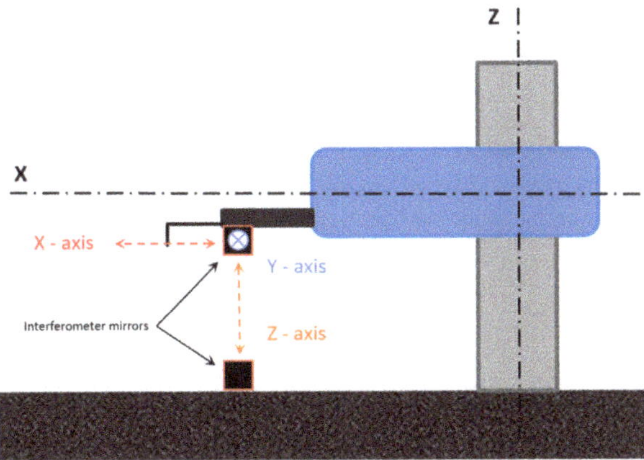

Figure 1. Scheme of the positions of interferometer mirrors for measuring displacements of X-, Y-, and Z-axes of profilometer, being under the influence of internal heat sources.

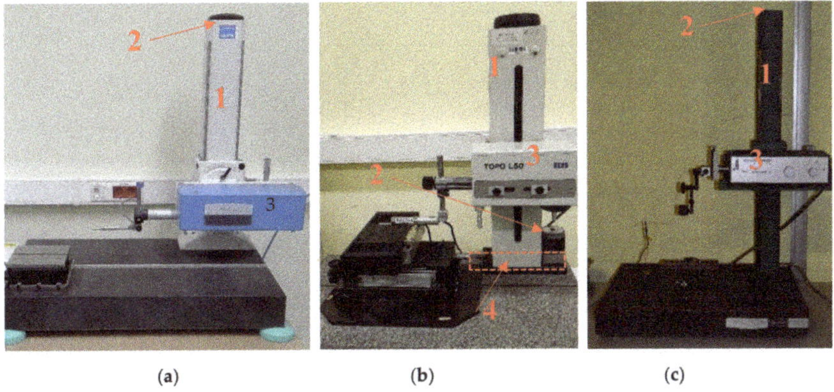

(a)   (b)   (c)

Figure 2. View of tested profilometers: 1—column, 2—drive Z, 3—traverse unit, 4—base. (**a**) Profilometer Hommel T8000; (**b**) Profilometer TOPO L50; (**c**) Profilometer Perthen S8P/PRK.

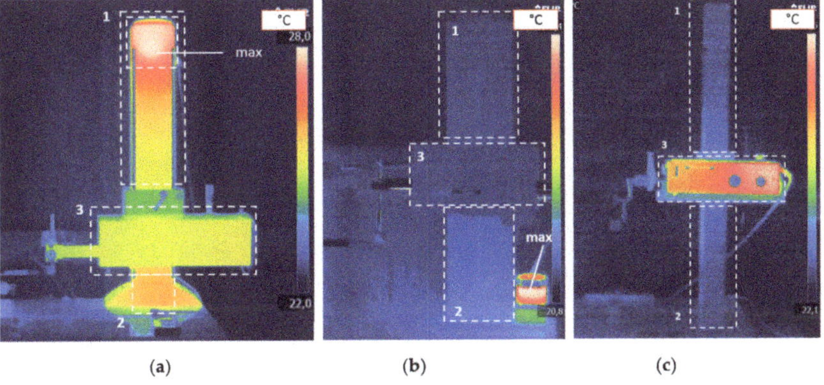

(a)   (b)   (c)

Figure 3. Distribution of thermal fields on the surface of profilometers: 1—part of the column above the traverse unit X, 2—part of the column below the traverse unit X, 3—traverse unit X. (**a**) Profilometer Hommel T8000; (**b**) Profilometer TOPO L50; (**c**) Profilometer Perthen S8P/PRK.

The temperature changes during testing of profilometer A are shown in the graph (Figure 4). The elongation of mechanical components for individual axes is illustrated in the next graph (Figure 5).

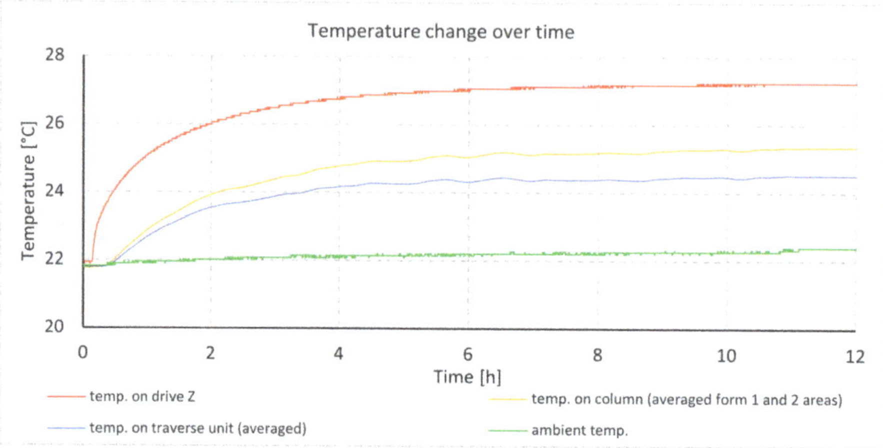

Figure 4. Temperature changes during heating of profilometer A drives.

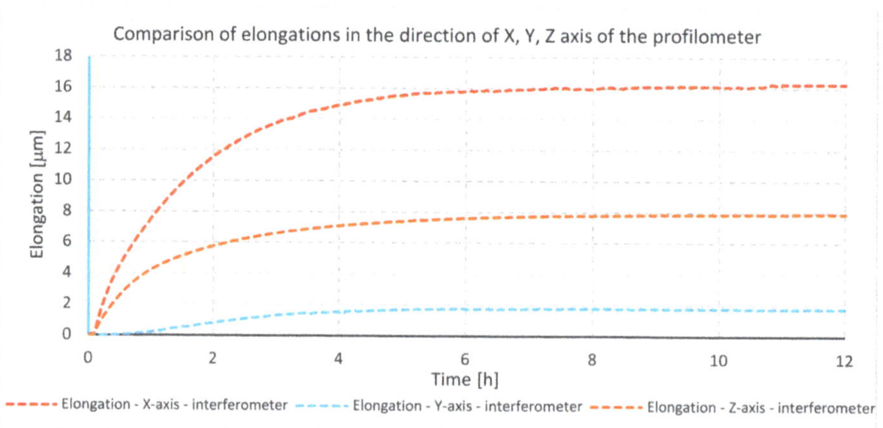

Figure 5. Comparison of elongation of individual axes of profilometer A during heating of its drives.

The graph (Figure 4) shows a rapid increase in the temperature of the stepper motor located in the Z column. Thermal stabilization of the drive and of the rest of the column can be observed after 6 h, counting from the moment the profilometer was turned on. The Z column increased its temperature by 4.3 °C and the traverse unit of the X-axis by 2.4 °C. At the same time, the growth of ambient temperature reached 0.4 °C.

The graph (Figure 5) shows displacements in the direction of individual axes due to elongation. The biggest change in the position during heating of the profilometer was observed in the direction of the X-axis. This is caused by a closed design of the traverse unit (X-axis), location of electronic modules inside it, and, due to this, practically no heat dissipation. The temperature inside the traverse unit directly affects the moving element of the measuring probe (pick-up), as well as the intermediary elements causing thermal expansion of all mentioned components. The smallest displacement was observed in the direction of the Y-axis of the profilometer. This is due to the design of the device. The traverse unit and its thermal deformability mostly affect moving parts and long elements (in accordance with the law of thermal expansion). There are no components in the Y-axis that can change their position, and the cross-section of the *Y*-axis has the smallest dimensions among all considered cross-sections (cross-section of the Z-column and traverse unit X).

The temperature changes during testing of the profilometer B are shown in the graph (Figure 6). The elongation of mechanical components along individual axes is illustrated in the next graph (Figure 7).

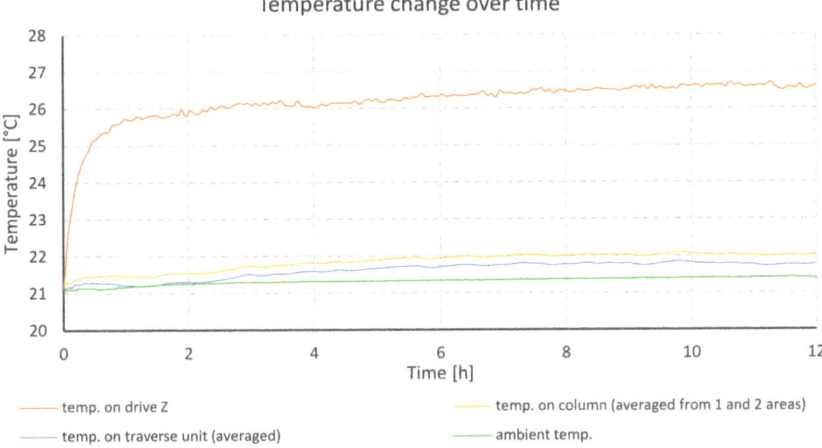

Figure 6. Temperature changes during heating of profilometer B drives.

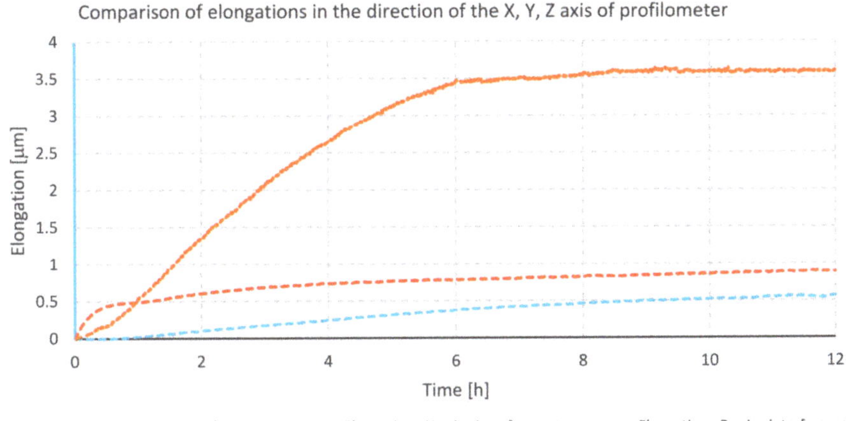

Figure 7. Comparison of elongation of individual axes of profilometer B during heating of its drives.

The graph (Figure 6) also shows a rapid increase in the temperature of the stepper motor located in the Z column. The temperature becomes stable after 2 h and increases 4.87 °C. However, the real thermal stability can be observed after 6 h from the moment the profilometer was turned on. A difference of 5.34 °C from the initial temperature was achieved. The Z column increased its temperature by 0.7 °C and the traverse unit of the *X*-axis by 0.4 °C. At the same time, the growth of ambient temperature reached 0.2 °C. The location of the drive outside the body of the profilometer limits the thermal influence on other components. Still, some heat is distributed through the metal base connecting the column with the drive.

The graph (Figure 7) illustrating the displacement of the measuring probe presents changes in position due to elongation, during the heating of the profilometer. The largest displacement reaching

3.6 µm was observed in the direction of the Z-axis. It is caused by heat transfer from the stepper motor located in the Z column, which is mounted on a steel connecting plate, which is attached to a granite plate and to the base of the column. The displacement of the measuring probe in the X- and Y-axes shows a much smaller value that does not exceed 1 µm. These displacements can be a consequence of thermal interaction of individual components occurring in particular axes, as well as an effect of Z column deformation.

The temperature changes during testing of the profilometer C are shown in the graph (Figure 8). The elongation of mechanical components toward individual axes is illustrated in the next graph (Figure 9).

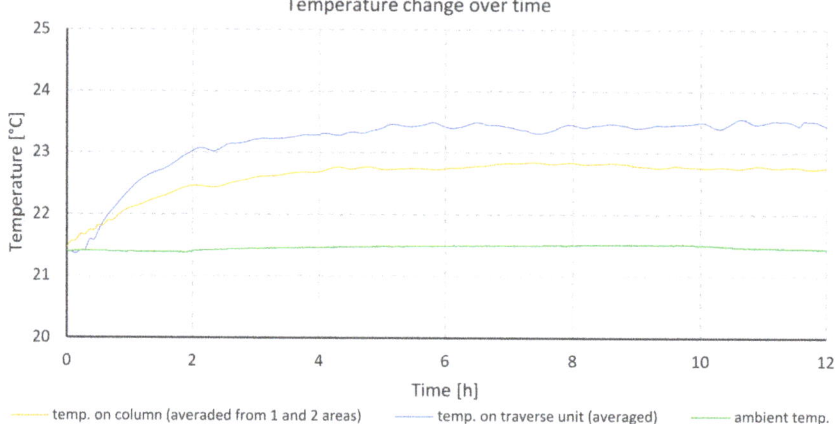

Figure 8. Temperature changes during heating of profilometer C drives.

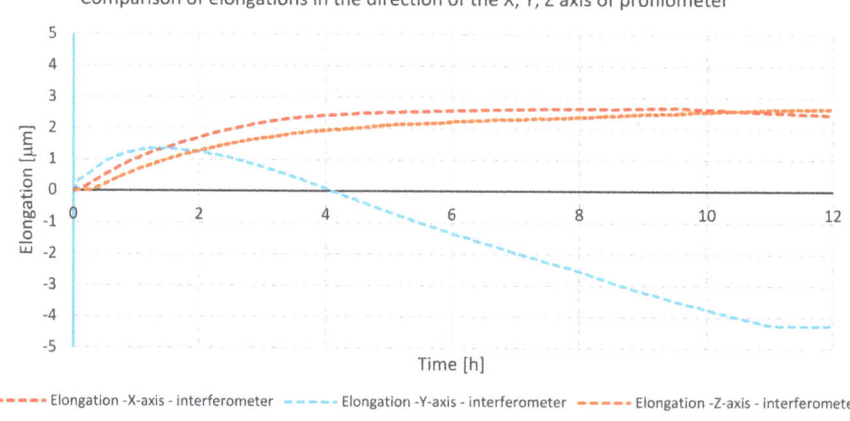

Figure 9. Comparison of elongation of individual axes of profilometer C during heating of its drives.

The temperature changes presented in the figure (Figure 8) show that the thermal stabilization of the traverse unit is reached about 4 h after switching the device on. The maximum temperature was recorded on the traverse unit X and was equal to 23.5 °C. The temperature increase of the Z column did not exceed 1 °C. The measurement was made in stable thermal environment conditions.

In the case of this device (profilometer C), column Z was not heated, which can be observed in the thermal image (Figure 3). For this reason, the temperature (drive Z) was not represented in the graph (Figure 8). It was measured by means of the semiconductor sensor located inside the housing, yet the readout was the same as ambient temperature. No heating effect is caused by a DC motor used to move the whole traverse unit in the Z direction. This kind of motor does not generate heat when the traverse unit is not moving vertically in contrast to stepper motors, where switching on the power supply already generates heat (as it was in profilometers A and B).

The results presented in the graph (Figure 9) show the change in the position of the measuring probe during the heating of the profilometer caused by its internal heat sources in relation to the starting point. The values of displacements in the X and Z are smaller than 3 µm and monotonic. The measuring probe shows the largest changes in position during the heating of the device in the direction of the Y-axis in a characteristic non-monotonic way. This situation is not typical and is caused by uneven heating of the component and its location on the positioning prisms (Figure 10a,b), which enabled displacement associated with the rotation around the apparent axis Z (Figure 10c).

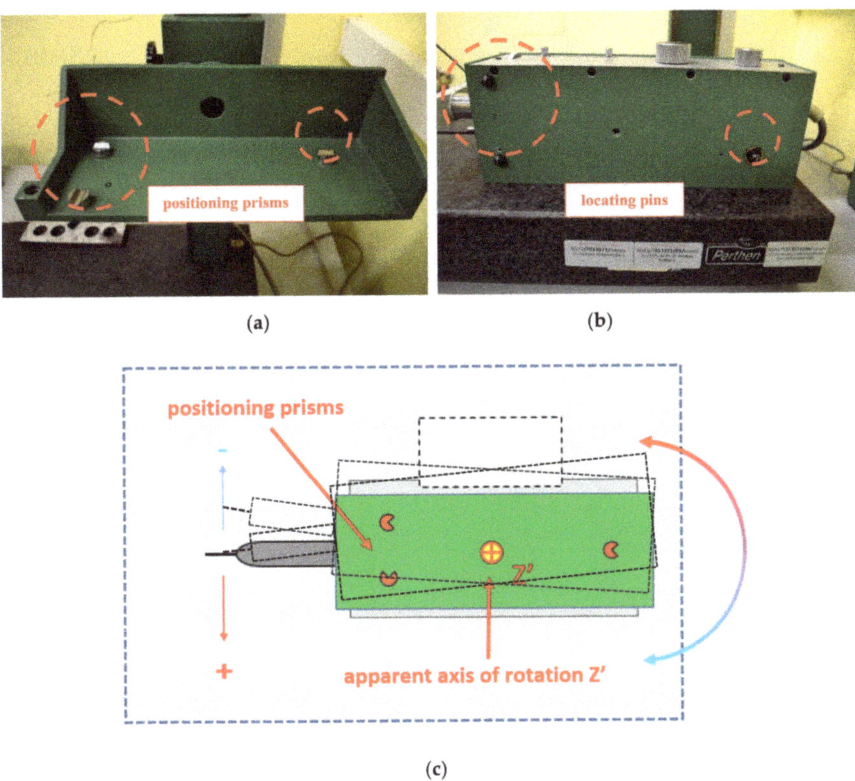

Figure 10. View of the supporting structure of X traverse unit of profilometer C: (**a**) Traverse unit cradle with positioning prisms marked; (**b**) traverse unit of the profilometer with locating pins marked; (**c**) deflection of the measuring probe of profilometer C relative to the apparent axis of rotation Z′.

Expanding elements—beginning from the location of the X-axis drive—cause this part of the traverse unit to expand and move on the positioning prism. The heat is distributed to the remaining part of the unit. This causes elongation due to thermal expansion and, consequently, displacement on the next prisms and rotation around the apparent Z-axis.

Data collected from each test were analyzed and presented in the form of a graph (Figure 11). The displacement in the measuring head in the Z-axis of the tested profilometers can have a direct impact on the correctness of spatial imaging of surface topography. Further analysis contained deformation of a column.

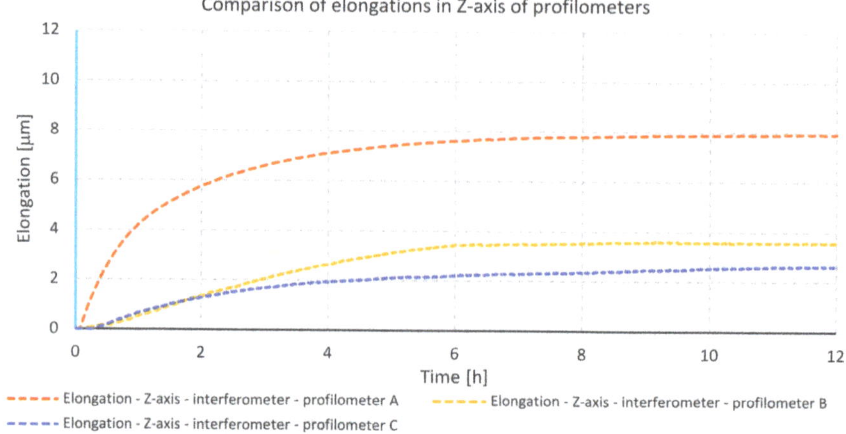

Figure 11. Comparison of elongation in Z-axes of profilometers during the heating of drives—internal heat sources.

Profilometer A has the highest Z-axis elongation value equal to 7.9 µm. Profilometer B has elongated by 3.58 µm, whereas profilometer C has elongated by 2.61 µm.

In general recommendations given by many manufacturers, it is suggested to turn the device on about 15 min before a measurement can be taken. Time determined experimentally, after which the 3D measurement of surface topography could be initialized, significantly varies from that information. It depends on the particular device and is about 6 h from the moment of turning the profilometer on (according to previously adopted criteria). After this time (in the case of tested devices), the temperature of the whole systems becomes stabilized, and no significant changes in geometric dimensions are expected.

In order to eliminate additional disturbances related to thermal aspects caused by, e.g., friction or traverse unit operation, it is recommended to perform a preliminary measurement, before starting a topography measurement. During this preliminary measurement, its time and conditions should be the same as during the real measurement. The only difference is a measuring tip that does not need to be in contact with the tested surface.

Based on the above presented results, it was found that profilometer A is characterized by the greatest susceptibility to the influence of internal heat sources. Therefore, further research should be concentrated on that device.

3.2. Dynamic Measurement

The research methodology of the dynamic measurement (measurement with movement of a measuring probe), similar to the static measurement, was based on monitoring the temperature of the profilometer components using two independent methods. These methods involved using temperature sensors located in places of heat sources occurrence and using a thermal imaging camera. The temperature measurement began from the moment the device was turned on and surface topography measurement (scanning the measured surface with the measuring tip) started. Turning on the power supply of the device caused passive power consumption, while movement of mechanical elements resulted in the heating of profilometer components located inside its structure, causing an expansion of the whole device structure. The increase in temperature was correlated with results of the simultaneous measurement of the distance between the measuring probe and the base on which the tested specimen was placed. The distance measurement was performed using a laser interferometer. One of the mirrors (reference one) was placed on a sliding table. The second mirror was placed on a rigid ABS (acrylonitrile butadiene styrene terpolymer) handle, attached to the housing of the device

(Figure 12a). The distance between the measuring head and the measured specimen was additionally verified during the measurement by reading the position of the measuring probe of the profilometer in relation to the sample. Information about the current position of the probe is registered and can be displayed in dedicated software.

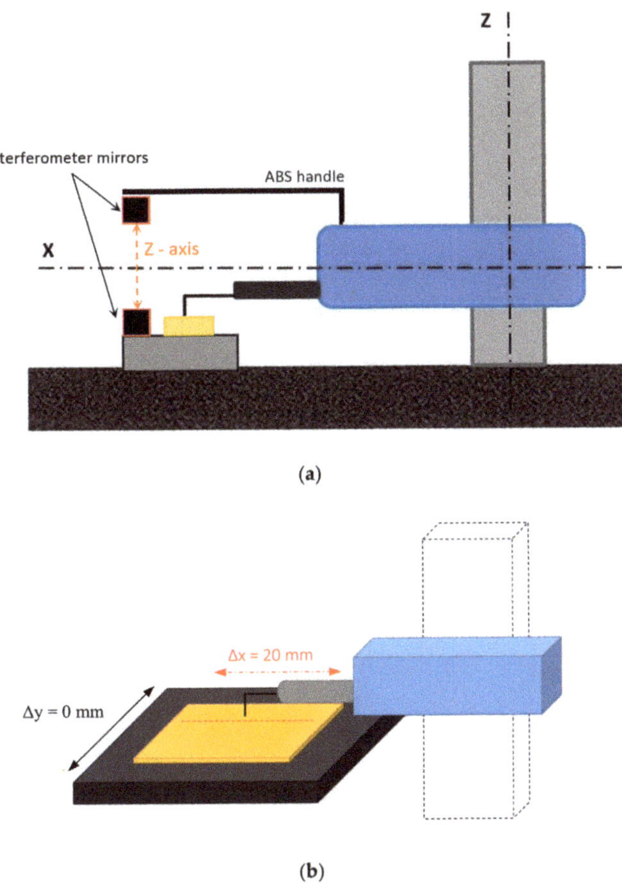

Figure 12. (**a**) Scheme of the interferometer mirrors during measurement of displacements in Z-axis; (**b**) graphic representation of measurement methodology.

The measured element was a standard optical flat—a parallel glass plate with known surface parameters and a flatness defined by the $\lambda/20$ parameter. In order to determine the influence of internal heat sources on the fidelity of surface topography imaging, the same profile was measured 600 times (to avoid influence of geometrical errors, the y table movement was switched off, $\Delta y = 0$). Then, the image of a surface was created from these profiles. This test, in the case of nearly ideal measuring conditions, was intended to provide an undisturbed, straight transverse profile from the generated surface. The measurement principle is presented in the figure (Figure 12b).

Before the beginning of the measurement, the measuring probe was levelled in relation to the plate being measured. The height deviation of a 20 mm-long distance from the starting point to the end point did not exceed 0.1 μm.

Two types of tests of the dynamic measurement were performed. The first involved measurement of the surface topography from the moment the device was switched on. During that measurement procedure, the profilometer and its components were heating up.

The second test of the dynamic measurement involved measurements after thermal stabilization of the profilometer. The measurement was made on the same surface (the measured specimen was fixed and the table motion was turned off).

The illustration (Figure 13) shows the surface generated from the measurements of the same track during heating of the device.

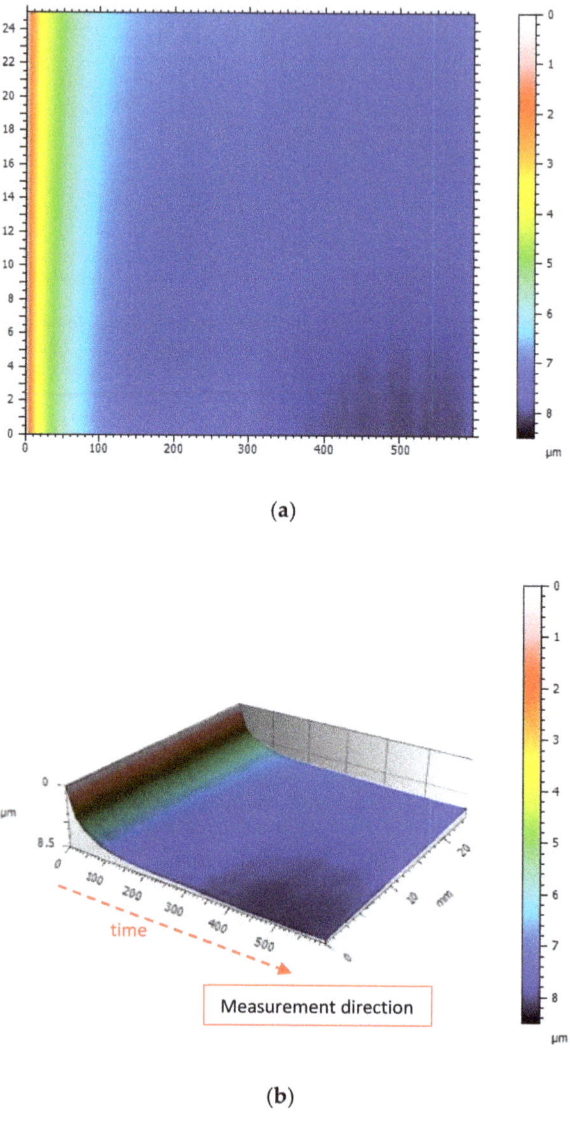

(a)

(b)

Figure 13. View of the surface generated from the measurements of the same track during heating of the profilometer: (**a**) A color map; (**b**) 3D image.

The graph (Figure 14) shows the correlation between the increase in temperature of the Z-axis drive and elongation of the profilometer column. The graph shows the displacement value of the measuring probe, which was measured with the laser interferometer and the measured surface transverse profile.

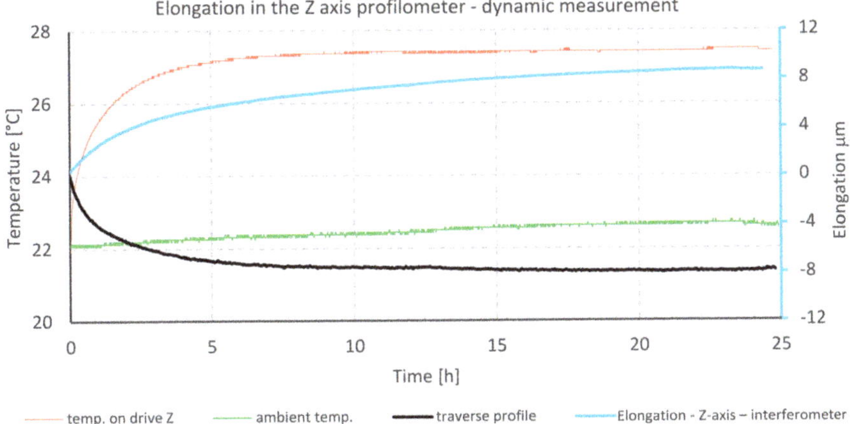

Figure 14. Elongation in Z-axis of profilometer A and temperature against time - dynamic measurement.

The cross-profile curves and the elongation values are inverted. This should be interpreted in such a way that the measuring tip, which was permanently in contact with the specimen, as an effect of elongation (due to the thermal expansion of the Z column), moved below the zero line, and the elongation of the column had a positive value in relation to the measured surface (Figure 15).

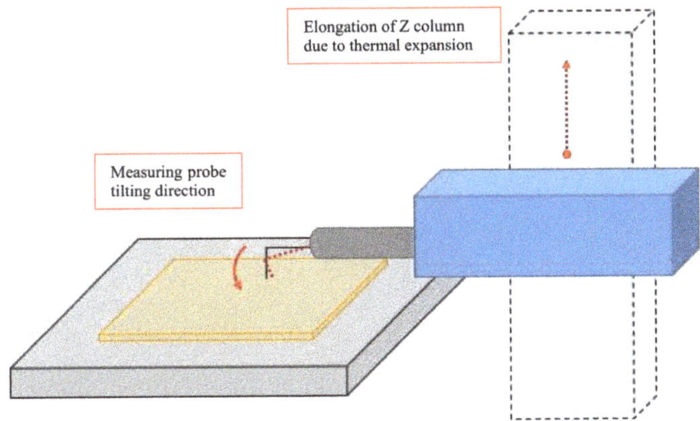

Figure 15. Probe movement due to heating of profilometer column.

The thermal conduction mechanism, concerning heat coming from power supplies or moving elements mounted inside a profilometer, influences the surface topography representation, particularly when changes in temperature take place within a column (Z direction). This results in displacement of the whole drive unit in relation to the base. Thus, the change in geometrical dimension due to the thermal expansibility of the column (elongation and shrinkage) results in the movement of a measuring tip being in contact with a measured surface. This movement is an error, which changes the representation of a surface and gives wrong values of topography parameters, a reason of workpiece malfunctioning and improper classification of manufactured parts.

The data presented above show the correlation of the temperature increase of the Z-axis drive (and the profilometer column) with the value of the Z-axis elongation. The maximum displacement of the measuring point placed on the measuring probe in the Z-axis reaches 8.3 µm in relation to the initial position, while the temperature of the Z column increased by 5.58 °C. These data correspond to the readings directly from the probe, presented in the form of a transverse profile. The maximum value of the probe displacement was 7.9 µm.

In the second test of the dynamic experiment, the profilometer performed the same profile measurement after thermal stabilization of its drives and structure. This test was intended to indicate differences in the fidelity of surface imaging when the influence of internal heat sources of the profilometer was significantly reduced.

The illustration (Figure 16) shows the surface generated from the measurements of the same profile after thermal stabilization of the profilometer.

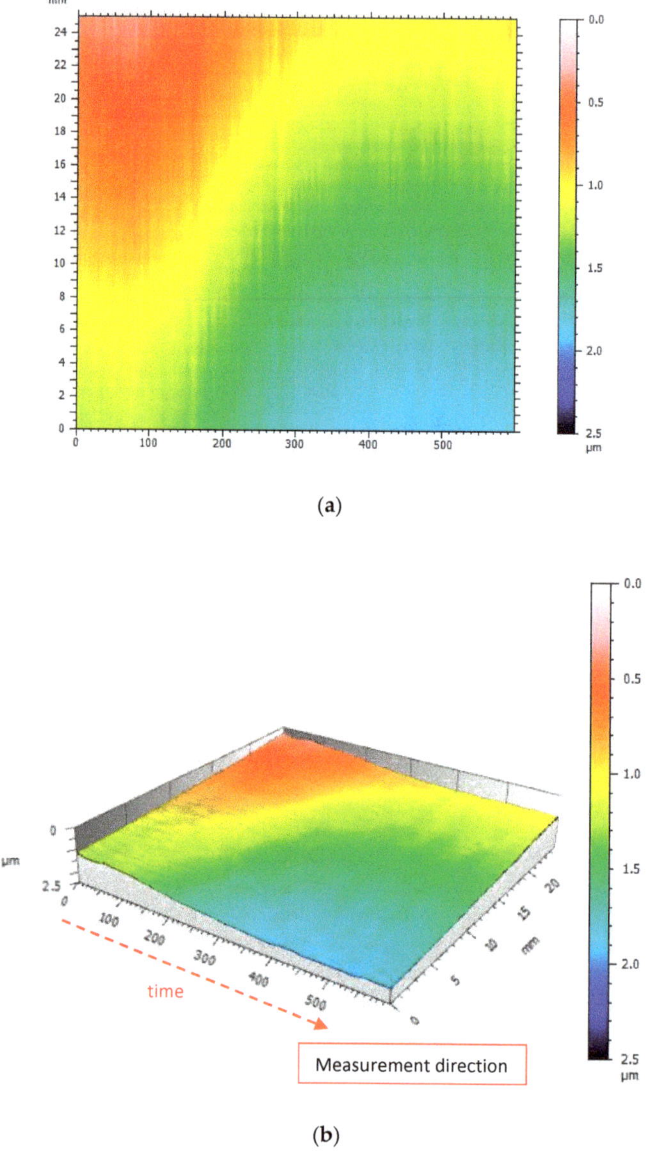

(a)

(b)

Figure 16. View of the surface generated from measurements of the same profile performed after thermal stabilization of the profilometer: (**a**) A color map; (**b**) 3D view.

The generated surface presented in the illustration (Figure 16) shows 90% less noise than in the case of an unstable profilometer. The signal amplitude illustrated in the figure is 0.8 µm, which is about 10 times less than in the case of a thermally unstable profilometer.

It is considered that regardless of the type of drives and electronic components used in a profilometer, if possible, they should be placed outside the profilometer housing so that the heat generated by them would be freely radiated to the environment. Motors should be separated from lead screws/linear guides to reduce heat transfer to essential elements of a measuring device, and connection with lead screws should be made with flexible toothed belts. This solution should protect the profilometer against vibrations coming from the drives while eliminating engine-lead screw clearances and, above all, reduce thermal conductivity between individual components.

4. Conclusions

After analyzing results of the research, a direct influence of thermal phenomena on surface imaging errors becomes clearly visible. These errors are especially evident during the initial phase of surface topography measurement, immediately after turning the device on.

There is a relation between the increase in temperature of drives and electronic components installed in contact profilometers (internal heat sources) and the deformation of the device structure in its individual axes. This directly affects the imaging of surface asperities, particularly when the measurement was started immediately after turning the profilometer on.

- The value of elongation in individual axes of the profilometer is different and it very much depends on the construction of the device, type of drives used, and their location. This might indicate that a change in design can limit the influence of thermal disturbances on the measurement results. Thus, it would improve the metrological characteristics of the device.
- The largest value of the displacement of the measuring tip occurs in the direction of the X-axis. This value (in the considered cases) reaches 16.2 µm.
- The largest impact on the imaging of the surface topography has the displacement of the probe in the direction of the Z-axis. This displacement directly translates into the obtained value of the height of the measured surface.
- The thermal and geometrical stabilization times should be precisely determined before beginning a 3D surface measurement. The stabilization time should be determined individually for a specific type of device in order to make a measurement correctly. Performing thermal stabilization of the tested device has reduced surface imaging errors by 90%.
- The comparison of analyzed constructions and drives of the contact profiler (based on Figures 4, 6 and 8) showed that DC motors working uniformly during the whole measurement are characterized by the best thermal properties. Change in feed should be executed by an electromagnetic clutch.
- Profilometers in which electronic systems and drives were located outside of the device body were characterized by lower values of displacement resulting from thermal deformation than the profilometer with drives inside its structure.

Author Contributions: K.G.—contributed to the study conception, experiment design, data analyses, and performed the experiments. M.W.—formal analysis and validation. P.P. and J.H.—review, editing, and contributed to the drafting of the manuscript. All authors have read and agreed to the published version of the manuscript and take responsibility for the integrity of the data and the accuracy of the data analysis.

Funding: This research was funded by Poznan University of Technology.

Conflicts of Interest: The authors declare no conflict of interest.

References

1. Trumpold, H. Process related assessment and supervision of surface textures. *Int. J. Mach. Tools Manuf.* **2001**, *41*, 1981–1993. [CrossRef]
2. Tonshoff, H.K.; Brinksmeier, E. Determination of the mechanical and thermal influences on machined surfaces by microhardness and residual stress analysis. *CIRP Annals* **1980**, *29*, 519–530. [CrossRef]
3. Mathia, T.; Pawlus, P.; Wieczorowski, M. Recent trends in surface metrology. *Wear* **2011**, *271*, 494–508. [CrossRef]
4. Pawlus, P.; Reizer, R.; Wieczorowski, M. Comparison of the results of surface texture measurement by stylus methods and optical methods. *Metrol. Meas. Syst.* **2018**, *3*, 589–602.
5. Brown, C.A.; Hansen, H.N.; Jiang, X.J.; Blateyron, F.; Berglund, J.; Senin, N.; Bartkowiak, T.; Dixon, B.; Le Goïc, G.; Quinsat, Y.; et al. Multiscale analyses and characterizations of surface topographies. *CIRP Ann.* **2018**, *67*, 839–862. [CrossRef]
6. Marteau, J.; Wieczorowski, M.; Xia, Y.; Bigerelle, M. Multiscale assessment of the accuracy of surface replication. *Surf. Topogr. Metrol. Prop.* **2014**, *2*, 44002. [CrossRef]
7. Elewa, I.; Koura, M.M. Importance of checking the stylus radius in the measurement of surface roughness. *Wear* **1986**, *109*, 401–410. [CrossRef]
8. Bodschwinna, H. Auswirkungen der Tastspitzengeometrie auf die industrielle Rauheitsmessung / Influences of the stylus tip geometry on industrial roughness measuring. *tm - Tech. Mess.* **1980**, *47*, 2–10. [CrossRef]
9. Trumpold, H.; Heldt, E. Influence of instrument parameters in the sub-micrometer range with stylus instruments. *Proc. X Coll. Surf.* **2000**, *1*, 106–121.
10. Smith, S.T.; Chetwynd, D.G. An Optimized Magnet-Coil Force Actuator and Its Application to Precision Elastic Mechanisms. *Proc. Inst. Mech. Eng. Part C: Mech. Eng. Sci.* **1990**, *204*, 243–253. [CrossRef]
11. Anbari, N.; Beck, C.; Trumpold, H. The Influence of Surface Roughness in Dependence of the Probe Ball Radius with Measuring the Actual Size *. *CIRP Ann.* **1990**, *39*, 577–580. [CrossRef]
12. Sherrington, I.; Smith, T. Performance assessment of stylus based areal roughness measurement systems. *Int. J. Mach. Tools Manuf.* **1992**, *32*, 219–226. [CrossRef]
13. O'Callaghan, P.W.; Babus'Haq, R.F.; Probert, S.D.; Evans, G.N. Three-dimensional surface-topography assessments using a stylus/computer system. *Int. J. Comp. Appl. Tech.* **1989**, *2*, 101–107.
14. McCool, J.I. Assessing the Effect of Stylus Tip Radius and Flight on Surface Topography Measurements. *J. Tribol.* **1984**, *106*, 202–209. [CrossRef]
15. Wieczorowski, M. Spiral sampling as a fast way of data acquisition in surface topography. *Int. J. Mach. Tools Manuf.* **2001**, *41*, 2017–2022. [CrossRef]
16. Senin, N.; Leach, R. Information-rich surface metrology. *Procedia CIRP* **2018**, *75*, 19–26. [CrossRef]
17. Pawlus, P.; Wieczorowski, M.; Mathia, T. *The Errors of Stylus Methods in Surface Topography Measurements*; ZAPOL: Szczecin, Poland, 2014.
18. *Optical Measurement of Surface Topography*; Leach, R. (Ed.) Springer: Berlin, Germany, 2011.
19. Pawlus, P.; Reizer, R.; Wieczorowski, M. Problem of Non-Measured Points in Surface Texture Measurements. *Metrol. Meas. Syst.* **2017**, *24*, 525–536. [CrossRef]
20. Pawlus, P.; Reizer, R.; Wieczorowski, M.; Żelasko, W. The effect of sampling interval on the predictions of an asperity contact model of two-process surfaces. *Bull. Pol. Acad. Sci. Tech. Sci.* **2017**, *65*, 391–398. [CrossRef]
21. Haitjema, H.; Morel MA, A. Noise bias removal in profile measurements. *Meas. J. Int. Meas. Confed.* **2005**, *38*, 21–29. [CrossRef]
22. Miller, T.; Adamczak, S.; Świderski, J.; Wieczorowski, M.; Łętocha, A.; Gapinski, B. Influence of temperature gradient on surface texture measurements with the use of profilometry. *Bull. Pol. Acad. Sci. Tech. Sci.* **2017**, *65*, 53–62. [CrossRef]
23. Krawiec, P.; Różański, L.; Czarnecka-Komorowska, D.; Warguła, Ł. Evaluation of the Thermal Stability and Surface Characteristics of Thermoplastic Polyurethane V-Belt. *Materials* **2020**, *13*, 1502. [CrossRef] [PubMed]
24. Zhou, Z.-F.; Hu, M.-Y.; Xin, H.; Chen, B.; Wang, G.-X. Experimental and theoretical studies on the droplet temperature behavior of R407C two-phase flashing spray. *Int. J. Heat Mass Transf.* **2019**, *136*, 664–673. [CrossRef]
25. Zha, J.; Zhang, H.; Li, Y.; Chen, Y. Geometric error measurement with high accuracy by ultra-precision CMM for closed hydrostatic guideways, European Society for Precision Engineering and Nanotechnology. In Proceedings of the 18th International Conference and Exhibition, EUSPEN 2018, Venice, Italy, 4–8 June 2018; pp. 137–138.

26. Ghidelli, M.; Sebastiani, M.; Collet, C.; Guillemet, R. Determination of the elastic moduli and residual stresses of freestanding Au-TiW bilayer thin films by nanoindentation. *Mater. Des.* **2016**, *106*, 436–445. [CrossRef]

27. Abdulshahed, A.M.; Longstaff, A.; Fletcher, S.; Myers, A. Thermal error modelling of machine tools based on ANFIS with fuzzy c-means clustering using a thermal imaging camera. *Appl. Math. Model.* **2015**, *39*, 1837–1852. [CrossRef]

28. Chenyang, Z.; Cheng, C.; Hang, Z. Study on nonlinear thermal error of the measurement machine. In Proceedings of the IEEE 2011 10th International Conference on Electronic Measurement & Instruments; Institute of Electrical and Electronics Engineers (IEEE), Chengdu, China, 16–19 August 2011; Volume 2, pp. 161–165.

29. Hao, S.; Liu, J.; Song, B.; Hao, M.; Zheng, W.; Tang, Z. Research on the Thermal Error of the 3D-Coordinate Measuring Machine Based on the Finite Element Method. *Computer Vision* **2008**, *5315*, 440–448. [CrossRef]

30. Muniz, P.; Magalhães, R.D.S.; Cani, S.P.N.; Donadel, C.B. Non-contact measurement of angle of view between the inspected surface and the thermal imager. *Infrared Phys. Technol.* **2015**, *72*, 77–83. [CrossRef]

31. Schwenke, H.; Knapp, W.; Haitjema, H.; Weckenmann, A.; Schmitt, R.; Delbressine, F. Geometrical Error Measurement and Compensation of Machines – An Update. *Ann. CIRP* **2008**, *57*, 660–675. [CrossRef]

32. Ma, C.; Liu, J.; Wang, S. Thermal error compensation of linear axis with fixed-fixed installation. *Int. J. Mech. Sci.* **2020**, *175*, 105531. [CrossRef]

33. Sładek, J.; Kupiec, R.; Gąska, A.; Kmita, A. Modeling the uncertainty changes caused by temperature with use of the Monte Carlo method. *Meas. Autom. Monit.* **2010**, *56*, 10–12.

34. Balsamo, A.; Pedone, P.; Ricci, E.; Verdi, M. Low-cost interferometric compensation of geometrical errors. *CIRP Ann.* **2009**, *58*, 459–462. [CrossRef]

35. Stejskal, T.; Kelemenová, T.; Dovica, M.; Demeč, P.; Štofa, M. Information Contents of a Signal at Repeated Positioning Measurements of the Coordinate Measuring Machine (CMM) by Laser Interferometer. *Meas. Sci. Rev.* **2016**, *16*, 273–279. [CrossRef]

36. Miao, E.; Liu, Y.; Liu, H.; Gao, Z.; Li, W. Study on the effects of changes in temperature-sensitive points on thermal error compensation model for CNC machine tool. *Int. J. Mach. Tools Manuf.* **2015**, *97*, 50–59. [CrossRef]

37. Lo, C.-H.; Yuan, J.; Ni, J. An application of real-time error compensation on a turning center. *Int. J. Mach. Tools Manuf.* **1995**, *35*, 1669–1682. [CrossRef]

38. Miao, E.-M.; Gong, Y.-Y.; Niu, P.-C.; Ji, C.-Z.; Chen, H.-D. Robustness of thermal error compensation modeling models of CNC machine tools. *Int. J. Adv. Manuf. Technol.* **2013**, *69*, 2593–2603. [CrossRef]

39. Baird, K.M. Compensation for Linear Thermal Expansion. *Metrol.* **1968**, *4*, 145–146. [CrossRef]

40. Kruth, J.; Vanherck, P.; Bergh, C.V.D. Compensation of Static and Transient Thermal Errors on CMMs. *CIRP Ann.* **2001**, *50*, 377–380. [CrossRef]

41. Sartori, S.; Zhang, G. Geometric Error Measurement and Compensation of Machines. *CIRP Ann.* **1995**, *44*, 599–609. [CrossRef]

42. Ge, Z.; Ding, X. Thermal error control method based on thermal deformation balance principle for the precision parts of machine tools. *Int. J. Adv. Manuf. Technol.* **2018**, *97*, 1253–1268. [CrossRef]

43. Tang, Y.; Xu, S.; Wang, Z. Prediction on the influence of ambient temperature and humidity to measuring instrument of thermal conductivity based on BP neural network. In Proceedings of the International Conference on Robotics, Intelligent Control and Artificial Intelligence, Shangai, China, 20–22 September 2019; pp. 562–567.

44. Milov, A.; Tynchenko, V.S.; Petrenko, V.E. Algorithmic and Software to Identify Errors in Measuring Equipment During the Formation of Permanent Joints. In Proceedings of the 2018 International Multi-Conference on Industrial Engineering and Modern Technologies (FarEastCon); Institute of Electrical and Electronics Engineers (IEEE), Vladivostok, Russia, 2–4 October 2018.

45. Hemming, B.; Esala, V.-P.; Laukkanen, P.; Rantanen, A.; Viitala, R.; Widmaier, T.; Kuosmanen, P.; Lassila, A. Interferometric step gauge for CMM verification. *Meas. Sci. and Technol.* **2018**, *29*, 7. [CrossRef]

46. Echerfaoui, Y.; El Ouafi, A.; Chebak, A. Experimental Investigation of Dynamic Errors in Coordinate Measuring Machines for High Speed Measurement. *Int. J. Precis. Eng. Manuf.* **2018**, *19*, 1115–1124. [CrossRef]

47. Schwenke, H.; Franke, M.; Hannaford, J.; Kunzmann, H. Error mapping of CMMs and machine tools by a single tracking interferometer. *CIRP Ann.* **2005**, *54*, 475–478. [CrossRef]
48. Jakubek, B.; Barczewski, R.; Rukat, W.; Różański, L.; Wrobel, M. Stabilization of vibro-thermal processes during post-production testing of rolling bearings. *Diagnostics* **2019**, *20*, 53–62. [CrossRef]

Article

Evaluation of Surface Roughness and Defect Formation after The Machining of Sintered Aluminum Alloy AlSi10Mg

Grzegorz Struzikiewicz [1,*] and **Andrzej Sioma** [2]

1   Production Engineering Institute, Mechanical Faculty, Cracow University of Technology, 31-155 Kraków, Poland
2   Department of Process Control, Faculty of Mechanical Engineering and Robotics, AGH University of Science and Technology, 30-059 Kraków, Poland; andrzej.sioma@agh.edu.pl
*   Correspondence: struzikiewicz@mech.pk.edu.pl

Received: 17 February 2020; Accepted: 31 March 2020; Published: 3 April 2020

Abstract: This article presents selected issues related to the workpiece surface quality after machining by the laser sintering of AlSi10MG alloy powder. The surfaces of the workpiece were prepared and machined by longitudinal turning with tools made of sintered carbides. The occurrence of breaches on the machined material surface was found, which negatively influence the values of 3D surface roughness parameters. The occurring phenomena were analyzed and proposals for their explanation were made. Guidelines for the machining of workpieces achieved by the laser sintering of powders were developed. The lowest value of the 3D roughness parameters was obtained for f = 0.06 mm/rev, a_p = 0.5–1.0 mm, and for the nose radius of cutting insert r_ε = 0.8 mm. The results of research on the effect of cutting parameters on the values of parameters describing the surface quality are presented. Topography measurements and 3D surface roughness parameters are presented, as well as the results of a microscopic 3D surface analysis. Taguchi's method was used in the research methodology.

Keywords: machining; sintered aluminum; 3D surface roughness parameters; surface defects

1. Introduction

The aim of currently observed directions and development trends visible in manufacturing techniques is to meet the requirements for dimension and shape accuracy and surface quality. One of the solutions used is a hybrid machining that combines subtractive manufacturing with additive machining. This approach is in accordance with the "All-in-One" and "Done-in-One" manufacturing philosophy used in multi-purpose numerically controlled machine tools produces by, e.g., DMG and Mazak. In this case, the hybrid manufacturing consists of the printing (additive manufacturing) of a workpiece or its piece, and then carrying out the final machining in the form of the burr removal (subtractive machining) of selected surfaces in order to ensure the required accuracy and quality level. The combination of subtractive machining with AM (Additive Manufacturing) is also beneficial in terms of manufacturing costs of machine parts [1]. This mainly concerns the manufacturing of large-size workpieces with thin-walled elements.

Both technologies have a number of advantages and disadvantages. In the case of AM, the advantage is the capability to achieve the workpiece of "any shape", while the disadvantage is still a lower surface quality and dimension and shape accuracy compared to workpieces made by the subtractive machining [2–4]. In addition, DMLS/SLM (Direct Metal Laser Sintering/Selective Laser Melting) technology allows for the manufacturing of complex and minute spatial structures, which cannot be achieved with casting and subtractive methods. Since the vast majority of manufactured workpieces is made of metal, the industry interest focuses on AM technology in terms of the production of fully functional metal workpieces.

Methods of additive manufacturing machine parts using metal alloy powders are innovative and still under research. They demonstrate a number of new problems requiring scientific and technological development. The most commonly used 3D printing technologies are FDM (Fused Deposition Modeling), SLS (Selective Laser Sintering), SLA (Stereolithergaphy), and others [1]. SLM technology, on the other hand, is a selective laser sintering and remelting of powdered metals, which are applied layer by layer until a fully durable workpiece is ready [5]. The aim of currently conducted research is to replace the casting technology with SLM and DMLS [2].

Metal materials obtained with the additive technology feature porosity and areas of varied material cohesion. AlSi10Mg has a good strength, corrosion resistance, low density, and high thermal conductivity compared with other alloys and is of tenfound in aerospace and automotive interior AM components, and in functional prototypes [2,6,7].

The surface quality obtained by DMLS/SLM is similar to that of castings by the lost model, with a surface roughness of Ra = 4–20 μm, depending on the alloy used and layer thickness. The characteristic property of workpieces manufactured with AM technology is the layered structure of the material. DMLS/SLM technologies most often use the ytterbium fiber laser operating in the infrared band (formerly weaker CO_2 lasers were used). The workpieces are manufactured by applying thin layers of metal powder (0.01–0.08 mm thickness). The process of material application usually includes the levelling out of the remelting roughness from the previously applied layer [2,8]. Recent publications on the additive manufacturing have mainly concerned the issues of surface metrology performed after additive manufacturing. As shown by Townsend et al. [9], in industrial conditions, the most commonly used parameters for surface quality assessment are 2D parameters describing the surface roughness (e.g., Ra, Rz etc.). As the material obtained by the additive method is porous and has a layered structure, the evaluation of the surface based on the above-mentioned parameters may not be insufficient [5,10]. Triantaphyllou et al. [11] presented the comprehensive literature analysis on surface texture metrology for metal additive manufacturing. They have shown that texture characterization is mainly based on the measurements of surface profiles and 2D parameters, of which Ra is the most commonly used.

However, the three-dimensional topography of a surface analysis is becoming more and more popular. In turn, Gao et al. [12] showed that there are still neither rules nor guidelines for additive manufacturing and indicated the need to standardize it. Based on the analysis of the surface measurement methods of workpieces made with SLM, Diatlov et al. [13] presented the concept of a roughness spectrum as an alternative to roughness value Ra. With reference to SLM, Rao et al. [10] demonstrated that, using optimal laser parameters, this technology can be implemented to manufacture workpieces made of A357 aluminum alloy, with the density and mechanical properties of the cast alloy standard and of a low porosity. However, different laser parameters also caused different melt pool size and morphology after SLM. Similar studies were conducted by Calignano et al. [14], who investigated the influence of process parameters, such as scan speed, laser power, and hatching distance (the perpendicular distance between successive laser scan lines), on the surface finish of direct metal laser sintered AlSi10Mg surfaces. Analyses of these issues available in the literature mainly concern the surface characteristics of workpieces obtained with additive technologies [1] rather than the subtractive machining. The authors of the published analyses focus mainly on the search for relationships between adjustable parameters of additive technology and parameters describing the surface quality, especially surface roughness. For example, Read et al. [7] described the influence of process parameters on the porosity of SLM-manufactured workpieces. He demonstrated that such surfaces have cracks and show the presence of significant amounts of unmelted powder, which results in the growth of material cracks. In turn, Strano et al. [15] analyzed the workpiece made of 316L steel alloy and carried out surface roughness tests. The analysis demonstrated an increasing density of spare particles positioned along the step edges, as the surface sloping angle increases. When the layer thickness is comparable to the diameter of the particle, the particles attached along the step edges can fill the gaps between the successive layers, thus affecting the actual surface roughness. Mumtaz and Hopkinson [16] carried out similar analyses, but for SLM of Inconel 625 alloy. They analyzed the processing parameters that simultaneously influence the roughness of the top and side surfaces of the manufactured workpiece. This demonstrated that higher peak powers tended to reduce top surface roughness and reduce side roughness as recoil pressures flatten out the melt pool and reduce balling formation by increasing the wettability of the melt. The increased repetition rate and reduced scan speed reduced top surface roughness but increased side roughness. Moreover, the authors (Grimm et al. [17]) analyze the correlation between the surface orientation of SLM parameters and *Sdr* (developed interfacial area ratio). In turn, [16] found that *Sa* and *Sq* were suitable measurement parameters for SLM.

The surface condition depends on the method and conditions of the workpiece material manufacturing (casting, forging, additive methods) and the methods and conditions of machining. Guo et al. [18] presents a theoretical and experimental investigation on the ultra-precision machining of V-groove structures on rapidly solidified aluminum RSA-905 using single-crystal diamond tools. The authors analyzed chip flow and material removal phenomena and effects of depth of cut and feed rate on cutting forces, surface quality, and form accuracy. The results showed that 15 nm Ra surface roughness were obtained on V-groove surface under the best machining condition. Guo et al. [19] carried out similar analyses on the surface integrity of rapidly solidified aluminum by magnetic field-assisted finishing. The effect of abrasive and polishing speed conditions on material removal and surface roughness was investigated. The results show that a low surface roughness was obtained under conditions using the SiC abrasive with a grit size of 12 μm at a polishing speed of 400 rpm or using the Al2O3 abrasive with a grit size of 5 μm at a polishing speed of 800 rpm. Therefore, there is a need to analyze the application of cutting the metal workpiece made by additive technology, and the impact of the subtractive machining (i.e., machining parameters such as feed, cutting speed, and cutting depth) on the surface quality of thus produced workpieces [3].

The authors of [20] compared the effect of aluminum alloy's cutting parameters in cast and sintered forms on values of the cutting force components and dimensional and shape accuracy. They showed that the burrs created during the passage of the cutting tool influence the dimensional and shape accuracy. Matras [21], on the other hand, analyzed the milling process of sintered aluminum and optimization of *Ra* and *Rz* surface roughness parameters. However, most of the research results described in the literature concern the machining of cast materials. Cutting forces in aluminum alloys are usually low and result from their lower mechanical strength [2]. The surface roughness during the subtractive machining of aluminum is significantly influenced by the hardness of the alloy and microstructural properties [22]. While machining alloys of a higher hardness, the values of the parameters describing surface roughness usually decrease [2,22], as hardness limits the adhesion of the material to the cutting edge of the tool. However, BUE (Build Up Edge) formation and the random extraction of hard particles from the material may occur. The high chemical affinity of aluminum alloys to cutting tool coating materials, such as TiC or Al_2O_3, causes the machined material to accumulate on the tool surface. This leads to deterioration of the material surface roughness due to the continuous adhesion of particles to the workpiece surface [2,22]. On the basis of the relationships obtained from the research, mathematical models are also developed to determine the values of selected surface topography parameters (Jayaraman et al. [23] Pawlus et al. [24] or Singh et al. [25]).

An important issue, also concerning aluminum alloys, is the optimization of the cutting process due to chip form and the choice of cutting parameters. The relationship between the chip form and cutting conditions was presented by Słodki et al. [4]. The authors presented investigations related to the effectiveness of selected chip breakers working in the local machining environment. Recommendations for cutting condition correction for the purpose of achieving an acceptable chip form were presented.

Currently, there are no dedicated procedures to optimize the cutting of workpieces produced by additive manufacturing. The authors attempted to establish a procedure for finding optimal cutting parameters for the finishing machining of laser sintered AlSi10Mg alloy, taking into account the criterion of the machined surface quality described by 3D surface roughness parameters. The analysis of the longitudinal turning of the cylinder made of sintered aluminum AlSi10Mg and the roughness, topography, and microscopic measurements of the machined surface were carried out.

2. Materials and Methods

In order to carry out the tests, a sample was prepared for testing by the selective sintering and remelting of powdered aluminum with a laser. The part was obtained using Renishaw's (Wotton-under-Edge, New Mills, UK) AM 250 with additive technology by selective laser sintering of AlSi10Mg aluminum powder. The properties of AlSi10Mg aluminum powder are presented in Table 1. The mechanical properties of the material and its chemical composition are presented in Tables 2 and 3. This alloy is used for large castings of complex shape and high strength, heavily and medium loaded, among others, in gearbox housings, steering gear housings, and blocks of internal combustion engines in motor vehicles.

Table 1. Properties of the powder AlSi10Mg.

Technical Data	
Recommended minimum layer thickness	30 μm
Accuracy for small items	+/− 20–50 μm
Accuracy for large items	+/− 0.2%

Table 2. Mechanical properties of the material AlSi10Mg.

Material	Tensile Strength Rm (MPa)	Elongation A5 (%)	Brinnel Hardness HB	Density (Sintered Part) (g/mm^3)
AlSi10Mg	193	2.5	68	0.064

Table 3. Chemical composition of aluminum AlSi10Mg (%).

Si	Fe	Cu	Mn	Mg	Cr	Ni	Zn	Ti	Be	Ca
9.738	0.312	0.011	0.436	0.202	0.0043	<0.0020	0.0096	0.0041	<0.0003	0.006

Cd	Co	Ga	Na	Pb	Sn	Sr	V	Zr	Al
<0.0005	<0.001	<0.002	0.001	<0.005	<0.005	<0.001	<0.002	<0.002	89.26

The following measurements were taken: surface roughness and topography and the microscopic measurements of the machined surface. The measurements were captured with Talysurf 50 surface profiler manufactured by Taylor Hobson (2 New Star Rd, Leicester, UK). The microscopic analysis of the machined surface was carried out using a VK-X1000 3D microscope by Keyence (Osaka, Japan) (Figure 1) with a resolution of 0.5 nanometers in the Z axis and 130 nanometers in the XY axis. The imaging field was 705 microns in the X axis and 528 microns in the Y axis.

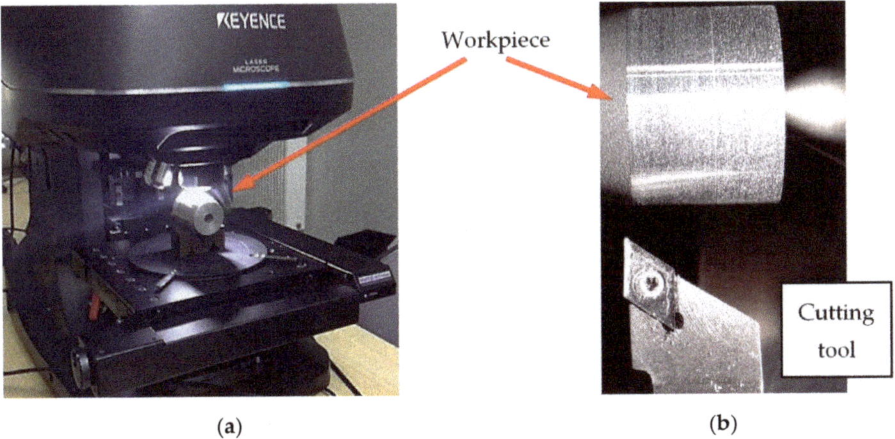

(**a**) (**b**)

Figure 1. Measuring station (**a**) for surface imaging with the Keyence VK-X1000 microscope and (**b**) the experimental station.

The analysis of the influence of cutting parameters on the surface roughness and dimension and shape accuracy of machined parts is often carried out on the basis of various methods, such as Taguchi [26,27] or its modifications [23,28]. The authors of the paper [3] presented the analysis of various optimization techniques used in manufacturing processes. The experimental research plan was developed according to the Taguchi method. The influence of variable cutting parameters, i.e., the feed rate, the speed and depth of cutting (f, v_c, a_p), and the nose radius of the cutting insert r_ε, on the values of the 3D surface roughness parameters was analyzed. In the statistical analysis of the test results, the model of the matching function according to Equation (1) was adopted.

$$Y_1 = y - \varepsilon = b_0 x_0 + b_1 x_1 + b_2 x_2 + b_3 x_3 + b_4 x_4, \tag{1}$$

where:
  —Y_1 is the estimated response based on first order equation;
  —y is the measured parameter (e.g., roughness parameter) on a logarithmic scale;
  —$x_0 = 1$ dummy variable;
  —x_1–x_4 are the logarithmic transformations of cutting speed and the feed and depth of cut;
  —ε is the experimental error;
  —b values are the estimates of corresponding parameters.

The S/N (signal-to-noise) ratio analysis strategy was adopted as "the lowest-the best" according to Equation (2):

$$\frac{S}{N} = -10 \cdot log\left(\frac{1}{n}\sum_{i=1}^{n} y_i^2\right) \tag{2}$$

where y_i is the respective characteristic and n is the number of observations.

In the cutting tests, DCGT 11T304-AS ($r_\varepsilon = 0.2$ mm) and DCGT 11T308-AS ($r_\varepsilon = 0.8$ mm) cutting inserts of type IC20 by ISCAR (Tefen, Israel) were used. The adopted ranges of the cutting parameter values are: $f = 0.06$; 0.12; 0.17; 0.25 mm/rev, $a_p = 0.5$; 1.0 mm and $v_c = 200$; 300 m/min. The values of the cutting parameters are within the range of cutting parameters recommended by the tool manufacturer.

Table 4 presents the test plan together with the actual values of the cutting parameters used in research.

Table 4. Research plan with real values.

No.	A	B	C	D	f (mm/rev)	v_c (m/min)	a_p (mm)	r_ε (mm)
1	1	1	1	1	0.06	200	0.5	0.2
2	1	2	2	2	0.06	300	1.0	0.8
3	2	1	1	2	0.12	200	0.5	0.8
4	2	2	2	1	0.12	300	1.0	0.2
5	3	1	2	1	0.17	200	1.0	0.2
6	3	2	1	2	0.17	300	0.5	0.8
7	4	1	2	2	0.25	200	1.0	0.8
8	4	2	1	1	0.25	300	0.5	0.2

3. Results and Discussion

In accordance with the adopted test plan, tests were carried out on the cutting (i.e., longitudinal turning) of the workpiece made by laser sintering. The microscopic observations and measurements of the selected 3D parameters of the surface roughness were performed afterwards.

The analysis of the microscopic measurement results show numerous breaches occurring after the machining of the surface of the workpiece made with additive technology. Figure 2 presents example microscopic images of the aluminum alloy surface after machining. The surface roughness was measured for each layout of the adopted test plan. The example measurement results are shown in Figure 3.

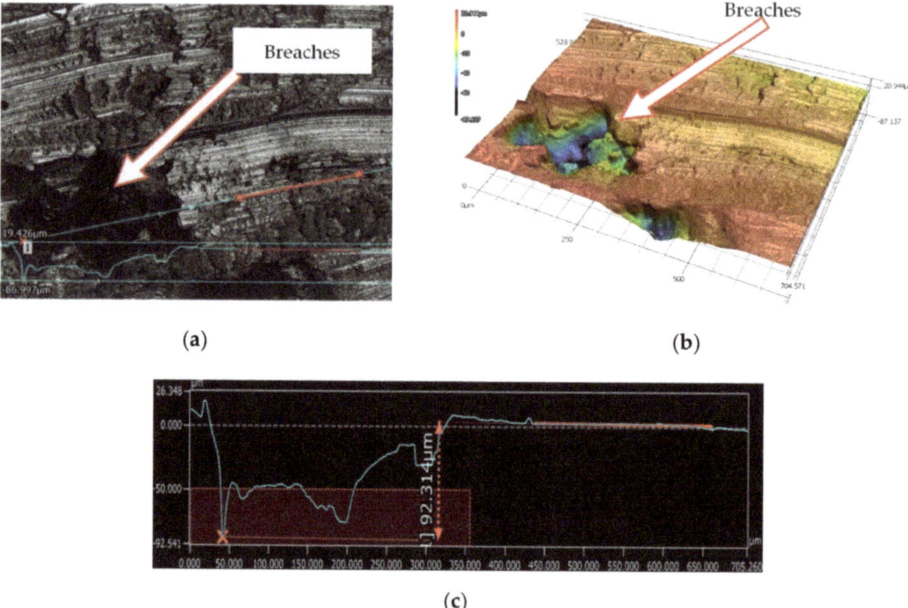

(a)   (b)

(c)

Figure 2. Example of surface measurement with a 3D microscope: (**a**) 2D surface view, (**b**) turned surface imaging (3D presentation), and (**c**) material breach measurement on a machined surface.

Surface	Profiles	Ra (μm)	Rz (μm)	Sa (μm)	Sz (μm)
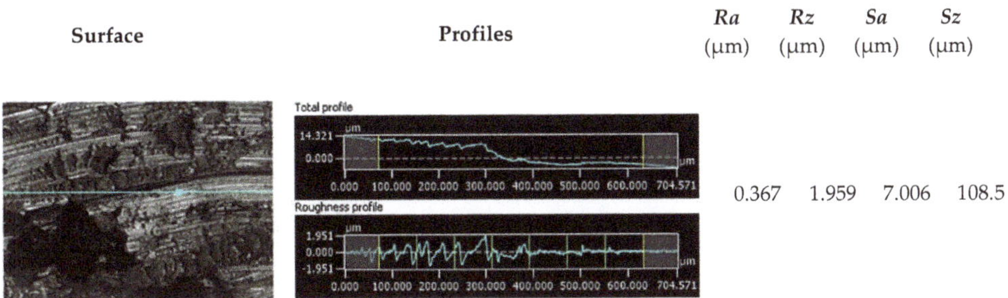		0.367	1.959	7.006	108.5

Figure 3. Example of measurement of 2 and 3D parameters of surface roughness with the Keyence microscope.

Figure 4 presents the selected topographies of the machined surface obtained from profilographical measurements. Table 5 presents the results of the measurements of the 3D parameters of surface roughness Sv (the maximum height of the surface pit), Sz (the maximum height of the surface), and Sa (the arithmetic mean height of the surface).

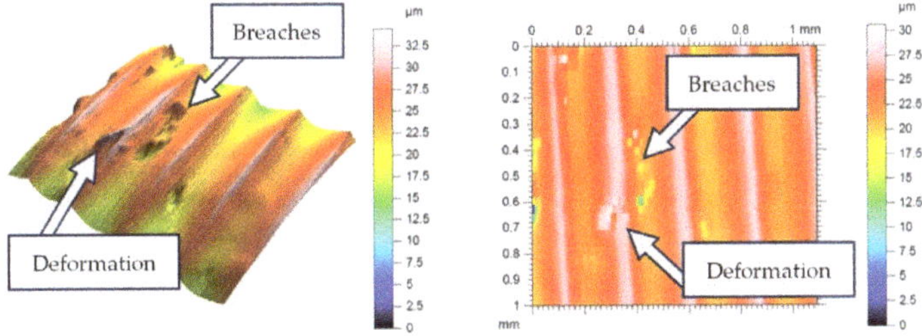

Figure 4. Examples of surface topography: f = 0.25mm/rev, ap = 0.5mm, v_c = 300m/min, r_ε = 0.8mm.

Figure 5 graphically shows the influence of the particular cutting data on the values of the 3D surface parameters Sv, Sz, and Sa. Figure 6 shows the surface roughness parameters Sv and Sz depending on the feed variables f and cutting depth a_p.

Table 5. Test results for the 3D roughness parameter measurements Sv, Sz, and Sa (μm).

No.	A	B	C	D	f (mm/rev)	v_c (mm/min)	a_p (mm)	r_ε (mm)	S/N Sv	$Sv_{\_mean}$ (μm)	S/N Sz	$Sz_{\_mean}$ (μm)	S/N Sa	$Sa_{\_mean}$ (μm)
1.	1	1	1	1	0.06	200	0.5	0.2	−18.8	8.73	−22.5	13.25	2.6	0.74
2.	1	2	2	2	0.06	300	1.0	0.8	−14.5	5.31	−26.1	20.15	0.8	0.91
3.	2	1	1	2	0.12	200	0.5	0.8	−23.4	14.71	−25.8	19.51	2.1	0.78
4.	2	2	2	1	0.12	300	1.0	0.2	−17.6	7.60	−26.6	21.39	−9.5	2.97
5.	3	1	2	1	0.17	200	1.0	0.2	−20.3	10.32	−31.1	35.67	−15.7	6.08
6.	3	2	1	2	0.17	300	0.5	0.8	−25.0	17.84	−20.4	10.38	−2.7	1.36
7.	4	1	2	2	0.25	200	1.0	0.8	−27.4	23.45	−29.7	30.43	−6.3	2.05
8.	4	2	1	1	0.25	300	0.5	0.2	−35.1	56.79	−40.7	107.98	−22.9	13.98

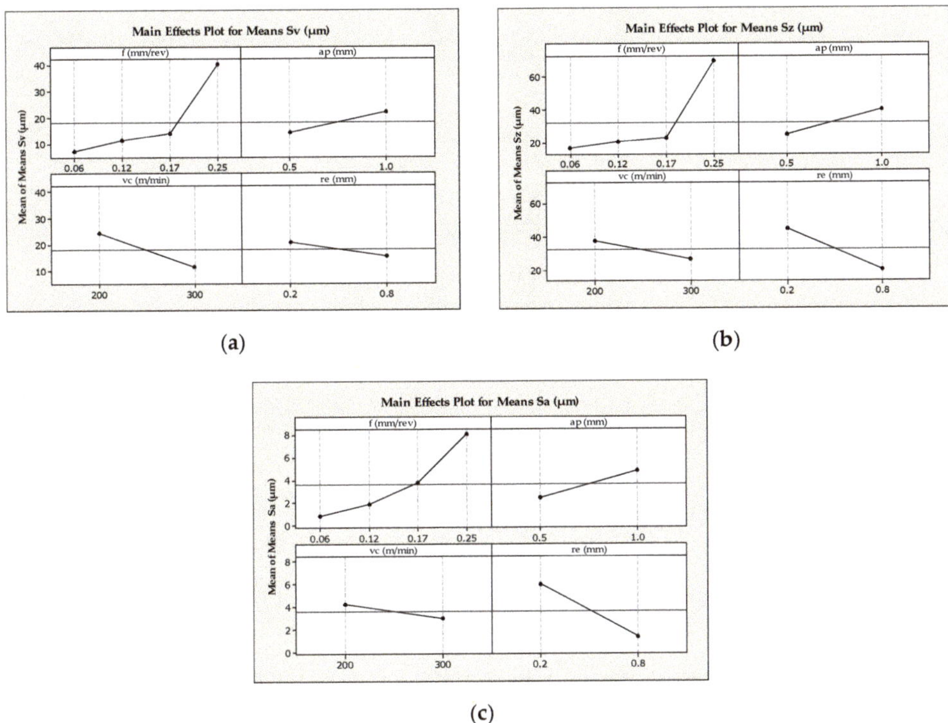

(a)

(b)

(c)

Figure 5. Influence of the cutting data on the values of the 3D surface parameters (**a**) *Sv* (μm), (**b**) *Sz* (μm), (**c**) *Sa* (μm).

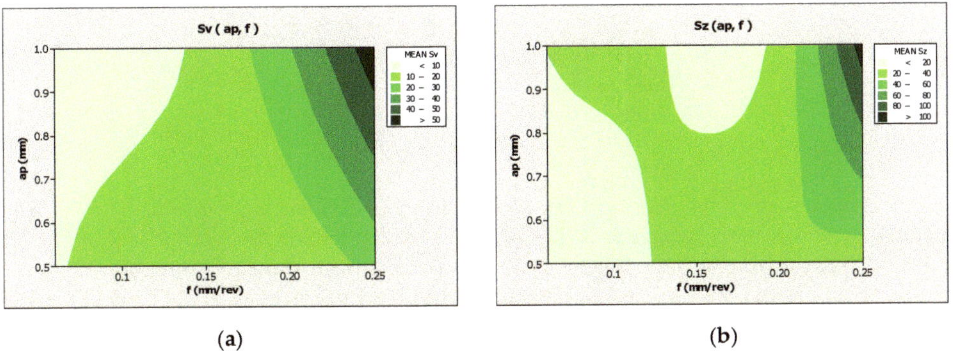

(a)

(b)

Figure 6. Three-dimensional surface parameters depending on the variable feed rate and depth of cut. (**a**) Parameter $Sv(a_p, f)$ (μm), and (**b**) $Sz(a_p, f)$ (μm).

Tables 6–8 show the ANOVA regression analysis results of the components for the *Sv*, *Sz*, and *Sa* parameters (where: *DF*—degrees of freedom, *Seq SS*—sums of squares, *Adj SS*—adjusted sums of squares, and *Adj MS*—adjusted means squares).

Table 6. Analysis of variance for means *Sv*.

Source	DF	Seq SS	Adj SS	Adj MS	F	P
A	3	1343.6	1343.6	447.9	4.11	0.344
B	1	115.0	115.0	115.0	1.06	0.491
C	1	329.9	329.9	329.9	3.03	0.332
D	1	61.3	61.3	61.3	0.56	0.590
Residual Error	1	108.9	108.9	108.9		
Total	7	1958.7				

Table 7. Analysis of variance for means *Sz*.

Source	DF	Seq SS	Adj SS	Adj MS	F	P
A	3	3662.9	3662.9	1221.0	0.84	0.645
B	1	465.8	465.8	465.8	0.32	0.672
C	1	236.2	236.2	236.2	0.16	0.756
D	1	1195.9	1195.9	1195.9	0.82	0.531
Residual Error	1	1454.4	1454.4	1454.4		
Total	7	7015.2				

Table 8. Analysis of variance for means *Sa*.

Source	DF	Seq SS	Adj SS	Adj MS	F	P
A	3	60.4	60.4	20.1	0.75	0.668
B	1	11.5	11.5	11.5	0.43	0.631
C	1	2.9	2.9	2.9	0.11	0.796
D	1	43.5	43.5	43.5	1.63	0.423
Residual Error	1	26.8	26.8	26.8		
Total	7	145.1				

Equation Sv (f, a_p, v_c, r_ε), Sz (f, a_p, v_c, r_ε) and Sa (f, a_p, v_c, r_ε) are described below (3–5).

$$Sv\left(f, a_p, v_c, r_\varepsilon\right) = 17.7 + 171f + 15.2a_p - 0.128v_c - 9.2r_\varepsilon, \tag{3}$$

$$Sz\left(f, a_p, v_c, r_\varepsilon\right) = 16.3 + 271f + 30.5a_p - 0.109v_c - 40.8r_\varepsilon, \tag{4}$$

$$Sa\left(f, a_p, v_c, r_\varepsilon\right) = 1.18 + 38.4f + 4.79a_p - 0.0121v_c - 7.77r_\varepsilon, \tag{5}$$

The analysis shows that the most important parameter influencing the values of 3D surface roughness parameters *Sv*, *Sz*, and *Sa* is the feed rate f and cutting depth a_p (Figures 5 and 6). The lowest value of the roughness pit height *Sv* was obtained for feed rate f = 0.06 mm/rev and depth a_p = 1.0 mm. In turn, the feed rate f = 0.17 mm/rev and the cutting depth a_p = 0.5mm at the same nose radius of cutting insert r_ε = 0.8 mm results in the lowest value of the roughness height *Sz* = 10.38 µm. In addition, the analysis of the results shows that the value of the cutting speed v_c and the nose radius of the cutting insert r_ε have an inversely proportional influence on the 3D values of the parameters characterizing the surface roughness. Additional microscopic analyses showed a number of deformations and breaches on machined surface. The largest number of breaches were observed inside the traces (grooves) caused by cutting tool edge passage. On the other hand, numerous deformations and burrs were observed on the tops of the machining tracks appearing on the machined surface after the passage of the cutting tool. The analysis of the results shows that the number and geometric dimensions of breaches, i.e., the breach width and depth, depend on the value of cutting parameters. Surface deformations and breaches expostulate into values of 3D parameters (i.e., *Sv* and *Sz*) describing the surface roughness. The structure and properties of the subsequent layers of laser sintered material depend on the conditions of metal powder remelting, and are characterized by porosity, as mentioned by, e.g., Kempen et al. [29] and Olakanmi et al. [30] and also defects (cracks), as mentioned by Read et al. [7]. This can lead to smaller forces between the particles of the combined material and can cause the material breach out of the machined surface by the cutting tool. The dynamics of the chip-forming process and material flow direction along cutting edges in turning could affect stress distributions, as mentioned by Guo et al. [18]. In addition, the torn-out material particles can be stretched and stuck on the surface machined by the cutting tool, which deteriorates the surface roughness parameters.

4. Conclusions

Based on the results obtained and the analyses carried out, the following conclusions, regarding the development of breaches formed during the turning of AlSi10MG aluminum parts made by the DMLS method, can be drawn:

(1) The values of 3D parameters describing surface roughness, i.e., Sv, Sz, and Sa, contain more useful information on the surface quality than 2D parameters (e.g., Ra, Rz).

(2) After lathe machining, there are numerous material breaches on the surface of sintered aluminum. The distribution of breaches and burrs is uneven. The number and dimensions of breaches, as well as material deformations and burrs on the machined surface influence the dimension and shape accuracy and performance properties of the workpieces. The depth and size of the breaches are determined by the feed rate of the cutting edge and the cutting depth. The lowest value of the 3D roughness parameters was obtained for $f = 0.06$ mm/rev, $a_p = 1.0$ mm, and for the nose radius of the cutting insert $r_\varepsilon = 0.8$ mm. Increasing the cutting speed value v_c causes a decrease in the 3D value of the parameters Sv, Sz, and Sa characterizing the surface roughness. The cause of breaches and deformations on the machined surface is probably the structure of the surface layer of the sintered aluminum, and the method and conditions of combining material particles during the laser sintering process. It is likely that there are areas with weaker material particle joints that were produced by melting and subsequently by combining metal powder particles. In the absence of the full melting of the material particles during laser sintering, the cohesive forces of the particles are smaller than those of the cast material, resulting in the easier breaching (removal) of particles and plastic strain. With an increased tool feed rate in the decohesion zone, conditions are created that promote the breaching of machined material particles. An additional factor may also be the fact that there are empty spaces (pores) in sintered materials.

Author Contributions: Conceptualization, G.S. and A.S.; methodology, G.S. and A.S.; formal analysis, G.S. and A.S.; investigation, G.S. and A.S.; resources, G.S.; writing—original draft preparation, G.S.; writing—review and editing, G.S. and A.S.; visualization, G.S. and A.S.; project administration, G.S.; funding acquisition, G.S. and A.S. All authors have read and agreed to the published version of the manuscript.

Funding: This research received no external funding.

Acknowledgments: The article is a part of scientific-research cooperation between Cracow University of Technology and AGH University of Science and Technology in Cracow.

Conflicts of Interest: The authors declare no conflict of interest.

References

1. Moreau, C. *The State of 3D Printin*; Sculpteo: San Francisco, CA, USA, 2016.
2. Santos, M.C.; Machado, A.R.; Sales, W.F.; Barrozo, M.; Ezugwu, E.O. Machining of aluminium: A review. *Int. J. Adv. Manuf. Technol.* **2016**, *86*, 3067–3080. [CrossRef]
3. Rao, R.V.; Kalyanakar, V.D. Optimization of modern machining processes using advanced optimization techniques: A review. *Int. J. Adv. Manuf. Technol.* **2014**, *73*, 1159–1188. [CrossRef]
4. Słodki, B.; Zębala, W.; Struzikiewicz, G. Correlation Between Cutting Data Selection and Chip Form in Stainless Steel Turning. *Mach. Sci. Technol.* **2015**, *19*, 217–235. [CrossRef]
5. Herzog, D.; Seyda, V.; Wycisk, E.; Emmelmann, K. Additive manufacturing of metals. *Acta Mater.* **2016**, *117*, 371–392. [CrossRef]

6. Zębala, W.; Kowalczyk, R.; Matras, A. Analysis and Optimization of Sintered Carbides Turning with PCD Tools. *Procedia Eng.* **2015**, *100*, 283–290. [CrossRef]

7. Read, N.; Wang, W.; Essa, K.; Attallah, M.M. Selective laser melting of AlSi10Mg alloy: Process optimisation and mechanical properties development. *Mater. Des.* **2015**, *65*, 417–424. [CrossRef]

8. Zhang, P.; Liu, Z. Modeling and prediction for 3D surface topography in finish turning with conventional and wiper inserts. *Measurement* **2016**, *94*, 37–45. [CrossRef]

9. Townsend, A.; Senin, N.; Blunt, L.; Leach, R.K.; Taylor, J.S. Surface texture metrology for metal additive manufacturing: A review. *Precision Eng.* **2016**, *46*, 34–47. [CrossRef]

10. Rao, H.; Giet, S.; Yang, K.; Wub, X.; Davies, C. The influence of processing parameters on aluminium alloy A357 manufactured by Selective Laser Melting. *Mater. Des.* **2016**, *109*, 334–346. [CrossRef]

11. Triantaphyllou, A.; Giusca, C.L.; Macaulay, G.D.; Roerig, F.; Hoebel, M.; Leach, R.K. Surface texture measurement for additive manufacturing. *Surf. Topogr. Metrol. Prop.* **2015**, *3*, 024002. [CrossRef]

12. Gao, W.; Zhang, Y.; Ramanujan, D.; Ramania, K.; Chen, Y.; Williams, C.B.; Wang, C.C.L.; Shin, Y.C.; Zhang, S.; Zavattieri, P.D. The status, challenges, and future of additive manufacturing in engineering, Comput. *Aided Des.* **2015**, *69*, 65–89. [CrossRef]

13. Diatlov, A.; Buchbinder, D.; Meiners, W.; Wissenbach, K.; Bültmann, J. Towards surface topography: Quantification of Selective Laser Melting (SLM) built parts. In *Innovative Developments in Virtual and Physical Prototyping*; London CRC Press Taylor and Francis Group: Leiria, Portugal, 2012; pp. 595–602. ISBN 978-0-415-68418-7.

14. Calignano, F.; Manfredi, D.; Ambrosio, E.P.; Iuliano, L.; Fino, P. Influence of process parameters on surface roughness of aluminum parts produced by DMLS. *Int. J. Adv. Manuf. Technol.* **2013**, *67*, 2743–2751. [CrossRef]

15. Strano, G.; Hao, L.; Everson, R.M.; Evans, K.E. Surface roughness analysis, modelling and prediction in selective laser melting. *J. Mater. Process. Technol.* **2013**, *213*, 589–597. [CrossRef]

16. Mumtaz, K.; Hopkinson, N. Top surface and side roughness of Inconel 625 parts processed using selective laser melting. *Rapid Prototyp. J.* **2009**, *15*, 96–103. [CrossRef]

17. Grimm, T.; Wior, G.; Witt, G. Characterization of typical surface effect in additive manufacturing with confocal microscopy. *Surf. Topogr. Metrol. Prop.* **2015**, *3*, 014001. [CrossRef]

18. Guo, J.; Zhang, J.; Wang, H.; Liu, K.; Kumar, A.S. Surface quality characterisation of diamond cut V-groove structures made of rapidly solidified aluminium RSA-905. *Precis. Eng.* **2018**, *53*, 120–133. [CrossRef]

19. Guo, J.; Wang, H.; Goh, M.H.; Liu, K. Investigation on Surface Integrity of Rapidly Solidified Aluminum RSA 905 by Magnetic Field-Assisted Finishing. *Micromachines* **2018**, *9*, 146. [CrossRef]

20. Struzikiewicz, G.; Zębala, W.; Słodki, B. Cutting parameters selection for sintered alloy AlSi10Mg longitudinal turning. *Measurement* **2019**, *138*, 39–53. [CrossRef]

21. Matras, A. Research and optimization of surface roughness in milling of SLM semi-finished parts manufactured by using the different laser scanning speed. *Materials* **2020**, *13*, 9. [CrossRef]

22. Singh, A.; Agrawal, A. Investigation of surface residual stress distribution in deformation machining process for aluminum alloy. *J. Mater. Process. Technol.* **2015**, *225*, 195–202. [CrossRef]

23. Jayaraman, P.; Kumar, L.M. Multi-response optimization of machining parameters of turning AA6063 T6 aluminium alloy using grey relational analysis in Taguchi method. *Procedia Eng.* **2014**, *97*, 197–204. [CrossRef]

24. Pawlus, P.; Reizer, R.; Wieczorowski, M. Reverse Problem in Surface Texture Analysis-One-Process Profile Modeling on the Basis of Measured Two-Process Profile after Machining or Wear. *Materials* **2019**, *12*, 4169. [CrossRef] [PubMed]

25. Singh, S.; Prakash, C.; Antil, P.; Singh, R.; Królczyk, G.; Pruncu, C.I. Dimensionless Analysis for Investigating the Quality Characteristics of Aluminium Matrix Composites Prepared through Fused Deposition Modelling Assisted Investment Casting. *Materials* **2019**, *12*, 1907. [CrossRef] [PubMed]

26. Asilturk, I.; Neseli, S.; Ince, M.A. Optimization of parameters affecting surface roughness of Co28Cr6Mo medical material during CNC lathe machining by using the Taguchi and RSM methods. *Measurement* **2016**, *78*, 120–128. [CrossRef]

27. Selvaraj, D.P.; Chandramohan, P.; Mohanraj, M. Optimization of surface roughness, cutting force and tool wear of nitrogen alloyed duplex stainless steel in a dry turning process using Taguchi method. *Measurement* **2014**, *49*, 205–215. [CrossRef]

28. Yadav, R.N. A hybrid approach of Taguchi-Response Surface Methodology for modeling and optimization of Duplex Turning process. *Measurement* **2017**, *100*, 131–138. [CrossRef]

29. Kempen, K.; Thijs, L.; Van Humbeeck, J.; Kruth, J.P. Mechanical properties of AlSi10Mg produced by Selective Laser Melting. *Phys. Procedia* **2012**, *39*, 439–446. [CrossRef]
30. Olakanmi, E.O.; Cochrane, R.F.; Dalgarno, K.W. Densification mechanism and microstructural evolution in selective laser sintering of Al–12Si powders. *J. Mater. Process. Technol.* **2011**, *211*, 113–121. [CrossRef]

 *materials*

Article

Evaluation of the Thermal Stability and Surface Characteristics of Thermoplastic Polyurethane V-Belt

Piotr Krawiec [1,*], Leszek Różański [2], Dorota Czarnecka-Komorowska [3] and Łukasz Warguła [1]

[1]   Faculty of Mechanical Engineering, Institute of Machine Design, Poznan University of Technology, Piotrowo 3 Str., 61-138 Poznan, Poland; Lukasz.Wargula@put.poznan.pl

[2]   Faculty of Mechanical Engineering, Division of Metrology and Measurement Systems, Institute of Mechanical Technology, Poznan University of Technology, Piotrowo 3 Str., 61-138 Poznan, Poland; Leszek.Rozanski@put.poznan.pl

[3]   Faculty of Mechanical Engineering, Institute of Materials Technology, Poznan University of Technology, Piotrowo 3 Str., 61-138 Poznan, Poland; Dorota.Czarnecka-Komorowska@put.poznan.pl

*   Correspondence: piotr.krawiec@put.poznan.pl; Tel.: +48-616-652-242

Received: 29 January 2020; Accepted: 23 March 2020; Published: 25 March 2020

Abstract: This article proposes thermography as a non-contact diagnostic tool for assessing drive reliability. The application of this technique during the operation of the belt transmission with a heat-welded thermoplastic polyurethane V-belt was presented. The V-belt temperature changes depending on the braking torque load at different values of the rotational speed of the active pulley, which were adopted as diagnostic characteristics. In this paper, the surface morphology of the polyurethane (PU) belts was assessed on the basis of microscopic and hardness tests. A surface roughness tester was used to evaluate the surface wear. The surface morphology and topography of the materials was determined by scanning electron microscopy (SEM) and optical microscopy. It was found that the most favorable operating conditions occurred when the temperature values of active and passive connectors were similar and the temperature difference between them was small. The mechanical and structure results indicate that the wear of the PU belt was slight, which provided stability and operational reliability for V-belt transmission. The microscopic images lacked clear traces of cracks and scratches on the surface, which was confirmed by the SEM observations.

Keywords: thermoplastic polyurethane; heat-welded V-belt; IR thermography; hardness; surface roughness; SEM morphology; optical microscopy

---

1. Introduction

The application of belts in different types of cars, machines and device drives is known and well described in the literature in the field of mechanical engineering [1–4]. A wide group of belts with round and V-sections as well as toothed and flat belts can be made of thermoplastic polymers [1,5] The condition for manufacturing belts made of this type of materials is their easy processing by extrusion at a temperature of about 200 °C, followed by cooling to obtain the desired shape of the belt. The advantage of this solution is the ability to process belts multiple times. The introduction to the widespread use of new technical solutions requires the testing of the behavior of the belts under various conditions of use to obtain the required reliability and safety over the assumed life of the belt transmission [4]. Gao et al. [4] discussed that the time-dependent reliability models, failure rate models and availability models of V-belt drive systems are developed based on the system's dynamic equations with the dynamic stress and the material property degradation taken into account.

Heat-welded V-belts are made of thermoplastic materials, such as polyesters and polyurethanes, with a hardness from 85 to 100 Shore A [5–7]. They are used both in classic drive technology and in various types of conveyors. The advantages of heat-welded V-belts include the possibility of welding their ends, which allows a belt of any length to be obtained, and its quick replacement in case of damage, excellent resistance to abrasion, action of fats, dirt and some chemicals, resistance to temperature influence from $-30\ °C$ up to $+80\ °C$ and considerable elasticity at a relatively low level of stretching. In addition, they show a high value of the friction coefficient and thus, very good anti-slip properties, even with load changes, quiet operation and safe use in contact with food (confirmed by an FDA certificate) [8].

Thanks to advantages such as maintaining smooth motion, ability to mitigate load changes, ensuring vibration damping, operation without lubrication, ability to transfer motion when the shafts are not parallel, low sensitivity to shaft axis spacing errors, possibility of obtaining variable gear ratios (connector variators, non-parallel transmission) and relatively low production costs, V-belts and serpentine belts have found wide application in many industries [9]. Therefore, this subject has become the basis for conducting scientific research towards their improvement [10–14]. Merghache and et al. [10] conducted an experimental and numerical thermal study of a toothed belt transmission type AT10. Then, Silva and et al. [11] described the state-of-the-art research on power losses (belt-hysteresis losses, pulley-belt slip losses) modelling in a front engine accessory drive and focused on internal and external losses of the belt.

The current progress of V-belts concerns research into the development of belt constructions, the development of new manufacturing techniques, and the use of new modified materials, e.g., polymer composites [12]. Amanow and et al. [1] investigated the friction and wear behavior of the vertical spindle and V-belt to improve the operation and to extend the service life of a cotton picker. The vertical spindle made of low-carbon steel was treated by the ultrasonic nanocrystal surface modification (UNSM) technique to control the friction and wear behaviour [1]. Almedia and et al. [13] studied the characteristics of the different belt types, with a particular emphasis on their energy efficiency, cost-effectiveness and field of application. Shim and et al. [14] presented a new shape optimization procedure to improve the fatigue life of the pulley in automotive applications and the shape control concept was introduced to reduce the shape design variables. Yu et al. [2] investigated V-ribbed belt design, wear, and traction capacity.

However, there are no accurate data on aspects of the reliability and safety of this type of construction. Krawiec and et al. reported the source of incorrect operation of the belt transmission, indicating inaccuracies in the produce and assembly of machines and devices [15,16], not balance of elements in rotational motion [17,18], wear of elements [19], low resistance to grease and pollutions [20] and especially, changes in belt temperature [21–23].

Krawiec and Grzelka [15] proposed a measurement methodology for nontypical cog belt pulleys, conducted experimental investigations, and presented a set of directions for process engineers manufacturing these pulleys. Krawiec [17] analyzed the generation capabilities of noncircular cog belt pulleys on the example of a cog belt pulley with an elliptical pitch line. Moreover, they described of selected dynamic features of a two-wheeled transmission system [18]. In another paper, the study results of the wear of the multiple grooved pulleys with the ATOS II optical method were presented. The evaluation of wear was made based on the comparison of the manufactured parts or assemblies by superimposing the CAD model and the surface model obtained from digitization [21].

The elements of transmission belts heat up to a temperature of about $82\ °C$ (pulleys). This value is close to the gear operating temperature of $70\ °C$ (shafts and housings in place of roller bearings and sealants) [24]. On the other hand, machine parts can warm up to lower temperatures—e.g., rolling bearings up to about 50–$60\ °C$ [25] and plain bearings up to $37\ °C$ [26]—thus, the heating of the transmission may be unfavorable for other parts of the machine. Other cases may be machine parts that heat up to higher temperatures and may adversely affect the transmission belts, e.g., the electric motor winding temperature is from 95 to $124\ °C$ [27,28] and electric motor housing is about $90\ °C$ [29].

Regarding transmission belt operation with low-power internal combustion engines, these are not the most heated parts of the machines, because the temperature of the exhaust system exceeds

300 °C [30]. The temperature of other components of the exhaust system—e.g., catalysts—in units of higher power may exceed 600 °C [31] and the optimum oil operating temperature is between 85 and 105 °C [32]. The effects of process conditions on the heat condition and safety of machine parts have been evaluated in many papers [33,34]. For example, one paper showed that convective interaction at a temperature of 400 °C may be the cause of fires in machines shredding wood waste or the surrounding infrastructure [33]. Wargula and et al. reported that a few minutes of exposure to the temperature can cause the emission of flammable gases and flame or flameless combustion [33]. It is known that a convection temperature from 100 to 150 °C can cause damage to the cover of electric cables used in machines [34]. It follows that belt transmission should have the lowest possible operating temperature [31–34]. According to Hakami and et al. [35], the process of conveyor belt wearing and damaging is a restraining condition which has a crucial impact on the operational service life of a conveyor belt.

In this paper, infrared (IR) thermography was used to evaluate the thermal condition of the belt transmission. IR thermography [36] is a technique which is suitable and very useful for recording the dynamics of the manufacturing process and operation and safety processes. The IR technique has also been applied to other mechanical applications, both for the diagnosis of faults and as a complement to other fault detection techniques [36–38]. This technique can be used for the diagnosis of kinematic pairs in the rotary mowers [39] or for the analysis of the extrusion process stability of microporous polyvinyl chloride (PVC) [40] and as a diagnostic tool for the assessment of the accumulation of discontinuities of the structure of polyester–glass pipes [41]. In pipe tests, the diagnostic characteristics of the composite were obtained and expressed as a change in temperature during the heating and cooling rates determined on the basis of the temperature distribution on the external surface of the heat-activated pipe [41]. Thermovision is increasingly used to evaluate the operation of belt transmission drives [42,43].

Thermographic systems indirectly measure the temperature of the tested object and directly measure the power of infrared radiation emitted by this object [44–52]. This radiation is converted by the system's detection structure into an electric signal, which is a carrier of information about the object's temperature. The temperature is determined using a thermograph and the thermometric characteristics are saved in the computer memory, designed, among others, to control the process of reproducing temperature fields visualized by a thermograph camera. The thermometric characteristic (calibration) is a function describing the dependence of the measured signal on the object temperature. Under calibration conditions, the object's emissivity is precisely determined. Measurement conditions often differ significantly from calibration conditions. These differences cause the signal at the detector output to be different than the signal under calibration conditions, which increases the uncertainty of temperature measurement. Temperature measurements using thermal cameras are affected by thermovision measurement method errors (emissivity estimation error, error caused by the influence of the ambient radiation reflected by the object and the influence of the ambient radiation, error caused by limited atmosphere permeability and its radiation, error due to the influence of ambient radiation and calibration, error caused by the inability to average the measurement results) [53,54]. Each temperature measurement made with the use of radiation methods is associated with the need to enter the values of parameters describing the emission properties of the tested object into the system, while knowledge of the object's emissivity is required. Three cases of the impact of emissivity on the results obtained using radiative temperature measurement methods can be considered, referring to three classes of thermal systems, i.e., systems operating in the entire spectrum, to band systems and to monochrome systems.

The simulation studies presented in [55–57] show that the impact of the factors mentioned above on the uncertainty of temperature measurement is negligible in the case of measurements carried out in laboratory conditions from a distance of about 1 m with 8–14 μm spectral band cameras, when the emissivity coefficient of the test object is greater than 0.9 (the emissivity coefficient of polyurethane is 0.95). Each of these conditions was met during measurements of the temperature distribution on the V-belt. This means that it can be assumed that for such favorable measurement conditions, calibration errors and errors of the electronic system are dominant—the total value of which is expressed by the maximum permissible error of the thermograph of ±2 °C. Thermography systems are band systems. According to typical procedures, correcting the impact of emissivity on a measurement signal does not require to enter its spectral characteristics into the computer system, but only one effective emissivity value. The effective object emissivity is defined as the emissivity of a grey body at the same temperature of the tested object, for which the signal at the output of the measuring system is identical to the signal emitted by the tested object.

In this study, the effect of thermal condition during the operating process of belt transmission with a heat-welded V-belt was investigated by the IR thermography technique. The investigation concerns the surface morphology, hardness and surface roughness and topography of the polyurethane V-belts.

2. Materials and Methods

2.1. Materials

For reliability testing, in order to determine significant wear indicators, a belt transmission with a heat-welded V-belt was investigated. The transmission pulleys were made of cast iron (EN 1561 standard) and the belts were made of thermoplastic polyurethane (type PU 75A), produced by Behabelts GmbH (Glottertal, Germany) [8]. Due to the dark color and matt surface of the pulleys, no paint coating was required for testing related to temperature distribution registration.

The selected physical properties of polyurethane V-belt are listed in Table 1.

Table 1. Physical properties of a red smooth polyurethane (PU) belt [8].

Properties	Profile Dimension	Hardness	Coefficient Friction Steel Topside	Recommended Min. Pulley	Working Tension
PU 75A	17 × 11 mm	80 °ShA	0.7 μ	100 mm	19 dN/belt

2.2. Experimental Test Stand

The tests of the V-belt transmission were carried out at room temperature at 21.2 °C. The drive pulley (AR 01) was mounted on a bearing shaft connected by a flexible coupling to the motor shaft. The gear load was applied by a magnetorheological brake, which was constantly cooled by water. The assessment of the thermal state of the belt transmission was recorded with a thermal imaging camera (Figure 1a) with a rotational speed of the active pulley (AR 02) of 500 rpm, 1000 rpm and 1500 rpm. The tests were carried out for various values of transmission braking torque load. The test stand for diagnostic thermovision tests was carried out as illustrated in Figure 1b.

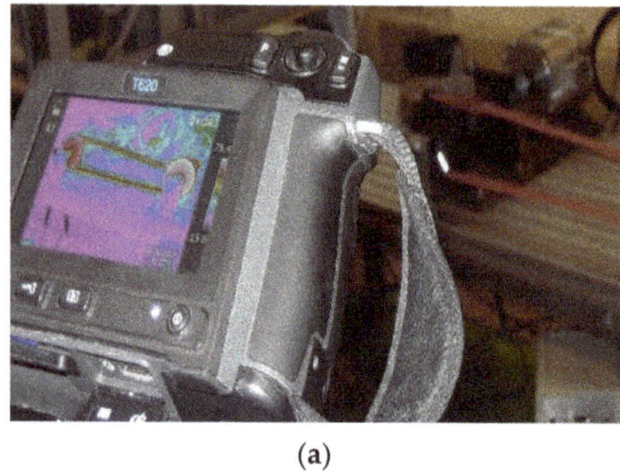
(a)

(b)

Figure 1. Thermovision set-up for testing heat concentration (**a**) and an experimental test stand (**b**) for the thermal condition of a belt transmission with a heat-welded V-belt.

2.3. Thermal Tests

Thermal tests of V-belt temperature distributions were carried out using a thermal imaging camera (see Figure 1a) type FLIR (Forward Looking InfraRed) T620 (FLIR Systems, Inc., Wilsonville, OR, USA), with a field of view (FOV) 25° × 19° lens, a spectral resolution understood as instantaneous field of view (IFOV) of 0.68 mrad and thermal resolution understood as noise equivalent temperature difference (NETD), determined for a temperature of 30 °C, ≤ from 0.05 °C. The maximum permissible error of this camera was ±2 °C or ±2% of the temperature reading—whichever was greater—at 25 °C.

2.4. Hardness Test

The hardness of polyurethane belts was also measured using a Shore hardness tester (Sauter HBD 100-0 GmbH (Balingen, Germany) according to the PN-EN ISO 868:2005. The hardness was indicative of an average penetration (Shore degrees on the A scale) value based on five readings from tests.

2.5. Optical Microscopy

The surface morphology of the belts was taken using a stereoscopic microscope type SK Opta-tech with an HDMI 6 Opta-tech RT16 Mpx camera (Warsaw, Poland), using 10× magnification. An optical polarized light microscope (Nikon Eclipse MA200, Kanagawa, Japan), equipped with the Nikon Imaging Software v.4.50 (NIS)—Elements Basic Research (BR) (Praha, Czech Republic) was used to study the topography surface of a polyurethane belt after wear.

2.6. Surface Roughness

The surface roughness model ART 3000 (Sunpoc CO., Guiyang, China) of polyurethane belts was measured using a surface roughness tester according to the PN-EN ISO 4287-1:1997.

2.7. Scanning Electron Microscopy (SEM)

A Tescan (model Mira 3 Tescan, Brno, Czech Republic) SEM at 23 °C, in a vacuum, and at a 15 kV accelerating voltage was used to examine the morphology of the PU belt surfaces used and obtained from the IR thermography test. Prior to the SEM analysis, the belts were coated with carbon powder. A magnification of 1000× was used.

3. Results and Discussions

3.1. Thermal Stability Analysis

As a result of the thermovision measurements, thermal images (thermograms) were obtained along with the surface temperature values of the examined objects and their distribution (profilograms). The profile in red for the active connector is marked in red and the passive connector in black. The obtained images were analyzed using the ThermaCAM Reporter 2000 Professional computer program. Figure 2a presents the temperature distribution (thermogram), and Figure 2b shows the profilograms recorded at a drive pulley speed of 500 rpm (15 min after starting the test stand) without external transmission braking torque load. The value of resistance to motion on the passive shaft measured at that time was 2.15 N·m. The direction of the rotation of the drive pulley (located on the right side of the transmission) was counterclockwise. During this time, using the camera, a temperature difference of 0.1 °C was recorded between the temperature in the active (lower) connector of 29.1 °C, and the temperature in the inactive (upper) connector of 29.2 °C. It was found that the obtained temperature difference $\Delta T = 0.1$ °C was insignificant.

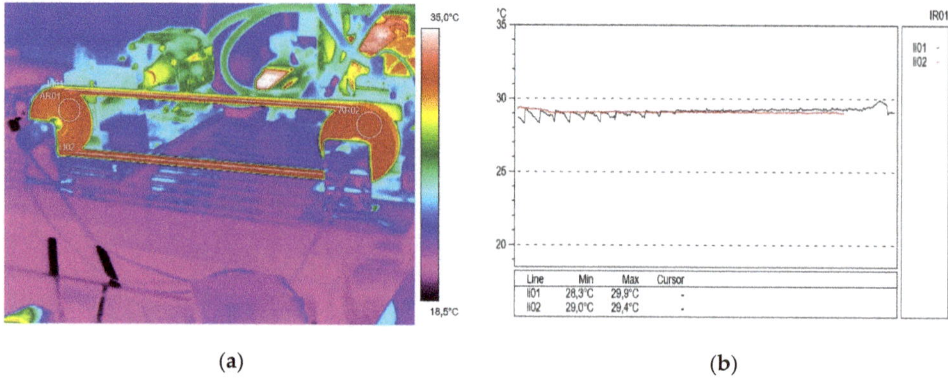

(a)　　　　　　　　　　　　　　　　　(b)

Figure 2. Thermogram (**a**) and profilogram (**b**): temperature distribution on the V-belt recorded at a drive pulley speed of 500 rpm (15 min after starting the test stand) and without external transmission braking torque load; red color—active connector and black color—passive connector.

Figure 3 shows the temperature distribution on the belt at a speed of 500 rpm and a braking torque load of 10 N m. In this case, an increase in the temperature was observed in the active connector (36.4 °C) in the bending phase and in the passive connector (36.9 °C) in the bending phase from the drive pulley. It should be noted that this was caused by a strong extension of the connector when the belt came into contact with the active pulley and immediately before leaving the cooperation, when the belt was compressed (Figure 3a).

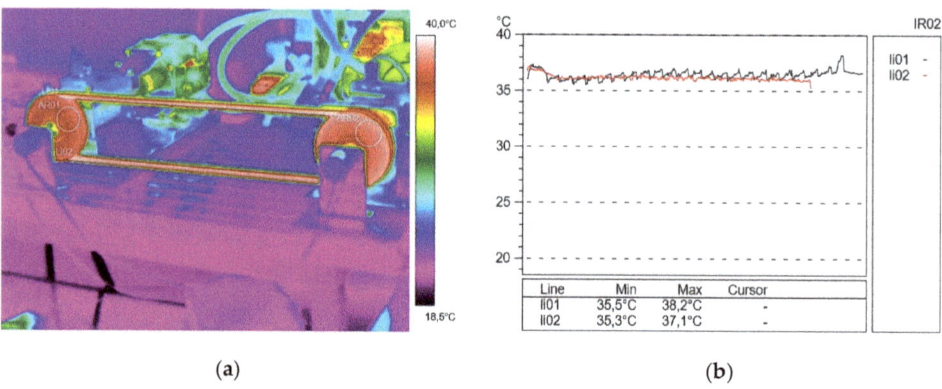

(a)　　　　　　　　　　　　　　　　　(b)

Figure 3. Thermogram (**a**) and profilogram (**b**): temperature distribution on the V-belt recorded at a rotational speed of 500 rpm and a braking torque load of 10 N·m, red color—active connector and black color—passive connector.

In the subsequent trials in the experiment, Figure 4 shows the temperature distribution on the V-belt and profilograms recorded at a rotational speed of 500 rpm and a braking torque load of 13 Nm. In this test, a significant temperature increase was noted both in the active (approximately 50.6 °C) and passive (50.8 °C) connector; the temperature difference was $\Delta T = 0.2$ °C. During the study, the formation of a so-called airbag occurred, which reduces the angle of wrap on the active pulley and the suction of the belt when it comes off the passive pulley, as a result of which the angle of pulley wrap increased. This phenomenon caused a change in the tension force in the active and passive connectors during gear operation and had an impact on its durability. In addition, significant vibrations in the active connector were visible. After these measurements, the transmission was relieved and blowers were used to lower the temperature.

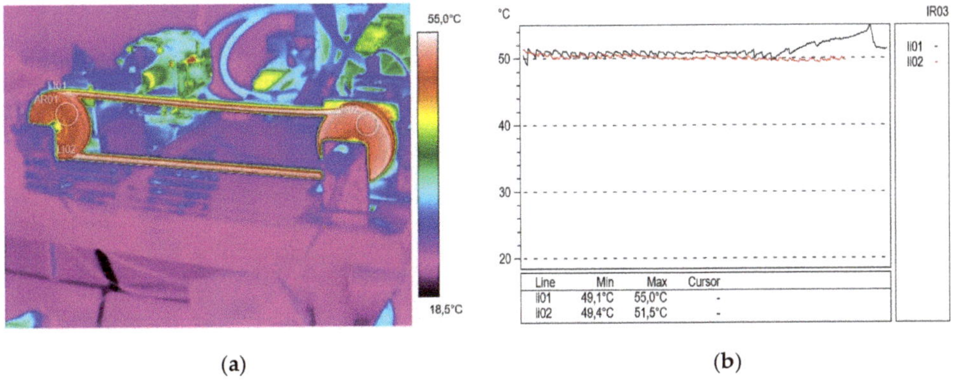

(a)   (b)

Figure 4. Thermogram (**a**) and profilogram (**b**): temperature distribution on the V-belt recorded at a rotational speed of 500 rpm and a braking torque load of 13 N·m (only transmission resistance), red color—active connector and black color—passive connector.

The next tests of the experiment were carried out at a rotational speed of 1000 rpm and a braking torque load of 2.15 N·m, the results of which are shown in Figure 5. On the basis of the registered PU belt temperature distributions and obtained profilograms, the phenomenon of temperature equalization in the active (approximately 38.2 °C) and passive (38.2 °C) connectors was observed, and the airbag phenomenon and vibration of the connector, recorded in the previous test, were not observed.

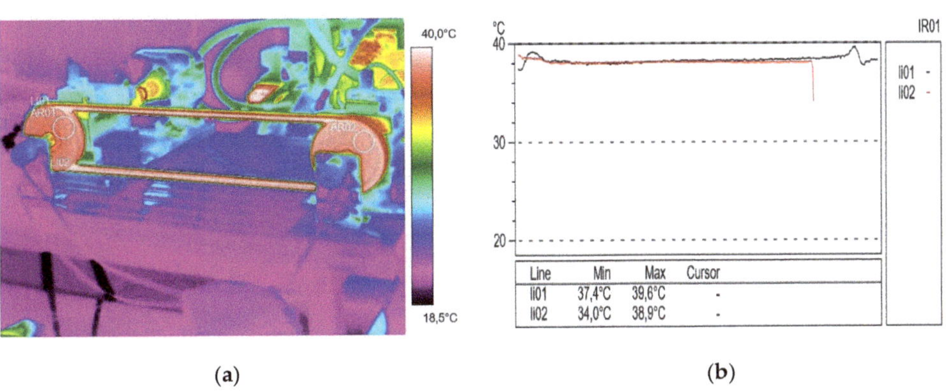

(a)   (b)

Figure 5. Thermogram (**a**) and profilogram (**b**): temperature distribution on the V-belt recorded at a rotational speed of 1000 rpm and a braking torque load of 2.15 N·m (only transmission resistance), red color—active connector and black color—passive connector.

On the other hand, Figure 6 shows the belt temperature distribution and profilograms recorded at an active shaft speed of 1000 rpm and a gear load of 10 Nm. In the studied case, in the given experiment conditions, an increase in the temperature difference was observed between the temperature values in active (52.1 °C) and passive (54.1 °C) connectors, corresponding to $\Delta T = 2$ °C.

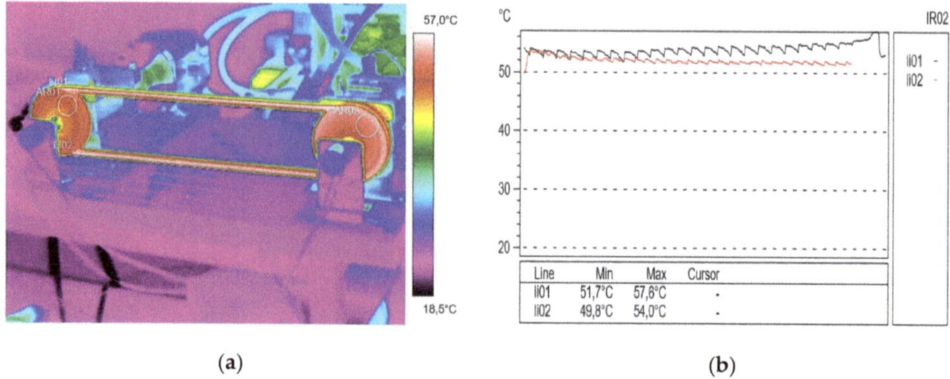

(a)   (b)

Figure 6. Thermogram (**a**) and profilogram (**b**): temperature distribution on the V-belt recorded at a rotational speed of 1000 rpm and a braking torque load of 10 N·m (only transmission resistance), red color—active connector and black color—passive connector.

Continuing the diagnostic measurements with the thermovision technique, tests were carried out at an active shaft speed of 1000 rpm and a braking torque load of 12.75 Nm, the thermal image and profilograms of which are illustrated in Figure 7. As a result of the presented experiment, the phenomenon of airbag formation and the occurrence of belt vibration was visible. According to Gao and et al. [58] the stiffness degradation of the belts has significant impacts on the reliability, failure rate, and lifetime distribution of belt drive systems. A significant difference was found between the temperature in active (55.7 °C) and passive (64.1 °C) connectors corresponding to $\Delta T = 9.4$ °C, and the obtained profilograms show the significant effect of the pulley temperature on the temperature of the polyurethane V-belt.

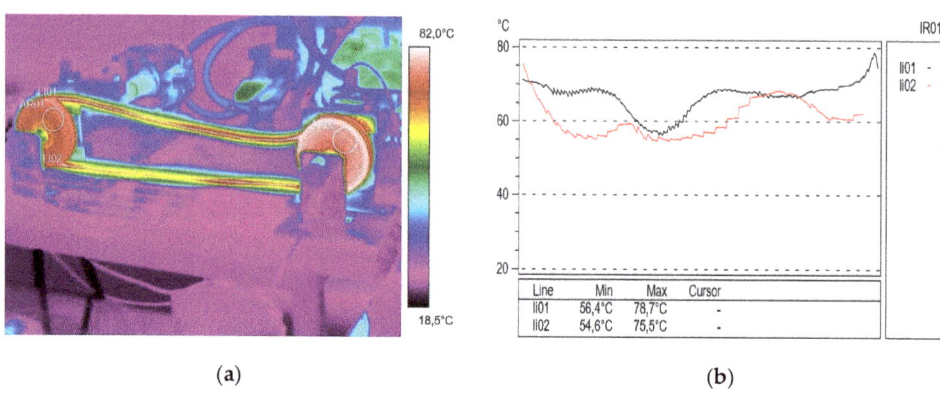

(a)   (b)

Figure 7. Thermogram (**a**) and profilograms (**b**): temperature distribution on the V-belt recorded at a rotational speed of 1000 rpm and a braking torque load of 12.75 N·m, red color—active connector and black color—passive connector.

Then, approximately 30 min after re-unloading the transmission and turning on the blowers, another thermogram was recorded (Figure 8a) at a rotational speed of the driving pulley of 1500 rpm and a transmission breaking torque load of 2.15 N·m. At that time, the difference between the temperature in active (30.4 °C) and passive (31.0 °C) connectors was small (approximately 0.6 °C). In the profilograms (Figure 8b), the stabilization of the V-belt temperature distribution is visible.

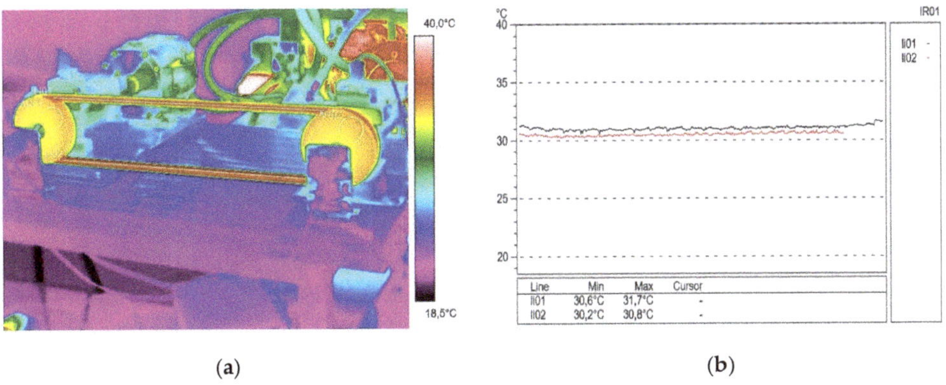

(a)   (b)

Figure 8. Thermogram (**a**) and profilogram (**b**): temperature distribution on the V-belt recorded at a rotational speed of 1500 rpm and a braking torque load of 2.15 N·m, red color—active connector and black color—passive connector.

It is worth noting that due to the aspects of operation analyzed, as well as in engineering practice, a thermographic image (Figure 9a) was observed during the experiment at a rotational speed of 1500 rpm and a braking torque of 8.20 N·m. The temperature distribution was measured at several V-belt measurement points and it was found that the passive connector temperature was lower than the active connector temperature. The difference was $\Delta T = 1.4\ °C$ (Figure 9b) between the temperature of active (44.1 °C) and passive (42.7 °C) connectors. This behavior of the transmission in the studied export conditions is caused by cooling down due to transverse V-belt vibrations.

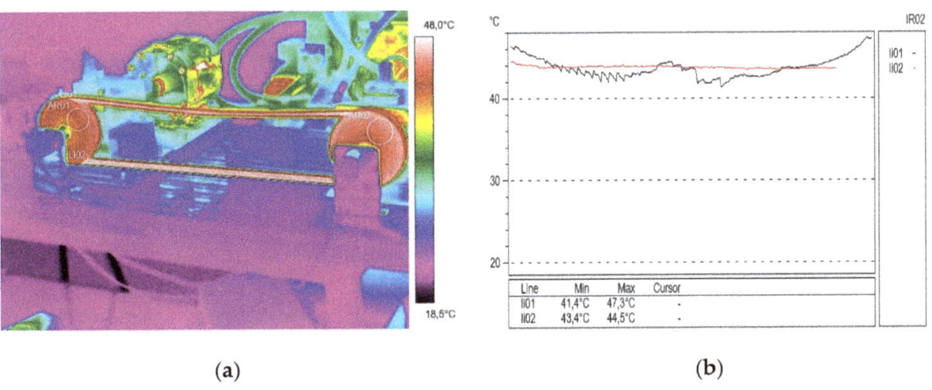

(a)   (b)

Figure 9. Thermogram (**a**) and profilogram (**b**): temperature distribution on the V-belt recorded at a rotational speed of 1500 rpm and a braking torque load of 2.15 N·m, red color—active connector and black color—passive connector.

Based on the obtained IR profilograms, an analysis of the temperature change of the belt in various experimental conditions was carried out. Figure 10 shows the dependence of the belt temperature change on the braking torque load at different rotational speeds for the passive and active pulleys. The obtained results of the belt heating temperature change as a function of rotational speed are described by a linear function with the correlation coefficient of the R function (Figure 10).

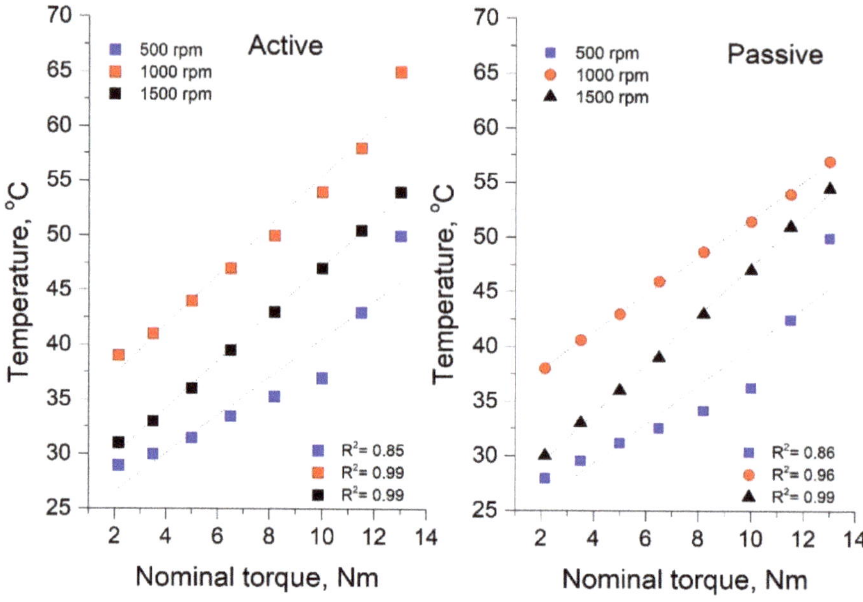

Figure 10. The dependence of the belt temperature change on the braking torque load at different rotational speeds for the active and passive pulleys.

Figure 10 shows that for a rotational speed of 500 rpm, the temperature difference between the connectors was 0.5 °C over the entire tested range; however, from a load value of 2.15 Nm to 9 Nm, the temperature change had a linear behavior, and above this value, it quickly increased to a value of approximately 51 °C. For the higher rotational speed of 1000 rpm, a temperature difference of 2 °C was recorded for a torque load of 10 N m. The results indicate that an increasing of the load caused an increase in the temperature difference to approximately 8 °C for a braking torque load of 13 Nm. For a rotational speed of 1500 rpm, both the value and the temperature difference were the smallest and assumed a stabilized value throughout the entire test ranging from 2.15–8.75 Nm. It is significant that there is a change in active and passive connector temperatures slightly above 4 Nm and from that moment, the passive connector has a slightly higher temperature than the active one. The difference is relatively small and for a load of 8.4 Nm, it is 1.4 °C. The reason for the differences is the introduction of the passive connector into the vibrations.

3.2. Hardness and Surface Roughness Analysis

The next stage of the study was to assess the hardness and roughness of the surface of new belts and after operation. The V-belts were given the cyclic loads described in Section 2.2. The belt loading cycles were repeated 50 times. The measurements were taken on both sides of the belt (designated as sides A and B) at 10 points. The obtained results were presented in conventional units, Shore degrees on the A scale (°ShA). The belt manufacturer states in his catalogues that the hardness of PU 75A belts is 80 °ShA [59]. The results of the hardness measurements for the original PU belt and used PU belt are shown in Figure 11a.

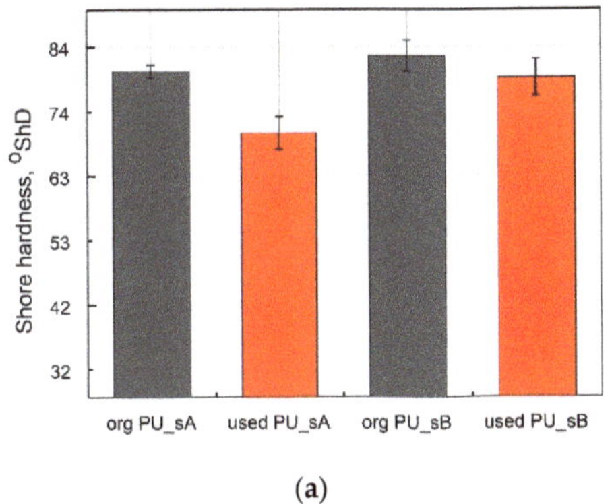
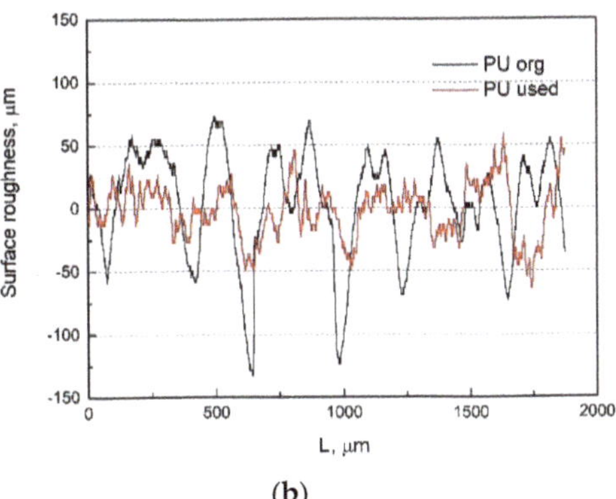

(a)   (b)

Figure 11. Shore hardness (**a**) and surface roughness profiles (R profiles) (**b**) of the polyurethane belt: original PU belt and used PU belt.

Based on the hardness tests (Figure 11a), it was found that the hardness of the belt after operating was 83 °ShA (side B), which indicates a slightly higher value than for the V-belt before operation. A significant change in hardness was observed on the A side. In this case, it was found that the hardness of the belt after operating cycles (used PU) was 70 °ShA, which means that it is lower by about 10% compared to the new polyurethane V-belt. The hardness distribution on the belt surface (A or B side) is not uniform, which indicates uneven belt wear under operating conditions. The decrease in surface hardness may be explained by the heat interaction generated during friction of the pulley–belt pair. Temperature changes diagnosed by the IR thermovision technique indicate low belt wear under given operating conditions. Slight changes in hardness on the belt surface confirm that polyurethane displays better abrasion and wear resistance than, for example, ultra-high-molecular-weight polyethylene (PE-UHMW) [60].

Figure 11b shows a typical roughness profile (R profile) for two types of polyurethane belt surfaces on side A: a new belt and used V-belt. The measured values of R_a, R_q and R_z are given in Table 2, where R_a is the arithmetic average height (µm), R_q is the root mean square roughness (µm) and R_z is the 10-point height (µm). R_a is the most widely known parameter and is the arithmetic mean deviation of the profile, but it does not describe contact surfaces completely since a totally different surface can have similar or even identical values of average surface roughness [1]. Therefore, R_z was provided to minimize the effect of the valleys which occasionally occur and can give an erroneous value [1]. The measurement results indicate that the R_a and R_z of the PU belt were reduced after wear from 0.349 µm to 0.179 µm, corresponding to about a 48% reduction, and from 1.576 to 0.216 µm, corresponding to a 86% reduction. The test results show significant changes in the PU V-belt roughness after the wear process as a result of the friction heat in belt drives. The reduction in surface roughness of a V-belt after wear may be attributed to the elimination of high peaks and valleys by the introduced plastic deformation, resulting in a reduction in the number of defects [1].

Table 2. Surface roughness measurements.

Profilometry Parameter	Original PU Belt	Used PU Belt
R_a (µm)	0.349 ± 0.15	0.179 ± 0.13
R_q (µm)	0.423 ± 0.12	0.894 ± 0.09
R_z (µm)	1.576 ± 0.11	0.216 ± 0.10

Additionally, the topography surface of a new polyurethane V-belt and a PU belt after wear are shown in Figure 12.

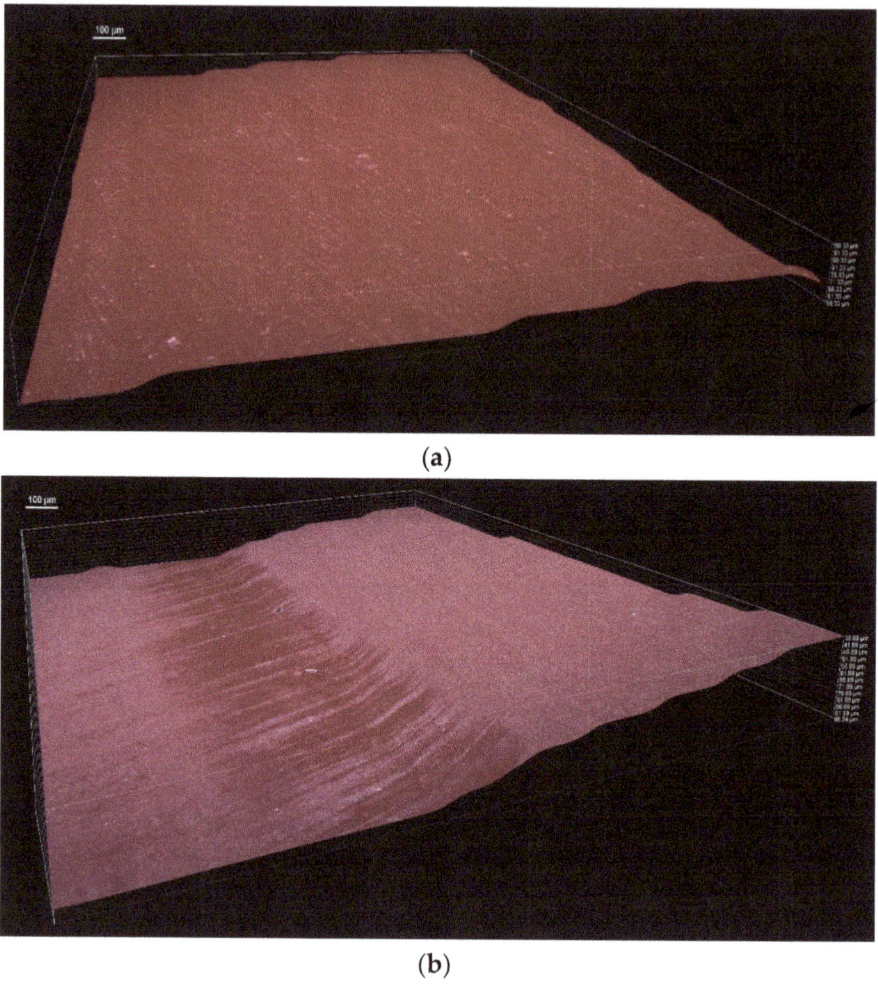

Figure 12. Comparison of 3D microscopic images of the polyurethane belt surface (**a**) with the used PU belt surface (**b**) (50× magnification).

Figure 12 presents the 3D images of surface profiles obtained with the Nikon Imaging Software (NIS)—Elements Basic Research (BR) microscopic images of the 3D surface system: the optical interferometer uses coherence scanning interferometry to create 3D surface maps for the investigation of the roughened surface topographies. The pictures show significant differences in the surface profile of the belt after operation and a new PU belt. The surface profile of the thermoplastic polyurethane belt is uneven and corrugated, probably due to heat from friction.

3.3. Optical and SEM Surface Morphology Analysis

In order to explain the mechanism of the abrasive process of the PU belt, a structural study of this surface before and after friction was carried out by means of a stereoscope and scanning electron microscope (SEM). The results of the optical microscopic analysis for a new V-belt and a used polyurethane (PU) belt are shown in Figure 13.

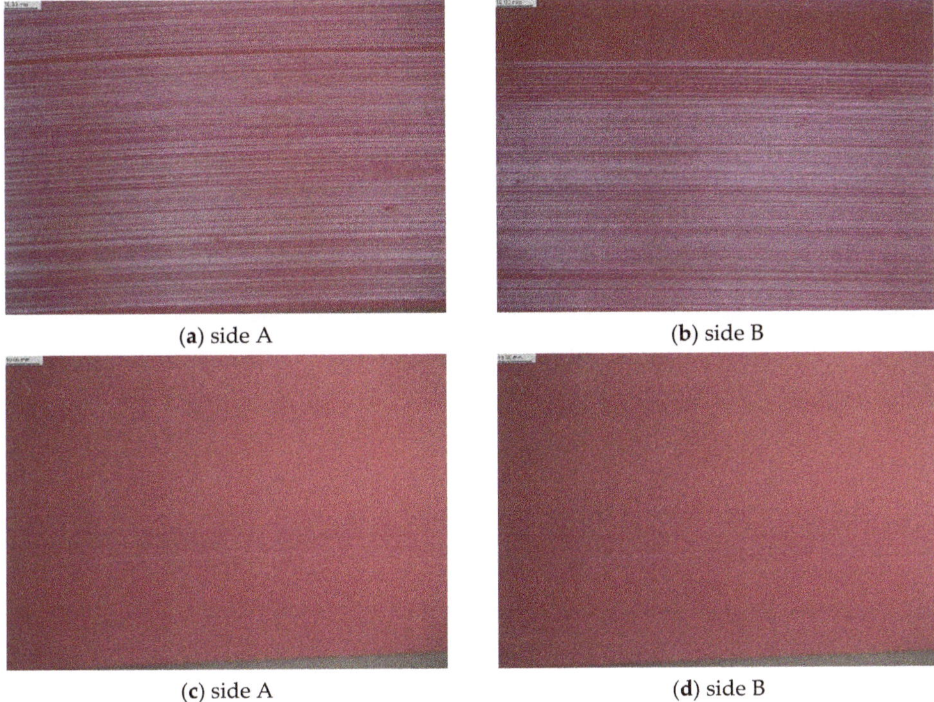

(**a**) side A   (**b**) side B

(**c**) side A   (**d**) side B

Figure 13. Microscopic images of a new belt surface (**a**,**b**) and a used PU belt (**c**,**d**), designated as sides A and B (×10 magnification).

Figure 13 shows the images of the damaged surface of the new V-belt in relation to the used belt on both sides of the belt A and B. As a result of the observations, significant differences were found in the quality of a new PU V-belt surface and after operation. Higher wear was observed for both V-belts on the A side, which correlates with the reduction of belt hardness after use. The occurrence of typical traces on the surface of the new V-belt, in the form of silvery stripes, formed when the material passed through the calibrator during the belt extrusion process. It was found that the surface of the used V-belt was clearly smoother—with no traces of adhesive wear—than the original polyurethane V-belt, which correlates with a decrease in the hardness of the belt due to the influence of friction heat. According to the works of Hook and Kukerek [61], the generated frictional heat has a huge impact on composite materials, used for work in helical gears, enabling a controlled temperature rise. Nachman and others [62] found that an increase in the annealing temperature leads to an increase in the PU molecular weight and degree of phase separation, which, in turn, improves the frictional wear resistance. Similar observations were reported in other papers in the case of some polymers, where, as the temperature increased, the coefficient of friction and elasticity of elastomers—e.g., nitril butadien rubber (NBR), polyurethane (PUR) as well as polytetrafluoroethylene (PTFE) and high-density polyethylene (PE-HD) thermoplastics, and polyoxymethylene (POM)—decreased [63–66]. After the thermal test of V-belt transmission with the polyurethane belts, the surface of the used PU belts was observed by scanning electron microscopy (SEM), which provided some information about the defects and location of failures. In this case, another methodology can also be used to assess the surface. Pawlus and others [67] developed the method of simulation of the one-process base (valley) profile on the basis of the two-process profile.

Figure 14 shows the SEM micrographs of the surface of the used PU belt and virgin PU belt on both sides of the belt: A (a) and B (b). Compared to a new PU belt, the surface of the damaged polyurethane belt showed a considerably smoother surface. The decreased surface roughness implied that the path of the crack tip was distorted because of heating treatment, making crack propagation more difficult [68].

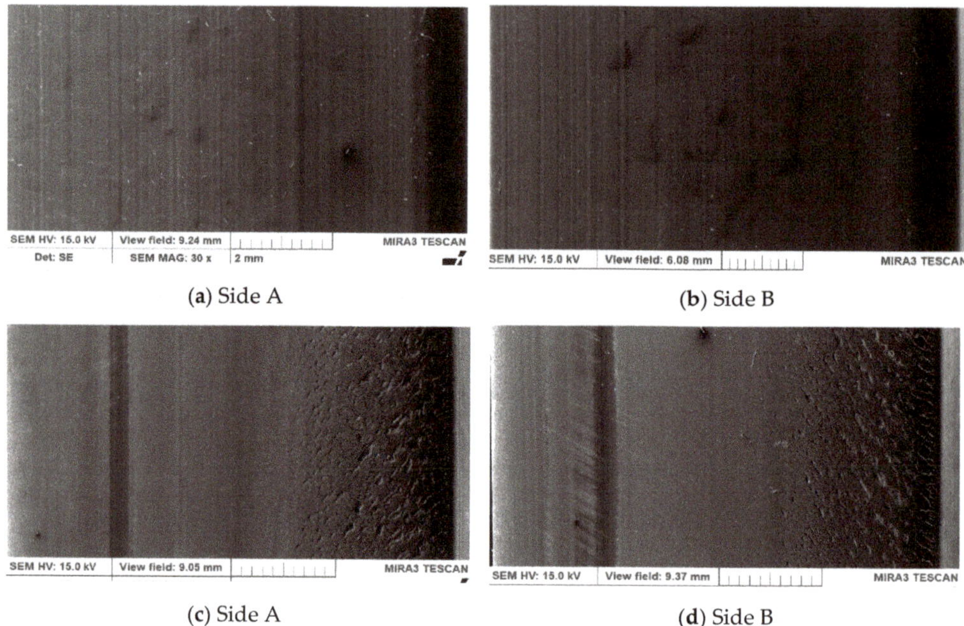

<table>
<tr><td align="center">(a) Side A</td><td align="center">(b) Side B</td></tr>
<tr><td align="center">(c) Side A</td><td align="center">(d) Side B</td></tr>
</table>

Figure 14. SEM images of surface for PU damage (**a**,**b**) and a new PU belt (**c**,**d**), designated as sides A and B at magnification of 30x.

The analysis of the SEM microphotographs shows that the friction surface of the tested belts had clear signs typical of abrasive wear. In the used PU belts with a lower hardness (Figure 14c,d), areas of intensive wear of the material can be observed. Particularly, numerous small cracks are visible at the outer edge of the belt. Those cracks may have resulted from the cyclic polyurethane deformation during the friction process [69]. Capanidis studied the effect of the hardness of the PUR in an elastomer foam used in various parts of machines on its abrasive wear resistance. He stated that the increase in the abrasive wear resistance directly determines an increase in the durability and operational reliability of machines [69].

4. Conclusions

In this paper, the reliability of the belt transmission operation was determined on the basis of the temperature distribution (thermogram) on the thermoplastic polyurethane belt surface, related to the characteristics of the friction process, transferred power, efficiency, load and type of belt material. The infrared thermography technique was applied to the assessment of temperature distribution of the polyurethane belt.

As a result of IR thermography tests, it was found that the polyurethane V-belt temperature increase (up to 80 °C) was directly related to slip and belt efficiency. It was noted that with the increase of rotational speed (from 500 to 1500 rpm) and braking torque load (to 13 Nm) of the active shaft, the slip value increased and the transmission efficiency decreased (about 30%).

The presented results indicate that taking into account the criterion of correct and long-term operation of belt drives, the most favorable working conditions (at a rotational speed of the driving pulley of 1500 rpm and transmission breaking torque load of 2.15 Nm) occur when the difference in the temperature of active (30.4 °C) and passive (31 °C) connectors is stabilized and small (approx. 0.6 °C).

In order to assess wear, the belts were subjected to hardness and roughness tests. The results showed that the surface roughness and the hardness of the used polyurethane belt were reduced, in comparison with those of the new belt. It was found that the hardness of the PU belt after operating cycles was decreased by 10% compared to the new polyurethane V-belt. This may be explained by the heat interaction generated during the friction of the pulley–belt pair.

The surface roughness results show significant changes in the PU V-belt roughness after the wear process as a result of the friction heat in belt drives. The R_a and R_z of the used PU belt were decreased by about 48% and 86%, respectively. As a result of SEM microscopic observations, it was found that the surface of the used polyurethane V-belt was clearly smoother—with no traces of wear—than the original PU belt, which correlates with a decrease in the hardness of the belt due to the influence of friction heat. The IR thermovision results, mechanical properties and surface structure indicate low V-belt wear under given the operating conditions.

The results confirm that the IR thermography technique may be a suitable tool for observing and registering the dynamics of the thermal processes of the polyurethane V-belts used in belt transmission.

Author Contributions: Conceptualization, P.K. and L.R.; methodology, P.K. and D.C.-K.; software, L.R.; validation, L.R., P.K. and Ł.W.; investigation, P.K., Ł.W. and D.C.-K.; resources, P.K, L.R. and D.C.-K.; data curation, D.C.-K.; writing—original draft preparation, P.K, L.R and D.C.-K.; writing—review and editing, D.C.-K.; project administration, D.C.-K.; funding acquisition, P.K and D.C.-K. All authors have read and agreed to the published version of the manuscript.

Funding: This research was funded by the Polish Ministry of Science and Higher Education, grants number 0613/SBAD/4630 and 05/53/SBAD/0101.

Acknowledgments: The authors would like to thank Lukasz Bernat, and Mikolaj Poplawski for SEM and NIS microscopic images (from the Poznan University of Technology, Poland).

Conflicts of Interest: The authors declare no conflict of interest.

References

1. Amanov, A.; Sembiring, J.P.B.A.; Amanov, T. Experimental Investigation on Friction and Wear Behavior of the Vertical Spindle and V-belt of a Cotton Picker. *Materials* **2019**, *12*, 773. [CrossRef] [PubMed]

2. Yu, D.; Childs, T.H.C.; Dalgarno, K.W. V-ribbed belt design, wear and traction capacity. *Proc. Inst. Mech. Eng. Part D J. Automob. Eng.* **1998**, *212*, 333–344. [CrossRef]

3. Kong, L.; Parker, R.G. Coupled Belt-Pulley Vibration in Serpentine Drives With Belt Bending Stiffness, ASME. *J. Appl. Mech.* **2004**, *71*, 109–119. [CrossRef]

4. Gao, P.; Xie, L.; Pan, J. Reliability and Availability Models of Belt Drive Systems Considering Failure Dependence. *Chin. J. Mech. Eng.* **2019**, *32*, 30. [CrossRef]

5. Czarnecka-Komorowska, D.; Struski, P. *Kompozyty Polimerowe Wytworzone z Odpadów Polioksymetylenu (rPOM) i Poliuretanu Termoplastycznego—Nowe Materiały. Nowoczesne Materiały Polimerowe i ich Przetwórstwo*; Klepka, T., Ed.; Wyd. Politechnika Lubelska: Lublin, Poland, 2017.

6. Thomas, S.; Yang, W. *Advances in Polymer Processing: From Macro- To Nano- Scales*; Woodhead Publishing: Cambridge, UK, 2009.

7. Whm_catalogue.pdf. Available online: https://www.whm.pl/oferta/technika-napedow/termozgrzewalne-pasy-okragle-i-klinowe-z-pu.html (accessed on 22 October 2019).

8. Behabelt_cataloque.pdf. Available online: http://www.behabelt.com/terms/agb.pdf (accessed on 22 October 2019).

9. Kong, L.; Parker, R.G. Steady Mechanics of Belt-Pulley Systems. *J. Appl. Mech.* **2005**, *72*, 25. [CrossRef]

10. Merghache, S.M.; Ghernaout, M.E.A. Experimental and numerical study of heat transfer through asynchronous belt transmission type AT10. *Appl. Therm. Eng.* **2017**, *127*, 705–717. [CrossRef]

11. Silva, C.A.F.; Manin, L.; Rinaldi, R.G.; Remond, D.; Besnier, E.; Andrianoely, M.A. Modeling of power losses in poly-V belt transmissions: Hysteresis phenomena (enhanced analysis). *Mech. Mach. Theory.* **2018**, *121*, 373–397. [CrossRef]

12. Czarnecka-Komorowska, D.; Mencel, K. Modification of polyamide 6 and polyoxymethylene with [3-(2-aminoethyl)amino]propyl-heptaisobutyl-polysilsesquioxane nanoparticles. *Przem. Chem.* **2014**, *93*, 392–396.

13. De Almeida, A.; Greenberg, S. Technology assessment: Energy-efficient belt transmissions. *Energy Build.* **1995**, *22*, 245–253. [CrossRef]

14. Shim, H.J.; Kim, J.K. Cause of failure and optimization of a V-belt pulley considering fatigue life uncertainty in automotive applications. *Eng. Fail. Anal.* **2009**, *16*, 1955–1963. [CrossRef]

15. Krawiec, P.; Grzelka, M.; Kroczak, J.; Domek, G.; Kołodziej, A. A proposal of measurement methodology and assessment of manufacturing methods of nontypical cog belt pulleys. *Measurement* **2019**, *132*, 182–190. [CrossRef]

16. Krawiec, P. Numerical Analysis of Geometrical Characteristics of Machine Elements Obtained Through CMM Scanning. In *Progress in Industrial Mathematics*; Springer: Berlin/Heidelberg, Germany, 2010.

17. Kujawski, M.; Krawiec, P. Analysis of Generation Capabilities of Noncircular Cog belt Pulleys on the Example of a Gear with an Elliptical Pitch Line. *J. Manuf. Sci. Eng. Trans. ASME* **2011**, *133*, 051006. [CrossRef]

18. Krawiec, P. Analysis of selected dynamic features of a two-wheeled transmission system. *J. Theor. Appl. Mech.* **2017**, *55*, 461–467. [CrossRef]

19. Krawiec, P.; Marlewski, A. Profile design of noncircular belt pulleys. *J. Theor. Appl. Mech.* **2016**, *54*, 561–570. [CrossRef]

20. Krawiec, P.; Waluś, K.; Warguła, Ł.; Adamiec, J. Wear evaluation of elements of V-belt transmission with the application of optical microscope. In Proceedings of the MATEC Web of Conferences, Proceedings of the Machine Modelling and Simulations, Sklené Teplice, Slovak Republic, 5–8 September 2017; Volume 157.

21. Krawiec, P.; Warguła, Ł.; Waluś, K.; Adamiec, J. Wear evaluation study of the multiple grooved pulleys with optical method. In Proceedings of the MATEC Web of Conferences, Proceedings of the XXIII Polish-Slovak Scientific Conference on Machine Modelling and Simulations, Rydzyna, Poland, 4–7 September 2018; Volume 254.

22. Krawiec, P.; Warguła, Ł.; Dziechciarz, A.; Małozięć, D.; Ondrušová, D. Evaluation of chemical compound emissions during thermal decomposition and combustion of V-belts. *Przem. Chem.* **2020**, *99*, 92–98.

23. Krawiec, P.; Różański, L. Usage of IR Thermography in Researching of Uneven Strand Transmissions with Cogbelts. *Pomiary Autom. Kontrola* **2009**, *11*, 966–969.

24. Walentynowicz, J.; Trawiński, G.; Wieczorek, M.; Polak, F.; Boruta, G. Methods of verification of main transmission gear boxes on test bench using vibrat;on measurement. *J. KONES Powertrain Transp.* **2013**, *20*, 471–478. [CrossRef]

25. Jakubek, B.; Barczewski, R.; Rukat, W.; Różański, L.; Wrobel, M. Stabilization of vibro-thermal processes during post-production testing of rolling bearings. *Diagnostyka* **2019**, *20*, 53–62. [CrossRef]

26. Lepiarczyk, D.; Tarnowski, J.; Uhrynski, A.; Gawedzki, W. Usage of thermo-vision in research concerning kinematic pair of friction in machines and mechanical devices. *a a aa a* **2012**, *4*, 1–7.

27. Kuchynkova, H.; Hajek, V. Measurement of temperature of electrical Machines using thermovision camera. *Zesz. Probl. Masz. Elektr.* **2010**, *87*, 139–144.

28. Barański, M.; Polak, A. Thermal diagnostic in electrical machines. *Przegląd Elektrotech. Electr. Rev.* **2011**, *87*, 305–308.

29. Rosca, D.; Rosca, A. Thermovision predictive maintenance for large capacity milling process in food industry. In *Annals of DAAAM for 2011 & Proceedings of the 22nd International DAAAM Symposium*; Katalinic, B., Ed.; DAAAM International: Vienna, Austria, 2011; pp. 1407–1408.

30. Wargula, Ł.; Waluś, K.J.; Krawiec, P. The problems of measuring the temperature of the small engines (SI) on the example of a drive for non-road mobile machines. In *MATEC Web of Conferences, Proceedings of the XXIII Polish-Slovak Scientific Conference on Machine Modelling and Simulations, Rydzyna, Poland, 4–7 September 2019*; EDP Sciences: Lez Ili, France, 2019; p. 254.

31. Worsztynowicz, B.; Uhryński, A. The analysis of heating process of catalytic converter using thermo-vision. *Combust. Engines* **2015**, *162*, 41–51.

32. Urzędowska, W.; Stępień, Z. Selected issues regarding the change of engine lubricating oil properties in operation (orginal title in Polish: Wybrane zagadnienia dotyczące zmiany właściwości silnikowego oleju smarowego w eksploatacji). *NAFTA-GAZ* **2012**, *12*, 1102–1110.

33. Warguła, Ł.; Kaczmarzyk, P.; Dziechciarz, A. The assessment of fire risk of non-road mobile wood chopping machines. *J. Res. Appl. Agric. Eng.* **2019**, *64*, 58–64.
34. Kaczmarzyk, P.; Małozięć, D.; Warguła, Ł. Research on electrical wiring used in the construction of working machines and vehicles in the aspect of fire protection. *J. Mech. Transp. Eng.* **2018**, *70*, 13–24.
35. Hakami, F.; Pramanik, A.; Ridgway, N.; Basak, A.K. Developments of rubber material wear in conveyer belt system. *Tribol. Int.* **2017**, *111*, 148–158. [CrossRef]
36. Osornio-Rios, R.; Antonino-Daviu, J.; de Jesus Romero-Troncoso, R. Recent industrial applications of infrared thermography: A review. *IEEE Trans. Ind. Inform.* **2018**, *15*, 615–625. [CrossRef]
37. Królczyk, G.; Li, Z.; Antonino Daviu, J.A. Fault Diagnosis of Rotating Machine. *Appl. Sci.* **2020**, *10*, 1961. [CrossRef]
38. Bagavathiappan, S.; Lahiri, B.B.; Saravanan, T.; Philip, J.; Jayakumar, T. Infrared thermography for condition monitoring—A review. *Infrared Phys. Technol.* **2013**, *60*, 35–55. [CrossRef]
39. Dudek, K.; Banasiak, J.; Bieniek, J. Diagnosis of kinematic pairs in the rotary mowers. *Maint. Reliab.* **2003**, *20*, 17–21.
40. Tor-Świątek, A.; Samujło, B. Wykorzystanie badań termowizyjnych do analizy stabilności procesu wytłaczania mikroporującego poli(chlorku winylu). *Maint. Reliab.* **2013**, *15*, 58–61.
41. Szymiczek, M. Ultrasonic and thermal testing as a diagnostic tool for the evaluation of cumulative discontinuities of the polyester-glass pipes structure. *Maint. Reliab.* **2017**, *19*, 1–7. [CrossRef]
42. Jia, S.S.; Song, Y.M. Elastic dynamic analysis of synchronous belt drive system using absolute nodal coordinate formulation. *Nonlinear Dyn.* **2015**, *81*, 1393–1410. [CrossRef]
43. Li, W.B.; Xin, Z.X. Flexural fatigue life prediction of a tooth V-belt made of fiber reinforced rubber. *Int. J. Fatigue* **2018**, *111*, 269–277. [CrossRef]
44. Różański, L.; Ziopaja, K. Applicability analysis of IR thermography and discrete wavelet transform for technical conditions assessment of bridge elements. *Quant. InfraRed Thermogr. J.* **2019**, *16*, 87–110. [CrossRef]
45. Sundararaman, S.; Hu, J.; Chen, J.; Chandrashekhara, K. Temperature dependent fatigue-failure analysis of V-ribbed serpentine belts. *Int. J. Fatigue* **2009**, *31*, 1262–1270. [CrossRef]
46. Chelladurai, V.; Jayas, D.; White, N. Thermal imaging for detecting fungal infection in stored wheat. *J. Stored Prod. Res.* **2010**, *46*, 174–179. [CrossRef]
47. Chrzanowski, K. *Błędy Metod Zdalnego Pomiaru Temperatury za Pomocą Urządzeń Podczerwieni*; Wyd. WAT: Warszawa, Poland, 1996.
48. Chrzanowski, K. *Non-Contact Thermometry—Measurement Errors*; Polski Oddział SPIE: Warszawa, Poland, 2000.
49. Fidali, M.; Urbanek, G. Application of evolutionary algorithm to limitation of a set of statistical features of thermovision images. *Comput. Assist. Mech. Eng. Sci.* **2007**, *14*, 601–609.
50. Gade, R.; Moeslund, T. Thermal cameras and applications: A survey. *Mach. Vis. Appl.* **2014**, *25*, 245–262. [CrossRef]
51. Gawdzińska, K.; L. Chybowski, L.; Przetakiewicz, W. Study of Thermal Properties of Cast Metal-Ceramic Composite Foams. *Arch. Foundry Eng.* **2017**, *17*, 47–50.
52. Jóźwik, J. Thermographic evaluation of temperature distribution on surface of workpiece during non-orthogonal turning. *Maint. Reliab.* **2001**, *5*, 58–61.
53. Minkina, W. *Pomiary Termowizyjne—Przyrządy i Metody*; Wyd. Politechniki Częstochowskiej: Częstochowa, Poland, 2014.
54. Stawicki, T.; Sędłak, P. Thermal Imaging Studies on Grain Harvester Belt Transmissions. *Diagnostyka* **2015**, *16*, 37–42.
55. Muniz, P.R.; de Araújo Kalid, R.; Cani, K.R.S.; da Silva Magalhães, R. Handy method to estimate uncertainty of temperature measurement by infrared thermography. *Opt. Eng.* **2014**, *53*, 074101. [CrossRef]
56. Minkina, W.; Dudzik, S. Infrared thermography. In *Errors and Uncertainty*; John Wiley & Sons, Ltd.: Chichester, UK, 2009.
57. Minkina, W.; Dudzik, S. Simulation analysis of uncertainty of infrared camera measurement and processing path. *Measurement* **2006**, *39*, 758–763. [CrossRef]
58. Gao, P.; Xie, L. Reliability and Lifetime Distribution Analysis of Belt Drive Systems considering Time-Dependent Stiffness of Belts. *Hindawi Shock Vib.* **2018**, *2018*, 1–9. [CrossRef]
59. Eurobelt_catalogue.pdf. Available online: http://www.eurobelt.pl (accessed on 22 October 2019).

60. Budinski, K.G. Resistance to particle abrasion of selected plastics. *Wear* **1997**, *203–204*, 302–309. [CrossRef]
61. Hooke, C.J.; Kukureka, S.N.; Liao, P.; Rao, M.; Chen, Y.K. The friction and wear of polymers in non-conformal contacts. *Wear* **1996**, *200*, 83–94. [CrossRef]
62. Nachman, M.; Kwiatkowski, K. The effect of thermal annealing on the abrasion resistance of a segmented block copolymer urethane elastomers. *Wear* **2013**, *306*, 113–118. [CrossRef]
63. Czarnecka-Komorowska, D.; Sterzynski, T.; Dutkiewicz, M. Polyoxymethylene/polyhedral oligomeric silsesquioxane composites: Processing, crystallization, morphology and thermo-mechanical behavior. *Int. Polym. Process.* **2016**, *31*, 598–606. [CrossRef]
64. Czarnecka-Komorowska, D.; Sterzynski, T. Effect of Polyhedral Oligomeric Silsesquioxane on the Melting, Structure, and Mechanical Behavior of Polyoxymethylene. *Polymers* **2018**, *10*, 203. [CrossRef]
65. Wieleba, W.; Dobrowolska, A. Frictional resistance of polymer materials during sliding on ice. *Tribologia* **2011**, *5*, 231–238.
66. Wróblewska-Krepsztul, J.; Rydzkowski, T.; Michalska-Pożoga, I.; Thakur, V.K. Biopolymers for Biomedical and Pharmaceutical Applications: Recent Advances and Overview of Alginate Electrospinning. *Nanomaterials* **2019**, *9*, 404. [CrossRef] [PubMed]
67. Pawlus, P.; Reizer, R.; Wieczorowski, M. Reverse Problem in Surface Texture Analysis—One-Process Profile Modeling on the Basis of Measured Two-Process Profile after Machining or Wear. *Materials* **2019**, *12*, 4169. [CrossRef] [PubMed]
68. Firas, A.; Zhang, S.; Tripathi, M.; Nikiforov, A.; Pugno, N. Cracks, microcracks and fracture in polymer structures: Formation, detection, autonomic repair. *Prog. Mater. Sci.* **2016**, *83*, 536–573.
69. Capanidis, D. The influence of hardness of polyurethane on its abrasive wear resistance. *Tribologia* **2016**, *4*, 29–39. [CrossRef]

 *materials*

Article

Abrasive Wear Behavior of Cryogenically Treated Boron Steel (30MnCrB4) Used for Rotavator Blades

Tejinder Pal Singh [1], Anil Kumar Singla [1], Jagtar Singh [1], Kulwant Singh [1], Munish Kumar Gupta [2], Hansong Ji [2], Qinghua Song [2,3], Zhanqiang Liu [2,3] and Catalin I. Pruncu [4,5,*]

[1] Mechanical Engineering Department, Sant Longowal Institute of Engineering and Technology, Punjab 148106, India; tejindergillz@gmail.com (T.P.S.); singlakiran1996@gmail.com (A.K.S.); jagtarsliet@gmail.com (J.S.); engrkulwant@yahoo.com (K.S.)

[2] Key Laboratory of High Efficiency and Clean Mechanical Manufacture, Ministry of Education, School of Mechanical Engineering, Shandong University, Jinan 250061, China; munishguptanit@gmail.com (M.K.G.); hansgji@163.com (H.J.); ssinghua@sdu.edu.cn (Q.S.); melius@sdu.edu.cn (Z.L.)

[3] National Demonstration Center for Experimental Mechanical Engineering Education, Shandong University, Jinan 250061, China

[4] Mechanical Engineering Department, University of Birmingham, Birmingham B15 2TT, UK

[5] Mechanical Engineering, Imperial College London, Exhibition Rd., London SW7 2AZ, UK

* Correspondence: c.pruncu@imperial.ac.uk

Received: 30 November 2019; Accepted: 14 January 2020; Published: 16 January 2020

Abstract: Rotavator blades are prone to significant wear because of the abrasive nature of sand particles. The aim of this research work is to investigate the effect of cryogenic treatment and post tempering on abrasive wear behavior, in the presence of angular quartz sand (grain size of 212–425 μm), of rotavator blade material of boron steel (30MnCrB4). Cryogenic treatment has caused an improvement in the abrasive wear resistance and microhardness of 30MnCrB4 by 60% and 260.73%, respectively, compared to untreated material due to enhancement in hardness, the conversion of retained austenite into martensite, and the precipitation of secondary carbides in boron steel after exposure to cryogenic temperature. Economic analysis justifies the additional cost of cryogenic treatment.

Keywords: austenitization; cryogenic; microstructure; microhardness; abrasive wear; tempering

1. Introduction

Rotavator and other earth engaging implements are needed in the farming industry, where they are used for seed bed preparation via the mixing of all crop residues to improve soil's organic health and yield [1]. Agricultural implements wear out frequently due to abrasive action by sand/soil particles. Rotavator also gets worn out frequently due to abrasive action of sand/soil, leading to repeated replacement of its blades, which is quite uneconomical. Poor abrasive wear behavior makes farmers less efficient and plays a major role in raising the cost of agriculture [2] by adversely affecting farmers' productivity, thereby making the whole process of agriculture economically unviable. Scientists and researchers have applied different techniques such as coatings [3–14], carburizing [15], nitriding and carbonitriding [15], boriding [15–20], and shot peening [21] to enhance the working life of rotavator blades. These techniques have shown promising results with regard to improving wear resistance in the short term, but limited data related to the long term success of these techniques is reported in the published work. These techniques largely improve a very thin layer of surface material, and chipping off this layer leads to severe wear. Therefore, there is need to explore material processing techniques for improving the abrasive wear resistance of rotavator blade material [22–25].

Conventional heat treatment (CHT) is simple, flexible, and economical, and is largely used for manipulating the microstructure, leading to alterations in the mechanical properties of the material used for earth engaging implements [26,27]. Suitable selection of the heat treatment temperature and subsequent cooling rate helps in attaining the desirable properties as per the requirements of application. Processing of materials at a low temperature in the range of −80 °C to −196 °C is another material processing technique used for achieving desirable properties in materials. Different review articles [28–33] pertaining to cryogenic processing have reported CHT's potential in improving the properties of materials, especially ferrous-based cutting tools/implements. The efficacy of cryogenic treatment in improving the properties of nonferrous materials has also been established by different research studies [34–39]. T Singla et al. [28] have extensively examined various cryogenic cycles and their parameters for different ferrous materials. The authors have recommended using tempering post cryogenic treatment for further improving the properties of materials. The optimum number of cycles of post tempering, temperature, and time may vary for different materials. These can be optimized only based on experimental observations for different materials. Cryogenic treatment also plays a significant role in eco-friendly machining of materials, especially difficult-to-cut materials, by replacing conventional cutting fluids with liquified nitrogen [40–42]. The cooling medium provided to the cutting zone during cryogenic cooling evaporates immediately and comes back to the atmosphere without any pollution [40].

In this work, comparative study has been conducted amongst different thermal treatments, with a major focus on cryogenic treatment, on rotavator blade material 30MnCrB4 vis-à-vis hardness, impact strength, and abrasive wear resistance. The alterations in mechanical behavior have been correlated with microstructural evolution. Economic analysis justifying the additional cost of cryogenic treatment due to its improvement in the performance of rotavator blade material has been reported.

2. Materials and Methods

2.1. Untreated Material

In the present work, low-carbon boron steel (DIN 30MnCrB4) is selected as an untreated material, which is widely used for manufacturing of rotavator blades [43,44]. Boron steel exhibits better hardness, sliding, and abrasive wear resistance due to the presence of hardenability enhancers like chromium and boron, as compared to high-carbon steel [25]. The chemical composition of as-received material is tested using optical emission spectrometer (GNR Solaris CCD-Plus, Torino, Italy). The obtained chemical composition along with standard composition are presented in Table 1.

Table 1. Chemical composition of boron steel (30MnCrB4) in weight (%).

Elements	As Received	As Per Standards
C	0.29	0.24–0.30
Mn	1.23	1.10–1.40
Si	0.221	0.40
P	0.032	0.030
Cr	0.306	0.30–0.60
B	0.002	0.0008–0.0050
Fe	balance	balance

2.2. Methods

The schematic plan depicting the various thermal treatments and mechanical and metallurgical investigations is given in Figure 1.

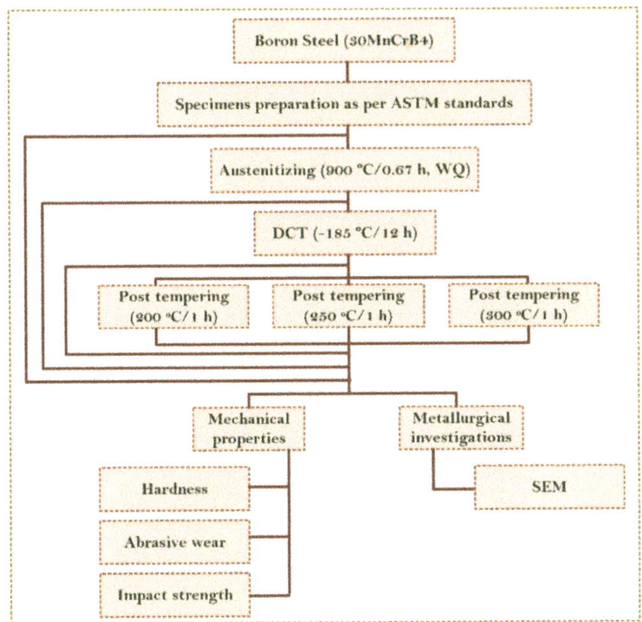

Figure 1. Schematic plan of thermal treatments and different investigations.

2.2.1. Austenitization

A pilot study [45,46] was undertaken to decide about the parameters of austenitization, such as austenitic temperature, soaking period, and quenching medium. L9 Taguchi DOE (design of experiments) was adopted for this purpose. The design factors along with three levels for each factor are shown in Table 2.

Table 2. Factor and level for DOE (design of experiments) of austenitization.

Factors	Levels		
	1	2	3
A: Medium	Water	Oil	Oil-Water
B: Temperature (°C)	800	850	900
C: Time (min)	20	30	40

The output parameters were abrasive wear volume loss, hardness, and impact toughness. It was observed that optimum results were attained corresponding to the austenitization temperature of 900 °C with a soaking period of 40 min and water as a quenching media. The complete details of the pilot study are given in research study by Singh et al. [46]. Accordingly, in this work, these optimized austenitization parameters have been chosen for further investigation of low-temperature treatment. Muffle furnace (Jupiter engineering works, New Delhi, India) has been used for austenitization, and all specimens have been properly cleaned ultrasonically using acetone before heat treatment to remove the oil, dust, foreign particles, and grease from the surface of specimens. The muffle furnace was allowed to reach 900 °C, and specimens were placed inside the muffle furnace. The specimens were held at that temperature for 40 min and then water-quenched.

2.2.2. Deep Cryogenic Treatment

Austenitization was immediately, without any time lag, followed by deep cryogenic treatment (DCT), and its parameters are reported in Table 3. All the specimens were again ultrasonically cleaned with acetone before deep cryogenic processing. DCT was conducted using cryogenic processor (3241-1-SAMSON, Super Cryogenic Systems Pvt. Ltd., Noida, India). Cryogenic treatment set up was brought to the desired temperature using computer controls in a well-insulated chamber with liquid nitrogen as a working medium.

Table 3. Description of deep cryogenic treatment (DCT) process parameters.

Sr. No.	Parameters	Level
1	Soaking temperature (°C)	−185
2	Cooling rate (°C/min)	0.5
3	Soaking period (h)	12
4	Heating rate (°C/min)	0.5

2.2.3. Post Tempering

Post tempering treatment was given to all the cryogenically treated specimens in three different groups at three tempering temperatures. The three sets of specimens were subjected to post tempering cycle at 200 °C, 250 °C, and 300 °C, respectively, for 1 h. The complete details of austenitization, DCT, and tempering cycles are presented in Figure 2. The various thermal cycles adopted for the present work and coding of specimens are detailed in Table 4.

Table 4. Thermal treatment schedules and coding of specimens.

Sr. No.	Type of Treatment	Specimen Coding
1	Untreated Material	UT
2	CHT (900 °C/0.67 h, WQ)	CHT
3	CHT (900 °C/0.67 h, WQ) + DCT (−185 °C/12 h)	CDCT-T0
4	CHT (900 °C/0.67 h, WQ) + DCT (−185 °C/12 h) + Tempering (200 °C/1 h, FC)	CDCT-T1
5	CHT (900 °C/0.67 h, WQ) + DCT (−185 °C/12 h) + Tempering (250 °C/1 h, FC)	CDCT-T2
6	CHT (900 °C/0.67 h, WQ) + DCT (−185 °C/12 h) + Tempering (300 °C/1 h, FC)	CDCT-T3

Note: WQ—water quenching, FC—furnace cooling, CHT—conventional heat treatment, UT—untreated.

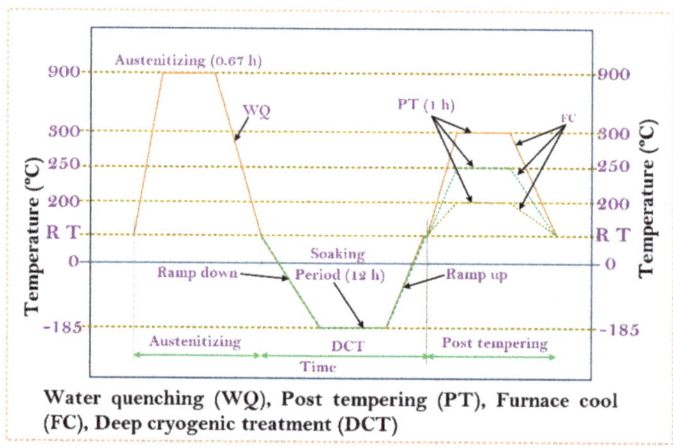

Water quenching (WQ), Post tempering (PT), Furnace cool (FC), Deep cryogenic treatment (DCT)

Figure 2. Thermal treatment cycles.

2.3. Metallurgical Investigation

Scanning electron microscopy, using JSM 6510LV (JEOL, Tokyo, Japan) was conducted for visualization of microstructure and study of fractography. For microstructure, the specimens were etched first with freshly prepared etchant 4% picric (picric acid 4 gm and ethanol 96 mL) by swabbing for 20–30 s followed by distilled water rinsing. They were then etched with 2% nital (nitric acid 2 mL and ethanol 98 mL) by swabbing for 15–25 s followed by distilled water rinsing. The specimens were dried in air and observed for obtaining SEM images.

2.4. Mechanical Properties

The micro hardness measurement was conducted as per ASTM E384 using digital micro Vickers hardness tester (RMHT-201, Radical, Ambala Cantt, India). Measurements were taken at five different locations of UT, CHT, CDCT-T0, CDCT-T1, CDCT-T2, and CDCT-T3 specimens. Four readings were taken at the periphery and one at the center of specimen. Vickers micro-hardness measurement was conducted using a 4.9 N load for a total cycle time of 40 s with a dwell time of 20 s. The average of five observations was computed and reported for this work.

Charpy V-notched impact test was performed as per ASTM E23-07a [47] using sub-size specimens (7.5 × 10 × 55) mm. Three specimens for each treatment CHT, CDCT-T0, CDCT-T1, CDCT-T2, and CDCT-T3 were tested, and observations with average value are reported in this work.

Specimens of size (76.2 × 25.4 × 6) mm, as shown in Figure 3a, were prepared using tabletop cutting equipment (Labotom 5, Struers, Ballerup, Denmark) for abrasive wear testing. Abrasive wear test was performed as per ASTM G65 [48] on the specimens of UT, CHT, CDCT-T0, CDCT-T1, CDCT-T2, and CDCT-T3 using abrasive wear test rig (TR-50-M7, Ducom, Bangalore, India) shown in Figure 3b. The parameters selected for abrasive wear testing are shown in Table 5. The values of speed, rotational speed of wheel, and sliding distance were selected to emulate the actual agricultural field conditions. Specimens were ultrasonically cleaned before the test to remove any foreign particles from the specimens. The weight of each specimen, before and after the experiment, was determined by using precision weighing balance (Citizen CY204, Denver Instruments GmbH, Denver, Germany) with a least count of 0.0001 g. The resulting weight loss was converted into abrasive wear loss in mm^3. Each experiment was performed three times to minimize the experimental errors.

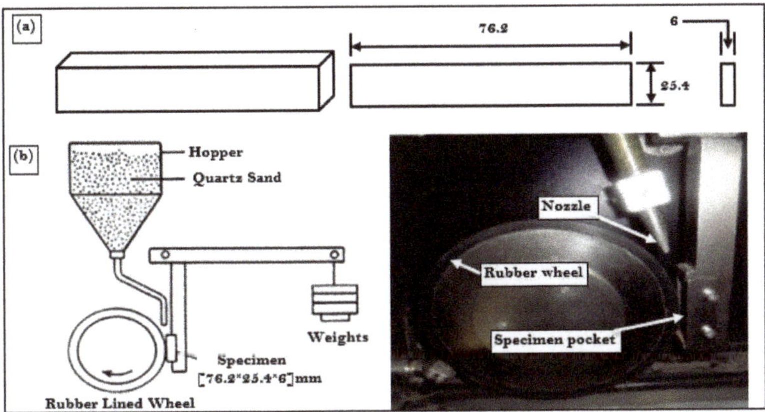

Figure 3. Schematic view of (**a**) abrasive wear test specimen and (**b**) dry sand abrasion test rig.

Table 5. Parameters used for three-body abrasion wear test (DSRW).

Parameters	Unit	Value
Load	N	130
Speed	m/s	3.17
Rotational speed of wheel	rpm	200
Abrasive flow rate	g/min	300
Abrasive size	μm	212–425
Sliding distance	m	1900

The abrasive wear study of the material was influenced by the size and morphology of abrasive sand. As per Avery [49], abrasive wear behavior of material was significantly affected by the sand particle size. The quartz sand was arranged for the test and standard sieve shaker (Kelsons Engineers and Fabricators, Kohlapur, India) was used to find out the particle size distribution. Particle size distribution for the used sand is presented in Figure 4. d10 and d90 of the quartz sand used for wear test were computed and marked on the particle distribution curve shown in Figure 4. It is clear from particle distribution curve that 80% of the quartz sand particles lie between 240 μm and 600 μm with an average particle size of 430 μm. The size of sand available in fields also lie in the same range as reported by Singh et al. [50].

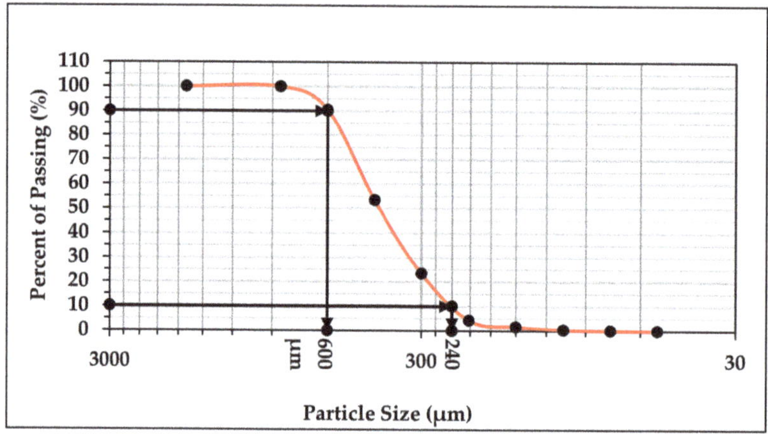

Figure 4. Particle size distribution curve.

To obtain morphology of the quartz sand particle, scanning electron microscopy (JSM 6510LV, JEOL, Tokyo, Japan) was used for the present study, and same is shown in Figure 5. The hardness value of quartz sand is determined based on past studies [50,51] and found to be in the range of 1070–1200 HV. It is clearly shown in the SEM image of quartz sand that particles are highly angular, which is responsible for aggressive wear [52,53].

Figure 5. The geometry of quartz sand.

3. Results and Discussions

3.1. Microstructure

The SEM micrograph of UT sample is characterized by presence of ferrite and pearlite structure, as shown in Figure 6a. The network of ferrite appeared as flat surface and pearlite as needle pattern surface [45]. During austenitization, ferrite and pearlite phases were transformed into austenite phase. Austenite is quite soft as compared to pearlite, and started transforming into martensite upon quenching in water. Formation of martensite has led to enhancement of hardness [54,55]. A sufficient fraction of retained austenite (RA) was also present upon quenching to room temperature, which is quite evident from microstructure of austenitized boron steel shown in Figure 6b. Few primary carbides and martensite formation have been identified and marked on SEM image shown in Figure 6b. During deep cryogenic treatment, retained austenite left after CHT was transformed into martensite. A significant fraction of martensite phase is lucidly visible in the SEM image shown in Figure 6c. Formation of secondary carbides on exposure to soaking temperature was also visible in microstructure of CDCT-T0 specimen. Soaking for long time helped in bringing more uniformity in distribution of secondary carbides, as shown in white spot encircled with red color, which led to improvement in hardness and wear resistance of the material [29,56–58]. The presence of martensite and secondary carbides in CDCT-T0 specimen were identified and marked in its SEM micrograph, as shown in Figure 6c. SEM micrograph of tempered post cryogenic treatment boron steel is shown in Figure 6d. It is evident from SEM image of CDCT-T1 specimen that post tempering resulted into grain coarsening and martensite decomposition. Transformation of martensite into tempered martensite and mechanical mixture of phases resulted in reduction of hardness and wear resistance of the CDCT-T1 specimen in comparison with that of CDCT-T0 sample. The results are in agreement with those reported by Kalsi et al. [32].

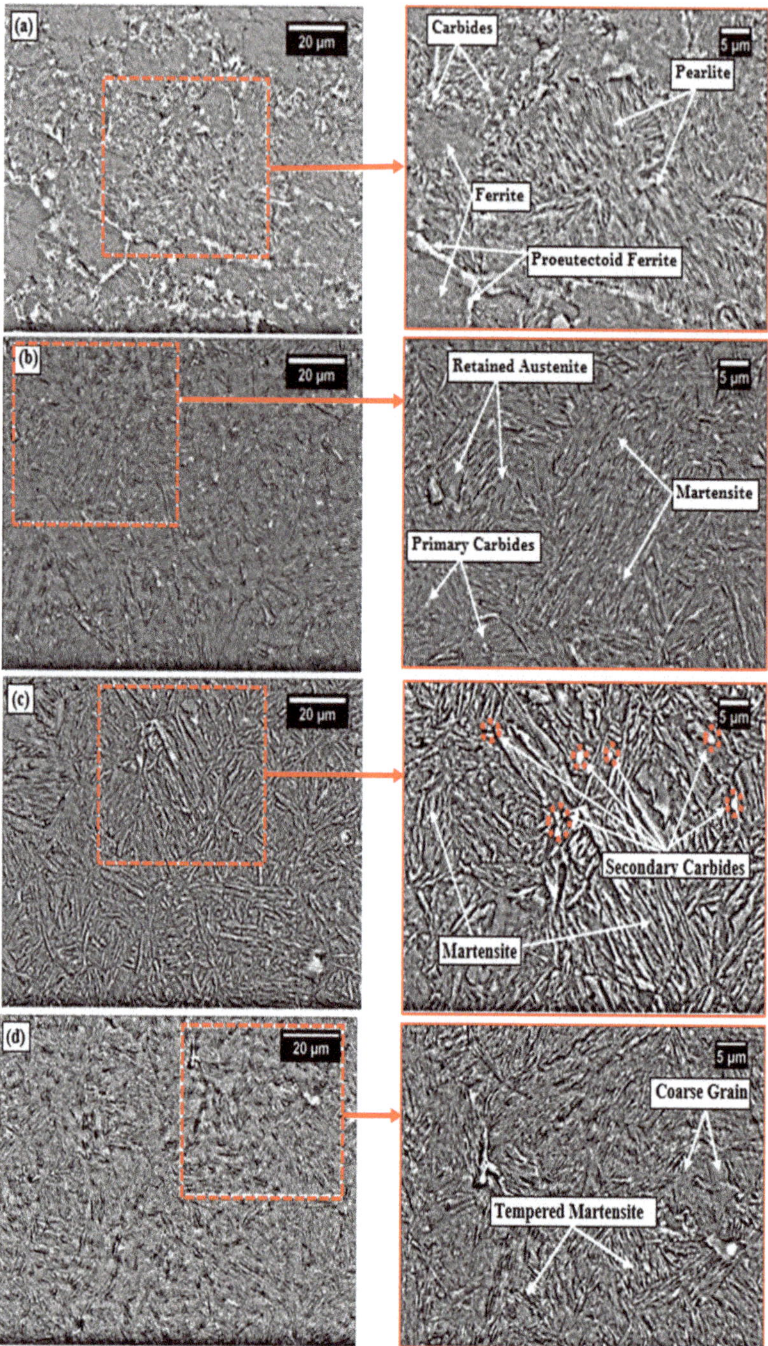

Figure 6. SEM microstructures of (**a**) untreated, (**b**) conventional heat-treated (900 °C/0.67 h, water-quenched), (**c**) conventional heat-treated (900 °C/0.67 h, water-quenched) + deep cryogenic treatment (−185 °C/12 h), and (**d**) conventional heat-treated (900 °C/0.67 h, water-quenched) + deep cryogenic treatment (−185 °C/12 h) + tempering (200 °C/1 h, furnace-cooled).

3.2. Hardness

The Vickers micro-hardness of UT, CHT, CDCT-T0, CDCT-T1, CDCT-T2, and CDCT-T3 specimens was measured at five different locations for each sample, and the mean value was computed and shown in the form of bar chart in Figure 7. From the results drawn in Figure 7, it is clear that hardness increased in the specimens CHT (215.76%), CDCT-T0 (260.73%), CDCT-T1 (216.60%), CDCT-T2 (197.42%), and CDCT-T3 (179.19%), as compared to UT sample. The maximum improvement in the hardness was observed in CDCT-T0 sample due to conversion of RA into martensite and formation of secondary carbides after deep cryogenic treatment [57,59–61].

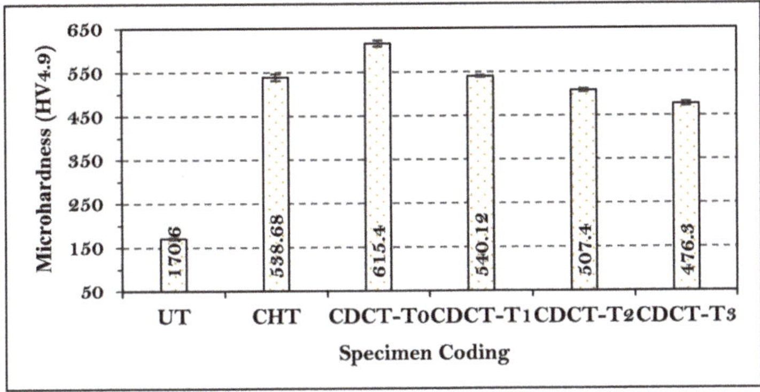

Figure 7. Microhardness of untreated and treated material.

3.3. Impact Strength

A set of three specimens were prepared for analyzing the impact strength of material, and the mean of these three values was computed. The average value has been drawn in the form of bar chart as shown in Figure 8. From the results shown Figure 8, it is clear that impact strength increased in the specimens CDCT-T0 (50%), CDCT-T1 (75%), CDCT-T2 (82.5%), and CDCT-T3 (95%), compared to CHT sample. The maximum increase in the impact strength was observed in CDCT-T3 sample due to transformation of martensite, which formed during hardening, into tempered martensite along with secondary carbide formation and removal of residual stresses due to tempering post cryogenic treatment. The nucleation of carbides and precipitation of finer carbides increase the impact strength of material after deep cryogenic treatment. During cryogenic treatment, fine platelets of martensite are formed from the retained austenite, and these platelets promote the precipitation of fine carbides by a diffusion mechanism during tempering [58].

SEM fractography of impact specimens was conducted to examine the nature of fractured surface. Impact strength of CHT samples was the lowest, and cryogenic treatment caused improvement in impact strength by 50%, which was further enhanced by tempering. The SEM fractography images of the CHT, CDCT-T0, and CDCT-T3 specimens taken at the central region of the Charpy V-notch impact test specimens are shown in Figure 9. Flat regions surrounded by the fibrous regions, which were formed by tearing ridges in the CHT specimen shown in Figure 9a. The presence of rippled morphology with very few voids and few visible cracks accounted for its low impact energy absorption capacity. In the CDCT-T0 specimen shown in Figure 9b, cracking was observed with severe bubble/dimples coalescence along the grain boundaries. The presence of micro dimples on the fracture facets shows that a considerable amount of plastic deformation occurred prior to the fracture, resulting in absorption of increased impact energy of the cryo-treated specimen. In the CDCT-T3 specimen shown in Figure 9c, more cracking was observed is evident, along with dense micro dimples coalescence formation. The presence of many microcracks and crack branching with dense micro dimples indicates more plastic deformation occurred before fracture, resulting in increase in the impact strength of post-tempered cryogenic-treated specimens [58,62].

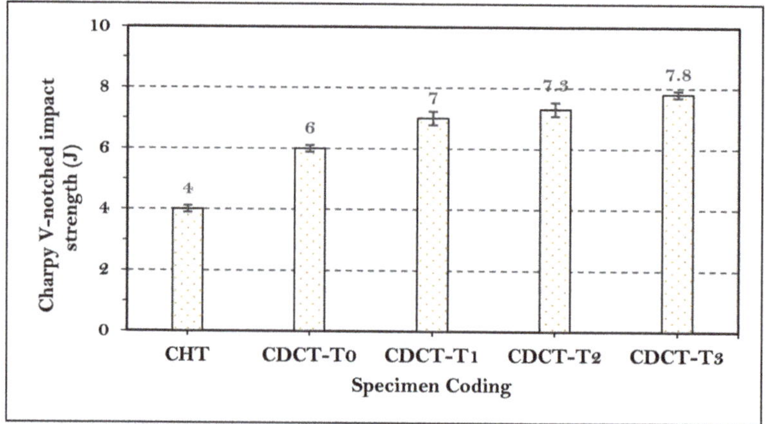

Figure 8. Charpy V-notch impact strength results of treated material.

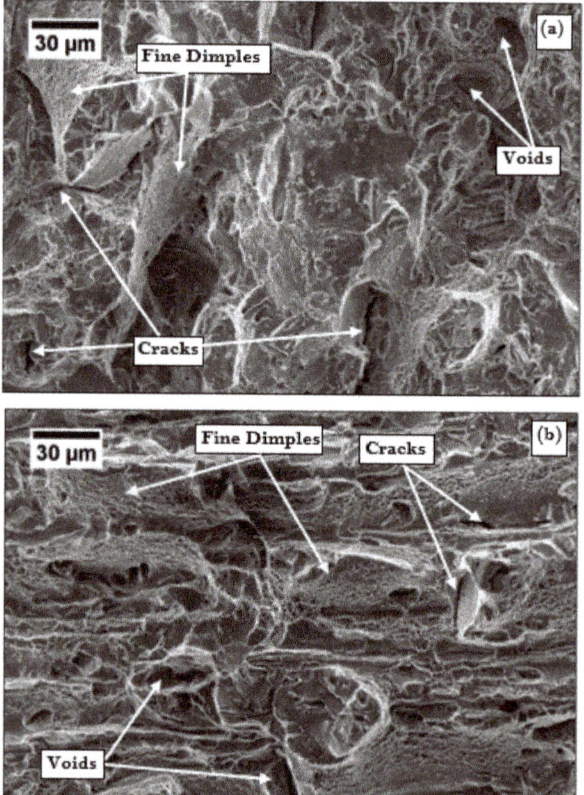

Figure 9. *Cont.*

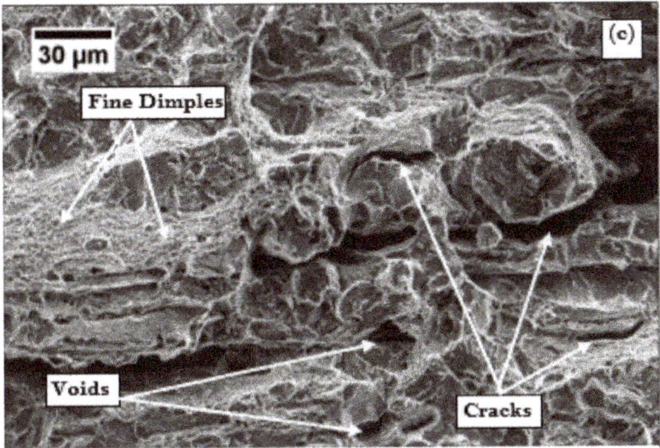

Figure 9. SEM fractography images of impact samples of boron steel: (**a**) conventional heat-treated (900 °C/0.67 h, water-quenched), (**b**) conventional heat-treated (900 °C/0.67 h, water-quenched) + deep cryogenic treatment (−185 °C/12 h), and (**c**) conventional heat-treated (900 °C/0.67 h, water-quenched) + deep cryogenic treatment (−185 °C/12 h) + tempering (300 °C/1 h, furnace-cooled).

3.4. Abrasive Wear

Abrasive wear volume loss was computed from the weight loss and density of material (7.85 g/cm^3) for each sample. The average abrasive wear volume loss was plotted in the form of bar chart as shown in Figure 10. It is clear from the results shown in Figure 10 that abrasive wear resistance increased in the specimens CHT (46.20%), CDCT-T0 (60.00%), CDCT-T1 (56.39%), CDCT-T2 (52.55%), and CDCT-T3 (48.87%), compared to UT sample. The maximum improvement in abrasive wear resistance was observed in CDCT-T0 sample due to conversion of RA into martensite and formation of secondary carbides after deep cryogenic treatment [57,61]. Post tempering increased wear volume loss with increase in tempering temperature attributable to decrease in hardness [63].

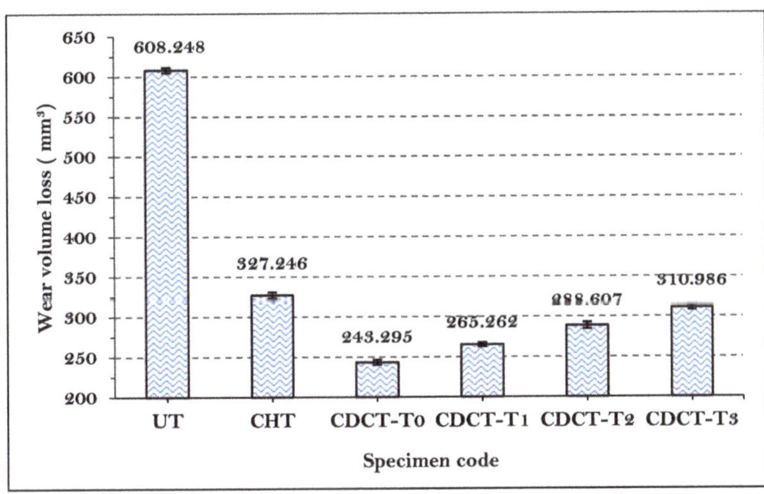

Figure 10. Comparison of abrasive wear loss of untreated and treated material.

Worn Surface Morphology

Since the abrasive wear volume loss was maximum in UT samples and minimum in CDCT-T0 boron steel, SEM images of worn surface of UT and CDCT-T0 specimens of boron steel are reported in Figure 11. The SEM image of UT specimen is marked by the pitting, deep, and broad grooves resulting from rolling and sliding motion of abrasive sand particles between the rubber wheel and the specimen. The sharped edge angular shaped abrasive particles pushed the material to both sides of abrasive groove, due to the repeated action of ploughing. Some material of specimen that drops out due to exhaustion of plasticity was responsible for pitting. These sharp-edge abrasive sand particles penetrated deep into the material of specimen, resulting in a large amount of material removal from specimens by changing the mode from rolling to sliding by abrasive sand particles. In steel with lower hardness, rolling mechanism is dominant while in steel having higher hardness, grooving is dominant feature on worn surfaces [51]. The continuous rubbing of abrasive sand against specimen surface also resulted in the formation of grooves. The results are in agreement with those reported by Chand et al. [64].

During sliding distance of 1900 m, some particles of specimen's material dropped out, generating pitting along with wide and deep grooves on the surface of the specimen due to their plastic deformation with the continuous and repeated ploughing action, which was responsible for large wear loss of the material [56,65]. It is clearly visible in the SEM image of UT specimen presented in Figure 11a. In SEM image of CDCT-T0 specimen, presented in Figure 11b, few shallow groove formations on the abraded surface of the specimen were observed. Number of grooves as well as depth of grooves was smaller than that in UT specimen, leading to significant improvement in the morphology of worn surface. Improvement in the hardness of the material reduced the ploughing depth, and offered stronger support for carbides to inhibit its spalling and could prevent large grooves forming during abrasive action of quartz sand. Improvement in abrasive wear behavior was due to increase in hardness, RA conversion into martensite, formation of secondary carbides, and their more uniform distribution, which ultimately led to reduction in abrasive wear volume loss [56,61,63,66,67].

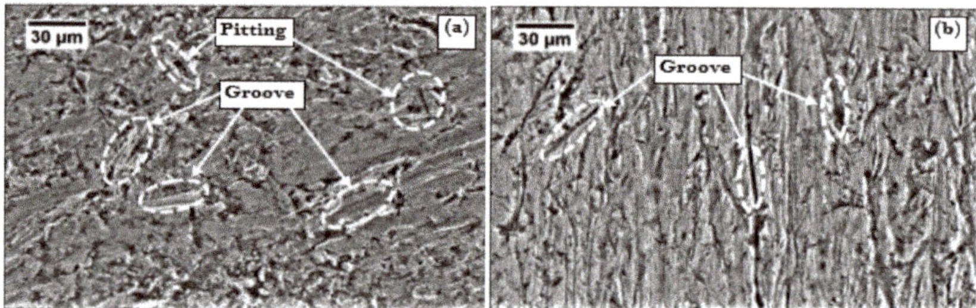

Figure 11. Abraded surfaces of: (**a**) untreated 30MnCrB4 Steel {UT} and (**b**) conventional heat-treated (900 °C/0.67 h, water-quenched) + deep cryogenic-treated (−185 °C/12 h) 30MnCrB4 steel {CDCT-T0}.

3.5. Economic Analysis

Economic analysis has been done to evaluate the expected improvement in performance of rotavator blade during field operation against the additional cost quantified in Indian Rupees (Rs.) due to cryogenic treatment. The details are as follows:

The cost of one blade (including cost of hardening)	=Rs. 215
Size of rotavator	=07 Feet
Total number of blades	=48
Weight of one blade	=1.04 Kg
Number of blades to be cryogenically treated in one lot	=5 × 48 = 240
Additional cost for proposed cryogenic treatment on one lot	=Rs. 7800
Additional cost/blade	=7800/240 = Rs. 32.50
Percentage improvement in wear resistance of blade after cryogenic treatment in comparison with CHT (hardened)	=25.7%
Percentage increase in cost per blade due to cryogenic treatment	=15.12%

It is evident from the economic analysis that improvement quantified in monetary terms is more that the additional cost due to cryogenic treatment. Therefore, deep cryogenic treatment processing of rotavator blades made of boron steel is recommended for improving the abrasive wear resistance and other mechanical properties.

4. Conclusions

The effect of cryogenic treatment and post-tempering on rotavator blade material were analyzed. The following conclusions were drawn:

(1) During deep cryogenic treatment, retained austenite left after conventional heat treatment was transformed into martensite. Cryogenic treatment also resulted in formation of secondary carbides and helped in bringing more uniformity in distribution of secondary carbides. Tempering post cryogenic treatment led to grain coarsening and martensite decomposition;

(2) Hardness of cryotreated (CDCT-T0) specimen was improved by 260.73% compared to UT material, due to the formation of martensite along with the precipitation of secondary carbides. Tempering post cryogenic treatment caused reduction in hardness due to grain coarsening and martensite decomposition;

(3) Impact strength of the cryotreated (CDCT-T0) specimen was augmented by 50% in comparison with CHT specimen, due to increasing the nucleation of carbides, which facilitated the precipitation of a higher number of fine carbides during cryogenic treatment, resulting in a higher impact strength of material. Post tempering enhanced the impact strength, which further increased with higher tempering temperature;

(4) Abrasive wear volume loss in cryotreated (CDCT-T0) specimens were reduced by 60% compared to UT samples, owing to improvement in hardness, RA conversion into martensite, and the formation of secondary carbides. Tempering post cryogenic treatment resulted in decline in abrasive wear resistance;

(5) The additional cost of 15.12% was incurred due to cryogenic treatment, whereas the expected augmentation in wear resistance of rotavator blade material was 25.70%. The economic analysis clearly justified the additional cost of cryogenic treatment.

Author Contributions: Conceptualization, T.P.S., A.K.S., and J.S.; methodology, T.P.S, Q.S., Z.L., and K.S.; validation, M.K.G., H.J.; formal analysis, M.K.G. and T.P.S.; investigation, C.I.P., Q.S., and Z.L.; resources, Q.S., Z.L.; data curation, C.I.P., A.K.S.; writing—original draft preparation, M.K.G., H.J., and C.I.P.; writing—review and editing, Q.S., H.J.; visualization, M.K.G.; supervision, Q.S., Z.L.; project administration, C.I.P; analysis and layout; Z.L. All authors have read and agreed to the published version of the manuscript.

Funding: This research received no external funding.

Conflicts of Interest: The authors declare no conflict of interest.

References

1. Saxena, A.C. Dushyant Singh Techno Economically Viable Production Package of Rotavator Blade for Entrepreneurs. *Agric. Eng. Today* **2010**, *34*, 23–25.

2. Ferguson, S.A.; Fielke, J.M.; Riley, T.W. Wear of cultivator shares in abrasive South Australian soils. *J. Agric. Eng. Res.* **1998**, *69*, 99–105. [CrossRef]

3. Salokhe, V.M.; Gee-Clough, D. Coating of cage wheel lugs to reduce soil adhesion. *J. Agric. Eng. Res.* **1988**, *41*, 201–210. [CrossRef]

4. Salokhe, V.M.; Gee-Clough, D.; Tamtomo, P. Wear testing of enamel coated rings. *Soil Tillage Res.* **1991**, *21*, 121–131. [CrossRef]

5. Kumar, S.; Mondal, D.P.; Jha, A.K. Effect of microstructure and chemical composition of hardfacing alloy on abrasive wear behavior. *J. Mater. Eng. Perform.* **2000**, *9*, 649–655. [CrossRef]

6. Singh, P.; Bansal, A.; Kumar Goyal, D. Erosion wear evaluation of HVOF sprayed WC-12Co coating on some pipeline materials using Taguchi approach. *Kovove Mater.* **2019**, *57*, 113–120. [CrossRef]

7. Bansal, A.; Singh, J.; Singh, H. Slurry Erosion Behavior of HVOF-Sprayed WC-10Co-4Cr Coated SS 316 Steel with and Without PTFE Modification. *J. Therm. Spray Technol.* **2019**, *28*, 1448–1465. [CrossRef]

8. Karoonboonyanan, S.; Salokhe, V.M.; Niranatlumpong, P. Wear resistance of thermally sprayed rotary tiller blades. *Wear* **2007**, *263*, 604–608. [CrossRef]

9. Bayhan, Y. Reduction of wear via hardfacing of chisel ploughshare. *Tribol. Int.* **2006**, *39*, 570–574. [CrossRef]

10. Chahar, V.K.; Tiwari, G.S. Wear characteristics of reversible cultivator shovels treated with different surface hardening processes. *J. Inst. Eng. Agric. Eng. Div.* **2009**, *90*, 42–45.

11. Raval, A.H.; Kaushal, O.P. Wear and tear of hard surfaced cultivator shovel. *Agric. Mech. Asia Afr. Lat. Am.* **1990**, *21*, 46–48.

12. Moore, M.A.; McLees, V.A.; King, F.S. Hard-facing soil-engaging equipment. *Agric. Eng.* **1979**, *34*, 15–19.

13. Buchely, M.F.; Gutierrez, J.C.; León, L.M.; Toro, A. The effect of microstructure on abrasive wear of hardfacing alloys. *Wear* **2005**, *259*, 52–61. [CrossRef]

14. Kumar, S.; Mondal, D.P.; Khaira, H.K.; Jha, A.K. Improvement in high stress abrasive wear property of steel by hardfacing. *J. Mater. Eng. Perform.* **1999**, *8*, 711–715. [CrossRef]

15. Moore, M.A. The abrasive wear resistance of surface coatings. *J. Agric. Eng. Res.* **1975**, *20*, 167–179. [CrossRef]

16. Atïk, E. Mechanical properties and wear strengths in aluminiumalumina composites. *Mater. Struct. Constr.* **1998**, *31*, 418–422. [CrossRef]

17. Meric, C.; Sahin, S.; Backir, B.; Koksal, N.S. Investigation of the boronizing effect on the abrasive wear behavior in cast irons. *Mater. Des.* **2006**, *27*, 751–757. [CrossRef]

18. Allaoui, O.; Bouaouadja, N.; Saindernan, G. Characterization of boronized layers on a XC38 steel. *Surf. Coatings Technol.* **2006**, *201*, 3475–3482. [CrossRef]

19. Béjar, M.A.; Moreno, E. Abrasive wear resistance of boronized carbon and low-alloy steels. *J. Mater. Process. Technol.* **2006**, *173*, 352–358. [CrossRef]

20. Bourithis, L.; Papadimitriou, G. Boriding a plain carbon steel with the plasma transferred arc process using boron and chromium diboride powders: Microstructure and wear properties. *Mater. Lett.* **2003**, *57*, 1835–1839. [CrossRef]

21. Sharma, M.C.; Mubeen, A. Effect of shot size on peening intensity for local peening. *J. Mech. Work. Technol.* **1983**, *8*, 155–160. [CrossRef]

22. Singh, D.; Saxena, A.C. Characterization of Materials Used for Rotavator Blades. *Agric. Eng. Today* **2011**, *35*, 10–14.

23. Yazici, A. Wear behavior of carbonitride-treated ploughshares produced from 30MnB5 steel for soil tillage applications. *Met. Sci. Heat Treat.* **2011**, *53*, 248–253. [CrossRef]

24. Singh, D.; Saha, K.P.; Mondal, D.P. Development of mathematical model for prediction of abrasive wear behaviour in agricultural grade medium carbon steel. *Indian J. Eng. Mater. Sci.* **2011**, *18*, 125–136.

25. Bhakat, A.K.; Mishra, A.K.; Mishra, N.S.; Jha, S. Metallurgical life cycle assessment through prediction of wear for agricultural grade steel. *Wear* **2004**, *257*, 338–346. [CrossRef]

26. Gupta, A.K.; Jesudas, D.M.; Das, P.K.; Basu, K. Performance evaluation of different types of steel for duck foot sweep application. *Biosyst. Eng.* **2004**, *88*, 63–74. [CrossRef]

27. Jha, A.K.; Prasad, B.K.; Modi, O.P.; Das, S.; Yegneswaran, A.H. Correlating microstructural features and mechanical properties with abrasion resistance of a high strength low alloy steel. *Wear* **2003**, *254*, 120–128. [CrossRef]

28. Singla, A.K.; Singh, J.; Sharma, V.S. Processing of materials at cryogenic temperature and its implications in manufacturing: A review. *Mater. Manuf. Process.* **2018**, *33*, 1603–1640. [CrossRef]

29. Kumar, T.V.; Thirumurugan, R.; Viswanath, B. Influence of Cryogenic Treatment on the Metallurgy of Ferrous Alloys - A Review. *Mater. Manuf. Process.* **2017**, *32*, 1789–1805. [CrossRef]

30. Akincioğlu, S.; Gökkaya, H.; Uygur, İ. A review of cryogenic treatment on cutting tools. *Int. J. Adv. Manuf. Technol.* **2015**, *78*, 1609–1627. [CrossRef]

31. Gill, S.S.; Singh, H.; Singh, R.; Singh, J. Cryoprocessing of cutting tool materials—A review. *Int. J. Adv. Manuf. Technol.* **2010**, *48*, 175–192. [CrossRef]

32. Kalsi, N.S.; Sehgal, R.; Sharma, V.S. Cryogenic Treatment of Tool Materials: A Review. *Mater. Manuf. Process.* **2010**, *25*, 1077–1100. [CrossRef]

33. Baldissera, P.; Delprete, C. Deep Cryogenic Treatment: A Bibliographic Review. *Open Mech. Eng. J.* **2008**, *2*, 1–11. [CrossRef]

34. Yong, J.; Ding, C.; Qiong, J. Effect of cryogenic thermocycling treatment on the structure and properties of magnesium alloy AZ91. *Met. Sci. Heat Treat.* **2012**, *53*, 589–591. [CrossRef]

35. Bhale, P.; Shastri, H.; Mondal, A.K.; Masanta, M.; Kumar, S.; Li, X.; Yao, Y.; Shaffer, D.; Reinstadtler, C.; Roth, J.T.; et al. Effect of Deep Cryogenic Treatment on Microstructure and Properties of AE42 Mg Alloy. *J. Mater. Eng. Perform.* **2016**, *25*, 3590–3598. [CrossRef]

36. Mohan, K.; Suresh, J.A.; Ramu, P.; Jayaganthan, R. Microstructure and Mechanical Behavior of Al 7075-T6 Subjected to Shallow Cryogenic Treatment. *J. Mater. Eng. Perform.* **2016**, *25*, 2185–2194. [CrossRef]

37. Steier, V.F.; Ashiuchi, E.S.; Reißig, L.; Araújo, J.A. Effect of a Deep Cryogenic Treatment on Wear and Microstructure of a 6101 Aluminum Alloy. *Adv. Mater. Sci. Eng.* **2016**. [CrossRef]

38. Singla, A.K.; Singh, J.; Kumar, P.; Kumar, A.; Sharma, V.S. Effect of cryogenic treatment on the wear resistance and microstructure of Ti-6AL-4V. *Indian J. Eng. Mater. Sci.* **2018**, *25*, 243–249.

39. Singla, A.K.; Singh, J.; Sharma, V.S. Impact of Cryogenic Treatment on Mechanical Behavior and Microstructure of Ti-6Al-4V ELI Biomaterial. *J. Mater. Eng. Perform.* **2019**, *28*, 5931–5945. [CrossRef]

40. Krolczyk, G.M.; Maruda, R.W.; Krolczyk, J.B.; Wojciechowski, S.; Mia, M.; Nieslony, P.; Budzik, G. Ecological trends in machining as a key factor in sustainable production—A review. *J. Clean. Prod.* **2019**, *218*, 601–615. [CrossRef]

41. Mia, M.; Gupta, M.K.; Lozano, J.A.; Carou, D.; Pimenov, D.Y.; Królczyk, G.; Khan, A.M.; Dhar, N.R. Multi-objective optimization and life cycle assessment of eco-friendly cryogenic N2 assisted turning of Ti-6Al-4V. *J. Clean. Prod.* **2019**, *210*, 121–133. [CrossRef]

42. Kopac, J.; Pusavec, F.; Krolczyk, G. Cryogenic machining, surface integrity and machining performance. *Arch. Mater. Sci. Eng.* **2015**, *71*, 83–93.

43. Eller, T.K.; Greve, L.; Andres, M.T.; Medricky, M.; Hatscher, A.; Meinders, V.T.; Van Den Boogaard, A.H. Plasticity and fracture modeling of quench-hardenable boron steel with tailored properties. *J. Mater. Process. Technol.* **2014**, *214*, 1211–1227. [CrossRef]

44. Namklang, P.; Uthaisangsuk, V. Description of microstructures and mechanical properties of boron alloy steel in hot stamping process. *J. Manuf. Process.* **2016**, *21*, 87–100. [CrossRef]

45. Naderi, M. *Hot Stamping of Ultra High Strength Steels*; Springer International Publishing: Cham, Switzerland, 2007.

46. Singh, T.P.; Singh, J.; Singh, K. An investigation on abrasive wear behaviour of rotary blade (AISI 30MnCrB4) under different quenching parameters. In Proceedings of the CPIE, Jalandhar, India, 19–21 December 2016.

47. ASTM. E 23-12c Standard Test Methods for Notched Bar Impact Testing of Metallic Materials. *Standards* **2013**, *i*, 1–25.

48. ASTM. International G65-16: Standard Test Method for Measuring Abrasion Using the Dry Sand/Rubber Wheel. *ASTM Stand.* **2013**, *4*, 1–12.

49. Avery, H.S. The measurement of wear resistance. *Wear* **1961**, *4*, 427–449. [CrossRef]

50. Singh, T.P.; Singh, J.; Singh, K. Enhancing the abrasive wear resistance of rotary blade material (AISI 30MnCrB4) by cryogenic treatment. *J. Sci. Ind. Res.* **2018**, *77*, 92–97.

51. Nahvi, S.M. Abrasive Wear Behaviour of Steels and Advanced HVOF-Sprayed WC-M Coatings. Ph.D. Thesis, University of Nottingham, Nottingham, UK, 2011.

52. Moore, M.A.; Swanson, P.A. The effect of particle shape on abrasive wear: A comparison of theory and experiment. In *Proceedings International Conference on Wear of Materials*; Ludema, K.C., Ed.; ASME: New York, NY, USA, 1983; pp. 1–11.

53. Kašparová, M.; Zahálka, F.; Houdková, S.; Ctibor, P. Abrasive wear of WC-NiMoCrFeCo thermally sprayed coatings in dependence on different types of abrasive sands. *Kovove Mater.* **2010**, *48*, 73–85.

54. Singh, D.; Mondal, D.P.; Modi, O.P.; Sethi, V.K. Low stress abrasive wear response of boron steel under three body abrasion: Effect of heat treatment and peening intensities. *Indian J. Eng. Mater. Sci.* **2010**, *17*, 208–212.

55. Güler, H.; Ertan, R.; Özcan, R. Influence of heat treatment parameters on the micrestructure and mechanical properties of boron-alloyed steels. *Mater. Test.* **2012**, *54*, 619–624. [CrossRef]

56. Liu, H.; Wang, J.; Yang, H.; Shen, B. Effects of cryogenic treatment on microstructure and abrasion resistance of CrMnB high-chromium cast iron subjected to sub-critical treatment. *Mater. Sci. Eng. A* **2008**, *478*, 324–328. [CrossRef]

57. Gola, A.M.; Ghadamgahi, M.; Ooi, S.W. Microstructure evolution of carbide-free bainitic steels under abrasive wear conditions. *Wear* **2017**, *376–377*, 975–982. [CrossRef]

58. Li, H.; Tong, W.; Cui, J.; Zhang, H.; Chen, L.; Zuo, L. The influence of deep cryogenic treatment on the properties of high-vanadium alloy steel. *Mater. Sci. Eng. A* **2016**, *662*, 356–362. [CrossRef]

59. Song, Z.; Zhao, S.; Jiang, T.; Sun, J.; Wang, Y.; Zhang, X.; Liu, H.; Liu, Y. Effect of nanobainite content on the dry sliding wear behavior of an Al-Alloyed high carbon steel with nanobainitic microstructure. *Materials* **2019**, *12*, 1618. [CrossRef]

60. Li, S.; Xiao, M.; Ye, G.; Zhao, K.; Yang, M. Effects of deep cryogenic treatment on microstructural evolution and alloy phases precipitation of a new low carbon martensitic stainless bearing steel during aging. *Mater. Sci. Eng. A* **2018**, *732*, 167–177. [CrossRef]

61. Chintha, A.R. Metallurgical aspects of steels designed to resist abrasion, and impact-abrasion wear. *Mater. Sci. Technol.* **2019**, *35*, 1133–1148. [CrossRef]

62. Jaswin, M.A.; Lal, D.M. Impact Behavior of Cryogenically Treated En 52 and 21-4N Valve Steels. *Int. J. Mech. Ind. Sci. Eng.* **2014**, *8*, 159–164.

63. Vuorinen, E.; Ojala, N.; Heino, V.; Rau, C.; Gahm, C. Erosive and abrasive wear performance of carbide free bainitic steels—Comparison of field and laboratory experiments. *Tribol. Int.* **2016**, *98*, 108–115. [CrossRef]

64. Chand, N.; Neogi, S. Mechanism of material removal during three-body abrasion of FRP composite. *Tribol. Lett.* **1998**, *4*, 81–85. [CrossRef]

65. Nadig, D.S.; Jacob, S.; Karunanithi, R.; Manjunatha, R.; Subramanian, D.; Prasad, M.V.N.; Sen, G.; Jha, A.K. Studies of cryotreatment on the performance of integral diaphragm pressure transducers for space application. *Cryogenics* **2010**, *50*, 561–565. [CrossRef]

66. Vuorinen, E.; Heino, V.; Ojala, N.; Haiko, O.; Hedayati, A. Erosive-abrasive wear behavior of carbide-free bainitic and boron steels compared in simulated field conditions. *Proc. Inst. Mech. Eng. Part J J. Eng. Tribol.* **2018**, *232*, 3–13. [CrossRef]

67. Hernandez, S.; Leiro, A.; Rodríguez, M.; Vuorinen, E.; Sundin, K.; Prakash, B. High temperature three-body abrasive wear of 0.25C 1.42Si steel with carbide free bainitic (CFB) and martensitic microstructures. *Wear* **2016**, *360–361*, 21–28. [CrossRef]

 *materials* 

Article

Reverse Problem in Surface Texture Analysis—One-Process Profile Modeling on the Basis of Measured Two-Process Profile after Machining or Wear

Pawel Pawlus [1], Rafal Reizer [2],* and Michal Wieczorowski [3]

[1] Faculty of Mechanical Engineering and Aeronautics, Rzeszow University of Technology, Powstancow Warszawy 8 Street, 35-959 Rzeszow, Poland; ppawlus@prz.edu.pl
[2] College of Natural Sciences, University of Rzeszow, Pigonia Street 1, 35-310 Rzeszow, Poland
[3] Faculty of Mechanical Engineering and Management, Poznan University of Technology, Piotrowo Street 3, 61-138 Poznan, Poland; michal.wieczorowski@put.poznan.pl
* Correspondence: rreizer@ur.edu.pl; Tel.: +48-17-8518582

Received: 13 November 2019; Accepted: 10 December 2019; Published: 12 December 2019

Abstract: The method of the base (valley) one-process profile modeling on the basis of the measured two-process profile was developed. The base one-process random profile of the Gaussian ordinate distribution is characterized by the standard deviation of the profile height and the correlation length. The problem of estimation of the correlation length of this one-process profile exists. In the procedure of the correlation length estimation, information about the averaged shape of the autocorrelation functions of many one-process profiles after the same type of machining is required. The correlation length of the base one-process profile can be obtained on the basis of the vertical truncation of the measured two-process profile. The average error of the correlation length estimation was not higher than 7%, while the maximum error was not larger than 14%. This method can be extended to simulate the one-process texture of 3D (areal) surface topography.

Keywords: profile; two-process surface; correlation length

---

1. Introduction

Many machined surfaces have an ordinate distribution similar to Gaussian distribution. They are called one-process surfaces because they contain tracks of only one machining process. Two-process surfaces can be created from initial one-process textures during a low wear (within the limits of the original surface topography). This type of the surface consists of smooth wear resistant plateau parts with deep valleys working as reservoirs for lubricant and traps for wear debris. This structure connects in an ideal way the good sliding property of a smooth surface with the great ability to maintain oil of a porous structure. Two-process textures are more functionally important than one-process topographies.

Therefore, attempts were made to obtain a two-process surface during the last stage of machining. This surface should resemble a texture created during running-in. Due to it, a duration of running-in and wear decreased. For this reason, two-process textures were created. A plateau-honed cylinder surface is the practical example of such structures [1–4]. It consists of two random portions—smooth plateau and rough valley parts. Because of the excellent tribological properties of such surfaces, many other two-process textures were created—random, deterministic or random-deterministic structures. They are called textured surfaces. Surface texturing is an option of surface engineering resulting in significant reductions of the frictional resistance in mixed and fluid lubrications, wear, and the inclination to seizure by creating oil pockets (dimples or cavities) on the sliding surfaces [5–10].

One-process surfaces are well described. The analysis of the measured one-process texture is simple. The digital filtration is an easy task. A typical Gaussian filter can be used. A one-process random profile can be easily described by only two parameters: height and horizontal. It can be completely characterized by the standard deviation of height (Pq), and a correlation length (CL), i.e., the distance at which the autocorrelation function slowly decays to a value of 0.1 [11,12]. Figure 1 presents the example of the random one-process profile and its autocorrelation function with the CL parameter.

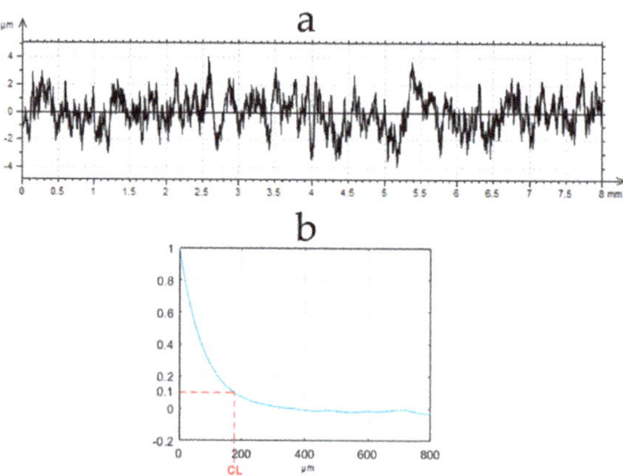

Figure 1. One-process random profile (**a**) and its autocorrelation function (**b**).

A correct characterization of contact between rough surfaces is substantial during the study of tribological problems like friction, wear or sealing. However, the contact of surfaces of Gaussian ordinate distribution was typically analyzed [13–15].

Simulation of surface texture offers many advantages. Computer generations of surfaces created in manufacturing or wear processes lead to decreases in both cost and time of experimental research. The modeled surfaces can be used in various problems, such as contact, friction, and wear. A one-process profile can be modeled based on the values of the Pq parameter (standard deviation of roughness height) and the correlation length CL using the ARMA (autoregressive-moving average) [16] or FFT (Fast Fourier Transform) [17] methods.

However, there are some problems when analyzing two-process textures. The application of the Gaussian filter causes profile distortions near the edges of deep valleys. It is necessary to increase the cut-off or use different filters, like a double Gaussian filter (ISO 13565–1 [18]) or robust filter [19]. Description of a two-process profile is more complicated compared to that of a one-process profile. Two-process profiles cannot be described only by two parameters. Therefore, the special two standards dedicated for two-process surfaces occur. The first of them, ISO 13565–2 [20], is based on the profile division into three parts: peak, core, and valley. There are five parameters describing the material ratio curve of a two-process profile: the core roughness height Pk, the reduced peak height Ppk, the reduced valley depth Pvk, and two material ratios of transitions points between profile parts [21,22]. The second standard ISO 13565–3 [23] divides the profile into only two parts: peak (plateau) and valley; therefore, three parameters characterize the material ratio curve: Ppq (the plateau root–mean square roughness), Pvq (the valley root–mean square roughness), and Pmq (the material ratio of plateau–to–valley transition) [24,25]. However, both presented standards do not include horizontal parameters, which are functionally important [11–15].

2. Formulation of the Problem

The standard ISO 13565–3 [23] is the base of two-process profile modeling. The parameters describing the two-process surface can be calculated from a probability plot of the material ratio curve. The two-process profile is represented by two straight lines describing peak (plateau) and valley portions. The slopes of these lines are the standard deviation of plateau (Ppq) and of valley (Pvq) parts, respectively [24]. This standard also includes the Pmq parameter (Figure 2).

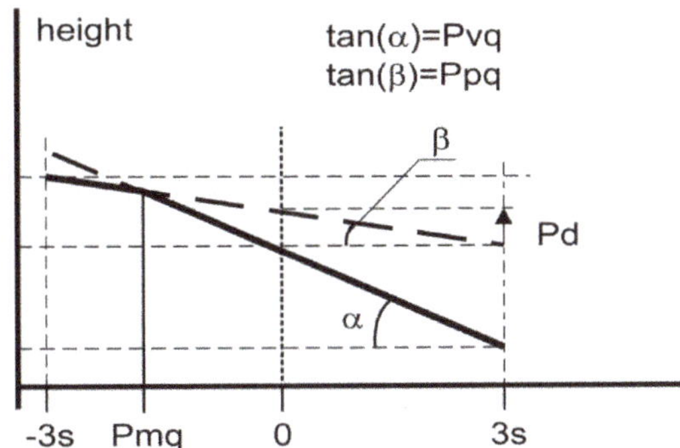

Figure 2. Graphical interpretation of the probability parameters.

During simulation of the two-process profile, one should generate two Gaussian profiles (plateau and valley) and take the point-wise minimum [26,27]. More precisely, two profiles of Gaussian ordinate distributions are superimposed [26,27] for a given vertical distance between their mean lines Pd shown in Figure 2. The standard deviations of the plateau and valley profiles are equal to the Ppq and Pvq parameters of the two-process profile, respectively. The distance Pd relates to the parameters Ppq, Pvq, and Pmq by the following formula [26]:

$$Pd = Pmq(Ppq - Pvq). \tag{1}$$

From ordinates of generated the two-process profile, the smaller ones are selected. Figure 3 presents the example of computer creation of the two-process profile.

Only amplitude parameters of plateau and valley profiles can be determined from the probability plot of the material ratio curve (Figure 3). However, each Gaussian profile is characterized not only by the height parameter (Sq), but also by the horizontal parameter, like the correlation length CL. They are necessary in the modeling procedure. The question arises of how horizontal parameters (correlation lengths) of two profiles of Gaussian ordinate distribution (plateau and valley) can be estimated. When these parameters are unknown, there would be a problem with obtaining the correct autocorrelation length of the two-process profile. The iterative procedure of selecting correlations lengths of plateau and valley Gaussian profiles can be a solution [28,29]. However, after its application, the correlation length of modeled two-process profile can be similar to that of the measured profile; nevertheless, the correlation lengths of plateau and valley parts can be incorrect.

Especially the proper estimation of the correlation length of the valley profile is the task of a primary importance, because this profile really exists, in contrast to the plateau profile. The profile after finish (one-step) honing during plateau honing and the one-process profile before wear are examples of valley profiles.

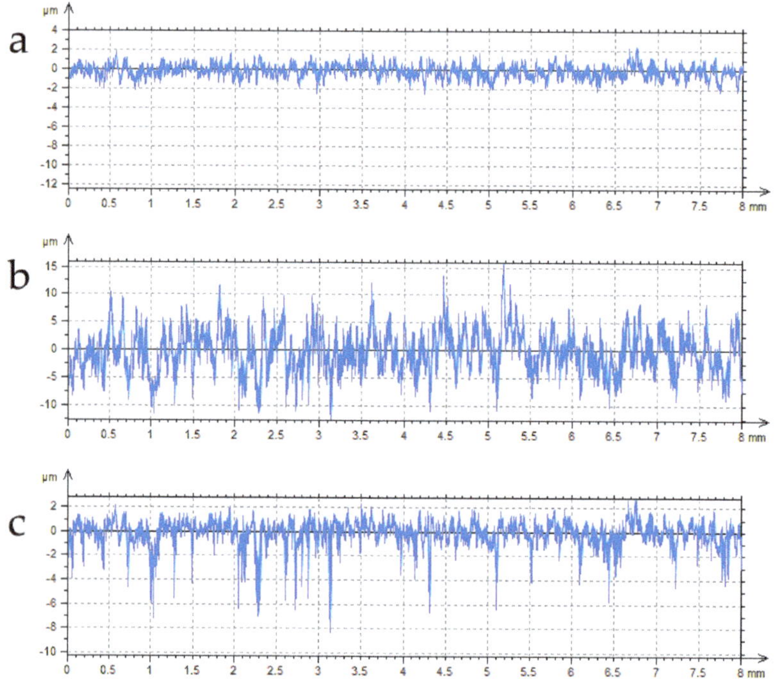

Figure 3. Computer creation of the two-process profile: plateau profile (**a**), valley profile (**b**), and two-process profile (**c**).

The following reverse problem should be solved: How can we simulate the valley profile (Figure 3b) when only the measured two-process profile is known (Figure 3c)? According to the knowledge of the authors of this paper, although many works in the field of surface topography modeling have been carried out, this problem has not been solved yet. Therefore, the solution to this problem is a novelty of this research work.

As mentioned above, each Gaussian profile can be described by the standard deviation of height (the Pq parameter) and the correlation length (CL). There is no problem with estimation of the Pq parameter of the valley profile. It is equal (close due to some measurement errors) to the Pvq parameter of the two-process profile. This parameter can be estimated from the probability plot of the material ratio curve (Figure 2).

The correct simulation of the valley profile is a problem of substantial importance. In some cases, one cannot measure the surface after machining, only after a low wear. It is necessary to obtain information about the shape of the one-process profile after machining to know how the roughness of the machined surface affects tribological properties of the sliding assembly. This problem is very important, since advanced machining methods have recently been used for surface topography creation [30–34]. Similar information is important also in a study of machining processes, like plateau honing. The question is how the initial surface roughness (after one-process finish honing) affects the surface obtained after the second process (plateau honing).

3. Solution of the Problem

The correlation length of the initial one-process profile can be estimated on the basis of analyses of (1) correlation lengths of many measured one-process profiles after the same type of machining treatment due to vertical truncation, and (2) the correlation length of the measured vertically truncated two-process profile, having only details belonging to the valley part.

First, one should know how autocorrelation functions of one-process profiles after the same kind of machining are changing with vertical truncations of heights. It was found that due to truncation, the correlation length decreased. This was connected with reductions in the widths of deep valleys affecting the correlation length and the cumulative spectrum [35]. Because for one-process texture, the profile probability plot has the shape of the straight line, one should know how the profile correlation length changes for the given material ratio. After analyses of the set of profiles, the average value of the correlation length for selected material ratios should be computed. This analysis should be conducted for many surfaces after the same treatment, like polishing, lapping, grinding or abrasive blasting. For instance, Figure 4a presents the curve obtained for one-process surfaces after one-process finish honing of more than 20 cylinders, which were honed using various methods including different kinds of stones (diamond and ceramic). It is important that shapes of the ordinate distributions of these textures should be approximately symmetric—the skewness Psk should be near 0 and the emptiness coefficient Pp/Pt near 0.5, the Pq/Pa ratio near 1.25 [36] (Pp—maximum peak height, Pt—total profile height, Pa—arithmetic mean deviation of the profile), without spikes or individual valleys. Then, reciprocals of the obtained correlation lengths' MF (magnification factors) for the given material ratios should be calculated (Figure 4b). The material ratios of truncated profiles were restricted to the 50–98% range.

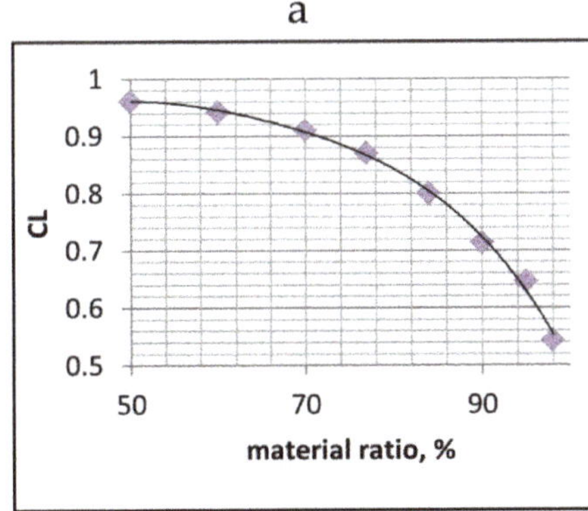
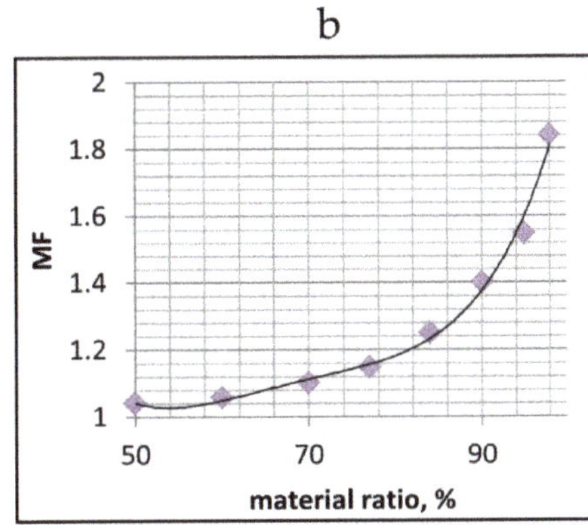

Figure 4. Correlation length (CL) of the truncated one-process profile after honing (**a**), and magnification factor (MF) (**b**) versus material ratio.

Determination of the curve shown in Figure 4b on the basis of series of profiles after the same type of machining is the initial step.

Then, for each measured two-process profile, it is necessary to extract details belonging only to the valley portion—Figure 5.

It is known that the part of the profile under the solid red line originates from the valley profile (Figure 5b). Therefore, the two-process profile (Figure 5a) should be vertically truncated and its lower part ought to be analyzed. In this case, the truncation level is the ordinate of crossing the approximate straight line characterizing the valley parts with the right vertical axis (in a profile study, it is a little larger than 3 s or 99.87%). However, sometimes, for a profile characterized by a high transition material ratio Pmq (Figure 2), the valley part could contain mainly individual nonstatistical valleys. In this case, the truncation level should lie at the top (see dotted line in Figure 5b).

However, one should know that near the transition point (of the abscissa Pmq), there can be a mixture of profile details originating from both the plateau and valley portions, which could be a source of the error. Therefore, the selection of the truncation level should be treated with great care.

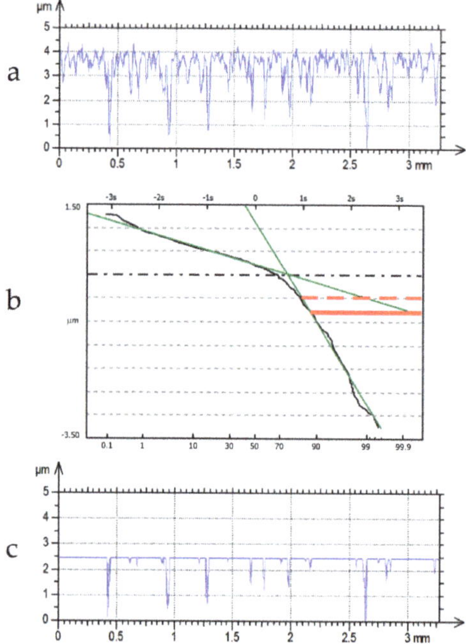

Figure 5. Two-process cylinder profile (**a**), its probability plot (**b**), and the truncated profile containing the valley part (**c**).

The correlation length of the one-process valley profile is equal to the correlation length of the truncated two-process profile (Figure 5c) multiplied by the magnification factor (MF) obtained from Figure 4b for the given material ratio of the truncation level (Figure 5b).

After determination of the curve presenting the magnification factor MF versus the material ratio (Figure 4b) on the basis of series of profiles after the same type of machining, there are the following steps in the procedure of simulation of the profile of the base (valley) one-process texture:

- Determination of the probability plot of the two-process measured profile;
- Determination of the Pq parameter of the one-process profile which is the Pvq parameter of the two-process profile;
- Vertical truncation of the two-process profile to extract profile details belonging only to the valley portion;
- Determinations of the autocorrelation function and correlation length of the truncated two-process profile;
- Calculation of the magnification factor MF (Figure 4b) based on the material ratio of the truncation level;
- Calculation of the correlation length of the one-process valley profile by magnification of the correlation length of the truncated two-process profile by the MF value;
- Simulation of the base (valley) profile using for example the FFT method [17].

Figure 6 presents a flow chart of research methodology. One can see that on the basis of a graph presenting the magnification factor MF versus the material ratio (Figure 4b), many profiles can be simulated.

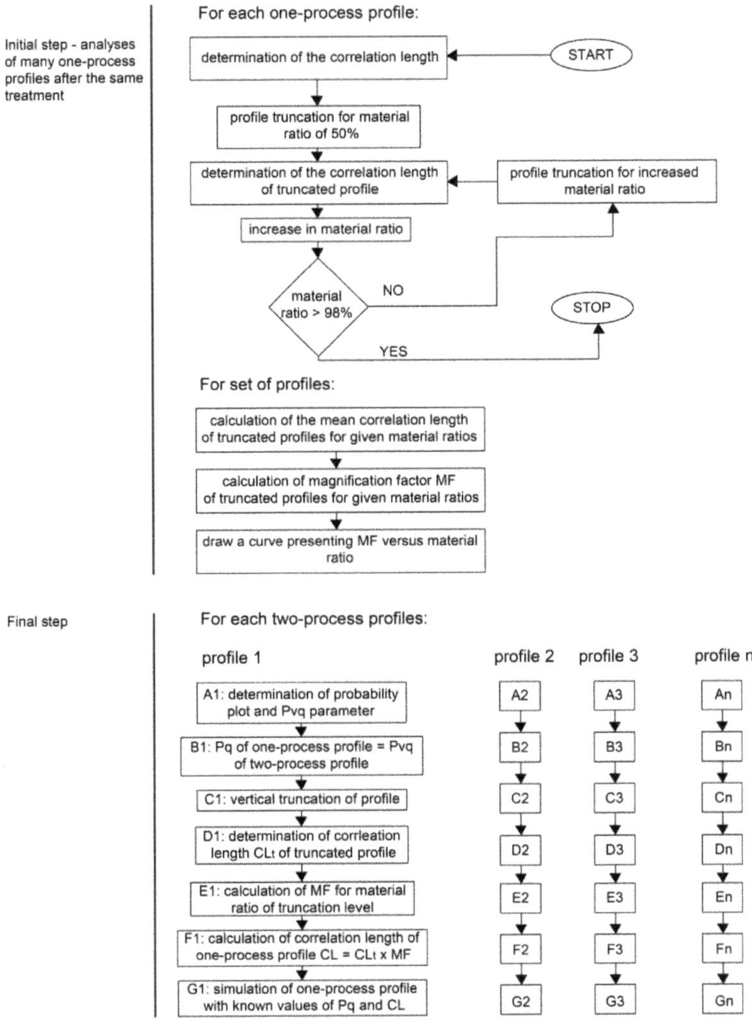

Figure 6. Methodology of one-process base (valley) profile simulation.

4. Validation of Method

This method was validated for two groups of cylinder liner surfaces made from grey cast iron. Cylinder liners from the first group were only finish honed with diamond or ceramic tools. Therefore, they had one-process random textures. These cylinders were subjected to tribological tests using an OPTIMOL SRV5 tester (Optimol Intruments, Munich, Germany). They co-acted with details of piston rings under lubricated conditions. During these tests, a low wear took place. Before and after test topographies of cylinder liner surfaces were measured in very similar places (the mechanical and then digital relocations were used) by a white light interferometer Talysurf CCI Lite—Figure 7. The initial measuring area of 3.3 × 3.3 mm contained 1024 × 1024 points. During the analysis, the form was removed by the polynomial of the second level; a digital filtration was not used. Spikes were eliminated by truncation.

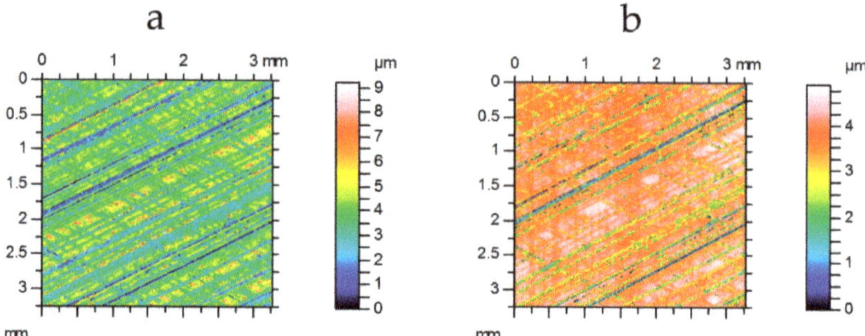

Figure 7. Surface of cylinder liner after finish honing (**a**) and after the tribological test (**b**).

From the surfaces of the cylinders, axial profiles were selected. For each profile after operating, the procedure described above was conducted for at least three truncation levels, starting from the upper level resulting from the probability plot of the material ratio curve (Figure 5b). The lowest level typically corresponded to material ratio of 98%; however, this level depended on the shape of the material ratio curve. For the one-process cylinder liner profile after finish honing, the correlation length was computed. The average correlation length of the profile after finish honing estimated from truncated cylinder liner profile after operating was compared with that from the measured (real) profile. About 20 different surface textures were analyzed. It was found that during operating, only the wear removal took place, without occurrence of the plastic deformation.

It was found that the average error of estimation of the correlation length of profiles after finish honing was 6.5%, while the maximum error was 12%. Figure 8 presents the example of one-process, two-process, and modeled one-process profiles. The one-process profile was simulated using the FFT method [17].

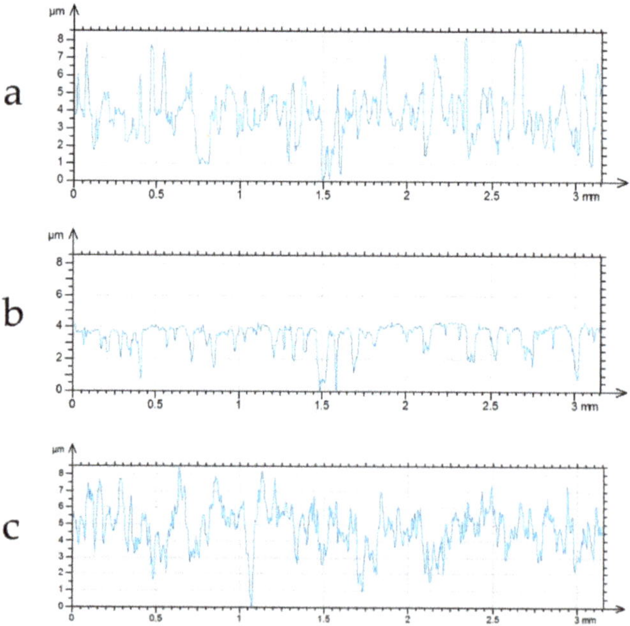

Figure 8. One-process cylinder liner profile after finish honing (**a**), after tribological test (**b**), and simulated one-process profile (**c**).

Similar tests were carried out for cylinders from the second group. There were finish honed (one-process) and plateau honed (two-process) surfaces of cylinder liners. In contrast to cylinders from the first group, the measurements of various one-process and two-process cylinder liners were conducted in similar places. However, the honing treatment of one-process and finish honing of two-process cylinder liners were carried out under the same conditions (see [37]). For each surface (also measured by the white light interferometer), three axial profiles were selected. The test procedure was similar to that mentioned above; however, the average value of three correlation lengths from one-process cylinder profiles (after finish honing) was compared with the average value of estimated correlation length obtained on the basis of measurement of three two-process profiles after plateau honing. It this part of study, about 20 different surface topographies were analyzed.

It was found that the mean error of estimation of correlation lengths of profiles after finish honing was 5.8%; however, the maximum error was 15%. Figure 9 shows the example of one-process, two-process, and simulated one-process profiles.

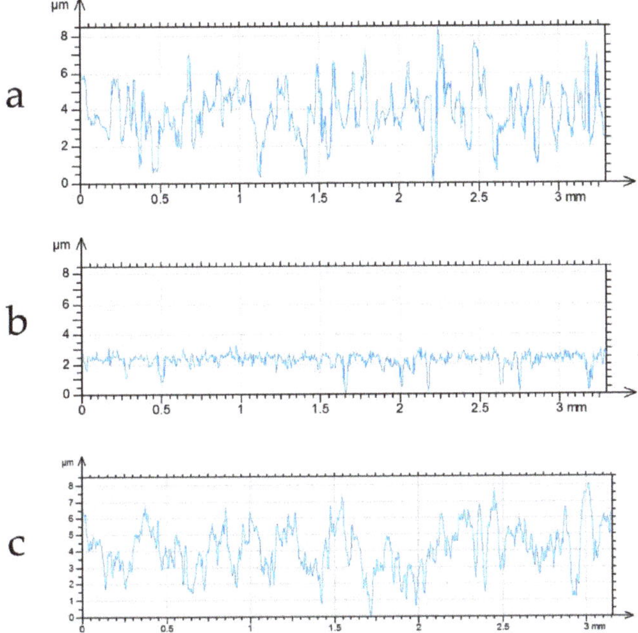

Figure 9. One-process cylinder liner profile after finish honing (**a**), after plateau honing (**b**), and the simulated one-process profile (**c**).

5. Discussion

One can see that the application of this method assured good results in both analyzed cases. The authors of this paper focused mainly on estimation of the correlation length, because the standard deviation of the roughness height can be obtained from the probability plot of the material ratio curve of two-process surfaces (Figure 5b). However, there can be errors connected with its estimation. For instance, the valley part in the roughness probability plot of a two-process surface sometimes has a deviation from the linear shape. However, these errors were analyzed elsewhere.

There are some limitations to the presented method application. The first of them is mentioned above: the necessity to exclude profile details related to individual nonstatistical grooves and the mixture of two processes. Therefore, the determination of the truncation higher level (Figure 5b) should be treated with great care. When it is too high (corresponding to a low material ratio), the valley part can include the mixture of the two processes. However, when it is too low (corresponding to a high material ratio), the individual grooves can disturb the results of one-process profile modeling. On the basis of analyses of many surfaces, the lowest truncation level for the material ratio of 98% was selected (Figures 4 and 6). The precise analysis of the probability plot of the material ratio curve can be helpful. The remaining valley part should not be changed during wear or machining. Therefore, the conditions of wear and machining processes should be carefully taken into consideration. Especially during wear, sometimes plastic deformation, not only wear removal, can take place. The correct determination of the correlation length can be the other problem. Although the effect of the sampling interval on the correlation length is low, the sampling interval should be small enough for the correct estimation of the correlation length (smaller than 0.4 of the correlation length) [38].

This method can be also used in 3D (areal) analysis of isotropic or anisotropic one-directional surfaces. In this case, the correlation lengths in perpendicular directions should be determined on the basis of analyses of several representative profiles.

The correlation length of the upper plateau profile can also be estimated using this method. However, this task is not such important as the analysis of the valley profile, since the plateau profile does not really exist. However, information of the correlation lengths of both plateau and valley parts can be helpful in the simulation of two-process profiles or 3D textures.

This method was dedicated to the random base profile. When the valley part of two-process surface has a deterministic character (piston skirt after a low wear [39] can be the example), it can be modeled more easily compared to the random valley portion [12].

6. Conclusions

The method of simulation of the one-process base (valley) profile on the basis of the two-process profile was developed. The one-process profile is characterized by the standard deviation of the height and the correlation length. In the simulation procedure, the probability plot of the two-process profile is helpful. After analysis of this plot, the height of the one-process profile can be directly estimated. The correlation length can be achieved based on the vertical truncation of the two-process profile. In the procedure of correlation length, estimation information about autocorrelation function shapes of many profiles after the same kind of machining is needed.

The proposed procedure was validated for two groups of surfaces. It was found that the average error of the correlation length of one-process profile estimation was not higher than 7%, while the maximum error was not larger than 14%.

This method can be easily extended to simulate the base one-process isotropic or one-directional anisotropic 3D (areal) surface topography. It would be helpful in two-process surface modeling.

Author Contributions: Conceptualization: P.P., R.R., and M.W.; methodology, investigation and formal analysis: P.P., R.R., and M.W.; writing—original draft preparation: P.P., R.R., and M.W.; writing—review and editing: P.P., R.R., and M.W.

Funding: This research received no external funding.

Conflicts of Interest: The authors declare no conflict of interest.

References

1. Grabon, W.; Pawlus, P.; Wos, S.; Koszela, W.; Wieczorowski, M. Effects of honed cylinder liner surface texture on tribological properties of piston ring-liner assembly in short time tests. *Tribol. Int.* **2017**, *113*, 137–148. [CrossRef]

2. Haasis, G.; Weigmann, U. New honing technique reduces oil consumption. *Ind. Diam. Rev.* **1999**, *3*, 205–211.

3. Mezghani, S.; Demirci, I.; Zahouani, H.; El Mansori, M. The effect of groove texture patterns on piston-ring pack friction. *Precis. Eng.* **2012**, *36*, 210–217. [CrossRef]

4. Anderberg, C.; Dimkovski, Z.; Rosén, B.G.; Thomas, T.R. Low friction and emission cylinder liner surfaces and the influence of surface topography and scale. *Tribol. Int.* **2019**, *133*, 224–229. [CrossRef]

5. Gachot, C.; Rosenkranz, A.; Hsu, S.M.; Costa, H.L. A critical assessment of surface texturing for friction and wear improvement. *Wear* **2017**, *372–373*, 21–41. [CrossRef]

6. Rosenkranz, A.; Grützmacher, P.G.; Gachot, C.; Costa, H.L. Surface Texturing in Machine Elements: A Critical Discussion for Rolling and Sliding Contacts. *Adv. Eng. Mater.* **2019**, *21*, 1900194. [CrossRef]

7. Koszela, W.; Dzierwa, A.; Galda, L.; Pawlus, P. Experimental investigation of oil pockets effect on abrasive wear resistance. *Tribol. Int.* **2012**, *46*, 145–153. [CrossRef]

8. Galda, L.; Dzierwa, A.; Sep, J.; Pawlus, P. The effect of oil pockets shape and distribution on seizure resistance in lubricated sliding. *Tribol. Lett.* **2010**, *37*, 301–311. [CrossRef]

9. Koszela, W.; Pawlus, P.; Reizer, R.; Liskiewicz, T. The combined effect of surface texturing and DLC coating on the functional properties of internal combustion engines. *Tribol. Int.* **2018**, *127*, 470–477. [CrossRef]

10. Mishra, P.; Ramkumar, P. Effect of additives on a surface textured piston ring–cylinder liner system. *Tribol.-Mater. Surf. Interfaces* **2019**, *13*, 67–75. [CrossRef]

11. Whitehouse, D.J.; Archard, J.F. The properties of random surface of significance in their contact. *Proc. R. Soc. (Lond.)* **1970**, *A316*, 97–121. [CrossRef]

12. Whitehouse, D.J. *Handbook of Surface Metrology*; Inst. of Physics: Philadelphia, PA, USA, 1994.

13. Greenwood, J.A.; Williamson, J.B.P. Contact of nominally flat surfaces. *Proc. R. Soc. (Lond.)* **1966**, *A295*, 300–319.

14. Pawlus, P.; Zelasko, W. The importance of sampling interval for rough contact mechanics. *Wear* **2012**, *276*, 121–129. [CrossRef]

15. An, B.; Wang, X.; Xu, Y.; Jackson, R.L. Deterministic elastic-plastic modelling of rough surface contact including spectral interpolation and comparison to theoretical models. *Tribol. Int.* **2019**, *135*, 246–258. [CrossRef]

16. Teja, S.R.; Jayasingh, T. Characterisation of ground surface profiles: A comparison of AR, MA and ARMA modelling methods. *Int. J. Mach. Tools Manuf.* **1993**, *33*, 103–109. [CrossRef]

17. Wu, J.J. Simulation of rough surfaces with FFT. *Tribol. Int.* **2000**, *33*, 47–58. [CrossRef]

18. *ISO 13565-1—Geometrical Product Specifications (GPS)—Surface Texture: Profile METHOD; SURFACES HAVING Stratified Functional Properties—Part 1: Filtering and General Measurement Conditions*; International Organization for Standardization: Geneva, Switzerland, 1996.

19. Li, H.; Jiang, X.; Li, Z. Robust estimation in Gaussian filtering for engineering surface characterization. *Precis. Eng.* **2004**, *28*, 186–193. [CrossRef]

20. *ISO 13565-2—Geometrical Product Specifications (GPS)—Surface Texture: Profile method; Surfaces having Stratified Functional Properties—Part 2: Height Characterization Using the Linear Material Ratio Curve*; International Organization for Standardization: Geneva, Switzerland, 1996.

21. Bohm, H.J. Parameters for evaluating the wearing behaviour of surfaces. *Int. J. Mach. Tools Manuf.* **1992**, *32*, 109–113. [CrossRef]

22. Schneider, U.; Steckroth, A.; Rau, N.; Hubner, G. An approach to the evaluation of surface profiles by separating them into functionally different parts. *Surf. Topogr.* **1988**, *1*, 71–83.

23. *ISO 13565-3—Geometrical Product Specifications (GPS)—Surface Texture: Profile Method; Surfaces Having Stratified Functional Properties—Part 3: Height Characterization Using the Material Probability Curve*; International Organization for Standardization: Geneva, Switzerland, 1998.

24. Malburg, M.C.; Raja, J. Characterization of surface texture generated by plateau-honing process. *CIRP Ann.* **1993**, *42*, 637–639. [CrossRef]

25. Whitehouse, D.J. Assessment of surface finish profiles produced by multiprocess manufacture. *Proc. Inst. Mech. Eng.* **1985**, *199*, 263–270. [CrossRef]

26. Pawlus, P. Simulation of stratified surface topographies. *Wear* **2008**, *264*, 457–463. [CrossRef]

27. Pérez-Rafols, F.; Almqvist, A. Generating randomly rough surfaces with given height probability distribution and power spectrum. *Tribol. Int.* **2019**, *131*, 591–604. [CrossRef]

28. Reizer, R.; Pawlus, P. Modeling of plateau honed surface topography. *Proc. Inst. Mech. Eng. Part B J. Eng. Manuf.* **2012**, *226*, 1564–1578. [CrossRef]

29. Hu, S.; Brunetiere, N.; Huang, W.; Liu, X.; Wang, Y. Bi-Gaussian surface identification and reconstruction with revised autocorrelation functions. *Tribol. Int.* **2017**, *110*, 185–194. [CrossRef]

30. Kishawy, H.A.; Hegab, H.; Umer, U.; Mohany, A. Application of acoustic emissions in machining processes: Analysis and critical review. *Int. J. Adv. Manuf. Technol.* **2018**, *98*, 1391–1407. [CrossRef]

31. Zhu, L.; Yang, Z.; Li, Z. Investigation of mechanics and machinability of titanium alloy thin-walled parts by CBN grinding head. *Int. J. Adv. Manuf. Technol.* **2019**, *100*, 2537–2555. [CrossRef]
32. Zhao, B.; Ding, W.; Chen, Z.; Ya, C. Processes Pore structure design and grinding performance of porous metal-bonded CBN abrasive wheels fabricated by vacuum sintering. *J. Manuf. Process.* **2019**, *44*, 125–132. [CrossRef]
33. Meng, B.; Yuan, D.; Xu, S. Coupling effect on the removal mechanism and surface/subsurface characteristics of SiC during grinding process at the nanoscale. *Ceram. Int.* **2019**, *45*, 2483–2491. [CrossRef]
34. Krolczyk, G.M.; Maruda, R.W.; Krolczyk, J.B.; Wojciechowski, S.; Mia, M.; Nieslony, P.; Budzik, G. Ecological trends in machining as a key factor in sustainable production: A review. *J. Clean Prod.* **2019**, *218*, 601–615. [CrossRef]
35. Pawlus, P.; Chetwynd, D.G. Efficient characterization of surface topography in cylinder bores. *Precis. Eng.* **1996**, *19*, 164–174. [CrossRef]
36. King, T.G.; Spedding, T.A. On the relationship between surface profile height parameters. *Wear* **1982**, *83*, 91–108. [CrossRef]
37. Pawlus, P.; Dzierwa, A.; Michalski, J.; Reizer, R.; Wieczorowski, M.; Majchrowski, R. The effect of selected parameters of the honing process on cylinder liner surface topography. *Surf. Topogr.-Metrol. Prop.* **2014**, *2*, 025004. [CrossRef]
38. Pawlus, P. Digitisation of surface topography measurement results. *Measurement* **2007**, *40*, 672–686. [CrossRef]
39. Krzyzak, Z.; Pawlus, P. 'Zero-wear' of piston skirt surface topography. *Wear* **2006**, *260*, 554–561. [CrossRef]